Seismic Design of Concrete Buildings to Eurocode 8

Seismic Design of Concrete Buildings to Eurocode 8

Michael N. Fardis

Eduardo C. Carvalho

Peter Fajfar

Alain Pecker

CRC Press
Taylor & Francis Group
Boca Raton London New York

CRC Press is an imprint of the
Taylor & Francis Group, an **informa** business

A SPON PRESS BOOK

CRC Press
Taylor & Francis Group
6000 Broken Sound Parkway NW, Suite 300
Boca Raton, FL 33487-2742

© 2015 by Taylor & Francis Group, LLC
CRC Press is an imprint of Taylor & Francis Group, an Informa business

No claim to original U.S. Government works

Printed on acid-free paper
Version Date: 20150102

International Standard Book Number-13: 978-1-4665-5974-5 (Paperback)

Visit the Taylor & Francis Web site at
http://www.taylorandfrancis.com

and the CRC Press Web site at
http://www.crcpress.com

Contents

Preface

The main aim of this book, published at the time Eurocode 8 is starting its course as the only seismic design standard in Europe, is to support its application to concrete buildings – the most common type of structure – through education and training. It is addressed to graduate or advanced undergraduate students who want to acquire the skills and knowledge that are necessary for the informed use of Eurocode 8 in their career, to practitioners wishing to expand their professional activity into seismic design with Eurocode 8, to instructors of such students or practitioners in University or professional training programmes, to researchers and academics interested in seismic analysis and design of concrete buildings, to software developers, code writers, to those with some official responsibility for the use and application of Eurocode 8, and so on. Besides its prime aim as support document for education and training in seismic design of concrete buildings with Eurocode 8, the book complements the currently available background documents for the present version of Eurocode 8 as far as RC buildings are concerned; as such, it will be useful for the coming evolution process of Part 1 of Eurocode 8.

The book puts together those elements of earthquake engineering, structural dynamics, concrete design and foundation/geotechnical engineering, which are essential for the seismic design of concrete buildings. It is not a treatise in any of these areas. Instead, it presumes that the reader is conversant with structural analysis, concrete design and soil mechanics/ foundation engineering, at least for the non-seismic case. Starting from there, it focuses on the applications and extensions of these subject areas, which are necessary for the specialised, yet common in practice, seismic design of concrete buildings. Apart from these fundamentals, which are only covered to the extent necessary for the scope of the book, the book presents and illustrates the full body of knowledge required for the seismic design of concrete buildings – its aim is to provide to the perspective designer of concrete buildings all the tools he/she may need for such a practice; the reader is not referred to other sources for essential pieces of information and tools, only for complementary knowledge.

A key component of the book is the examples. The examples presented at the end of each chapter follow the sequence of its sections and contents, but often gradate in length and complexity within the chapter and from Chapter 2 to 6. Their aim is not limited to illustrating the application of the concepts, methods and procedures elaborated in the respective chapter; quite a few of them go further, amalgamating in the applications additional pieces of information and knowledge in a thought-provoking way. More importantly, Chapter 7 is devoted to an example of a close-to-real-life multistorey concrete building; it covers in detail all pertinent modelling and analysis aspects, presents the full spectrum of analysis results with two alternative methods and highlights the process and the outcomes of detailed design. Last but not least, each chapter from 2 to 6 includes problems (questions) without giving the answers to the reader. The questions are, in general, more challenging and complex than the examples; on average they increase in difficulty from Chapters 2 to 6 and – like most of the examples – often extend the scope of the chapter. Unlike the complete example in Chapter 7, which relies on calculations by computer for the analysis and the detailed design, the questions – and most of the examples – entail only hand calculations, even for the analysis. They are meant to be solved with help and guidance from an instructor, to whom the complete and detailed answers will be available. Moreover, the questions have been formulated in a way that provides flexibility to the instructor to tune the requirements from students to their background and skills, and possibly to extend them according to his/her judgement.

Acknowledgements

The authors would like to acknowledge the technical contributions and support of Dr. M. Kreslin, Dr. I. Peruš, Mr. K. Sinkovič and Dr. G. Tsionis, especially to Chapters 3 and 7. Moreover, sincere thanks are due to Ms. V. Vayenas for her editorial support.

Authors

Michael N. Fardis is Professor and director of Structures Laboratory, Civil Engineering Department, University of Patras, Greece. He is Honorary President of the International Federation for Structural Concrete (*fib*), after serving as its President between 2009 and 2010, as Deputy President between 2007 and 2008 and as Presidium member between 2002 and 2012. He is vice chairman of the CEN Committee TC250 'Structural Eurocodes' (2013–2016) and one of the Directors of the International Association for Earthquake Engineering (2012–2016). He holds MSc in civil engineering (1977) and nuclear engineering (1978) and a PhD in structural engineering (1979), all from the Massachusetts Institute of Technology (MIT), where he taught at the Civil Engineering Department to the rank of Associate Professor. As chairman of the CEN Committee for Eurocode 8: 'Design of Structures for Earthquake Resistance' (1999–2005), he led the development of its six parts into European standards. He is a editorial board member of *Earthquake Spectra, Earthquake Engineering & Structural Dynamics, Structural Concrete, Bulletin of Earthquake Engineering* and *Journal of Earthquake Engineering*. He authored *Seismic Design, Assessment and Retrofitting of Concrete Buildings* (Springer, 2009), and was the lead author of *Designers' Guide to EN1998-1 and EN1998-5: Eurocode 8* (Thomas Telford, 2005) and co-author of *Designers' Guide to Eurocode 8: Design of Bridges for Earthquake Resistance* (ICE Publishing, 2012). He has written seven books in Greek and was the editor of two books published by Springer and co-editor of two more (2010–2013). He has published more than 300 papers in international journals or conference proceedings. He received the 1993 Wason Medal of the American Concrete Institute for the best paper in materials.

Eduardo C. Carvalho graduated at the Technical University of Lisbon in 1974 as a civil engineer and obtained the specialist degree in structures by the National Laboratory for Civil Engineering (LNEC) in Lisbon in 1980. He received the Manuel Rocha Research award in 1982. He is a member of the Portuguese Academy of Engineering. He was a principal researcher at LNEC for 30 years, where he headed the Structural Analysis Division between 1981 and 1995 and the Centre for Earthquake Engineering between 1994 and 2004. Since 2005, he is the chairman of the CEN committee for Eurocode 8: 'Design of Structures for Earthquake Resistance' (CEN/TC250/SC8), after being its Secretary from 1990 to 2005. During this period he participated intensely in the preparation of Eurocode 8. He chairs the Portuguese mirror group for the implementation of Eurocode 8 in Portugal. He was a member of the Administrative Council of CEB-Comité Euro-Internationale du Béton (1995/1998) and is currently member of the Technical Council of federation internationale du béton (*fib*), having received the *fib* medal in 2009. He co-authored the book *Designers' Guide to EN1998-1 and EN1998-5: Eurocode 8* (Thomas Telford, 2005) and the book *Sismos e Edifícios* (*Earthquakes and Buildings*) (Edições Orion, 2008). He was one of the reviewers of the book *Seismic Design, Assessment and Retrofitting of Concrete Buildings* (Springer, 2009) by Michael N. Fardis. He is the chairman of Gapres, a structural design office in Lisbon.

Peter Fajfar is Professor of structural and earthquake engineering at the Faculty of Civil and Geodetic Engineering, University of Ljubljana, Slovenia. He is a member of the Slovenian Academy of Sciences and Arts, of the Slovenian Academy of Engineering and of the European Academy of Sciences (Belgium). He obtained his PhD from University of Ljubljana (1974). He was a visiting professor at McMaster University (1994), Stanford University (1995), University of Bristol (2006) and University of Canterbury (2009). He is the editor of the international journal *Earthquake Engineering and Structural Dynamics* (Wiley) since 2003 and a member of the editorial boards of four international journals. He is a honorary member of the European Association of Earthquake Engineering and was one of the Directors of the International Association of Earthquake Engineering (2004–2012). He was the president of the Yugoslav Association of Earthquake Engineering (1984–1988) and the founding president of the Slovenian Association of Earthquake Engineering (1988–1990). He authored several books in Slovenian and co-authored the first comprehensive book on earthquake engineering in former Yugoslavia. He is the co-editor of three books published by international publishers. In journals and conference proceedings, he has published more than 200 scientific papers. He has been active in the Technical committee TC250/SC8, responsible for the development of Eurocode 8. He was the leader of the implementation process of Eurocode 8 in Slovenia, the first country to implement Eurocode 8. As designer, consultant and reviewer, he has participated in more than 100 design projects, which have mainly dealt with static and dynamic analysis of buildings and civil engineering structures and the determination of seismic design parameters.

Alain Pecker is the president of Géodynamique et Structure, Professor of Civil Engineering at Ecole des Ponts ParisTech in France. He is a member of the French National Academy of Technologies, Honorary President of the French Association for Earthquake Engineering, former President of the French Association for Soil Mechanics and Geotechnical Engineering, one of the former Directors of the International Association of Earthquake Engineering (1996–2004) and is presently member of the Executive Committee of the European Association for Earthquake Engineering. He holds a degree in civil engineering from Ecole des Ponts ParisTech (formerly Ecole Nationale des Ponts et Chaussées) and an MSc from the University of California at Berkeley. He is president of the French committee for the development of seismic design codes and was a member of the drafting panel of EN 1998-5. He is an editorial board member of *Earthquake Engineering & Structural Dynamics*, *Bulletin of Earthquake Engineering*, *Journal of Earthquake Engineering* and *Acta Geotechnica*. He has 130 papers in international journals or conference proceedings. He has been a consultant to major civil engineering projects in seismic areas worldwide, most notably the Vasco de Gama bridge in Lisbon, the Rion Antirion bridge in Greece, the Athens metro, the Second Severn bridge in the United Kingdom, the Chiloe bridge in Chile and several nuclear power plants in France, South Africa and Iran. He has authored two books and co-authored three books, including *Designers' Guide to Eurocode 8: Design of Bridges for Earthquake Resistance* (ICE Publishing, 2012). He received the Adrien Constantin de Magny award from the French National Academy of Sciences (1994).

Chapter 1

Introduction

1.1 SEISMIC DESIGN OF CONCRETE BUILDINGS IN THE CONTEXT OF EUROCODES

As early as 1975, the European Commission launched an action programme for structural Eurocodes. The objective was to eliminate technical obstacles to trade and harmonise technical specifications in the European Economic Community. In 1989, the role of Eurocodes was defined as European standards (European Norms (EN)) to be recognised by authorities of the Member States for the following purposes:

- As a means for enabling buildings and civil engineering works to comply with the Basic Requirements 1, 2 and 4 of the Construction Products Directive 89/106/EEC of 1989, on mechanical resistance and stability, on safety in case of fire and on safety in use (replaced in 2011 by the Construction Products EU Regulation/305/2011 (EU 2011), which also introduced Basic Requirement 7 on the sustainable use of natural resources)
- As a basis for specifying public construction and related engineering service contracts; this relates to Works Directive (EU 2004) on contracts for public works, public supply and public service (covering procurement by public authorities of civil engineering and building works) and the Services Directive (EU 2006) on services in the Internal Market – which covers public procurement of services
- As a framework for drawing up harmonised technical specifications for construction products

It is worth quoting from EU Regulation/305/2011 of the European Parliament and the European Union (EU) Council (EU 2011), given its legal importance in the EU, which deals with the basic requirement for buildings and civil engineering works (called 'Construction works' in the following text) which the Eurocodes address:

Construction works as a whole and in their separate parts must be fit for their intended use, taking into account in particular the health and safety of persons involved throughout the life cycle of the works. Subject to normal maintenance, construction works must satisfy these basic requirements for construction works for an economically reasonable working life.

1. Mechanical resistance and stability
   Construction works must be designed and built in such a way that the loadings that are liable to act on them during their construction and use will not lead to any of the following:
   (a) collapse of the whole or part thereof;
   (b) major deformations to an inadmissible degree;

      (c)   damage to other parts of the construction work or to fittings or installed equipment as a result of major deformation of the load-bearing construction;

      (d)   damage by an event to an extent disproportionate to the original cause.

2. Safety in case of fire

Construction works must be designed and built in such a way that in the event of an outbreak of fire:

      (a)   the load-bearing capacity of the construction work can be assumed for a specific period of time;

      (b)   the generation and spread of fire and smoke within the construction work are limited;

      (c)   the spread of fire to neighbouring construction works is limited;

      (d)   occupants can leave the construction work or be rescued by other means;

      (e)   the safety of rescue teams is taken into consideration.

[...]

4. Safety and accessibility in use

Construction works must be designed and built in such a way that they do not present unacceptable risks of accidents or damage in service or in operation such as slipping, falling, collision, burns, electrocution, injury from explosion and burglaries. In particular, buildings must be designed and built taking into consideration accessibility and use for disabled persons.

[...]

7. Sustainable use of natural resources

Construction works must be designed, built and demolished in such a way that the use of natural resources is sustainable and in particular ensure the following:

      (a)   reuse or recyclability of the construction works, their materials and parts after demolition;

      (b)   durability of the construction works;

      (c)   use of environmentally compatible raw and secondary materials in the construction works.

Totally, 58 EN Eurocode Parts were published between 2002 and 2006, to be adopted by the CEN members and to be fully implemented as the sole structural design standard by 2010. They are the recommended European codes for the structural design of civil engineering works and of their parts to facilitate integration of the construction market (construction works and related engineering services) in the European Union and enhance the competitiveness of European designers, contractors, consultants and material and product manufacturers in civil engineering projects worldwide. To this end, all parts of the EN Eurocodes are fully consistent and have been integrated in a user-friendly seamless whole, covering in a harmonised way practically all types of civil engineering works.

In 2003, the European Commission issued a 'Recommendation on the implementation and use of Eurocodes for construction works and structural construction products' (EC 2003). According to it, EU member states should adopt the Eurocodes as a suitable tool for the design of construction works and refer to them in their national provisions for structural construction products. The Eurocodes should be used as the basis for the technical specifications in the contracts for public works and the related engineering services, as well as in the water, energy, transport and telecommunications sector. Further, according to the 'Recommendation', it is up to a Member State to select the level of safety and protection (which may include serviceability and durability) offered by civil engineering works on its national territory. To allow Member States to exercise this authority and to accommodate geographical, climatic and geological (including seismotectonic) differences, without sacrificing the harmonisation of structural design codes at the European level, nationally determined parameters (NDPs) have been introduced in the Eurocodes (and have been adopted by the Commission in its Recommendation) as the means to provide the necessary flexibility in their application across and outside Europe. Therefore, the Eurocodes allow national choice in all key parameters or aspects that control safety, durability, serviceability and economy of civil engineering works designed and built by them. As a matter of fact, the same approach has been followed when consensus could not be reached for some aspects not related to safety, durability, serviceability or economy. The NDPs in Eurocodes are:

- Symbols (e.g. safety factors, the mean return period of the design seismic action, etc.).
- Technical classes (e.g. ductility or importance classes).
- Procedures or methods (e.g. alternative models of calculation).

Alternative classes and procedures/methods considered as NDPs are identified and described in detail in the normative text of the Eurocode. For NDP symbols, the Eurocode may give a range of acceptable values and will normally recommend in a non-normative note a value for the symbol. It may also recommend a class or a procedure/method among the alternatives identified and described in the Eurocode text as NDPs.

National choice regarding the NDPs is exercised through the National Annex, which is published by each Member State as an integral part of the national version of the EN-Eurocode. According to the Commission's 'Recommendation', Member States should adopt for the NDPs the choices recommended in the notes of the Eurocode, so that the maximum feasible harmonisation across the EU is achieved (diverging only when geographical, climatic and geological differences or different levels of protection make it necessary). National Annexes may also contain country-specific data (seismic zoning maps, spectral shapes for the various types of soil profiles foreseen in Eurocode 8, etc.), which also constitute NDPs. A decision to adopt or not to adopt an Informative Annex of the Eurocode nationally may also be made in the National Annex. If the National Annex does not exercise national choice for some NDPs, the choice will be the responsibility of the designer, taking into account the conditions of the project and other national provisions.

A National Annex may also provide supplementary information, non-contradictory to any of the rules of the Eurocode. This may include references to other national documents to assist the user in the application of the EN. What is not allowed is modifying through the National Annex any Eurocode provisions or replacing them with other rules, for example, national rules. Such deviations from the Eurocode, although not encouraged, are allowed in national regulations other than the National Annex. However, when national regulations are used, allowing deviation from certain Eurocode rules, the design cannot be called 'a design according to the Eurocodes', as by definition this term means compliance with all EN Eurocode provisions, including the national choices for the NDPs.

A National Annex is not required for a Eurocode part, if that part is not relevant to the Member State concerned. This is the case for Eurocode 8 in countries of very low seismicity.

The approved Eurocodes were given to the National Standardisation Bodies (NSB) in English, French and German. NSBs have adopted one of these three official versions, or have translated them into their national language, or have adopted the translation by another NSB. This national version is supreme in the country over those in any other language (including the original three-language version). NSBs have also published the National Annexes, including the national choice for the NDPs, after calibrating them so that, for the target safety level, structures designed according to the national version of the Eurocodes do not cost significantly more than those designed according to National Standards that were applicable hitherto.

Member States are expected to inform the Commission of the national choices for the NDPs. The impact of differences in the national choices upon the end product of the design (the works or their parts), as far as the actual level of protection and the economy provided is concerned, will be assessed jointly by the Member States and the Commission. According to the European Commission's 'Recommendation', on the basis of the conclusion of such an evaluation, the Commission may ask Member States to change their choice, so that divergence within the internal market is reduced.

National Standards competing or conflicting with any EN Eurocode part have been withdrawn, and the Eurocodes have become the exclusive standards for structural design in the European Union.

In the set of 10 Eurocodes, two cover the basis of structural design and the loadings ('actions'), one covers geotechnical and foundation design and five cover aspects specific to concrete, steel, composite (steel-concrete), timber, masonry or aluminium construction. Instead of distributing seismic-design aspects to the Eurocodes on loadings, materials or geotechnical design, all aspects of seismic design are covered in Eurocode 8: 'EN1998: Design of Structures for Earthquake Resistance'. This is for the convenience of countries with very low seismicity, as it gives them the option not to apply Eurocode 8 at all.

Seismic design of concrete buildings is covered in EN1998-1 'General rules, seismic actions, rules for buildings', also called (including throughout this book) Part 1 of Eurocode 8. However, this part of Eurocode 8 is not sufficient for the seismic design of concrete buildings. Therefore, it is meant to be applied as part of a package, which includes all Eurocodes needed for the package to be self-sufficient, namely:

- EN1990: 'Basis of structural design'
- EN1991-1-1: 'Actions on structures – General actions – Densities, Self-weight and Imposed loads for buildings'
- EN1991-1-2: 'Actions on structures – General actions – Actions on structures exposed to fire'
- EN1991-1-3: 'Actions on structures – General actions – Snow loads'
- EN1991-1-4: 'Actions on structures – General actions – Wind actions'
- EN1991-1-5: 'Actions on structures – General actions – Thermal actions'
- EN1991-1-6: 'Actions on structures – General actions – Actions during execution'
- EN1991-1-7: 'Actions on structures – General actions – Accidental actions'
- EN1992-1-1: 'Design of concrete structures – General – General rules and rules for buildings'
- EN1992-1-2: 'Design of concrete structures – General – Structural fire design'
- EN1997-1: 'Geotechnical design – General rules'
- EN1997-2: 'Geotechnical design – Ground investigation and testing'
- EN1998-1: 'General rules, seismic actions, rules for buildings'
- EN1998-3: 'Assessment and retrofitting of buildings'
- EN1998-5: 'Foundations, retaining structures, geotechnical aspects'

Besides Part 1 of Eurocode 8 (CEN 2004a), four other Eurocodes from the package are important for the seismic design of concrete buildings: EN1992-1-1 (CEN 2004b), EN 1997-1 (CEN 2003), EN1998-5 (CEN 2004c) and EN1998-3 (CEN 2005), which are referred to in this book as Eurocode 2, Eurocode 7, and Parts 5 or 3 of Eurocode 8, respectively.

1.2 SEISMIC DESIGN OF CONCRETE BUILDINGS IN THIS BOOK

This work is addressed to graduate or advanced undergraduate students, researchers and academics interested in the seismic response, behaviour or design of concrete buildings, seismic-design professionals, software developers and other users of Eurocode 8, even code writers. Familiarity and experience of the reader in structural dynamics, earthquake engineering or seismic design are not presumed: although the book does not go in depth in each one of these topics, it is self-sufficient in this respect. However, a background in structural analysis and in design of concrete structures and foundations, be it without reference to seismic loading, is necessary. Familiarity with the notation which has become the international standard and is currently used in Europe is also desirable.

In order to define the target of design at the outset, according to the objective of Chapters 2–7, this chapter presents two of the requirements of Eurocode 8 (namely protection of life in a rare earthquake and protection of property in a more frequent one) for the performance of buildings of different material types and the way they are implemented. A general overview of the physics and the mechanics of earthquakes and of their typical effects on concrete buildings and their foundations is provided in Chapter 2, along with an overview of the effects on other geotechnical works. Pictures and descriptions of typical damage help the reader to understand and appreciate the specific objectives of Eurocode 8 and the means it uses to achieve them.

Chapter 3, after presenting the fundamentals of structural dynamics, with emphasis on dynamic loading due to seismic ground motions, gives a fairly detailed and complete description of the methods adopted in Eurocode 8 for the linear or non-linear analysis of buildings under seismic loading, alongside the appropriate modelling. The fundamental concept of the reduction of elastic forces by a factor, which derives from the deformation capacity as well as from the ability of the structure to dissipate energy and links linear analysis with non-linear response ('behaviour factor' in Eurocode 8), is introduced. Three short analysis examples of simple structures illustrate the basic points of the chapter.

Chapter 4 covers the principles of sound conceptual seismic design of concrete buildings, emphasising its importance and the challenges it poses. It presents the available system choices for the superstructure and the foundation, their advantages and disadvantages, along with ways to profit the most from the former and minimise the impact of the latter. It then proceeds with the fundamentals of capacity design, which is the main means available to the designer according to Eurocode 8 to control the inelastic seismic response of the building. It is to be noted that although the concept of capacity design originated and was first introduced into seismic-design codes in New Zealand, it is in Eurocode 8 that it has found its widest scope of application in its purest and most rigorous form, with very little empirical additions or interventions. The specifics of practical application are described in Chapter 5. Chapter 4 closes with the choices offered by Eurocode 8 for trading deformation capacity and ductility for strength, alongside the values prescribed for the 'behaviour factor' under the various possible circumstances. A good number of short examples illustrate various aspects of conceptual design, as well as the use of the 'behaviour factor' to reduce seismic design loads.

Chapters 5 and 6 cover all aspects of detailed design of the superstructure of concrete buildings and their foundation. Although they deal primarily with the base case, leaving aside special applications or cases, they go into significant depth, presenting everything a designer may need for the complete seismic design of a concrete building. Numerous short examples are given, with transparent hand calculations.

The book culminates in the design of a real-life building, complete with analysis using the two main methods as per Eurocode 8, capacity design across the board and sample detailed design calculations for all types of elements:

- Design of beams (and deep foundation beams) at the ultimate limit state (ULS) in flexure and serviceability limit state (SLS) design for crack and stress control under service loads.
- Check of columns for second-order effects under the factored gravity loads ('persistent and transient design situation' in EN1990) and dimensioning of their vertical reinforcement for the ULS in flexure with axial force.
- Capacity design of beams and columns in shear, with ULS design of their shear reinforcement, including detailing for confinement.
- Dimensioning of the vertical reinforcement of walls for the ULS in flexure with axial force and of their horizontal reinforcement for capacity-design shears, with detailing for ductility.
- Capacity design of footings at the ULS in flexure, shear or punching shear, with capacity-design verification of the bearing capacity of the soil.

Outcomes are illustrated through diagrams of internal forces from the two types of analysis, full construction drawings of the framing and detailing, and representative examples of all sorts of design/dimensioning calculations.

1.3 SEISMIC PERFORMANCE REQUIREMENTS FOR BUILDINGS IN EUROCODE 8

1.3.1 Life safety under a rare earthquake: The 'design seismic action' and the 'seismic design situation'

The main concern in Eurocode 8 for buildings subjected to earthquake is safety of the public – occupants and users of the facility. Eurocode 8 pursues safety of life under a specific earthquake, called 'design seismic action', whose choice is left to the National Authorities, as an NDP. The 'design seismic action' should be a rare event, with low probability of being exceeded during the conventional design life of the building. For 'ordinary' buildings, Eurocode 8 recommends setting this probability to 10% in 50 years. This is equivalent to a mean return period of 475 years for earthquakes at least as strong as the 'design seismic action'. The performance requirement is then to avoid failure ('collapse') of structural members or components under this 'design seismic action'.

Member integrity under the 'design seismic action' is verified as for all other types of design loadings: it is ensured that members possess a design resistance at the Ultimate Limit State (ULS), R_d, which exceeds the 'action effect' (internal force or combinations thereof), E_d, produced by the 'design seismic action', acting together with the long-term loadings expected to act when this seismic action occurs:

$$R_d \geq E_d \tag{1.1}$$

These long-term loadings are the arbitrary-point-in-time loads, or, in Eurocode terminology, the 'quasi-permanent combination' of actions, $\sum_j G_{k,j} + \sum_i \psi_{2,i} Q_{k,i}$, that is, the loads acting essentially all the time. The Eurocode 1990 'Basis of Structural Design' (CEN 2002) defines the quasi-permanent value of the other actions as:

- The nominal value (subscript: k) of permanent loads, $G_{k,j}$ (where index j reflects the possibility of having several types of permanent loads: dead loads, earth or water pressure, etc.)
- The expected value of variable actions, such as the imposed (i.e. live) gravity loads or snow at an arbitrary-point-in-time ('quasi-permanent value'); if $Q_{k,i}$ is the nominal value (i.e. the characteristic, hence the subscript k) of variable action i, its 'quasi-permanent value' is taken as $\psi_{2,i}Q_{k,i}$

The values of $\psi_{2,i}$ are given in Normative Annex A1 of Eurocode EN 1990 as an NDP, with recommended values as follows:

- $\psi_{2,i} = 0.3$ on live loads in residential or office buildings and traffic loads from vehicles of 30–160 kN
- $\psi_{2,i} = 0.6$ on live loads in areas of public gathering or shopping, or on traffic loads from vehicles less than 30 kN
- $\psi_{2,i} = 0.8$ on live loads in storage areas
- $\psi_{2,i} = 0$ for live loads on roofs
- $\psi_{2,i} = 0$ for snow on the roof at altitudes less than 1000 m above sea level in all CEN countries except Iceland, Norway, Sweden and Finland, or $\psi_{2,i} = 0.2$ everywhere in these four countries and at altitudes over 1000 m above sea level everywhere else
- $\psi_{2,i} = 0$ for wind or temperature

The combination of the 'design seismic action' and the 'quasi-permanent combination' of actions, $\sum_j G_{k,j} + \sum_i \psi_{2,i}Q_{k,i}$, is called 'seismic design situation' in the Eurocodes. In common language, it is the design earthquake and the concurrent actions.

The 'seismic design situation' is the condition for which the local verifications of Equation 1.1 are carried out; the 'quasi-permanent combination' comprises the loads acting at the instant of the 'design seismic action' on a limited part of the building and directly affects the local verification. These loads are always taken into account in E_d, regardless of whether they are locally favourable or unfavourable for the verification of Equation 1.1. However, the inertia forces are considered to be produced not by the full mass corresponding to $\psi_{2,i}Q_{k,i}$, but by a fraction thereof. This is because it is considered unlikely to have 100% of the 'quasi-permanent value' of variable action i, $\psi_{2,i}Q_{k,i}$, applied throughout the building. Moreover, some masses associated to live loads may be non-rigidly connected to the structure and can vibrate out of phase to their support, or with smaller amplitude.

The fraction of $\psi_{2,i}Q_{k,i}$ considered to produce inertia forces through its mass is an NDP. Its recommended value is 0.5, for all storeys (except the roof) of residential or office use, or those used for public gathering (except shopping), provided that these storeys are considered as independently occupied. In storeys of these uses which are considered to have correlated occupancies, the recommended fraction is 0.8. There is no reduction of the masses corresponding to $\psi_{2,i}Q_{k,i}$ for uses other than the above, or on roofs.

The 10% probability of exceedance in 50 years, or the mean return period of 475 years are recommended in Eurocode 8 for the 'design seismic action' of 'ordinary' buildings. To offer better protection of life to facilities with large occupancy and to reduce damage to facilities critical for the post-disaster period (e.g. hospitals, power stations, etc.), the 'design seismic action' is multiplied by an 'importance factor' γ_I (cf. Section 4.1). By definition, for buildings of ordinary importance $\gamma_I = 1.0$; for facilities other than 'ordinary', the importance factor γ_I is an NDP, with recommended values as in Table 1.1.

Table 1.1   Importance classes and factors for buildings in Eurocode 8

Importance Class and Type of Facility	γ_I
I: Not occupied by people; temporary or auxiliary buildings	0.8
II: Ordinary	1.0
III: High consequences (large occupancy, congregation areas, etc.), cultural facilities	1.2
IV: Critical, essential for civil protection (hospitals, fire stations, power plants, etc.)	1.4

1.3.2 Limitation of damage in occasional earthquakes

In addition to life safety under the 'design seismic action', Eurocode 8 aims at protecting property, by minimizing structural and non-structural damage in occasional, more frequent, earthquakes. A specific occasional earthquake, called 'serviceability seismic action' or 'damage limitation seismic action', is selected for that purpose by national authorities, as NDP. The recommendation in Eurocode 8 is to choose an earthquake with 10% probability of being exceeded in 10 years, which corresponds to a mean return period of 95 years. Specifically for buildings, Eurocode 8 introduces the ratio of the 'serviceability seismic action' to the 'design seismic action', v, and considers it an NDP. For buildings of ordinary or lower importance ('Importance Classes' I and II in Table 1.1), it recommends $v = 0.5$; for importance above ordinary ('Importance Classes' III and IV) a value of 0.4 is recommended for v. In the end, this gives about the same level of property protection to Importance Classes II and III; property protection is 15%–20% lower for 'Importance Class' I and 15% higher for Class IV, compared to Classes II and III.

After the occurrence of the 'serviceability seismic action', the structure itself is meant to be free of permanent deformations, not to need any repair and to retain its full strength and stiffness. Non-structural elements, notably partition walls, may have suffered some damage, which should be easily and economically repairable later.

The verification required as per Eurocode 8 for buildings is carried out in terms of the inter-storey drift ratio (i.e. the relative horizontal displacement of the mass centres of two successive floors due to the 'serviceability seismic action', Δu, divided by the storey height, h_{st}). For a partition wall, this corresponds to an average shear strain in the plane of a wall panel. If the partitions are in contact or attached to the structure and follow its deformations, the limits to be met by this average shear strain are:

$$\Delta u/h_{st} \leq 0.5\%, \text{ for brittle partitions;} \tag{1.2a}$$

$$\Delta u/h_{st} \leq 0.75\%, \text{ for ductile partitions (uncommon in practice)} \tag{1.2b}$$

For buildings without partitions, or with partitions not attached to the structure in a way that imposes on them horizontal relative deformations, the limit for the inter-storey drift ratio is:

$$\Delta u/h_{st} \leq 1\% \tag{1.2c}$$

Equation 1.2c refers to the structure itself and aims to protect its members from large excursions in the inelastic range under the 'serviceability seismic action'.

If the structure is a frame, Equation 1.2 may govern the size of the cross sections of its members.

Chapter 2

Earthquakes and their structural and geotechnical effects

2.1 INTRODUCTION TO EARTHQUAKES

An earthquake is a sudden rupture along a fault. The slip or offset (i.e. relative displacement) along the fault due to the rupture may reach several meters over a surface on the fault that may exceed 10,000 km². The sudden slip generates seismic waves which propagate in the earth, inducing, in turn, vibration of the ground in all three directions. In addition, under certain circumstances and close to the fault, permanent displacements of several hundreds of millimetres to several meters may affect the ground surface. Examination of the distribution of seismicity on the earth surface during the period 1900–2012 (Figure 2.1), for instance, shows that earthquake occurrence is not uniformly distributed over the Earth's surface, but tends to concentrate along well-defined lines, which are known to be associated with the boundaries of 'plates' of the Earth's crust (Figure 2.2).

Plate tectonics (Wegener 1915) is nowadays recognised as the general framework to explain the distribution of seismicity over the Earth. The elastic rebound theory (Reid 1910) provides the most satisfactory explanation for the types of earthquakes causing potentially damaging surface motions. This concept is displayed in Figure 2.3a represents the slip that takes place over time along a fault plane; slip accumulates during long intervals at a slow, but constant, rate until the strains and stresses that have developed along the fault plane exhaust the material strength; a rupture will then start at a critical location in the fault zone, producing a sudden slip jump. Figure 2.3b depicts a plan view of the ground surface, with the fault trace and a dotted line representing a fence crossing the fault; during the long period of slow strain accumulation the fence gently deforms, but when the fault rupture takes place the fence breaks. Figure 2.4 shows an example of the phase of straining at the San Andreas Fault in the town of Hollister (California). When the rupture takes place, the accumulated strain energy is suddenly released and is converted into heat and radiated energy carried by the elastic waves. Depending on the state of stress in the rock that leads to the rupture, relative displacements on the fault plane may be mainly horizontal; in that case, the fault is called a 'strike-slip' fault. If the relative displacement on the fault is mainly vertical, the fault is called 'normal' or 'reverse', depending on the relative movement of the two parts separated by the fault ('walls'). Figure 2.5 shows the different fault types.

With the development of satellite imagery (GPS, radar interferometry, etc.), it is nowadays possible to measure ground displacement of the order of 1 mm/year and therefore to know the slip rates along fault planes. From that information, the recurrence period of major earthquakes can be estimated. So, slip rate measurements represent an alternative to the more traditional estimation of earthquake recurrence intervals from historical data.

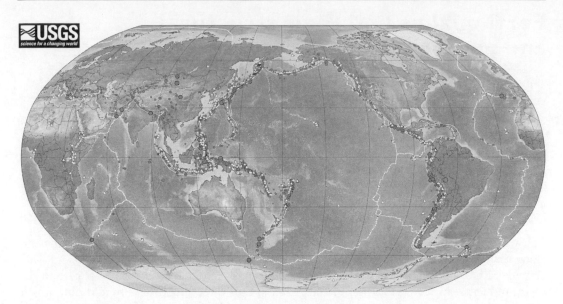

Figure 2.1 World seismicity between 1900 and 2012. (From United State Geological Survey – USGS.)

2.1.1 Measure of earthquake characteristics: Magnitudes

Having estimated the location of a possible future earthquake, it is necessary from an engineering standpoint to characterise the strength of that earthquake. In the old days, earthquakes were classified according to their effects on structures: depending on the amount of damage caused to a building of a given typology (masonry structure, wooden structure, etc.), and/or on the effects on the soil and people, the earthquake was assigned an intensity. Intensity is usually denoted by a Roman numeral. Several intensity scales exist: the MSK scale in the USA, the JMA scale in Japan and so on. In Europe the most recent intensity scale is the European Macroseismic Scale (EMS): it distinguishes six classes of vulnerability for buildings and 12 degrees (I–XII) – the latest revision was in 1998. Such intensity scales are the only means to characterise historical earthquakes for which no instrumental records are available.

In order to advance beyond the somewhat subjective characterisation of earthquakes with intensity scales, a quantitative parameter, the magnitude M_L, was introduced by Richter in 1935. It is an instrumental measurement based on the amplitude of seismic waves, intended to quantify the energy released by the fault rupture. Since the amplitude depends not only on the strength of the earthquake, but also on the distance of the recording station from the source and on the recording instrument, various corrections have been introduced in the calculation of M_L. As the decay of seismic waves with distance may vary from one region to another, different magnitude scales have been introduced, which, although still based on the logarithm of the displacement amplitude, measure the energy radiated in different frequency bands: the body wave magnitude, m_B, measures the energy at 1 Hz, the surface wave magnitude, M_S, measures the energy at 0.05 Hz. Other scales have been calibrated locally; an example is the magnitude M_{JMA} of the Japanese Meteorological Agency. There seems to be a consensus, nowadays, among seismologists to use the same definition of the magnitude, namely, one based on the seismic moment M_0. The moment magnitude, M_w, is defined as

$$M_w = \frac{2}{3} M_0 - 6 \qquad (2.1)$$

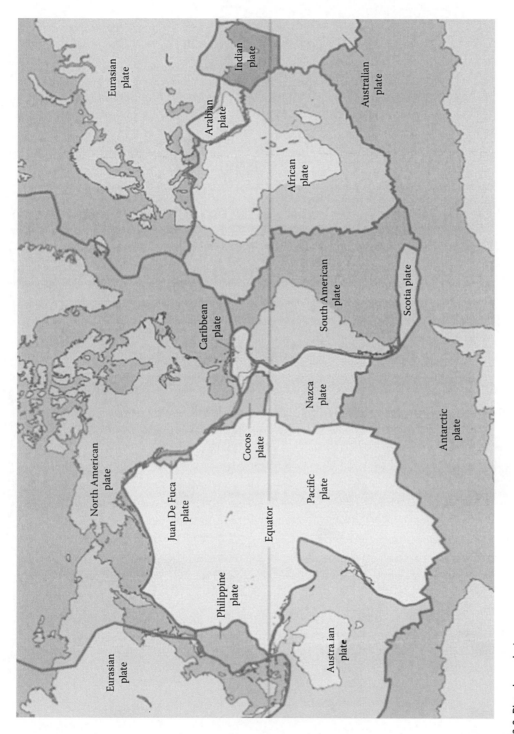

Figure 2.2 Plate boundaries.

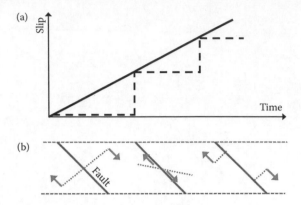

Figure 2.3 Elastic rebound theory: (a) slip as a function of time; (b) from left to right: initial stage, straining before earthquake, after earthquake.

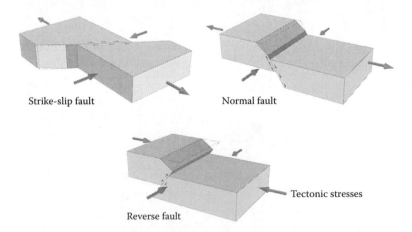

Figure 2.4 Fence offset in Hollister, California.

Figure 2.5 Fault types.

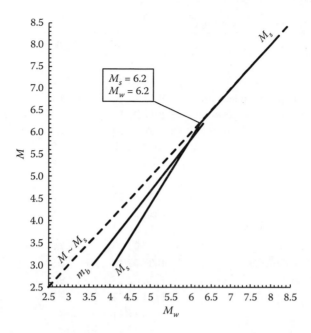

Figure 2.6 Relationship between magnitude scales.

with the seismic moment, M_0, defined as

$$M_0 = G\bar{u}A \tag{2.2}$$

where G is the material shear modulus, A is the rupture area, and $\bar{u}$ is the average slip; note that M_0 has the dimension of energy.

Not all magnitude scales are equivalent; empirical correlations have been developed to give the relationship between them. Figure 2.6, adapted from Scordilis (2006), depicts a recent correlation based on 2000 records from around the world.

Since magnitudes are computed as the logarithm of a displacement amplitude, they can be negative for small earthquakes, not perceived. In addition, the magnitude has no upper bound and can theoretically reach large values. In reality, the rock strength and physical limits of fault and rupture lengths set an upper limit on the amount of energy that can be radiated from a source. The largest magnitude value that has been assigned so far to a seismic event is 9.5, for the Chile earthquake of 1960. Table 2.1 lists the largest earthquakes ever recorded worldwide. In Europe, the largest expected intra-plate events may reach magnitudes in the order of 7.0. However, for inter-plate events in the fault between the Eurasian and the African plates (see Figure 2.2), magnitudes may reach 8 to 8.5, as was the case of the great Lisbon earthquake in 1755.

Note that the increase of one unit in the magnitude corresponds to an energy release multiplied by 31.6. Hence, in terms of its energy release, a magnitude 8 event is 1.000 times larger than a magnitude 6 event.

2.1.2 Characteristics of ground motions

Magnitude by itself is not an indicator of the damaging potential of an earthquake: damage also depends on the distance from the rupture area to the site, on local soil conditions and

Table 2.1 Largest seismic events by magnitude

	Location	Date UTC	Magnitude	Latitude	Longitude
1.	Chile	1960 05 22	9.5	−38.29	−73.05
2.	Prince William Sound, Alaska	1964 03 28	9.2	61.02	−147.65
3.	Off the west coast of northern Sumatra	2004 12 26	9.1	3.30	95.78
4.	Near the east coast of Honshu, Japan	2011 03 11	9.0	38.322	142.369
5.	Kamchatka	1952 11 04	9.0	52.76	160.06
6.	Offshore Maule, Chile	2010 02 27	8.8	−35.846	−72.719
7.	Off the coast of Ecuador	1906 01 31	8.8	1.0	−81.5
8.	Rat Islands, Alaska	1965 02 04	8.7	51.21	178.50
9.	Northern Sumatra, Indonesia	2005 03 28	8.6	2.08	97.01
10.	Assam–Tibet	1950 08 15	8.6	28.5	96.5
11.	Off the west coast of northern Sumatra	2012 04 11	8.6	2.311	93.063
12.	Andreanof Islands, Alaska	1957 03 09	8.6	51.56	−175.39
13.	Southern Sumatra, Indonesia	2007 09 12	8.5	−4.438	101.367
14.	Banda Sea, Indonesia	1938 02 01	8.5	−5.05	131.62
15.	Kamchatka	1923 02 03	8.5	54.0	161.0
16.	Chile–Argentina border	1922 11 11	8.5	−28.55	−70.50
17.	Kuril Islands	1963 10 13	8.5	44.9	149.6

Source: United State Geological Survey (USGS).

on other less important factors. For instance, a low magnitude earthquake ($M < 6$) occurring just below a town at shallow depth, like the Agadir earthquake in Morocco in 1960, might well be more damaging than a large magnitude earthquake ($M \sim 7.5–8.0$) occurring at large distance (~ 70 km).

The most direct measurement of ground motion has for long been the peak ground acceleration (PGA), as it was the only quantity accessible from analog records. For destructive earthquakes, it is larger than 2 m/s^2 and may reach values above 10 m/s^2. However, it is recognised that PGA is far from sufficient to characterise the response of a given structure to an earthquake. Furthermore, it is poorly correlated to the observed damage. In order to approximately account for the spectral content of ground motions, but still preserving the simplicity of using maximum values, the peak ground velocity (PGV) or the peak ground displacement (PGD) may be used: PGA characterises the high frequency content of the motion (>5 Hz), PGV the intermediate frequency range (0.5 to a few Hertz) and PGD the low frequency range (<0.1 Hz). For destructive earthquakes, PGV varies typically from a few cm/s to more than 1 m/s. However, neither PGV nor PGD correlates well with the observed damage (i.e. the macroseismic intensity scale). Furthermore, their determination is less accurate than that of PGA: it requires integration of the time-history of the acceleration record (see Figure 2.7 for examples of such records), a process which, for analog records and even early digital ones, is sensitive to low-frequency noise.

Quantification of the frequency content can be achieved more accurately through the Fourier response spectrum, or the single-degree-of-freedom (SDOF) response spectrum (see Section 3.1.2). Seismologists prefer the Fourier spectra, as they are related to the physics of wave propagation and emission of energy at the source. From an engineering standpoint, Fourier spectra are not convenient; SDOF response spectra are commonly used instead. However, the response spectrum does not convey all the information about a seismic motion. For instance, it does not provide information on the duration of the motion, which may be a key parameter when the structure behaves inelastically. For that reason, other parameters

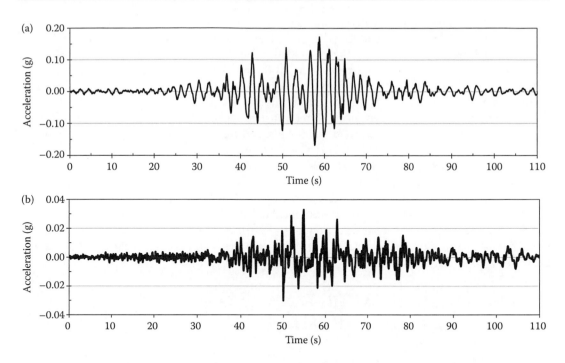

Figure 2.7 Records of the 1985 Michoacán Guerrero earthquake in Mexico City: (a) SCT (soft soil); (b) Tacubaya (rock).

may be introduced to characterise a ground motion. These parameters, although still lacking direct engineering applications, may be used, for instance, for the selection of natural ground motion records representing a given seismic scenario. Two of them are

- The cumulative absolute velocity $CAV = \int_0^{T_f} |a(t)| dt$ (2.3)

- The Arias intensity $I_A = \dfrac{\pi}{2g} \int_0^{T_f} a^2(t) dt$ (2.4)

  with the notation:
  $a(t)$ ground motion acceleration (m/s²)
  T_f total duration of ground motion (s)
  g acceleration of gravity (m/s²)

- Duration, defined as the time interval necessary for the Arias intensity to build up from 5% to 95% of its full value

2.1.3 Determination of ground motion parameters

Based on the statistical analyses of recorded ground motions, ground motion prediction equations (GMPEs) have been developed that allow prediction of one of the parameters characterising the motion as a function of several independent parameters characterising the earthquake. GMPEs have been developed initially for PGA, but nowadays GMPEs exist for almost every parameter listed in the previous subsection: PGV, PGD, duration, Arias intensity, cumulative absolute velocity (CAV) and pseudo-spectral acceleration (PSA, ordinate of the one-degree of freedom response spectrum). The independent parameters were originally the earthquake magnitude, the distance to site (epicentral or focal distance, distance to surface projection of fault, etc.) and local geotechnical conditions. The first published GMPEs exhibited a large scatter, due to the poor definition of the independent parameters: the magnitude scale was not homogeneous, several definitions of the distance were used without distinction and soil classification was rather crude (rock, stiff soil, soft soil). With the increasing number of records, and in an attempt to reduce the scatter in the prediction, the independent parameters are now better constrained (moment magnitude is uniformly accepted, distances are better defined, local soil conditions are quantitatively assessed with a measurable parameter, like shear wave velocity) and new independent parameters, which are not always known before the earthquake, are introduced, like depth to bedrock, tectonic environment, fault mechanism and so on.

The most popular GMPEs in use are the so-called NGA West GMPEs (Power et al. 2008), which are valid for an active tectonic context. Such GMPEs have been recently developed for Europe, for example, Akkar and Bommer (2010). A comprehensive overview of recent GMPEs can be found in Douglas (2010). To illustrate the format of a typical GMPE, the one derived by Akkar and Bommer (2010) is reproduced as follows:

$$\log(\text{PSA}) = b_1 + b_2 M + b_3 M^2 + (b_4 + b_5 M)\log\sqrt{R_{jb}^2 + b_6^2}$$
$$+ b_7 S_S + b_8 S_A + b_9 F_N + b_{10} F_R + \varepsilon\sigma \tag{2.5}$$

where PSA is the pseudo-spectral acceleration (denoted in Chapter 3 as A), M is the moment magnitude, R_{jb} is the Joyner–Boore distance; S_S and S_A take the value 1 for soft (mean shear wave velocity in the upper 30 metres, $V_{s30} < 360$ m/s) and stiff soil sites (360 m/s < $V_{s30} < 750$ m/s), respectively, or zero for rock sites defined by $V_{s30} > 750$ m/s. Similarly, F_N and F_R take the value of unity for normal and reverse faulting earthquakes respectively, otherwise, they are equal to zero; ε is the number of standard deviations above or below the mean value of log(PSA); σ represents the (inter-event and intra-event) variability. All coefficients b_i ($i = 1$–10) are period-dependent parameters, tabulated by Akkar and Bommer for periods between 0 and 3.0 s. Users of GMPEs must realise that, although considerable improvement has been achieved to better constrain the independent parameters, there still exists a large scatter in the prediction, with typical standard deviations of 0.3 for the logarithm of PSA (a factor of 2 for PSA).

2.1.4 Probabilistic seismic hazard analyses

Design of buildings for earthquake loading first requires quantification of the possible ground motions that would affect the structure during its life time. The goal of probabilistic seismic hazard analyses (PSHA) is to quantify the rate (or the probability) of exceeding various ground motion levels at a site, given all possible earthquakes that can affect the site.

The approach to PSHA was first formalised by Cornell (1968) and is now commonly implemented in earthquake engineering (Abrahamson 2000). Originally, PGA has been used to quantify ground motions, but today, with the emergence of sophisticated GMPEs, PSA is preferred. In the following, the general framework of PSHA will be presented for one parameter of the ground motion, Y. This parameter can be PGA, PGV and PSA at any period and so on.

PSHA involves three steps: (1) the definition of the seismic hazard source model(s); (2) the specification of the GMPE(s), and (3) probabilistic calculations.

1. *Seismic hazard source model.* The seismic hazard source model is a description of all earthquake scenarios that can affect the site; each earthquake scenario has its own magnitude, M, location, L, and annual rate of occurrence, r. There are an infinite number of earthquake scenarios. For example, the magnitude may be a single value associated to a specific fault (called the characteristic magnitude) or may have a continuous distribution of possible values. Observations have shown that the distribution of magnitude usually follows a simple relationship, called the Gutenberg–Richter law. If $n(M)$ represents the number of events with a magnitude between $M - \Delta M$ and $M + \Delta M$ occurring in a given area during a given time interval, n varies with M as

$$\log(n) = a - bM \tag{2.6}$$

Coefficient a varies from one area to another and characterises the seismicity of the area; coefficient b is always close to 1, indicating that the number of occurrences of earthquakes with magnitude $M + 1$ in a region is ten times less than the number of occurrences of earthquakes with magnitude M. The Gutenberg–Richter law for Earth as a whole indicates that, on average, one earthquake with magnitude above 8 occurs per year, one with magnitude at least 7 every month and two with magnitudes greater than 6 every week.

The rate of occurrence of earthquakes unfortunately does not obey simple rules, although simple mechanical interpretations have been attempted, like a gradual stress accumulation at faults with imposed displacements at boundaries. Real physics seems more complicated, and despite the fact that sophisticated time-dependent models have been proposed, a Poisson model, in which occurrences are random, still prevails in PSHA. According to this model, the probability of having more than one earthquake on a given source in T years is given by

$$P = 1 - \exp(-rT) \tag{2.7}$$

2. *Ground motion prediction equations.* GMPEs have been discussed in Section 2.1.3. Care should be exercised to choose an appropriate GMPE for each seismic source model.
3. *Probabilistic calculations.* Suppose that the seismic source model has provided N earthquake scenarios for a particular site, each of them characterised by a given magnitude M_i, location and rate r_i. From the scenario location one can define the distance to the site, D_i, the tectonic regime, fault mechanism, the soil conditions of the site and so on; the GMPE then provides the value of the ground motion parameter of interest $Y = g(M_i, D_i, \ldots)$. The probability of Y exceeding a value Y_0 is then given by

$$P_i\left(Y > Y_0\right) = \frac{1}{\sigma_{\ln Y}\sqrt{2\pi}} \int_{Y_0}^{\infty} \exp\left[-\frac{1}{2}\left(\frac{\ln Y - g(M,D,\ldots)}{\sigma_{\ln Y}}\right)^2\right] dY \tag{2.8}$$

The annual rate at which Y is exceeded due to this particular scenario is $r_i\, P_i(\ln Y > \ln Y_0)$. Summing up all possible scenarios, the annual rate of exceeding Y_0 at the site is obtained as

$$R_{\text{tot}}(Y > Y_0) = \sum_{i=1}^{N} r_i P_i(\ln Y > \ln Y_0) \qquad (2.9)$$

Finally, using the Poisson distribution, the probability of exceeding the ground motion Y_0 in the next T years is

$$P(Y > Y_0, T) = 1 - \exp(-R_{\text{tot}} T) \qquad (2.10)$$

This result is known as a hazard curve. Examples of hazard curves at a site are depicted in Figure 2.8 for several seismic sources: each curve corresponds to one seismic source. The total hazard at the site is calculated using Equation 2.9.

Note that the calculations have been presented earlier, as done in numerical calculations, for a finite number of discrete earthquake scenarios. In practice, their number is infinite: each discrete location on the fault plane is capable of producing an earthquake of magnitude M with a continuous distribution. Therefore, the finite discrete summations are replaced by continuous integrals over the fault area and magnitude distribution.

In performing PSHA it is mandatory to account for uncertainties. Today's practice classifies the uncertainties into epistemic uncertainties and aleatory ones. Epistemic uncertainties arise from a lack of knowledge, and they can theoretically be reduced with additional studies, investigations and so on. For instance, determination of the shear wave velocity profile at a site may be improved by increasing the number of measurements; a large number of measurements will help bracket more accurately the velocities and, therefore, reduce the uncertainty. Aleatory uncertainties are due to the variability inherent in nature and cannot

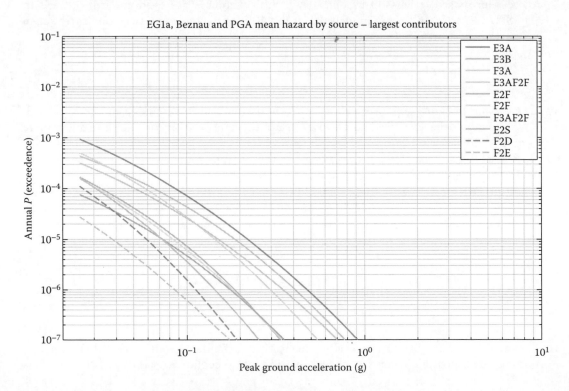

Figure 2.8 Hazard curves: each curve corresponds to a given seismic source.

be reduced with additional information. For instance, the depth of an earthquake on a fault plane can be viewed as an aleatory variable (in a given depth range). Epistemic uncertainties are best handled by considering different alternatives to which different weights are assigned; this results in a logic tree approach, while aleatory uncertainty is treated with a distribution around the mean value. As an illustration of this concept, a set of GMPEs could be considered viable for our PSHA; which one is the best is uncertain, until more and more records become available; to handle this epistemic uncertainty the hazard curve is computed for all GMPEs, each result being assigned a degree of confidence through a weight in the logic tree. On the other hand, each GMPE has its own uncertainty reflected by the range (mean, fractiles) of the predicted ground motion Y for a given magnitude, distance and so on. This uncertainty shall be carried along the calculations of the hazard curve. For a more thorough discussion of uncertainties, one can refer to Bommer and Scherbaum (2008).

2.2 EFFECTS OF EARTHQUAKES ON CONCRETE BUILDINGS

2.2.1 Global seismic response mechanisms

A structure supported on the ground follows its motion during an earthquake, developing, as a result, inertial forces. A typical concrete building is neither stiff enough to follow the ground motion as a rigid body, nor sufficiently flexible to stay in the same absolute position in space, while its base adheres to the shaking ground. As we will see in Sections 3.1.1, 3.1.2 and 3.1.4, the building will respond to the seismic inertial forces by developing its own oscillatory motion. The amplitude, frequency content and duration of that motion depend on both the corresponding characteristics of the ground shaking and on the dynamic properties of the structure itself (see Section 3.1.1).

The base of the structure will follow all three translational and all three rotational components of the motion of the ground it is supported on; accordingly, its dynamic response will be in 3D, with displacements and rotations in all three directions. However, for a typical concrete building, only the structural effects of the two horizontal translational components of the ground motion are worth considering. The – by and large poorly known – rotational components are important only for very tall and slender structures, or those with twisting tendencies very uncommon in buildings designed for earthquake resistance. Concerning the vertical translational component, its effects are normally accommodated within the safety margin between the factored gravity loads (e.g. the 'persistent and transient design situation' of the Eurocodes, where the nominal gravity loads enter amplified by the partial factors on actions) for which the building is designed anyway, and the quasi-permanent ones considered to act concurrently with the 'design seismic action' (see Section 1.3.1). Important in this respect is the lack of large dynamic amplification of the vertical component by the vibratory properties of the building in the vertical direction.

As we will see in detail in Chapters 3 and 4, a concrete building is expected to respond to the horizontal components of the ground motion with inelastic displacements. It is allowed to do so, provided that it does not put at risk the safety of its users and occupants by collapsing. Very important for the possibility of collapse are the self-reinforcing second-order $(P - \Delta)$ effects produced by gravity loads acting through the lateral displacements of the building floors: if these displacements are large, the second-order moments (i.e. the overlying gravity loads times the lateral displacements) are large and may lead to collapse.

Because the major part of lateral structural displacements are inelastic and, besides, they tend to concentrate in the locations of the structural system where they first appeared, very important for the possibility of collapse is the 'plastic mechanism', which may develop in the building under the horizontal components of the ground motion. Inelastic seismic deformations in concrete buildings are flexural; they concentrate as plastic rotations wherever members yield in flexure (normally at member ends). Once the yield moment is reached at such a location, a 'plastic hinge' forms and starts developing plastic rotations with little increase in the acting moment. The 'plastic hinges' may form at the appropriate locations and in sufficient numbers to turn the building structure into a 'mechanism', which can sway laterally under practically constant lateral forces (plastic mechanism). The two extreme types of mechanism in concrete buildings are shown in Figure 2.9. Of the two mechanisms, the one that can lead to collapse is the 'column-sway' or 'soft-storey' mechanism in Figure 2.9a. If the ground storey has less masonry infills or other components with significant lateral stiffness and strength than the storeys above, a 'soft-storey' mechanism is more likely to develop there.

Mixed situations are very common, with plastic hinges forming at column ends at a number insufficient for a 'soft-storey' mechanism, and in fewer beams than in a full-fledged 'beam-sway' mechanism (see Example 5.2 in Chapter 5). Strictly speaking, a mixed distribution of plastic hinges does not give a 'mechanism' that kinematically allows sway of the building at little additional lateral force. Therefore, normally it does not lead either to collapse or to notable residual horizontal drifts. A full mechanism of the types shown in Figure 2.9 (especially the one in Figure 2.9a) may lead to collapse, or to demolition because of large, irreversible residual drifts.

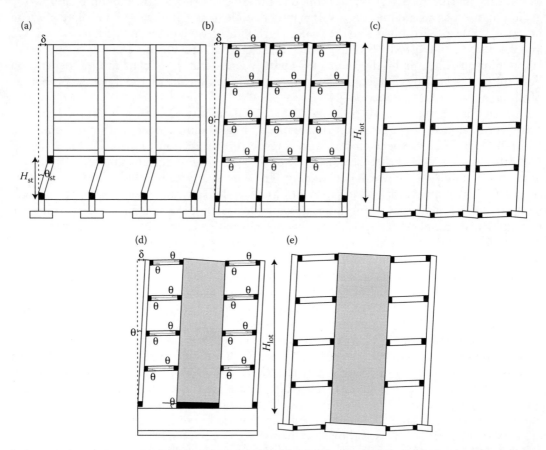

Figure 2.9 Side-sway plastic mechanisms in concrete buildings: (a) soft-storey mechanism in weak column–strong beam frame; (b), (c) beam-sway mechanisms in strong column/weak beam frames; (d), (e) beam-sway mechanisms in wall-frame systems.

2.2.2 Collapse

Collapses of 'open ground storey' buildings are depicted in Figures 2.10 and 2.11. Figure 2.11 shows on the left a very common type of collapse in multi-storey concrete buildings: the so-called 'pancake' collapse, with the floors falling on top of each other, trapping or killing the occupants.

As we will see in detail in Sections 4.5.2 and 5.4.1, a stiff vertical spine of strong columns or large concrete walls promotes 'beam-sway' mechanisms of the type illustrated in Figures 2.9b to 2.9e and helps avoid 'soft-storey' ones per Figure 2.9a. Walls are quite effective in that respect: in Figure 2.12a the walls in the middle of the lateral sides and at the corners with the

Figure 2.10 (a) Collapse of open ground storey building; (b) collapsed building shown at the background; similar building at the foreground is still standing with large ground storey drift.

Figure 2.11 Typical collapses of frame buildings with open ground storey; 'pancake' type of collapse shown on the right.

Figure 2.12 Role of walls in preventing pancake collapse of otherwise condemned buildings.

back side (shown inside dark-coloured frames) have failed at the ground storey (one is shown inside a light-coloured frame), but have prevented the collapse of columns all along the front side from triggering 'pancake' collapse; in Figure 2.12b perimeter walls (shown inside dark-coloured frames) may have failed terminally, but have prevented collapse of the building.

The dismal performance of walls in the earthquake of February 2010 in Chile has shown that walls are not a panacea. Wall buildings were a success story in past Latin American earthquakes, leading designers to extremes in their use in high-rise construction: in recent practice, very narrow, long walls, bearing the full gravity loads, are used in tall buildings, in lieu of columns and non-load bearing partitions. These walls were subjected to very high axial stresses due to gravity loads and failed at the lowest level in flexure-cum-compression, sometimes with lateral instability. A typical case is that of the building on the cover of this book, depicted in more detail in Figure 2.13.

In all the examples shown so far, as well as in Figure 2.14, the ground storey was critical. Figure 2.14c depicts the typical case of a concrete frame building with masonry infills, which have suffered heavy damage at the ground storey but may have saved the building from collapse. Figures 2.10 to 2.14 may be contrasted to Figure 2.15, where the top floors or an intermediate one have collapsed, but the underlying ones withstood both the earthquake and the collapse of the floors above. Such exceptions to the rule are most often due to an abrupt reduction in the lateral resistance of a floor, because that floor and those above were thought to be non-critical. Higher modes of vibration (see Sections 3.1.4 and 3.1.5), which are more taxing on certain intermediate floors than on the ground storey, may have played a role as well.

Twisting of the building about a vertical axis is more often due to the horizontal eccentricity of the inertia forces with respect to the 'centre of stiffness' of the floor(s) than to the rotational component of the motion itself about the vertical. In such cases, twisting takes place about a vertical axis passing through the 'centre of stiffness' which is closer to the 'stiff side' in plan and produces the maximum displacements and the most severe damage to the perimeter elements on the opposite, 'flexible side'. The example in Figure 2.16 is typical of such a response and its consequences – twisting about the corner of the building plan where the stiff and strong elements were concentrated (including a wall around an elevator shaft, the staircase, etc.) – caused the failure of the elements of the 'flexible side'. The seismic displacements on that flexible side, as increased by twisting, exceeded the – otherwise

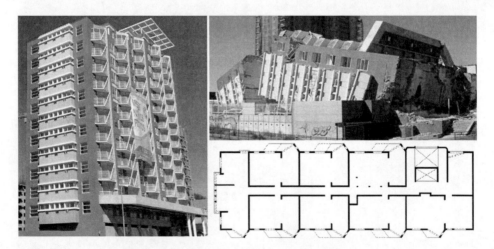

Figure 2.13 Collapse of Alto Rio wall building in Concepción, Chile; February 2010 earthquake (structural walls are shown in black in the framing plan).

(a)

(b)   (c)

Figure 2.14 Typical concentration of failures or damage in ground storey (a), (b) with role and damage to infills shown in (c).

Figure 2.15 Collapse of top floors in Mexico City (1985) or of an intermediate one in Kobe (1995).

Figure 2.16 Collapse of flexible sides in torsionally imbalanced building with stiffness concentrated near one corner.

Figure 2.17 Shear failure of short columns on stiff side (inside rectangle) causes collapse of flexible side as well.

ample – ultimate deformation of these columns. The collapse of the strongly asymmetric one-storey building in Figure 2.17 demonstrates the opposite effect: calling the side in Figure 2.17a as front, the vertical elements of the back side were shear-critical 'short columns', developing higher shear forces than the columns on the front, owing to their much larger stiffness and short length. However, they did not have sufficient shear strength to resist these forces. They collapsed, pushing out the columns of the front side as well.

The remark about 'short columns' brings up the effects of earthquakes on typical concrete members: columns, beams, the connections between them ('joints') and walls.

2.2.3 Member behaviour and failure

Typical seismic damage or failures of columns, joints, beams and walls are shown in Figures 2.18 to 2.23 and are commented in the following.

Figure 2.18 Flexural damage (a) or failure (b, c) at column ends.

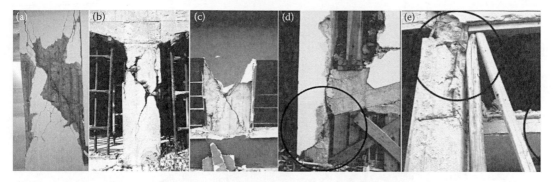

Figure 2.19 Shear failure of columns, (a)–(e), including a captive one between the basement perimeter wall and the beam (c) and short columns due to mid-storey constraint by a stair (d) or a landing (e) supported on the column.

Figure 2.20 Despite complete failure of columns across the ground storey, their residual axial load capacity still supports gravity loads.

2.2.3.1 Columns

Columns may be damaged or fail in flexure, as shown in Figure 2.18. Flexural damage or failure phenomena are concentrated in horizontal bands at the very top or bottom of a column in a storey (where the bending moments are at maximum). Such regions are the physical manifestation of flexural 'plastic hinges', where the plastic rotations take place. It is clear from Figure 2.18 that 'plastic hinging', although essential for the seismic design of the building for ductility and energy dissipation (see Sections 3.2.2, 3.2.3 and 4.6.3), is not painless:

Figure 2.21 Shear failure of beam–column joints.

Figure 2.22 Typical features of beam behaviour: (a) pullout of beam bars from narrow corner column, due to short straight anchorage there; (b) wide crack in slab at right angles to the beam at the connection with the columns shows the large participation of the slab as effective flange width in tension; (c) failure, with concrete crushing and bar buckling at bottom flange next to the column.

it implies damage, normally reparable, but sometimes not (especially if it is accompanied by irreversible residual horizontal drifts). Flexural damage always includes a visible horizontal crack and loss of concrete cover, often accompanied by bar buckling, opening of stirrups or partial disintegration of the concrete core inside the cage of reinforcement; sometimes one or more vertical bars rupture, or the concrete core completely disintegrates. The cyclic and

Figure 2.23 Typical failures of concrete walls: (a) flexural, with damage in shear; (b) in shear; (c) by sliding shear.

reversed nature of the deformation imposed on concrete elements by the earthquake plays an important role on its response: the opposite sides of the element are cyclically subject to tension and compression – when in tension, transversal cracking occurs but, then, when the force changes to compression the crack closes and the concrete cover may be lost (if the compressive strain is too large). Additionally, if the lateral restraint of the longitudinal bars is insufficient, the bars on the compressed face may buckle outwards, rupturing the stirrups and accelerating the loss of the concrete cover. Note that the Bauschinger effect decreases very sharply the buckling resistance of bars that have yielded previously in tension.

A column may fail in shear anywhere between its two ends, the end regions included (since the shear force is essentially constant along the height of the column). The signature of a shear failure is a diagonal crack or failure zone (Figure 2.19); sometimes such cracks or zones form in both diagonal directions and cross each other. If the column carries a low axial load relative to its cross-sectional area, the inclination of the shear failure plane to the horizontal is about 45°; it is steeper, sometimes over 60°, if the column is heavily loaded. In columns engaged in two-way frame action, the shear failure plane may be at an inclination to both transverse directions of the column. Stirrups intersected by the diagonal failure band(s) may open or break. The concrete may disintegrate all along the diagonal failure zone or across the full core inside the reinforcement cage (especially if failure is not due to one-way shear, parallel to a single transverse direction of the column). For shear, the cyclic and reversed nature of the earthquake effects on the elements is even more important than for flexure. In fact, as the direction of the shear alternates, two 'families' of diagonal cracks form, intersecting each other and leading to a very fast disintegration of the concrete. Additionally, since the horizontal stirrups are in tension for both directions of shear, diagonal cracks do not close upon reversal of the force; hence, the cracks become wider ever more, causing a very fast degradation of the lateral stiffness and strength of the column, denoting a so-called brittle failure.

Cases (c) to (e) in Figure 2.19 are 'short columns', which develop very high shear force demands and are very vulnerable to shear; the one in (c) is made 'short' by design: those in (d) and (e) unintentionally, as the secondary elements supported by the column between its two ends split its free height into two shorter ones. The back side columns in Figure 2.17, whose failure triggered the global collapse of the building, were also short.

Except for the one in Figure 2.18a, all columns in Figures 2.18 and 2.19 have essentially lost their entire lateral resistance and stiffness: they will not contribute at all against an aftershock or any other future earthquake. However, except for the column in Figure 2.18c, they all retain a good part of their axial load capacity. Note that the 'quasi-permanent' gravity loads normally exhaust only a small fraction of the expected actual value of the axial load capacity of the undamaged column. Moreover, the overlying storeys, thanks, among others, to their masonry infills, can bridge over failed columns working as deep beams. So, buildings with many failed columns or a few key ones in a storey are often spared from collapse. For example, very few columns were left in the building of Figure 2.20 with some axial load capacity. Another example are the six storeys above the failed corner column in Figure 2.21a, which survived by working as a 6-storey-deep multilayer-sandwich cantilever beam, with the concrete floors serving as tension/compression flanges or intermediate layers and the infills as the web connecting them.

2.2.3.2 Beam–column joints

As explained in Section 4.4.3.1 with the help of Figure 4.12, an earthquake introduces very high shear stresses to the core of a beam–column joint. These stresses are parallel to the plane of frame action. Effects of such shear stresses are shown in Figure 2.21: in (a), complete diagonal failure of an unreinforced joint; in (b), (c), diagonal cracking in reinforced joints. These effects are clearly manifested in exterior joints, especially corner ones (Figures 2.21a and 2.21b). Interior joints profit from the confinement by the slab on all four sides and by the beams in any direction they frame into the joint.

The joints also provide the anchorage zone of beam bars, whether they terminate there (as in corner joints; see Figure 2.22a), or continue into the next beam span across the joint. The next subsection addresses this issue.

2.2.3.3 Beams

Beam bars with insufficient anchorage in a joint may pull out in an earthquake. Such a failure of bond and anchorage shows up at the end section as a crack through the full depth of the beam (Figure 2.22a). A characteristic feature of a pull-out crack is its large width, well in excess of the residual crack width typical of yielding of the steel (which is a fraction of a millimetre or around 1 mm). The impact of this type of bond failure on the global behaviour is not dramatic: the beam cannot develop its full moment resistance at the end section and the force resistance and stiffness of the frame it belongs to drops accordingly. The damage is reparable, although the original deficiency, namely the poor anchorage of beam bars in the joint, cannot be corrected easily.

Beams are designed to develop flexural plastic hinges at the ends and are expected to do so in an earthquake. The loss of beam anchorage highlighted previously is part of such flexural action (although it prevents a proper plastic hinge from forming). A standard feature of a flexural plastic hinge in a beam is its through-depth crack at the face of the supporting beam or column, with a residual width indicative of yielding of the beam bars; that crack often extends into the slab and travels a good distance at right angles to the beam, sometimes joining up with a similar crack from a parallel beam (Figure 2.22b). The length and the sizeable residual crack width of such an extension show that the slab fully participates in the flexural action with its bars which are parallel to the beam, serving as a very wide tension flange.

Flexural damage is mostly associated with cracking and spalling of concrete and yielding of the reinforcement. By contrast, flexural failure comes with disintegration of concrete beyond the cover, often with buckling (or even rupture) of bars. Such effects (demonstrated in Figure 2.22c) happen only at the bottom flange of a beam, because the slab provides the top flange with abundant cross-sectional areas of concrete and steel reinforcement. Larger amounts of top reinforcement at the supports also result from the design for the hogging moments due to the factored gravity loads (the 'persistent and transient design situation' of EN 1990 (CEN 2002)). Note that a bottom reinforcement smaller than the top one is unable to close the crack at the top face (as it is unable to yield the top reinforcement in compression): the vertical crack at the face of the support, across the full depth of the beam, tends to remain open and increase in width for each cycle of deformation; bottom bars may buckle and then rupture under the large cyclic excursions of strain across the open crack.

2.2.3.4 Concrete walls

Flexural or shear damage and failure phenomena in walls (Figures 2.23a and 2.23b) are similar to those in columns, but take place almost exclusively right above the base of the wall, and very rarely in storeys higher up. One difference concerning flexure is that spalling and disintegration of concrete are normally limited to the edges of the wall section (Figure 2.23a). Owing to the light axial loading of the wall section by gravity loads, diagonal planes of shear failure are normally at about 45° to the horizontal (Figure 2.23b).

Walls have lower friction resistance than columns, owing to their lower axial stress level and vertical reinforcement ratio; so, they may slide at their through-cracked base section, which happens to coincide with a construction joint (Figure 2.23c).

2.3 EFFECTS OF EARTHQUAKES ON GEOTECHNICAL STRUCTURES

Earthquakes may affect geotechnical structures in several ways. The incoming motion may be altered by the local geotechnical conditions; the subsurface or surface topography may give rise to significant amplification of the seismic motion. These effects are known as site effects. In addition, soil instability, like liquefaction, flow failures, lateral spreading or slope instability, may be induced.

2.3.1 Site effects

Seismic motions may be significantly altered by the geotechnical conditions close to the ground surface. Typical wave lengths, λ, of the incoming motion vary between some meters to few hundred meters ($\lambda = V_S/f$ where V_S is the wave velocity ranging from 100 m/s to 2 km/s and f the predominant frequency of the motion, typically in the range 1–10 Hz); these values are of the same order of magnitude as surface or subsurface heterogeneities. As these heterogeneities might be very pronounced, interference between the incoming and diffracted wave fields may be important, possibly leading to significant modifications of the frequency content and amplitudes of the incoming wave field. These modifications are broadly referred to as site effects and often lead to very significant amplification or de-amplification. They are affected by topographic reliefs, sedimentary basins and so forth.

Observations have shown that a factor of 3–4 on PGA or PGV between the crest and the foot of a relief is common. Numerical analyses of such configurations invariably predict amplification for convex topographic reliefs, albeit with a very high sensitivity to the incoming wave field characteristics (wave type, azimuth, incidence angle). The present state of knowledge can be summarised as follows:

- Theory and observations are qualitatively in good agreement: convex topography induces amplification, while concave topography causes de-amplification.
- Amplification is more pronounced for the horizontal component of the motion than for the vertical one; furthermore, for 2D geometries amplification is more important in the component perpendicular to the slope than in the component parallel to it.
- The magnitude of amplification depends on the aspect ratio of the slope (height/width); the higher the aspect ratio, the larger is the amplification.
- Amplification is strongly frequency dependent; maximum effects are associated with wave lengths comparable to the horizontal dimensions of the relief.

However, the qualitative agreement between observations and numerical analyses is not confirmed quantitatively: more often than not, numerical analyses overpredict amplification, although the reverse might also be true. This explains why in engineering practice, and especially in seismic design codes, topographic amplification is handled with a rather crude approach. Eurocode 8, Part 5 in particular, simply gives a frequency-independent amplification factor, ranging from 1.0 to 1.4, depending only on the slope geometry; this factor is uniformly applied to the whole spectrum.

For a complete description and characterisation of topographic effects, one can refer to Bard and Riepl-Thomas (1999).

Even more important than topographic amplification, it has long been recognised that damage is more important in sedimentary basins than on rock outcrops. One of the most famous examples is provided by the records of the 1985 Michoacán-Guerrero earthquake in Mexico City. Records on stiff soil outcrops exhibit PGAs of the order of 0.04 g, while on the lake bed deposits (very soft clay deposits) PGAs reach 0.18 g with a totally different frequency content (Figure 2.7). Mexico City does represent a prime example; several others are available worldwide. The physical reason for this amplification stems from the incoming waves being trapped in the superficial layers of low rigidity. For horizontally layered profiles, seismic waves are reflected back and forth between the ground surface and the interface located at the soil–rock interface (Figure 2.24), leading to resonance of the layer. In 2D or 3D geometries, reflection of waves also occurs at the side boundaries, possibly giving rise to surface waves travelling back and forth horizontally between the edges of the valley; they not only induce resonance of the valley, but are also responsible for an increased duration of the seismic motion (Figure 2.24).

Amplification, or de-amplification, strongly depends on the predominant frequencies of the incoming signal and on the soil characteristics: for a given incoming motion, some frequencies may be amplified while others are de-amplified; a deep deposit with a low natural frequency may de-amplify a nearby, moderate earthquake with a high-frequency content, while it will strongly amplify a long distance earthquake with a low-frequency content. The example of the records in San Francisco during the 1957 Daly City earthquake (close by $M = 5.3$ event) and the 1989 Loma Prieta earthquake (70 km distant $M = 7.1$ event) illustrates this statement: although the recorded rock PGAs were similar in San Francisco (~0.10 g), ground surface PGAs recorded at Alexander Building Station on top of 45 m of clayey silt and sand were respectively 0.07 g (Daly City) and 0.17 g (Loma Prieta).

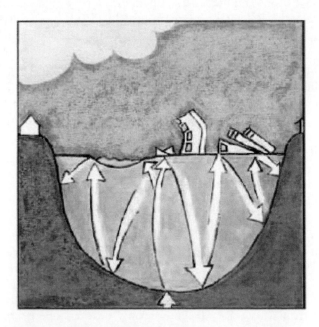

Figure 2.24 Illustration of wave trapping in sedimentary basins.

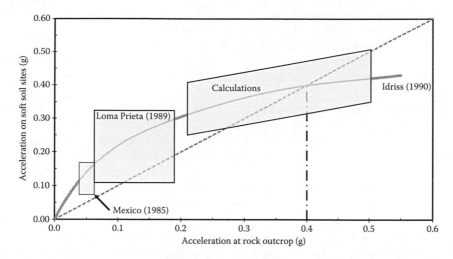

Figure 2.25 Relationship between PGA on rock and PGA at ground surface.

The interpretation and observation of site effects is further complicated by the non-linear behaviour of soils. It is well recognised, since the pioneering work of H.B. Seed and his co-workers in the 1970s, that soils are highly non-linear, even at very small strains (of the order of 10^{-5} to 10^{-4}). This is reflected by a decrease in the apparent shear modulus and an increase in the material energy dissipation capacity (traditionally called damping ratio). For strong motions, the decrease in material rigidity causes a shift in the natural frequency of the soil profile towards lower values and, as a consequence, a decrease of PGA as the rock incoming motion increases. This is portrayed in Figure 2.25 (adapted from Idriss 1990). This figure should not be taken at face value: it only relates to PGA, a high-frequency characteristic parameter of the ground motion; when looking at PGD, or pseudo-spectral acceleration at low frequencies, a reverse phenomenon is observed: ground surface PGD increases, as the rock incoming motion increases.

2.3.2 Soil liquefaction

Liquefaction is a process by which a solid is transformed into a liquid. In saturated cohesionless soils this phenomenon happens under undrained conditions, that is when the material is loaded at a rate high enough to prevent dissipation of excess pore water pressures. Liquefaction has been identified since the 1964 Niigata (Japan) and Valdez (Alaska) earthquakes as the cause of major damage. The state of the art is nowadays well developed; reliable predictions can be made and effective countermeasures implemented (Youd and Idriss 2001; Idriss and Boulanger 2008).

Liquefaction is caused by the tendency of dry sands to densify under cyclic loading. Progressive densification occurs by repeated back-and-forth straining of dry sand samples; each cycle causes further densification, at a decreasing rate, until the sand reaches a very dense state. The densification is the result of the soil particles being rearranged during straining. Actually, if the sand becomes dense enough, each half cycle may cause dilation of the sand sample, as particles roll or slide upon each other, but in the end they attain a still denser packing. Densification is mainly a function of the past history of loading, of the current density and stresses, and of the amplitude of shear strain. If the soil sample is saturated and water prevented from draining, the reduction in volume caused by cyclic loading cannot occur (assuming incompressible water). Instead, the tendency to decrease in volume is counteracted by a reduction in effective stress; for a constant total stress, a decrease in effective stress means an increase in pore water pressure. Hence, part of the applied grain-to-grain contact stress is transferred to the water. A zero effective stress condition is eventually reached, triggering liquefaction. This corresponds to the first step in most engineering treatments of soil liquefaction: assessment of 'liquefaction potential', or the risk of triggering liquefaction.

Once it is concluded that occurrence of liquefaction is a potentially serious risk, the next step should be the assessment of the consequences of the potential liquefaction. This involves assessment of the available post-liquefaction strength and resulting post-liquefaction overall stability (of a site, and/or of a structure or other built facilities, etc.). There has been considerable progress in the evaluation of post-liquefaction soil strength and stability over the past 20 years. If stability after liquefaction is found to be critical, then the deformation/ displacement potential is large; engineered remediation is typically warranted in such cases, because the development and calibration/verification of engineering tools and methods to estimate liquefaction-induced displacements are still at a research stage. Similarly, very few engineering tools and guidelines are available regarding the effects of liquefaction-induced deformations and displacements on the performance of structures and other engineered facilities; moreover, criteria for 'acceptable' performance are not well established. The ongoing evolution of new methods for the mitigation of liquefaction hazards provides an ever-increasing suite of engineering options, but the efficiency and reliability of some of them remain debatable. Accurate and reliable engineering analysis of the improved performance provided by many of these mitigation techniques continues to be difficult. Despite these difficulties, Mitchell and Wentz (1991) provide evidence of good performance of some mitigation techniques during the Loma Prieta earthquake.

The effect of liquefaction on a site or a built environment may consist of flow failures, where large volumes of earth are displaced over several tens of meters (as in Valdez, during the 1964 Alaska earthquake), or lateral spreading, which is similar to flow failures but involves much smaller volumes of soil and displacements of few meters (Figure 2.26). For flow failure and lateral spreading to take place, gentle slopes and the presence of a free surface, like a river bank, are necessary; in horizontal layers vertical settlements take place upon pore water pressure dissipation (Figure 2.27). If a foundation is on top of a liquefied

Figure 2.26 Lateral spreading (El Asnam, 1980).

Figure 2.27 Liquefaction-induced settlement in Marina district (Loma Prieta earthquake, 1989).

layer, loss of bearing capacity takes place (Figure 2.28). On the contrary, buried structures, which are usually lighter than the surrounding soil, may float due to buoyancy.

In engineering practice, liquefaction assessment of a site is carried out using empirical correlations between the cyclic undrained shear strength and some index parameter, like the standard penetration test (SPT) blow count, the cone penetration test (CPT) point resistance or the shear wave velocity. Use of laboratory tests is not recommended, except for important civil engineering structures, because, to be meaningful, they must be performed on truly undisturbed samples. Retrieving truly undisturbed samples from loose saturated cohesionless deposits is a formidable task, of a cost beyond the budget of any 'common' project.

Figure 2.28 Bearing capacity failure due to liquefaction (Hyogo-ken Nambu earthquake, 1995).

Techniques for retrieving undisturbed samples may call for in-situ soil freezing, or sampling with very large diameter samplers (of at least 300 mm).

2.3.3 Slope stability

An earthquake may cause landslides, including debris avalanches from volcanoes. Earthquake-induced acceleration can produce additional downslope forces, causing otherwise stable or marginally stable slopes to fail. In the 1964 Alaska earthquake, for instance, most rockfalls and debris avalanches were associated with bedding plane failures in the bedrock, probably triggered by this mechanism. In addition, liquefaction of sand lenses or changes in pore pressure in sediments trigger many coastal bluff slides.

Pseudo-static analysis is used for simple slope design to account for seismic forces. Displacement analysis is used to estimate the amount of permanent displacement suffered by a slope due to strong ground shaking. Simplified charts are developed for displacement analyses, to estimate the amount of permanent displacement of a slope due to strong ground shaking.

Failure of a slope may cause damage to buildings located on the slope itself or at its foot, but may also interrupt transportation systems: in the Loma Prieta earthquake, a large landslide in the Santa Cruz Mountains disrupted State Highway 17 (Figure 2.29), the only direct high-capacity route between Santa Cruz and the San Jose area, which was closed for about one month for repair.

Natural slope stability is difficult to assess, as it strongly depends on the initial state of the slope before the earthquake: the water regime, pre-existing fractures, previous slides, tectonic stresses and so on. For man-made slopes, the situation may be easier, provided that information on the construction method and materials constituting the slope is available.

Figure 2.29 Slope failure on State Highway 17, California (Loma Prieta earthquake, 1989).

2.4 EARTHQUAKE EFFECTS ON SHALLOW FOUNDATIONS

The discussion will be restricted to shallow foundations, although deep foundations can also suffer from earthquakes. Most often, damage to the foundation is the result of liquefaction of the underlying layers and the subsequent loss of bearing capacity (Figure 2.28). However, severe damage may also occur without liquefaction, due to bearing capacity failure or excessive settlements.

Bearing capacity failures of shallow foundations were seldom observed until 1985, which may explain why this topic did not attract much research. Furthermore, it may be difficult to make a clear classification between bearing capacity failure and excessive settlement during an earthquake since the loads do not act permanently. The situation changed significantly after the 1985 Michoacán Guerrero earthquake, when several buildings founded on individual footings or basemats undoubtedly failed due to loss of bearing capacity in Mexico City (Figure 2.30). Research carried out around and since the end of the century has shown that the most sensitive structures are those with low initial safety factor against gravity loads, especially if they also have large load eccentricity, which is a design deficiency.

In areas of high seismicity, the inertial forces developed in the supporting soil by the passage of the seismic waves also contribute to the reduction of the overall safety; it has been shown that neglecting these inertial forces may lead one to conclude that the foundation is stable, while taking them into account points to the opposite conclusion, confirmed by observations.

The state of the art allows determining the pseudo-static foundation-bearing capacity, taking into account the eccentricity and inclination of the load at the foundation level, as well as the inertial forces developed in the soil. Such a verification has been included, as an informative annex, in Eurocode 8 – Part 5 (see Section 6.2 in Chapter 6).

Less spectacular than bearing capacity failures, earthquake-induced settlements may also cause damage to foundations and supported structures. Settlements mainly occur in loose, dry, or partially saturated, cohesionless deposits as a consequence of densification under cyclic loading (see Section 2.3.2). In cohesive soils and saturated cohesionless deposits, settlements are not observed during the earthquake but may occur later on, upon dissipation

Figure 2.30 Bearing capacity failure in Mexico City (Michoacán Guerrero earthquake, 1985).

Figure 2.31 Earthquake-induced foundation settlement (Michoacán Guerrero earthquake, 1985).

of the earthquake-induced excess pore water pressure. Dense sands are less sensitive to settlements. Figure 2.31 is an example of an intermediate column of a steel frame structure of a factory in Lazaro Cardenas, which settled during the earthquake, resulting in loss of support of the column. Poorly compacted backfill is prone to large densification and settlements (Figure 2.32). Earthquake-induced settlements may reach dozens or hundreds of millimetres. Predicting settlements is a challenging task: they depend on the initial density of the soil, the amplitude of the induced shear strains, the number of cycles of loading and so on. Empirical charts have been developed to estimate them and can be used as a guideline (Pyke et al. 1975, see Section 6.2.3).

Figure 2.32 Settlement of a poorly compacted backfill (Moss Landing, Loma Prieta earthquake, 1989).

2.5 EARTHQUAKE EFFECTS ON LIFELINES

Although most of this book (notably Chapters 4 to 7) is devoted to concrete buildings, this chapter provides more general background on various aspects of earthquake engineering and dynamics. In this context, for completeness, this section highlights the role and performance of lifelines in earthquakes.

Unlike individual buildings, lifelines have a distinguishing characteristic: they cover large geographical areas and are composed of many diverse components interacting with each other. Lifelines provide cities with services and resources necessary to security, commerce and communications. When earthquakes strike urban areas, they can disrupt lifeline systems, threatening life and property in the short term and postponing economic recovery during post-earthquake rehabilitation (O'Rourke 1996).

A review of the lifeline performance during earthquakes reveals that electric power systems generally perform well; in the vast majority of cases, restoration requires less than a few days. However, electric power is critical for other lifelines; power loss reflects directly in reduced serviceability of water supplies, wastewater facilities, telecommunications and transportation. There are many examples, including loss of sewage and water pumping capacity, loss of rapid transit services during the Loma Prieta earthquake and loss of power for telecommunications during the Northridge and Hyogo-ken Nambu earthquakes. Even though power losses are of short duration, their consequences can be important; for example, electric power affects remote control of water supply and thereby influences fire protection.

The behaviour of buried pipelines is controlled by ground movement; therefore the geotechnical characteristics are critically important for lifeline systems. Earthquakes, like Northridge, have shown that damage to buried pipelines can be caused by transient motion; however such damage is mainly related to pipeline deterioration (corrosion, characteristics of welds) or past construction practices, which reduce the capacity compared to that achieved with modern materials and procedures. Simplified analytical procedures are available for assessing such effects: they model the seismic excitation as a traveling wave and consider that the pipeline strain is equal to the ground strain, unless slippage takes place at the interface between the pipeline and the surrounding soil; at slippage, the pipeline stress is limited by the frictional force that can be transmitted to it. Special detailing, like coating, may contribute to the reduction of friction and therefore pipe stresses.

Even though modern pipelines in more or less homogeneous soil profiles are not very sensitive to transient motions, special attention must be paid to singular points; that is, at transitions between two layers with a sharp rigidity contrast, or at the connection with a building, whose motion is different from the free-field motion, creating transient differential displacements. These differential displacements can usually be accommodated through special detailing, providing enough flexibility at the connection.

Instead, buried pipelines are more sensitive to permanent ground deformation caused by settlements, slope instability, fault offsets or liquefaction. Settlements of pipelines are typically encountered close to buildings, where pipelines are constructed in open narrow trenches; those trenches are backfilled afterwards; however, heavy compaction is difficult to achieve and soil densification may take place during the earthquake (Figure 2.33). Differential settlements between the building and the pipeline may cause damage to the connection.

Liquefaction may induce lateral spreading, but also large transient shear strain in liquefied layers; these strains may reach 1.5%–2%; so, when integrated over the thickness of the liquefiable soil, they may impose large lateral deformation on the buried pipelines. As an illustration of the complexity and interdependence of lifeline systems, Figure 2.34, adapted

Figure 2.33 Settlement of a pipeline trench adjacent to a building (Mexico, 1985).

from O'Rourke (1996), presents a bar chart of the number of Kobe reservoirs that emptied as a function of time after the Hyogo-ken Nambu earthquake. Only one of the 86 reservoirs supplying Kobe was structurally damaged; in all other cases, loss of water was the result of ruptures of water pipelines, mostly caused by liquefaction-induced ground movements. Within 24 h after the main shock, all reservoirs were empty, impairing firefighting and contributing to the destruction of part of the town by fire.

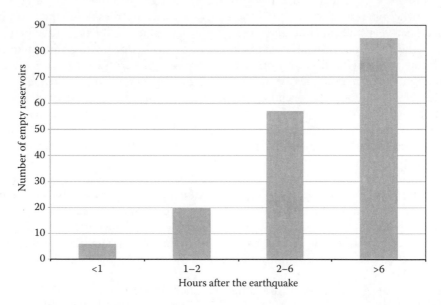

Figure 2.34 Loss of reservoirs after the 1995 Hyogo-ken Nambu earthquake. (Modified from O'Rourke, T.D. 1996. Lessons learned for lifeline engineering from major urban earthquakes. Paper no. 2172. *Eleventh World Conference on Earthquake Engineering.* Acapulco, Mexico.)

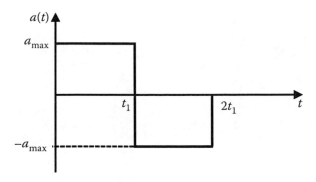

Figure 2.35 Ground acceleration for Question 2.1.

QUESTION 2.1

Consider the pulse-like excitation depicted in Figure 2.35, with $a_{max} = 0.1$ g and $t_1 = 0.15$ s. Calculate the CAV and the Arias intensity I_A.

QUESTION 2.2

Figure 2.8 gives the annual probability of exceedance of peak ground acceleration (PGA) at a given site, from several individual seismic sources. Calculate the annual probability that a PGA of 0.1 g will be exceeded for a building located at that site. For a building structure designed for a lifetime of 50 years, what is the probability that a PGA of 0.1 g will be exceeded during the lifetime of the structure?

QUESTION 2.3

What is the mode of failure or damage of the beams in Figure 2.36? Would you characterise the case as damage or as failure?

QUESTION 2.4

What is the mode of failure or damage of the columns in Figure 2.37? Would you characterise the case as damage or as failure?

(a)   (b)   (c)

Figure 2.36 (a–c) Beams of Question 2.3.

Figure 2.37 (a–l) Columns of Question 2.4.

QUESTION 2.5

What is the mode of failure or damage of the concrete walls in Figure 2.38? Would you characterise the case as damage or as failure?

Figure 2.38 (a–f) Walls of Question 2.5.

Chapter 3

Analysis of building structures for seismic actions

3.1 LINEAR ELASTIC ANALYSIS

3.1.1 Dynamics of single degree of freedom systems

3.1.1.1 Equation of motion

When loads or displacements are applied very slowly to a structure, inertia forces, which are equal to the mass times the acceleration, are negligible and may be disregarded in the equation of force equilibrium. This corresponds to what is normally referred to as the static response of the structure. By contrast, if the loads or displacements are applied quickly, the inertia forces may not be disregarded in the equilibrium equation and the structure responds dynamically to those excitations.

Furthermore, damping forces may also develop and should also be considered in the equilibrium.

To better understand this concept, consider the very simple system shown in Figure 3.1. It depicts a single degree of freedom (SDOF) system, with constant parameters, that is subject to a ground displacement $u_g(t)$ and an applied force $p(t)$ varying with time.

Under this excitation, the system deforms, developing:

- A restoring force that (in the simpler case of linear behaviour) is proportional to the relative displacement u and the stiffness of the system k
- A damping force that may be assumed to be proportional to the relative velocity $\dot{u}$ (rate of deformation of the system) and a damping constant c (in which case the system is considered to have viscous damping)
- An inertia force that is proportional to the (absolute) acceleration $\ddot{u}^t$ of the mass m

All these forces should be in equilibrium as is represented by Equation 3.1 where, for simplicity, we omit the dependence on time of the acceleration, velocity and displacement of the system as well as of the applied force:

$$m\ddot{u}^t + c\dot{u} + ku = p \tag{3.1}$$

It should be noticed that u^t, $\dot{u}^t$ and $\ddot{u}^t$ correspond to the *absolute* displacement, velocity and acceleration, whereas u, $\dot{u}$ and $\ddot{u}$ correspond to the *relative* (to the ground) displacement, velocity and acceleration.

The response of the system is thus governed by a linear differential equation and the system is represented by its three properties:

1. m, Mass
2. c, Viscous damping constant
3. k, Stiffness of the system

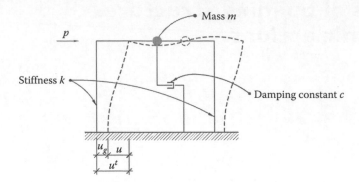

Figure 3.1 Single degree of freedom system.

Considering that $\ddot{u}^t = \ddot{u}_g + \ddot{u}$, Equation 3.1 may be re-written in a more convenient form, where the excitation terms are grouped in the right-hand side of the equation:

$$m\ddot{u} + c\dot{u} + ku = p - m\ddot{u}_g \qquad (3.2)$$

If we divide Equation 3.2 by m, we obtain:

$$\ddot{u} + 2\zeta\omega_n\dot{u} + \omega_n^2 u = \omega_n^2 u_{st} - \ddot{u}_g \qquad (3.3)$$

where we have replaced the proportionality coefficients as:

$$2\zeta\omega_n = c/m$$

$$\omega_n^2 = k/m$$

$u_{st} = p/k$ (i.e. u_{st} is the static displacement of the system under the lateral force p).

The left-hand term of Equation 3.3 represents the characteristics of the system, whereas the right-hand term represents the excitation (either as an applied force or as an applied motion at the base).

In Equation 3.3 the characteristics of the system are represented by the quantities ω_n and ζ, which shall be discussed later.

The equation covers all cases of interest with regard to the dynamic response of the system: *free vibration*; *forced vibration* and *transient disturbance*.

3.1.1.2 Free vibration

The simplest case of dynamic response corresponds to the *free vibration*, in which the base is motionless ($\ddot{u}_g = 0$) and there is no external force applied ($p = 0$).

Let us consider additionally a further simplification in which there is no damping in the system (i.e. the system is conservative, meaning that there is no dissipation of energy associated to the motion).

Under these assumptions, Equation 3.3 becomes:

$$\ddot{u} + \omega_n^2 u = 0 \tag{3.4}$$

and its general solution is:

$$u(t) = a \sin\left[\omega_n(t - t_1)\right] \tag{3.5}$$

where a is an arbitrary constant (with units of length) and t_1 is an arbitrary value of time t. This equation represents a simple harmonic motion with amplitude a and circular frequency ω_n, which is called the *undamped natural circular frequency* of the system. In other words, if such a system is displaced from its resting position and released, it will remain oscillating indefinitely. It is worth noticing that this occurs, since there is no dissipation of energy in the system, because the energy which is input into the system to start the motion is conserved–hence the denomination of the system as 'conservative'.

The *undamped natural circular frequency* of the system (expressed in radians per second) may be converted into the *natural frequency of the system* by the expression:

$$f_n = \omega_n/2\pi \tag{3.6}$$

The *undamped natural frequency* of the system f_n is expressed in cycles per second or the corresponding unit hertz (Hz). The inverse of f_n is the *undamped natural period* of the system T_n which is given by:

$$T_n = 1/f_n = 2\pi/\omega_n \tag{3.7}$$

and is expressed in seconds. It corresponds to the duration of each cycle of oscillation.

As mentioned earlier, during the undamped free vibration, the energy in the system is kept constant and corresponds, at each moment, to the sum of the deformation energy and the kinetic energy.

For a harmonic motion with amplitude a the *deformation energy* in the system is maximal at maximum displacement ($u = a$) with a value of:

$$E_{s\,max} = \frac{ka^2}{2} \tag{3.8}$$

On the other hand, the maximum *kinetic energy* is attained when the velocity (of the mass) is maximal. For a harmonic motion this occurs when the displacement is zero ($u = 0$). If the amplitude is a and the oscillatory frequency is ω_n, such maximal velocity is $\dot{u}_{max} = a\omega_n$. Hence the maximal kinetic energy in the system is:

$$E_{k\,max} = \frac{ma^2\omega_n^2}{2} \tag{3.9}$$

Equating these two maximal values of the deformation and kinetic energies, we obtain $k = m\omega_n^2$ or $\omega_n^2 = k/m$ which is precisely the value of the square of the *undamped natural circular frequency* of the system as derived above. This means that the energy in the undamped free vibration is kept constant only if the oscillation occurs with a frequency equal to the natural frequency of the system.

Still within the framework of the free vibration of the system, we consider now the case where the viscous damping is not zero, that is, the system dissipates energy during the oscillatory motion. In such a case, Equation 3.4 is replaced by:

$$\ddot{u} + 2\zeta\omega_n\dot{u} + \omega_n^2 u = 0 \tag{3.10}$$

and its general solution is:

$$u(t) = a\,\exp\left[-\zeta\omega_n(t - t_1)\right]\sin\left[\omega_D(t - t_1)\right] \tag{3.11}$$

This equation represents a damped harmonic motion, but the equation is only valid if the damping constant c is smaller than a limiting value known as the *critical damping* given by:

$$c_{cr} = 2\sqrt{km} \tag{3.12}$$

If we normalise the damping constant of the system c and consider the definition of ζ adopted in Equation 3.3, by simple substitution, we obtain:

$$\zeta = \frac{c}{c_{cr}} \tag{3.13}$$

This quantity is called the *damping ratio* and is a measure of the damping in the system.

If c is smaller than c_{cr}, the system, when released from a displaced position, tends to the resting position, oscillating with a circular frequency ω_D which is called the *damped natural circular frequency* of the system and is given by:

$$\omega_D = \omega_n\sqrt{1 - \zeta^2} \tag{3.14}$$

For the values of damping normally applicable in structural dynamics, the difference between the damped (ω_D) and the undamped (ω_n) natural frequencies is very small. For instance, for 5% damping ($\zeta = 0.05$) the difference is negligible (0.1%), whereas for 20% damping ($\zeta = 0.20$) the difference is still only 2%.

It is apparent that the pace at which the oscillation tends to the resting position increases with the damping ratio. To illustrate this effect, in Figure 3.2 the free vibration oscillation

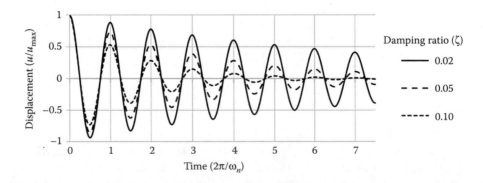

Figure 3.2 Free vibration of systems with different damping ($\zeta = 2\%$, 5% and 10%).

of systems with different damping is depicted. The influence of the damping ratio is clear: after 5 cycles of oscillation, the system with 2% damping is still presenting amplitude of the order of one-half the initial amplitude, whereas the system with 10% damping is practically at rest showing just a residual amplitude of oscillation.

If c is equal to or greater than c_{cr} (i.e. $\zeta \geq 1$), the system does not oscillate when it is released after having been displaced from its resting position. In that case, the system comes back to the resting position always with the displacement on the same side and taking infinite time to rest.

3.1.1.3 Forced vibration

We consider now the case in which the ground is at rest and there is a dynamic excitation of the system caused by the application of a force varying harmonically with a circular frequency ω and amplitude p_0. This corresponds to introducing $\ddot{u}_g = 0$ and $p = p_0 \sin \omega t$ in the general Equation 3.2. In this case, the (homogeneous) equation has a general solution given by Equation 3.11 and a particular solution given by:

$$u = u_{st}\, B_d \sin(\omega t - \phi) \tag{3.15}$$

where $u_{st} = p_0/k$ (i.e. u_{st} is the static displacement of the system under the lateral force at its maximum p_0) and

$$B_d = \left[\left(1 - \frac{\omega^2}{\omega_n^2} \right)^2 + \left(2\zeta \frac{\omega}{\omega_n} \right)^2 \right]^{-1/2} \tag{3.16}$$

$$\phi = \tan^{-1} \frac{2\zeta(\omega/\omega_n)}{1 - (\omega^2/\omega_n^2)} \tag{3.17}$$

B_d is a dimensionless *response factor*, equal to the ratio of the dynamic to the static displacement amplitudes and ϕ is a phase shift between the excitation and the response.

The variation of the *response factor* B_d with ω/ω_n is depicted in Figure 3.3 for five values of the damping ratio ζ (notice that the vertical axis is in logarithmic scale). Immediately apparent from Figure 3.3 is the fact that the *response factor* is influenced by the value of the

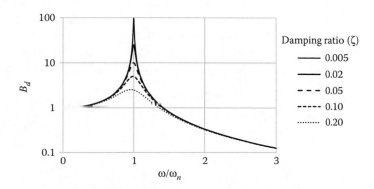

Figure 3.3 Values of the response factor B_d for systems with different damping ($\zeta = 0.5\%$, 2%, 5%, 10% and 20%) acted by a dynamic force.

damping ratio only in the vicinity of $\omega/\omega_n = 1$. As we depart from this value (either descending or ascending), the response factor becomes practically independent of the damping ratio ζ.

When ω/ω_n is close to zero, the value of B_d is close to unity. This means that, in such a case, there is practically no dynamic amplification. In fact, ω/ω_n close to zero means that the frequency of excitation ω is so small (with regard to the natural frequency of the system ω_n) that the system takes it as a static action.

For $\omega/\omega_n = 1$, the response factor is $B_d = 1/(2\zeta)$, which is very close to its maximum that occurs at *resonance*, as described below.

For large values of ω/ω_n the value of B_d tends to zero, meaning that there is very little deformation in the system and the system equilibrates the external force with the inertia force in its mass. In fact large values of ω/ω_n indicate that the excitation is very fast in comparison with the natural frequency of the system ω_n and so it is the mass (by its inertia) that fully resists the applied force.

In what concerns the phase angle ϕ, it is apparent from Equation 3.17 that it goes from $\phi = 0$ at $\omega/\omega_n = 0$ to $\phi = \pi/2$ at $\omega/\omega_n = 1$ and then tends to $\phi = \pi$ as ω/ω_n tends to infinity. This means that for excitations that are 'slow' with regard to the natural frequency of the system, the response is practically in phase with the excitation (and with very little amplification as seen before). The response is essentially static and the dynamic effects are negligible. By contrast, for 'fast' excitations (i.e. for large values of ω/ω_n) the displacement response is opposite to the excitation, that is, the direction of the displacement is contrary to the direction of the applied force.

It should be noticed that in such case, as seen before, the deformation of the system is very small and thus the force resulting from its stiffness is very small. The external force is resisted essentially by the inertia of the mass and hence the phase of the displacement is not really very relevant.

At $\omega/\omega_n = 1$ the phase shift is $\phi = \pi/2$, meaning that in that case the external force is essentially equilibrated by the damping force developed in the system.

The case where the response factor is maximal is normally called *resonance*. If we consider the response factor for displacement B_d, the maximum occurs for $\omega = \omega_n\sqrt{1 - 2\zeta^2}$ with a value of $B_d = (1/2\zeta)(1 - \zeta^2)^{-1/2}$.

For the small values of ζ normally associated to structural dynamics (damping ratios in the range of 0.01 to 0.2) this value is practically equal to the value indicated before ($B_d = 1/(2\zeta)$) for the excitation with a frequency equal to the undamped natural frequency ($\omega = \omega_n$). It may also be noticed that for damping ratios up to $\zeta = 0.1$, the difference is less than 0.5%. Likewise, for small damping, the excitation frequency leading to resonance is practically equal to the undamped natural frequency (up to $\zeta = 0.1$, the difference is less than 1%).

We consider now the other case of forced excitation, in which there is no external force applied and the ground moves harmonically with a circular frequency ω and amplitude a. In this case the ground displacement is described by $u_g = a \sin \omega t$, which in terms of ground acceleration is described by $\ddot{u}_g = -a\,\omega^2 \sin\omega t$. Moreover, in this case we have $p = 0$. Then the (homogeneous) equation has a general solution given by Equation 3.11 and a particular solution given by:

$$u = a\,B_d \sin(\omega t - \phi) \tag{3.18}$$

with

$$B_d = \frac{(\omega/\omega_n)^2}{\left[\left(1 - (\omega^2/\omega_n^2)\right)^2 + \left(2\zeta(\omega/\omega_n)\right)^2\right]^{1/2}} = \left[\left(1 - \frac{\omega_n^2}{\omega^2}\right)^2 + \left(2\zeta\frac{\omega_n}{\omega}\right)^2\right]^{-1/2} \tag{3.19}$$

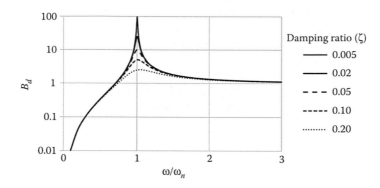

Figure 3.4 Values of the response factor B_d for systems with different damping ($\zeta = 0.5\%$, 2%, 5%, 10% and 20%) acted by ground motion.

In this case B_d is a dimensionless *response factor* equal to the ratio of the relative displacement in the system to the amplitude of the base motion. It is worth noticing that the response factor is then equal to the one for the force excitation case (Equation 3.16) but with ω and ω_n interchanged. The phase shift ϕ between the excitation and the response is given by Equation 3.17.

The variation of the *response factor* B_d for ground excitation with ω/ω_n is depicted in Figure 3.4 for five values of the damping ratio ζ (notice that the vertical axis is in logarithmic scale).

As for the force excitation case, it is apparent from Figure 3.4 that the *response factor* is influenced by the value of the damping ratio only in the vicinity of $\omega/\omega_n = 1$.

When ω/ω_n is close to zero, the value of B_d is very small. In such case the frequency of excitation ω is very small and there is practically no deformation of the system. Hence the relative displacement in the system (i.e. its deformation) is negligible and the system moves essentially as a rigid body. By contrast, for large values of ω/ω_n the value of B_d tends to one. In that case the system is so flexible (in relation to the frequency of excitation) that the mass remains motionless. Hence the relative displacement in the system (i.e. its deformation) is equal to the amplitude of the ground motion.

3.1.1.4 Numerical evaluation of dynamic response

In the case of earthquake ground motion, which represents a *transient disturbance*, the equation of motion, Equation 3.2, can be solved only by numerical step-by-step methods for integration of differential equations. A large number of methods have been presented in the literature. Only a very brief presentation of one of them is presented here for illustration. The method is usually called the *Average acceleration method*, also known as *Newmark's method* with $\gamma = 0.5$ and $\beta = 0.25$.

In the equation of motion, Equation 3.2, we will consider, for convenience, only the applied force $p(t)$ on the right-hand side. (It will be easy to replace it with the ground motion at the end of the derivation). The full duration of the motion is divided into a number of short-time intervals Δt, taken to be constant, although this is not necessary. In each interval, it is assumed that the acceleration is constant and equal to the average value of the accelerations at the beginning and at the end of the interval. This is the only assumption in the integration of the equation of motion; on the basis of it, it is possible to transform the differential equation of motion into a number of algebraic equations, which can be easily solved.

The assumption of constant acceleration is an approximation which violates the principles of physics, since it requires a sudden jump in acceleration at the boundaries of the time intervals. Nevertheless, it produces quite accurate results, provided that the time step is short enough, as discussed later in this Section.

Constant acceleration implies a linear variation of the velocity and a quadratic variation of displacements within the time interval. The following relations between the quantities at the end and at the beginning of the interval are obtained:

$$\dot{u}_{i+1} = \dot{u}_i + (\ddot{u}_i + \ddot{u}_{i+1})\frac{\Delta t}{2} \tag{3.20}$$

$$u_{i+1} = u_i + \dot{u}_i \Delta t + (\ddot{u}_i + \ddot{u}_{i+1})\frac{\Delta t^2}{4} \tag{3.21}$$

where the index i applies to the beginning of the interval and index $i + 1$ to its end. Equations 3.20 and 3.21 can be rearranged into the form:

$$\ddot{u}_{i+1} = \frac{4}{\Delta t^2}(u_{i+1} - u_i) - \frac{4}{\Delta t}\dot{u}_i - \ddot{u}_i \tag{3.22}$$

$$\dot{u}_{i+1} = \frac{2}{\Delta t}(u_{i+1} - u_i) - \dot{u}_i \tag{3.23}$$

Introducing Equations 3.22 and 3.23 into the equation of motion at the end of the interval

$$m\ddot{u}_{i+1} + c\dot{u}_{i+1} + ku_{i+1} = p_{i+1} \tag{3.24}$$

the following equation is obtained, after some rearrangement:

$$\left(\frac{4}{\Delta t^2}m + \frac{2}{\Delta t}c + k\right)u_{i+1} = p_{i+1} + \left(\frac{4}{\Delta t^2}u_i + \frac{4}{\Delta t}\dot{u}_i + \ddot{u}_i\right)m + \left(\frac{2}{\Delta t}u_i + \dot{u}_i\right)c \tag{3.25}$$

or

$$\bar{k}u_{i+1} = \bar{p}_{i+1} \tag{3.26}$$

where the effective stiffness $\bar{k}$ is defined as

$$\bar{k} = \frac{4}{\Delta t^2}m + \frac{2}{\Delta t}c + k \tag{3.27}$$

and the effective applied force $\bar{p}_{i+1}$ as

$$\bar{p}_{i+1} = p_{i+1} + \left(\frac{4}{\Delta t^2}u_i + \frac{4}{\Delta t}\dot{u}_i + \ddot{u}_i\right)m + \left(\frac{2}{\Delta t}u_i + \dot{u}_i\right)c \tag{3.28}$$

In order to apply the above procedure to the integration of the equation of motion, the following computational steps are needed:

1. The initial displacement u_1 and the initial velocity $\dot{u}_1$ are known. The initial acceleration $\ddot{u}_1$ is calculated from the equation of motion at the beginning of the first step.
2. The effective stiffness $\bar{k}$ is determined from Equation 3.27.

At each time step:

3. The initial values of the displacement, velocity and acceleration, u_i, $\dot{u}_i$ and $\ddot{u}_i$, respectively, are either the initial values determined in step 1 (only at the beginning of the calculation), or the values determined in the previous step (at the end of the interval).
4. The effective applied force $\bar{p}_{i+1}$ is determined from Equation 3.28.
5. The displacement at the end of the interval u_{i+1} is determined from Equation 3.26.
6. The velocity $\dot{u}_{i+1}$ is determined from Equation 3.23.
7. The acceleration $\ddot{u}_{i+1}$ is determined from Equation 3.22, or from the equation of motion (Equation 3.24).

Equation 3.26 is equivalent to the equilibrium equation used in statics. Formally, the step-by-step integration procedure transforms the dynamic problem into several static ones. For each time-step, the equilibrium equation, Equation 3.26, has to be solved. Dynamic effects are included in the effective stiffness $\bar{k}$ and the effective loading $\bar{p}_{i+1}$.

In the case of the excitation in the form of ground motion, the applied force p_{i+1} is replaced by $-m\ddot{u}_{i+1}$.

Note that in the case of multiple degrees of freedom (MDOF) systems, the same procedure can be used for the numerical integration of a system of differential equations. The same equations apply, only the scalar values are replaced by matrices.

The developed method for step-by-step integration of the equation of motion can be formulated also in another form, where instead of displacements the incremental displacements $\Delta u_i = u_{i+1} - u_i$ are calculated. Both forms are equivalent in the case of elastic analysis, whereas in the case of inelastic structural behaviour only the second form is applicable. For the development of equations in this form, the equation of motion at the beginning of the interval

$$m\ddot{u}_i + c\dot{u}_i + ku_i = p_i \tag{3.29}$$

is subtracted from the equation of motion at the end of the interval (Equation 3.25) resulting in

$$\left(\frac{4}{\Delta t^2}m + \frac{2}{\Delta t}c + k\right)(u_{i+1} - u_i) = p_{i+1} - p_i + \left(\frac{4}{\Delta t}\dot{u}_i + 2\ddot{u}_i\right)m + 2c\dot{u}_i \tag{3.30}$$

or

$$\bar{k}\Delta u_i = \bar{p}_{i+1} \tag{3.31}$$

where $\bar{k}$ is the same as in the first variant (Equation 3.27) and $\bar{p}_{i+1}$ is defined as:

$$\bar{p}_{i+1} = p_{i+1} - p_i + \left(\frac{4}{\Delta t}\dot{u}_i + 2\ddot{u}_i\right)m + 2c\dot{u}_i \tag{3.32}$$

Computational errors occur in all numerical methods. In the case of time-stepping procedures for the solution of the equation of motion, the error decreases when the size of the time interval is reduced. When using the average acceleration method presented in this Section (and other similar methods), reasonable accuracy can be achieved if the time interval is not larger than one-tenth of the period of the structure ($\Delta t \leq 0.1 T_n$). In the case of MDOF systems, this condition applies to all modes that exhibit a significant contribution to the structural response. Of course, in the case of earthquake ground motion, the time interval should not be shorter than the interval of the input acceleration data, which is usually 0.01 s or 0.005 s. This condition is typically decisive in the case of SDOF systems subjected to earthquake ground motion.

The average acceleration method is an unconditionally stable procedure and leads to bounded solutions regardless of the length of the time interval.

In the case of non-linear analysis, the linear force–deformation relation does not apply. The stiffness k depends on the displacement and changes with time. Consequently, $\bar{k}$ is not a constant as in the case of elastic analysis, but changes with the time steps.

3.1.2 Seismic response of SDOF systems – Response spectrum

3.1.2.1 Response spectra

The seismic response of SDOF systems can be obtained by solving Equation 3.2 by means of a numerical procedure, for example, the method presented in Section 3.1.1.4. The analysis which determines the structural response as a function of time is called 'time-history' or 'response-history' analysis. Several numerical methods and computer codes are available in the literature on structural dynamics, which provide a numerical solution in terms of the displacement u as a function of time. In practice, the whole response history is usually not needed. In most cases, the analyst is interested only in the maximum response values, which may be obtained from a 'response spectrum'.

A *response spectrum* gives, by definition, the maximum absolute values of a response quantity (in seismic analyses this is typically acceleration, velocity and/or displacement) as a function of the period, T_n (or a related quantity such as the frequency ω_n), for a fixed damping ratio and for a given ground motion. An example of response spectrum is shown in Figure 3.5. It shows the maximum (relative) displacements, u_o, of an SDOF system subjected to a ground motion recorded at Ulcinj (Albatros, N–S direction) during the 1979 Montenegro earthquake. The spectrum was obtained by performing a response-history analysis of several SDOF systems with different natural periods, but always with the same accelerogram (Figure 3.5). The damping ratio ζ was in all cases equal to 0.05 (i.e. 5%).

The displacement spectrum u_o represents the absolute values of the maximum (relative) displacements. In a similar way, spectra for the (relative) velocity, $\dot{u}_o$, and the (absolute) acceleration $\ddot{u}_o^t$, which represent the absolute values of the maximum relative velocity and absolute acceleration, respectively, can be obtained.

The spectral values are defined as

$$u_o\left(T_n,\ \zeta\right) \equiv \max\left|u\left(t,\ T_n,\ \zeta\right)\right| \tag{3.33}$$

$$\dot{u}_o\left(T_n,\ \zeta\right) \equiv \max\left|\dot{u}\left(t,\ T_n,\ \zeta\right)\right| \tag{3.34}$$

$$\ddot{u}_o^t\left(T_n,\ \zeta\right) \equiv \max\left|\ddot{u}^t\left(t,\ T_n,\ \zeta\right)\right| \tag{3.35}$$

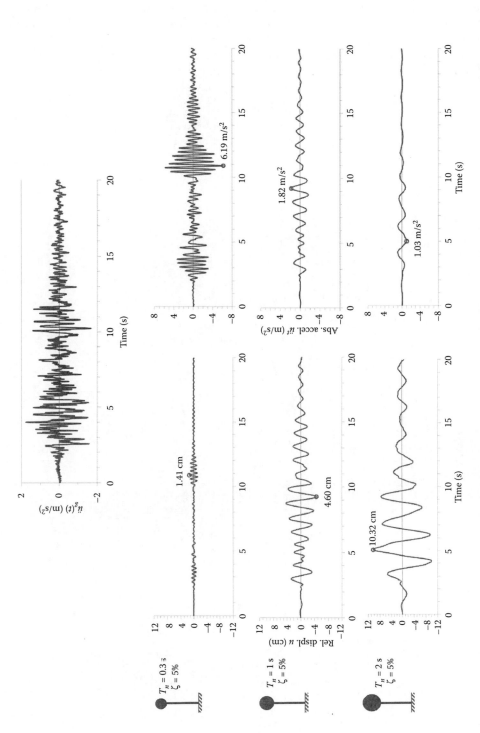

Figure 3.5 Construction of relative displacement and absolute acceleration spectra for the Ulcinj (Albatros, N–S direction) record of ground motion. The peak values of relative displacements and absolute accelerations represent the spectral displacements and spectral accelerations, respectively (see Figure 3.6).

The spectra for the relative displacement u_o, relative velocity $\dot{u}_o$ and absolute acceleration $\ddot{u}_o^t$ for the Ulcinj – Albatros ground motion and 5% damping are shown in Figure 3.6.

If the analyst is only interested in the maximum response of a structure subjected to a specific ground motion, and the displacement response spectrum u_o (T_n, ζ) is known, the maximum displacement u_0 can be obtained as the ordinate of the spectrum (corresponding to damping ζ) at the natural period of the system T_n. The same applies to the velocity and acceleration spectra.

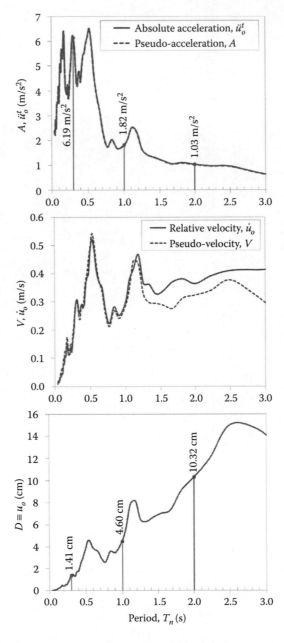

Figure 3.6 Spectra for the relative displacement $u_o \equiv D$, relative velocity $\dot{u}_o$ and absolute acceleration $\ddot{u}_o^t$, as well as the pseudo-velocity V and pseudo-acceleration spectra A for the Ulcinj (Albatros, N–S direction) ground motion and 5% damping.

3.1.2.2 Pseudo-spectra and seismic force

In seismic analyses, the velocity spectrum and the acceleration spectrum are, for convenience, usually replaced by the so-called *pseudo-velocity spectrum*, *V*, and *pseudo-acceleration* spectrum, *A*. They are defined as (using the notation $D \equiv u_o$):

$$V = \omega_n D = \frac{2\pi}{T_n} D \tag{3.36}$$

$$A = \omega_n^2 D = \frac{4\pi^2}{T_n^2} D \tag{3.37}$$

or

$$V = \frac{A}{\omega_n} = \frac{T_n}{2\pi} A \tag{3.38}$$

$$D = \frac{A}{\omega_n^2} = \frac{T_n^2}{4\pi^2} A \tag{3.39}$$

For small damping values, the pseudo spectra are very similar to the actual spectra, with the exception of pseudo-velocity spectra for very flexible structures with a long natural period, T_n.

$$A \approx \ddot{u}_o^t \tag{3.40}$$

$$V \approx \dot{u}_o \tag{3.41}$$

For zero damping, the pseudo-acceleration spectrum becomes exactly equal to the acceleration spectrum.

A comparison of pseudo-velocity and pseudo-acceleration spectra with actual spectra is shown in Figure 3.6. It can be seen that, for small damping, the pseudo-acceleration spectrum is practically equal to the acceleration spectrum. Some differences between the pseudo-velocity and velocity spectra occur in the intermediate and long-period ranges.

The use of pseudo-spectra instead of actual spectra simplifies the analysis. First, the three spectra (D, V and A) are related by simple equations. So, seismic standards typically provide only one spectrum, that is, the pseudo-acceleration spectrum, A, whereas the other two spectra can be determined, if needed, from Equations 3.38 and 3.39. Using Equations 3.36 to 3.39, it is also possible to plot several spectra in a single plot, for example, in the acceleration–displacement (AD) format, in which spectral accelerations are plotted against spectral displacements, with the periods T_n represented by radial lines (Figure 3.7). Secondly, the pseudo-acceleration spectrum is directly related to the *seismic force*, as shown next.

The absolute maximum value of the elastic force in the spring, also called the restoring force, is defined as:

$$f_{S0} = ku_0 = kD = m\omega_n^2 D = mA \tag{3.42}$$

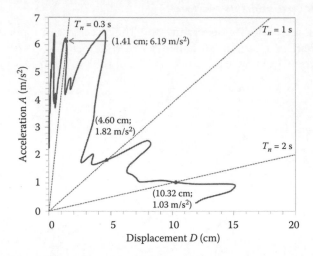

Figure 3.7 Spectrum for the Ulcinj (Albatros, N–S direction) ground motion and 5% damping in AD format.

Equation 3.42 shows that the maximum restoring force can be calculated either as the product of stiffness and maximum displacement (i.e. the value in the displacement spectrum D), or as the product of mass and the value in the pseudo-acceleration spectrum A. The latter option is simpler in the case of MDOF systems and is typically used in seismic standards for the determination of seismic actions.

The idea of a typical seismic analysis according to standards is to perform a usual elastic static analysis (as for the other types of actions). This idea can be realised by using an equivalent static loading, which can be called 'seismic forces'. The equivalent static loading must, in the static analysis, produce the same displacements as those determined in the dynamic analysis. In the case of an SDOF system, the maximum displacement can be obtained from the displacement spectrum. Moreover, in static analyses the external force is equal to the internal force, that is, the elastic force in the spring (the restoring force). Consequently, considering Equation 3.42, the *seismic force*, that is, the equivalent static external force, f_n, which produces the maximum dynamic displacement in a static analysis, is defined as

$$f_n = mA \tag{3.43}$$

In an SDOF system, f_n is equal to the seismic base shear force V_{bn}.

Equation 3.43 shows that the seismic force (i.e. the seismic action) can be determined as the product of the mass and the value in the pseudo-acceleration spectrum, $A(T_n, \zeta)$, which depends on the natural period and the damping of the system. Static analysis of a system subjected to the seismic force will produce the same displacement as obtained in a dynamic analysis of the same system, when subjected to a ground motion represented by the pseudo-acceleration spectrum. The same concept can be used for MDOF systems, as shown in Section 3.1.5.

The values of the acceleration and displacement spectra at periods $T_n = 0$ and $T_n = \infty$ follow physical constraints. For infinitely rigid structures ($T_n = 0$), the spectral acceleration $A(T_n = 0)$ is equal to the peak ground acceleration PGA, whereas the spectral displacement $D(T_n = 0)$ is equal to zero. The structure moves with the ground without any deformation. For infinitely flexible structures, however, the spectral acceleration is equal to zero, whereas the spectral displacement is equal to the maximum ground displacement PGD. The support moves with the ground but the mass remains at rest (Figure 3.8).

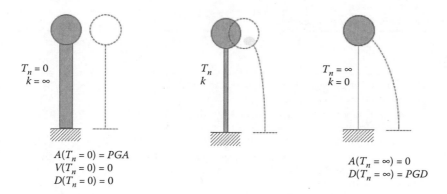

Figure 3.8 Seismic response of structures with different natural periods of vibration. Limit cases of infinitely rigid and infinitely flexible structures.

In the following, pseudo-acceleration spectra will be used. It is assumed that the pseudo-acceleration is equal to the acceleration and the prefix 'pseudo' is omitted.

3.1.3 Elastic response spectra according to Eurocode 8

Seismic standards and codes use highly idealised spectra, which follow the physical constraints and the characteristic features of actual spectra and are intended to represent average characteristics of ground motions in the region where the standard and/or code is implemented.

The shape of spectra in codes follows the typical characteristics of spectra. The spectral accelerations are the largest in the short-period range, the spectral velocities are the largest in the medium period range and the spectral displacements are the largest in the long-period range. The three ranges are also called acceleration-, velocity- and displacement-controlled regions. Typically, with the exception of the extreme cases, the absolute elastic spectral accelerations (and forces) decrease with increasing period (and flexibility) of the structure, whereas the relative spectral displacements increase.

In order to define the design seismic action according to Eurocode 8, the following parameters need to be defined:

1. The reference return period for the design seismic action.
2. PGA on rock, defined as a material with an equivalent shear wave velocity larger than 800 m/s.
3. The Importance Class of the building.
4. The representative ground type.
5. The predominant surface wave magnitude of earthquakes that contribute to the seismic hazard.

The reference return period is typically chosen by the National Authorities (see Section 1.3). The importance classification results from the use and occupancy of the building (see Section 1.3, Table 1.1). PGA on rock and the earthquake magnitude are results of a probabilistic seismic hazard analysis (PSHA – see Section 2.1.4); the ground type depends on the local soil conditions.

In Eurocode 8, the ground type is defined in terms of the average shear wave velocity, $V_{S,30}$, in the top 30 m below the ground surface. When $V_{S,30}$ is not available, other

Table 3.1 Ground type classification per EC8

Ground type	Description	$V_{s,30}$ (m/s)	N_{SPT} blows/0.3 m	c_u (kPa)
			Parameters	
A	Rock or similar geological formation, with at most 5 m of weaker material at the surface.	>800	–	–
B	Deposits of very dense sand, gravel, or very stiff clay, at least several tens of m thick, with gradual increase of mechanical properties with depth.	360–800	>50	>250
C	Deep deposits of dense or medium-dense sand, gravel or stiff clay, from several tens to many hundreds metres thick.	180–360	15–50	70–250
D	Deposits of loose-to-medium cohesionless soil (with or without some soft cohesive layers), or of predominantly soft-to-firm cohesive soil.	<180	<15	<70
E	A 5–20 m thick surface alluvium layer with V_s values of type C or D, underlain by stiffer material with V_s > 800 m/s			
S_1	Deposits consisting, or containing an at least 10 m thick layer of soft clays/silts with high plasticity index (PI > 40) and high water content	<100 (indicative)	–	10–20
S_2	Deposits of liquefiable soils, sensitive clays, or any other soil profile not included in types A to E or S_1			

parameters like the Standard Penetration Test (SPT) blow count, N, or the soil undrained shear strength, c_u, may be used as proxies. Modelling the top 30 m of the soil profile as a stack of horizontal layers, each with thickness h_i and shear wave velocity V_{Si}, the average shear wave velocity is defined as:

$$V_{S30} = \frac{\sum_{i=1}^{n} h_i}{\sum_{i=1}^{n} h_i/V_{Si}} = \frac{30 \text{ m}}{\sum_{i=1}^{n} h_i/V_{Si}} \tag{3.44}$$

Table 3.1 presents the different ground types defined in Eurocode 8.

For soil types S_1 and S_2, special studies for the definition of the seismic action shall be undertaken as too few sites entering that category are available to define a design response spectrum. For the other site categories, the horizontal component of the seismic action, the (pseudo-) acceleration elastic response spectrum A, which in Eurocode 8 is denoted as $S_e(T)$, is defined by the following equations:

$$
\begin{aligned}
0 \le T \le T_B \qquad & S_e(T) = a_g S\left[1 + \frac{T}{T_B}(2.5\eta - 1)\right] \\[2mm]
T_B \le T \le T_C \qquad & S_e(T) = 2.5\, a_g S\eta \\[2mm]
T_C \le T \le T_D \qquad & S_e(T) = 2.5\, a_g S\eta\left[\frac{T_C}{T}\right] \\[2mm]
T_D \le T \le 4 \text{ s} \qquad & S_e(T) = 2.5\, a_g S\eta\left[\frac{T_C T_D}{T^2}\right]
\end{aligned}
\tag{3.45}
$$

where:

- T is the vibration period of a linear SDOF oscillator (denoted as T_n in other parts of this book)
- a_g is the design acceleration on rock (equal to the reference peak ground acceleration on rock times the importance factor)
- T_B is the lower limit of the constant acceleration branch
- T_C is the upper limit of the constant acceleration branch
- T_D is the value defining the beginning of the constant displacement branch
- S is the soil factor
- η is the damping correction factor given by $\eta = \sqrt{10/(5 + \zeta)} \geq 0.55$ with the damping ratio of the structure (ζ) expressed as a percentage

Values of S, T_B, T_C and T_D are given in Table 3.2. Because statistical analyses of recorded events show that the spectral shape is magnitude-dependent and practically distance-independent (from source to site), the parameters are defined for two different spectral shapes that depend on the seismicity of the area: Type 1 defines areas of high intensity characterised by earthquakes with a surface wave magnitude larger than 5.5; Type 2 defines areas of low intensity characterised by earthquakes with a surface wave magnitude (see Section 2.1.1) smaller than 5.5.

The shape of the acceleration spectrum according to Eurocode 8, normalised to 1 g peak ground acceleration, is shown in Figure 3.9 for soil categories A to E and both earthquake types. The displacement spectrum is defined by Equation 3.39, but only for periods $T \leq T_E$, where T_E depends on the soil type ($T_E = 4.5$ s for soil type A). For periods $T > T_F$, where $T_F = 10$ s, the values in the Eurocode 8 displacement spectrum are equal to an estimate of the maximum ground displacement $u_g = \text{PGD}$ (see also Figure 3.8).

The spectral shapes defined by Equation 3.45 are valid so long as no near source effects are expected; base-isolated structures are very sensitive to near source effects, which create a large velocity pulse in the motion. Therefore, Eurocode 8 requires the development of a site-specific response spectrum for base-isolated structures of importance category IV (the highest one) located within 15 km of an active fault that can generate an earthquake with a magnitude larger than 6.5. The resulting spectrum shall not, however, fall below the Eurocode 8 spectrum.

Since observations of recorded motions have shown that the frequency content of the vertical component is different from the frequency content of the horizontal motion, the traditional way of defining the vertical spectrum as a fraction of the horizontal one is abandoned and the vertical component of the seismic action is defined independently.

Table 3.2 Values of horizontal elastic spectrum parameters recommended in EC8

Ground type	Spectrum Type 1				Spectrum Type 2			
	S	T_B (s)	T_C (s)	T_D (s)	S	T_B (s)	T_C (s)	T_D (s)
A	1.00	0.15	0.4	2.0	1.0	0.05	0.25	1.2
B	1.20	0.15	0.5	2.0	1.35	0.05	0.25	1.2
C	1.15	0.20	0.6	2.0	1.50	0.10	0.25	1.2
D	1.35	0.20	0.8	2.0	1.80	0.10	0.30	1.2
E	1.40	0.15	0.5	2.0	1.60	0.05	0.25	1.2

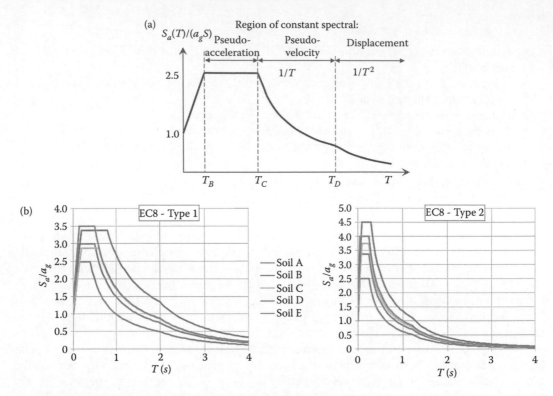

Figure 3.9 Elastic spectral shapes according to Eurocode 8; (a) generic shape of acceleration spectrum; (b) recommended 5%-damped horizontal acceleration spectrum of type 1 or 2.

The pseudo-acceleration vertical elastic response spectrum $S_{ve}(T)$ is defined by the following equations:

$$0 \leq T \leq T_B \qquad S_{ve}(T) = a_{vg}\left[1 + \frac{T}{T_B}(3.0\eta - 1)\right]$$

$$T_B \leq T \leq T_C \qquad S_{ve}(T) = 3.0\,a_{vg}\eta$$

$$T_C \leq T \leq T_D \qquad S_{ve}(T) = 3.0a_{vg}\eta\left[\frac{T_C}{T}\right] \tag{3.46}$$

$$T_D \leq T \leq 4\text{ s} \qquad S_{ve}(T) = 3.0\,a_{vg}\eta\left[\frac{T_C T_D}{T^2}\right]$$

where a_{vg} is the vertical acceleration defined as a fraction of the horizontal one. The recommended value of a_{vg} and those of the controlling periods T_B, T_C and T_D are:

- $T_B = 0.05$ s
- $T_C = 0.15$ s
- $T_D = 1.0$ s
- $a_{vg} = 0.9a_g$, if the Type 1 spectrum is used
- $a_{vg} = 0.45a_g$, if the Type 2 spectrum applies

Note that the spectral shape no longer depends on the ground classification; the rationale behind this comes from the fact that vertical motions are mainly induced by the propagation of *P*-wave (dilatational waves), while horizontal motions are mainly induced by propagation of *S*-waves (shear waves); *S*-wave velocities are much more sensitive to ground type than *P*-wave velocities.

3.1.4 Dynamics of multiple degrees of freedom systems

3.1.4.1 Equation of motion

Very rarely, real structures may be described by an SDOF system: the ways in which a real structure may deform are very complex and cannot be described by a single coordinate. The complexity of the deformation of a structure depends on the distribution of its mass and stiffness, as well as on the characteristics of the loading (distribution in space and variation in time); in most cases, the response includes a variation with time of the shape and amplitude. Hence, the response of the structure may only be described well if the deformation is defined by more than one degree of freedom. Hence, the need to consider MDOF systems.

The degrees of freedom in a discrete parameter system may be the displacements of certain selected points of the structure. In principle, these points may be chosen arbitrarily in the structure; in reality, it is convenient to choose the points in connection to the specific features of the structure, so that they are appropriate to best describe the way in which the structure deforms in response to the loading.

In many cases of discretisation of engineering structures, it is acceptable, without major loss of accuracy, to consider the mass of the system lumped at the points where the degrees of freedom are defined.

In a spatial system, at each discretisation point there are six degrees of freedom, three corresponding to translations and three to rotations; however, in many cases the structural model may be simplified and the number of degrees of freedom may be reduced. This is, for instance, the case of the so-called 'rigid diaphragm' buildings where the in-plane stiffness of the floor slabs is much higher than the stiffness of the lateral resisting elements. In such cases the deformation of the building under earthquake loading may be described just by two horizontal degrees of freedom at each storey, plus the twisting rotation around a vertical axis, with normally the discretisation point at each storey established at the centroid of the storey mass (see Section 3.1.10).

In any case, the configuration of the system (displacements and rotations) is described by as many linearly independent quantities as there are degrees of freedom. These quantities are called generalised displacements, denoted here by u_r. We may regard the generalised displacement as the product of a vector by a scalar, the former being a generalised coordinate. Then the modification of the amplitude of the displacement corresponds to modification of the scalar. In general we may choose as coordinates at each lumped mass the three displacements of the centroid of the mass and the rotation around the principal axis of inertia.

The displacements scalars describe the configuration of the system and may be arranged in a column (in any order). For a system with N degrees of freedom, the displacements scalars constitute an N-dimensional column vector, denoted by u and called the generalised configuration of the system.

In a similar way we may organise the internal and external generalised forces (forces and moments) in column vectors. Each term in the vector stands for the component of the force (or moment) on the corresponding coordinate, arranged in the same order as used for the terms describing the generalised configuration.

As in SDOF systems, the dynamic equilibrium condition requires that at each degree of freedom the restoring force, the damping force and the inertia force equilibrate the applied external force. This corresponds to an N-dimensional set of equations that may be expressed by:

$$\mathbf{f}_I + \mathbf{f}_D + \mathbf{f}_S = \mathbf{p}(t) \tag{3.47}$$

where
    $\mathbf{f}_I$ is the vector of the inertia forces
    $\mathbf{f}_D$ is the vector of the damping forces
    $\mathbf{f}_S$ is the vector of the restoring (stiffness) forces
    $\mathbf{p}(t)$ is the vector with the external (applied) forces (varying with time)

Let's consider first the case in which there is no motion of the ground, meaning that absolute and relative displacements, velocities and accelerations are the same. The restoring forces (in a linearly elastic system) depend on the generalised displacements $\mathbf{u}$ through the so-called stiffness matrix $\mathbf{k}$ with an $N \times N$ dimension, where each coefficient k_{ij} is defined as the force corresponding to coordinate i due to a unit displacement of coordinate j. Accordingly the vector of the restoring (stiffness) forces may be expressed by:

$$\mathbf{f}_S = \mathbf{k}\,\mathbf{u} \tag{3.48}$$

Now, if we assume, as before, that the damping forces are proportional to the velocity $\dot{\mathbf{u}}$ at each coordinate (i.e. the viscous damping assumption) we may express the damping forces vector through the damping matrix $\mathbf{c}$ as:

$$\mathbf{f}_D = \mathbf{c}\,\dot{\mathbf{u}} \tag{3.49}$$

Similarly to the definition of the stiffness matrix coefficients, the coefficients c_{ij} of the damping matrix are defined as the force corresponding to coordinate i due to a unit velocity of coordinate j.

Finally, the inertia forces depend on the acceleration at each coordinate and the corresponding mass. In matrix form this is expressed by:

$$\mathbf{f}_I = \mathbf{m}\,\ddot{\mathbf{u}} \tag{3.50}$$

where the coefficients m_{ij} of the mass matrix $\mathbf{m}$ are defined as the force corresponding to coordinate i due to a unit acceleration of coordinate j.

Replacing Equations 3.48 to 3.50 in Equation 3.47, the complete dynamic equilibrium of the system is given by a set of equation represented in matrix form by:

$$\mathbf{m}\ddot{\mathbf{u}} + \mathbf{c}\dot{\mathbf{u}} + \mathbf{k}\mathbf{u} = \mathbf{p} \tag{3.51}$$

where, for simplicity, we have omitted again the dependence on time of the accelerations, velocities and displacements of the system, as well as of the applied forces.

Let's consider now the case where the base is not fixed but moves, as for earthquake action. In that case, for the computation of the restoring and the damping forces we have to consider the relative displacements and the relative velocities, whereas for the computation of inertia forces the absolute accelerations apply.

The simplest case is when the base is rigid and hence all points of the system fixed to the base have the same motion $\mathbf{u}_g$, $\dot{\mathbf{u}}_g$ and $\ddot{\mathbf{u}}_g$. In that case the relations between relative and absolute displacements, velocities and accelerations are given by:

$$\mathbf{u} = \mathbf{u}^t - \iota\,\mathbf{u}_g \tag{3.52}$$

$$\dot{\mathbf{u}} = \dot{\mathbf{u}}^t - \iota\,\dot{\mathbf{u}}_g \tag{3.53}$$

$$\ddot{\mathbf{u}} = \ddot{\mathbf{u}}^t - \iota\,\ddot{\mathbf{u}}_g \tag{3.54}$$

where ι is the influence vector representing the displacements of the masses resulting from static application of a unit ground displacement in the direction of ground excitation. In a special case of a planar system with all degrees of freedom in the same direction as the ground motion (e.g. a planar model of a multi-storey building with concentrated masses at the floor levels) the influence vector becomes a unit column vector **1**.

Considering that inertial forces are determined as the mass matrix **m** multiplied by the vector of absolute accelerations $\ddot{\mathbf{u}}^t$ and using Equation 3.54, Equation 3.51 becomes:

$$\mathbf{m}\ddot{\mathbf{u}} + \mathbf{c}\dot{\mathbf{u}} + \mathbf{k}\mathbf{u} = \mathbf{p} - \mathbf{m}\,\iota\,\ddot{\mathbf{u}}_g \tag{3.55}$$

This is the basic equation for MDOF systems; it is similar to the basic Equation 3.2 presented before, with regard to SDOF systems.

3.1.4.2 Free vibration

Let's consider the simplest case of dynamic response of the MDOF system that corresponds to its *free vibration* response when the base is still ($\ddot{u}_g = 0$) and there is no external force applied ($\mathbf{p} = 0$). Additionally, neglecting the term of damping in Equation 3.55 the equation representing free vibration is:

$$\mathbf{m}\ddot{\mathbf{u}} + \mathbf{k}\mathbf{u} = 0 \tag{3.56}$$

To solve this equation, let us assume that the system vibrates harmonically with a circular frequency ω_n. Such motion is given by:

$$\mathbf{u} = \Phi_n \sin\!\left[\omega_n(t - t_n)\right] \tag{3.57}$$

where **u** is a vector with the shape of the deformed configuration of the system Φ_n (which does not change with time) and t_n is an arbitrary value of time t. Taking the second derivative of time of this expression and replacing in Equation 3.56, we obtain successively:

$$-\omega_n^2\mathbf{m}\,\Phi_n \sin\!\left[\omega_n(t - t_n)\right] + \mathbf{k}\,\Phi_n \sin\!\left[\omega_n(t - t_n)\right] = 0 \tag{3.58}$$

$$(\mathbf{k} - \omega_n^2\mathbf{m})\Phi_n = 0 \tag{3.59}$$

Besides the trivial solution of Equation 3.59, $\Phi_n = 0$, non-trivial solutions are possible only when:

$$\det(\mathbf{k} - \omega_n^2 \mathbf{m}) = 0 \tag{3.60}$$

For a system with N degrees of freedom, this is an Nth degree equation in ω_n^2, called the characteristic equation of the system. Its solution corresponds to the determination of the eigenvalues and eigenvectors of the system.

The square roots of the eigenvalues are the *natural frequencies* of the system ω_n. Replacing ω_n^2 in Equation 3.59 and solving the resulting matrix equation provides the shape of the *n*th *natural mode* Φ_n. It should be noticed that the vector Φ_n may have any arbitrarily chosen scale, that is, what matters is the configuration of the mode and not its size.

In any case, usually the mode shapes are normalised according to certain criteria, one of the most common being choosing the scale of Φ_n so that $\Phi_n^T \mathbf{m} \Phi_n = 1$.

Conventionally, the results are ordered in increasing order of the natural frequencies ω_n, with n varying from 1 to N and the lowest frequency, that is, ω_1, called *fundamental frequency* of the system.

The natural modes form a complete orthogonal set with $\mathbf{m}$ or $\mathbf{k}$ as weighting matrix. This implies that:

$$\Phi_m^T \mathbf{m} \Phi_n = 0 \quad \text{if } \omega_m \neq \omega_n \tag{3.61}$$

and

$$\Phi_m^T \mathbf{k} \Phi_n = 0 \quad \text{if } \omega_m \neq \omega_n \tag{3.62}$$

where the superscript T indicates that the matrix or vector is transposed.

The orthogonality between mode shapes means that the inertia forces associated with the *m*th mode of vibration do not perform work when displaced with the configuration of the *n*th mode.

On the other hand, the fact that the natural modes constitute a complete set means that any deformed configuration $\mathbf{u}$ of the system may be represented by a linear combination of the natural mode shapes.

$$\mathbf{u} = \sum_n q_n \Phi_n \tag{3.63}$$

In Equation 3.63 q_n are (dimensionless) weighting coefficients of the contribution of each mode to the global deformed configuration.

Note that the calculation of eigenvalues and eigenvectors by solving the characteristic equation of the system, Equation 3.60, is practical for two or mostly three degrees of freedom. In the literature, a number of numerical methods are available for solving the eigenvalue problem (see, e.g. Chopra 2007).

As pointed out before, Equation 3.57 is a solution of the general free vibration, Equation 3.56, provided that ω_n is one of the *natural frequencies* of the system. Hence, considering the linearity of the system, any linear combination of the modal vibration

$$\mathbf{u} = \sum_N q_n \Phi_n \sin[\omega_n(t - t_n)] \tag{3.64}$$

is also a solution of Equation 3.56 for the free vibration motion of the system.

This means that the system, when disturbed from its resting position, will respond with an oscillatory motion, that is, a combination of all of its natural modes. The relative importance of the different modes for the global response is reflected by the values of the weighting coefficients q_n.

Also, the relation in time of the response of the various modes is reflected by the values of the time shift t_n. Both q_n and t_n depend on the initial conditions that trigger the motion of the system.

In the case of damped systems the free vibration equation becomes:

$$\mathbf{m\ddot{u} + c\dot{u} + ku} = 0 \qquad (3.65)$$

In such case, only under particular conditions the system possesses classical natural modes (i.e. in the real domain). The necessary and sufficient condition for the existence of natural modes in the real domain is that the transformation that diagonalises the mass and stiffness matrices $\mathbf{m}$ and $\mathbf{k}$ also diagonalises the damping matrix $\mathbf{c}$. In this case the natural modes of the damped system are the same as those for the corresponding undamped system. The condition above is satisfied if the damping matrix is a linear combination of the mass and stiffness matrices:

$$\mathbf{c} = a\mathbf{m} + b\mathbf{k} \qquad (3.66)$$

In this case, the damping ratios of the various natural modes are given by:

$$\zeta_n = \frac{a}{2\omega_n} + \frac{b\omega_n}{2} \qquad (3.67)$$

From Equation 3.67 it is apparent that the damping ratio shall be different for the various natural modes. In fact, in line with that equation, it is only possible to fix damping ratio for two natural modes considering (from Equation 3.67) a system of two linear equations and solving it for a and b.

In spite of this limitation, that only allows us to use in damped systems the classical natural modes if severe restrictions on the values of the damping of different modes are accepted, this is not, in practice, an important limitation for the evaluation of the dynamic response of structural systems. In most cases, it is acceptable, with no significant loss of accuracy, to assume the system as undamped and perform the modal analysis and then correct the results reducing the response of each mode based on approximate coefficients that reflect the effect of the damping ratio assigned to each mode.

The concept of vibration modes of MDOF systems and their features are illustrated in Example 3.1 at the end of this chapter, for an oscillator with 3 degrees of freedom.

3.1.5 Modal response spectrum analysis

3.1.5.1 Modal analysis

The system of coupled differential equations, Equation 3.55, can be solved by so-called *modal analysis*. This approach is based on a transformation into a new coordinate system defined by:

$$\mathbf{u}(t) = \mathbf{\Phi}\,\mathbf{q}(t) \qquad (3.68)$$

where $\mathbf{q}$ is the vector of generalised displacements in the new coordinate system and Φ is the matrix of eigenvectors whose columns are the eigenvectors corresponding to individual modes Φ_n. $\mathbf{u}$ and $\mathbf{q}$ are functions of time, whereas Φ is independent of time. Equation 3.68 can be written also in the form of Equation 3.63, which is repeated here for convenience

$$\mathbf{u}(t) = \sum_{n=1}^{N} \Phi_n\, q_n \tag{3.63}$$

This shows the physical meaning of the transformation defined by Equation 3.68: the vector of displacements $\mathbf{u}$ is expressed as a linear combination of N mode shapes Φ_n. The elements of the vector of the generalised displacements $\mathbf{q}$, q_n, represent the amplitudes of the mode shapes.

An approximation can be made which can substantially reduce the computational effort. Usually, the influence of different vibration modes decreases with increasing number of the mode. From a certain mode, for example, from mode M upwards, the influence becomes negligible. In such a case, only the first M modes can be taken into account in Equation 3.63:

$$\mathbf{u}(t) = \sum_{n=1}^{M} \Phi_n\, q_n \quad M \ll N \tag{3.69}$$

In the simplest case, it is sufficient to consider only the fundamental mode (Section 3.1.9). Note that, among the relevant response parameters, displacements are the least susceptible to the effects of higher modes. The contribution of higher modes is greater in the case of more local quantities, for example, storey drifts or deformations at the element level, and internal forces.

Using Equation 3.68, the system of coupled equations Equation 3.55 can be transformed into a system of N uncoupled equations:

$$\mathbf{M}\,\ddot{\mathbf{q}} + \mathbf{C}\,\dot{\mathbf{q}} + \mathbf{K}\,\mathbf{q} = \mathbf{P}(t) \tag{3.70}$$

where

$$\mathbf{M} = \Phi^T \mathbf{m}\, \Phi \tag{3.71}$$

$$\mathbf{C} = \Phi^T \mathbf{c}\, \Phi \tag{3.72}$$

$$\mathbf{K} = \Phi^T \mathbf{k}\, \Phi \tag{3.73}$$

$$\mathbf{P} = \Phi^T (\mathbf{p} - \mathbf{m}\iota\ddot{u}_g) \tag{3.74}$$

Thanks to the orthogonality of the mode shapes (Equations 3.61 and 3.62), the transformed mass matrix $\mathbf{M}$ given by Equation 3.71 and the transformed stiffness matrix, $\mathbf{K}$, given by Equation 3.73, are diagonal. The damping matrix $\mathbf{C}$, given by Equation 3.72 is, in general, not diagonal, since the mode shapes (eigenvectors) are determined for the undamped and not for the damped system (see also Section 3.1.4.2). Nevertheless, for practice, it can be assumed that the transformed damping matrix $\mathbf{C}$ is also diagonal. An example of a damping matrix $\mathbf{c}$, which becomes diagonal after the transformation as per Equation 3.72, is shown in Equation 3.66.

With all three matrices on the left-hand side of Equation 3.70 being diagonal, the system of equations (Equation 3.70) becomes uncoupled. Each of the N equations can be written in terms of the diagonal terms of matrices $\mathbf{M}$, $\mathbf{C}$ and $\mathbf{K}$ as:

$$M_n \ddot{q}_n + C_n \dot{q}_n + K_n q_n = P_n(t) \tag{3.75}$$

The differential equation, Equation 3.75, has the same form as Equation 3.2 of the case of SDOF systems. To each vibration mode corresponds one equation and can be treated independently. If only the first M vibration modes are taken into account (the approximation represented by Equation 3.69), the number of independent equations is reduced to M. In the extreme case, when only the fundamental mode is considered, there is only one equation.

After dividing by M_n (see Equation 3.80, below), Equation 3.75 becomes

$$\ddot{q}_n + 2\zeta_n \omega_n \dot{q}_n + \omega_n^2 q_n = \frac{P_n(t)}{M_n} \tag{3.76}$$

Neglecting the applied forces $\mathbf{p}$, that is, considering only excitation by ground motion, the right-hand side of Equation 3.76 can be transformed as follows:

$$\frac{P_n(t)}{M_n} = -\ddot{u}_g(t) \frac{\Phi_n^T \mathbf{m} \, \iota}{M_n} = -\ddot{u}_g(t) \frac{L_n}{M_n} = -\ddot{u}_g(t) \, \Gamma_n \tag{3.77}$$

where

$$\Gamma_n = \frac{L_n}{M_n} \tag{3.78}$$

$$L_n = \Phi_n^T \mathbf{m} \, \iota \tag{3.79}$$

and (see Equation 3.71)

$$M_n = \Phi_n^T \mathbf{m} \Phi_n \tag{3.80}$$

Γ_n is called *modal participation factor*. It is a measure of the degree to which the nth mode participates in the response.

In order to determine the coefficients in Equation 3.76, the results of free vibration analysis, that is, the dynamic characteristics of the structure defined by the frequencies ω_n and the mode shapes Φ_n, have to be known for all vibration modes which will be taken into account, in addition to the dimensionless damping coefficients ζ_n and the mass matrix of the system $\mathbf{m}$.

By solving the differential equation, Equation 3.76, the generalised displacement q_n is obtained as a function of time. It is related to vibration in the nth mode, which represents a part of the total response. Any method, appropriate for the solution of the differential equation for an SDOF system, can be used.

3.1.5.2 Elaboration for the seismic action

If one is interested only in the maximum value of q_n, that is, q_{n0}, it can be obtained from the response spectrum, similarly as in the case of an SDOF system. By comparing Equation 3.76

and its counterpart for the SDOF system, Equation 3.3, it can be seen that for the MDOF system the ground acceleration on the right-hand side of the equation is multiplied by Γ_n. Consequently, when determining the maximum response by using the response spectrum, the spectra for ground motion are multiplied by the participation factor Γ_n for the vibration corresponding to the nth mode of the MDOF system. Thus, q_{n0} can be obtained as

$$q_{n0} = \Gamma_n D(T_n, \zeta_n) = \Gamma_n D_n \tag{3.81}$$

where Γ_n is defined by Equation 3.78 and $D(T_n, \zeta_n) = D_n$ is the value in the displacement spectrum at the period of the nth vibration mode, T_n, for the damping in this mode ζ_n.

The next step in the computational procedure is back-transformation from the generalised displacements q_n to the displacements in the original coordinate system $\mathbf{u}_n$, which is performed by means of Equation 3.63, by considering only one vibration mode, that is, mode n:

$$\mathbf{u}_{n0} = \Phi_n q_{n0} = \Phi_n \Gamma_n D_n = \Phi_n \Gamma_n \frac{T_n^2}{4\pi^2} A_n \tag{3.82}$$

Equation 3.82 is the expression for the determination of the maximum displacements for vibration in the nth mode. In Equation 3.82, A_n represents the value in the pseudo-acceleration spectrum at the period of the nth vibration mode, T_n, considering the damping in this mode, ζ_n.

Knowing the displacements, all other relevant quantities, that is, local deformations and internal forces, can be obtained, with the methods of static analysis. However, for convenience, it is usual to perform static analysis by applying equivalent external forces, called seismic forces (see also Section 3.1.2.2), rather than by imposing displacements. Moreover, the mathematical model used in static analysis for the determination of internal forces is typically much more complex than the condensed model used for dynamic analysis. *Seismic forces* (representing the seismic action), $\mathbf{f}_n$, are the external forces which, in the case of a static analysis, produce the displacements $\mathbf{u}_{n0}$, determined in the dynamic analysis via Equation 3.82. In static analysis the following equation applies:

$$\mathbf{f}_n = \mathbf{k}\,\mathbf{u}_{n0} \tag{3.83}$$

Considering Equations 3.82 and 3.60, the right-hand side of Equation 3.83 can be transformed into a form which can be used for the determination of the seismic forces corresponding to vibration mode n:

$$\mathbf{f}_n = \mathbf{k}\,\Phi_n \Gamma_n D_n = \omega_n^2 \mathbf{m}\,\Phi_n \Gamma_n D_n = \mathbf{m}\Phi_n \Gamma_n A_n \tag{3.84}$$

Equation 3.84 shows that the seismic forces corresponding to an individual vibration mode are proportional to the shape of this mode, weighted by the masses. A comparison of Equations 3.84 and 3.82 reveals that the displacements are less influenced by vibration in higher modes than the forces. The period T_n, which enters in Equation 3.82 squared, has the largest value in the first (fundamental) mode and decreases with increasing mode.

The shear force at the base of a structure (V_{bn}, called the *base shear force*) in the direction of the applied excitation is equal to the sum of all the lateral seismic forces for the vibration mode n in this direction. It can be calculated as:

$$V_{bn} = \mathbf{f}_n^T \iota = \Phi_n^T \mathbf{m}\iota \frac{L_n}{M_n} A_n = \frac{L_n^2}{M_n} A_n = M_n^* A_n \tag{3.85}$$

where M_n^* is the *effective modal mass* for vibration mode n, defined as:

$$M_n^* = \frac{L_n^2}{M_n} \tag{3.86}$$

The effective mass has dimensions of mass, and can be interpreted as the part of the total mass responding to the ground excitation in mode n. In general, the effective mass of a mode has different values for different directions of application of the seismic action. Indeed, its value is a good indicator of the 'direction' of the mode shape or, in other words, the direction of the seismic action that excites this mode most. The sum of the effective modal masses for all modes in a given direction is equal to the total mass of the structure. This is a very important feature of the effective modal masses, which can be used to determine the contribution of different vibration modes to the total response. Sometimes, the effective modal mass is presented normalised by the total mass of the structure excited in the relevant direction. Then it is denoted as the *participating mass ratio*. The sum of the participating mass ratios (for all modes and for a given direction) is equal to 1.

Equation 3.85 shows that the base shear corresponding to the vibration mode n can be determined as the product of the effective modal mass and the spectral acceleration corresponding to this vibration mode.

Equations 3.82, 3.84 and 3.85 are general equations, based on the dynamics of structures, explicitly or implicitly included in seismic guidelines, standards and codes, including Eurocode 8, where this type of analysis is called *Modal response spectrum analysis*. The equations are general and apply to any structural model. The base shear notion is common mostly in the analysis of building structures.

3.1.5.3 Combination of modal responses

Equations 3.82, 3.84 and 3.85 apply to vibration mode n. In general, several vibration modes contribute to the structural response. The question is how to determine the peak value of the response which is represented by the combined contribution of all relevant vibration modes. Equation 3.63 can be used if a time-history analysis is performed. In the analysis by means of response spectra, only the maximum values of the response for individual modes are known, whereas the time at which these maxima occur is not known. However, it is highly improbable that the maximums would occur simultaneously in all vibration modes. Thus, if Equation 3.63 is used in response spectrum analysis to combine the maximum values for different modes, the result represents an upper limit and would be typically highly conservative.

Several other approaches for the combination of responses in different vibration modes exist, providing more realistic results than the sum of maximum values according to Equation 3.63. Among them, the most widely used is the Square Root of the Sum of Squares (SRSS) combination rule. According to the SRSS rule, the resulting value of any response quantity E_E is obtained as the square root of the sum of the squared values of this response quantity for all the relevant modes:

$$E_E = \sqrt{\sum_{n=1}^{M} E_{En}^2} \tag{3.87}$$

The SRSS combination rule is based on random vibration theory and is intended to represent the expected value of the peak response for a set of ground motions, which are typically defined by a smooth response spectrum. The SRSS combination rule does not take into account the correlation between different modes of vibration. It provides very close estimates of peak responses if the natural frequencies of the relevant modes are well separated. However, in the case of closely spaced frequencies, the SRSS combination rule is not applicable. According to Eurocode 8, the SRSS rule can be used if the periods of two relevant vibration modes T_n and T_r satisfy (with $T_n \leq T_r$) the condition $T_n \leq 0.9 T_r$.

One of the combination rules, which does take into account the correlation between different vibration modes and is applicable also for closely spaced natural frequencies, is the Complete Quadratic Combination (CQC) (Der Kiureghian 1981; Wilson et al. 1981):

$$E_E = \sqrt{\sum_{n=1}^{M}\sum_{r=1}^{M}\rho_{nr}\, E_{En}\, E_{Er}} \tag{3.88}$$

where n and r are indexes of vibration modes. The value of the correlation coefficient ρ_{nr} is between 0 and 1. In a special case, where the damping ratio ζ is the same in all vibration modes, ρ_{nr} can be determined as:

$$\rho_{nr} = \frac{8\zeta^2 \left(1 + \beta_{nr}\right)\beta_{nr}^{3/2}}{\left(1 - \beta_{nr}^2\right)^2 + 4\zeta^2\,\beta_{nr}\left(1 + \beta_{nr}\right)^2} \tag{3.89}$$

where β_{nr} is the ratio of the frequencies of the modes n and r:

$$\beta_{nr} = \frac{\omega_n}{\omega_r} = \frac{T_r}{T_n} \tag{3.90}$$

Equation 3.89 cannot be used in the case of zero damping and equal frequencies ($\zeta = 0$, $\beta_{nr} = 1$). In this limit case: $\rho_{nr} = 1$.

Note that in the case of well-separated natural frequencies the CQC rule reduces to the SRSS combination rule.

The question arises, however, as to how many vibration modes have to be taken into account, in order to obtain reasonably accurate results. In seismic analyses, the influence of higher modes is typically small for displacements and increases for more local response quantities. According to Eurocode 8, 'the response of all modes of vibration contributing significantly to the global response shall be taken into account'. Eurocode 8 further considers that this principle is deemed to be satisfied if either the sum of the effective modal masses for the modes taken into account amounts to at least 90% of the total mass of the structure, or all the modes with effective modal masses greater than 5% of the total mass are taken into account. The effective modal mass is defined according to Equation 3.86.

It is important to emphasise that the combination rule should be applied to the final response quantities, that is, to the deformations and internal forces in structural elements, and not to intermediate quantities, like seismic forces. In a usual procedure, the seismic forces are first determined for each relevant mode. A static analysis of the structure is then performed separately for each vector of seismic forces, that is, for each relevant mode. Finally, the results of static analyses for each mode are combined by a combination rule.

The procedure would be simplified if the combination rule were applied at the level of forces because, in such a case, only one static analysis would be necessary. However, the results of such an analysis would be (overly) conservative. For illustration, let us consider a planar building structure. For such a structure, the seismic forces corresponding to the higher modes change sign along the height of the building. These changes have a beneficial effect on the amplitude of the response quantities. When combining the seismic forces, for example, by the SRSS method, the beneficial effect is lost, since the negative signs are lost by squaring.

The modal analysis of the example building of Chapter 7, complete with the periods and participating mass ratios of the 10 natural modes that are needed to capture at least 90% of the total mass in both horizontal directions, and the shapes of the three lower modes, can be found in Section 7.3.5, Table 7.2 and Figure 7.6. Moreover, Example 3.2 at the end of this chapter, extends Example 3.1 to compute the participating masses and participation factors of the same oscillator with 3 degrees of freedom.

3.1.5.4 Special case: Planar building models

If a building structure is doubly symmetrical in plan, there is no torsional effect and two planar (two-dimensional, 2D) models can be used for the analysis, one in each horizontal direction. According to Eurocode 8, a planar model can be used as an approximation, for all structures that are regular in plan. In the case of a planar model (see Section 3.1.10), the equations presented in the previous chapter can be simplified. The masses are concentrated at the levels of individual storeys, j. The N degrees of freedom correspond to the horizontal displacements of the masses. For such a model, the mass matrix $\mathbf{m}$ is diagonal, with the floor masses m_j along the diagonal, and all the elements of the influence vector ι are equal to 1 ($\iota = 1$). Equations 3.79 and 3.80 can be written as:

$$L_n = \sum_j \Phi_{nj}\, m_j \tag{3.91}$$

$$M_n = \sum_j \Phi_{nj}^2\, m_j \tag{3.92}$$

where n stands for the vibration mode and j for the storey. From Equations 3.82 and 3.84 the expressions for the displacement u_{nj} and seismic force f_{nj} in storey j are obtained

$$u_{nj} = \Phi_{nj}\, \Gamma_n\, D_n = \Phi_{nj}\, \frac{\sum_j \Phi_{nj}\, m_j}{\sum_j \Phi_{nj}^2\, m_j}\, D_n \tag{3.93}$$

$$f_{nj} = \Phi_{nj}\, m_j\, \Gamma_n\, A_n = \Phi_{nj}\, m_j\, \frac{\sum_j \Phi_{nj}\, m_j}{\sum_j \Phi_{nj}^2\, m_j}\, A_n \tag{3.94}$$

The base shear force V_{bn} is the sum of the seismic forces in all the storeys

$$V_{bn} = \sum_j f_{nj} = \frac{\left(\sum \Phi_{nj}\, m_j\right)^2}{\sum \Phi_{nj}^2\, m_j}\, A_n = \frac{L_n^2}{M_n}\, A_n = M_n^*\, A_n \tag{3.95}$$

Considering Equation 3.95, another form of Equation 3.94 can be obtained:

$$f_{nj} = V_{bn} \frac{\Phi_{nj}\, m_j}{\sum_j \Phi_{nj}\, m_j} \tag{3.96}$$

3.1.6 Lateral force method

The *lateral force method* of analysis is a simplified approach widely used for simple structures in seismic standards and codes. It is based on the assumption that the influence of higher vibration modes is negligible. According to Eurocode 8, 'this type of analysis may be applied to buildings whose response is not significantly affected by contributions from modes of vibration higher than the fundamental mode in each principal direction'. Eurocode 8 considers that this requirement is deemed to be satisfied in buildings that are regular in elevation and do have a fundamental period less than 2 s and four times the corner period T_C of the applicable design spectrum (see Section 3.1.3 and Table 3.2).

In the lateral force method, the starting point is the base shear force, determined as follows:

$$V_b = M\, A_1 \tag{3.97}$$

where M is the total mass of the structure and A_1 is the spectral acceleration for the period of the fundamental mode of vibration. Equation 3.97 is the same as Equation 3.43 of SDOF systems. In the modal analysis of MDOF systems, a similar equation, namely Equation 3.85, applies. However, the base shear force in Equation 3.97 is determined using the total mass, whereas according to Equation 3.85 the base shear force is related to the effective mass M^*. As stated earlier, the sum of effective masses for all the vibration modes is equal to the sum of all the masses. Thus, the effective mass for a single mode is always less than the total mass (except in a SDOF system, where it is the same). So, the base shear force in the lateral force method according to Equation 3.97 is always greater than the base shear force for the first mode in modal analysis, Equation 3.85. This conservatism of the approximate lateral force method can be considered as a reasonable compensation, usual when simplified methods are used. However, in Eurocode 8 the conservatism has been intentionally removed by multiplying the base shear force according to Equation 3.97 with a correction factor λ equal to 0.85, except in buildings with up to two storeys or flexible ones (those with a fundamental period longer than twice the corner period, T_C, of the design spectrum), for which $\lambda = 1.0$.

If the simple lateral force method of analysis is used, it is reasonable to determine also the fundamental period of vibration by a simplified method, rather than performing a rigorous free eigenvalue analysis. A practical approach, which yields quite accurate values of the fundamental period, is the Rayleigh method (described in Section 3.1.9). Some standards and codes, including Eurocode 8, also allow the use of purely empirical formulas for the estimation of the fundamental period.

The base shear force, which represents the sum of all lateral seismic forces, has to be distributed along the height of the building. This can be done by means of Equation 3.96 ($n = 1$), provided that the mode shape of the fundamental mode Φ_1 is known from free vibration analysis. If not, an approximation can be used for the first mode shape. Eurocode 8 allows a height-wise linear one; then, the seismic force acting at floor j is determined as:

$$f_{sj} = V_b \frac{z_j\, m_j}{\sum_j z_j\, m_j} \tag{3.98}$$

where z_j is the height of storey j above the base, that is, the distance from a rigid base to storey j.

The analysis of the example building of Chapter 7 with the lateral force method and the estimation of the fundamental period in the two horizontal directions can be found in Section 7.3.4 and Table 7.2.

3.1.7 Combination of seismic action components

All of the analyses so far presented apply to ground excitation in one direction. Typically, the analysis is carried out also for the seismic excitation in the orthogonal direction (rarely also in the vertical direction). Since the analyses are linear, the superposition law applies and the results can be superposed in order to obtain the total response for the structure when subjected to ground motion in two (or three) directions. In response spectrum analysis, where only peak response values are known, a problem similar to the combination of the peak effects of different vibration modes arises in the combination of the peak effects of different directions of ground motion. It is highly improbable that peak values from different directions will occur at the same time. A good approximation of the final value of any response quantity E_E can be obtained (Smebby and Der Kiureghian 1985) by using the SRSS combination rule:

$$E_E = \sqrt{E_X^2 + E_Y^2 \left(+E_Z^2\right)} \qquad (3.99)$$

where E_X, E_Y and E_Z represent the total values (considering all the relevant modes) of the response quantity of interest due to the application of the seismic action along the chosen horizontal axes x and y, and the vertical axis z of the structure, respectively. (E_Z is in brackets, since the vertical direction is only exceptionally taken into account.)

According to Eurocode 8, as an alternative to Equation 3.99, the action effects due to combination of the three components of the seismic action may be computed using all of the following combinations:

$$E_X \;\text{'+'}\; \lambda E_Y \;\text{'+'}\; \lambda E_Z \qquad (3.100a)$$

$$\lambda E_X \;\text{'+'}\; E_Y \;\text{'+'}\; \lambda E_Z \qquad (3.100b)$$

$$\lambda E_X \;\text{'+'}\; \lambda E_Y \;\text{'+'}\; E_Z \qquad (3.100c)$$

where $\lambda = 0.3$ and '+' means 'to be combined with', but in this case with the same sign. Again, the vertical component is used only exceptionally.

Equation 3.99 captures in a single load combination all seismic action components. As a matter of fact, when modal response spectrum analysis is used, computationally this combination can be carried out in the same phase as the SRSS or CQC combination of modal response as per Section 3.1.2.3. By contrast, when Equation 3.100 is used, two separate combinations are needed for the two horizontal components, or three, when the vertical one is considered as well. These separate combinations should be superimposed separately with the gravity load effects in the 'seismic design situation', each combination with alternating sign.

An example of application of the combination rules, Equations 3.99 and 3.100, is presented in Figure 3.10, where plan views of two simple single-storey building structures are shown. In both cases, only the horizontal components are taken into account. The structure with three walls is symmetric with respect to the x-axis and asymmetric with respect to the y-axis.

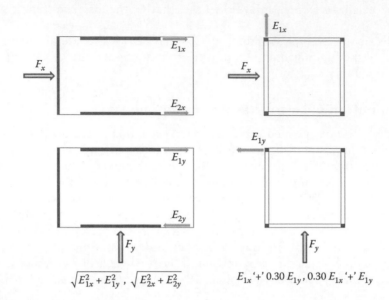

Figure 3.10 Illustration of the use of different combination rules for the effects of the two seismic action components.

In the two walls in the x-direction, action effects (e.g. shear forces, moments) are present for seismic action in the x-direction and also, due to torsion, in the y-direction. Any response quantity in each of the walls in the x-direction can be obtained by using the SRSS combination rule, Equation 3.99. Alternatively, according to Eurocode 8, the combination rules in Equation 3.100 can also be used. In the frame structure on the right-hand side of Figure 3.10, we consider one of the columns. In the corner column, there are two bending moments: one with respect to the y-axis due to seismic action in the x-direction, E_{1x}, and one around the x-axis due to loading in the y-direction, E_{1y}. For dimensioning (or checking) of the column, both bending moments have to be taken into account, but not with their maximum values. In this case, it is not possible to use the SRSS combination for the two bending moments representing two different response quantities; it is possible to use the combination according to Equation 3.100 and to dimension the column with the maximum bending moment around one axis and, simultaneously, 30% of the maximum bending column around the other axis (see also Section 5.8.1 and Example 5.12).

Figures 7.8 to 7.25 depict the moment, shear, and axial force diagrams of the example building, from modal analysis and the SRSS combination of the effects of the two horizontal components, Equation 3.99; they also compare them to the outcome of the lateral force method, this time using Equations 3.100a and 3.100b for the combination. Discussion and comments on these results are summarised in Section 7.5.1.

3.1.8 Accidental torsion

3D structural models take into account coupling between translational and torsional vibrations. If a building is plan-wise fully symmetric with respect to both axes, the horizontal components of the ground motion do not produce any torsional response. However, conventional seismic response analysis cannot capture possible variations in the stiffness and/or mass (and/or strength in the case of non-linear analysis) distributions from their nominal values. Moreover, there are possible components of torsional ground motion, which are not taken into account in the seismic analyses. Such effects may produce torsional response even in nominally fully symmetric buildings. In order to account for all these uncertainties, and to ensure a minimum of torsional resistance and stiffness, as well as to limit the possible consequences of an unforeseen torsional response, the concept of 'accidental eccentricity' has been used in seismic codes, including Eurocode 8.

Accidental torsional effects can be introduced by shifting the masses from their nominal positions by a distance equal to the accidental eccentricity, e_{aj}, which, in Eurocode 8, is taken to be equal to 5% of the dimension of the floor in storey j:

$$e_{aj} = \pm 0.05 L_j \tag{3.101}$$

where L_j is the floor dimension perpendicular to the direction of the seismic action.

In Eurocode 8, the accidental eccentricity takes twice the value from Equation 3.101, if it is considered in a simplified way on a separate 2D model for each horizontal component of the seismic action, or if masonry infills have a moderately irregular and asymmetric distribution in plan.

Shifting the masses is possible in dynamic analyses of 3D structural models (either modal response spectrum or response-history analysis). However, such an approach requires, in general, four different models and is very inconvenient for practical application. For this reason, the accidental torsional effects are usually taken into account through a static analysis of a 3D structural model subjected to storey torsional moments about the vertical axis. These torsional moments are equal to the storey lateral loads due to the horizontal component in question multiplied by the accidental eccentricity at the storey. In such a way, the accidental eccentricity of the masses from their nominal positions is replaced by an accidental eccentricity of the lateral seismic forces with respect to the nominal position of the masses. The resulting action effects are then superimposed to those determined by an analysis, which does not take into account accidental torsion.

The approach with torsional moments does not, in general, produce the same results as the shifting of masses. However, it is much more convenient for application. It does not make sense to try to 'accurately' predict the effects of accidental torsion, which is a highly uncertain phenomenon, while the magnitude of accidental eccentricities as per Equation 3.101 is just postulated. For a more detailed description of the treatment of accidental torsion in the different analysis procedures in Eurocode 8, see Fardis (2009).

Section 7.3.6 highlights the analysis of the example building of Chapter 7 for the accidental eccentricities in X and Y.

3.1.9 Equivalent SDOF systems

If the structural response is dominated by vibration in the fundamental mode, which is often the case for simple, not very flexible structures, the seismic analysis can be simplified by transforming an MDOF system into an equivalent SDOF system, and performing dynamic analyses on this SDOF system. In the present Section, a planar model is used and only the determination of the natural frequency based on the equivalent SDOF system is discussed.

Development of the procedure can start from Equation 3.70. It is assumed that the structure vibrates in the fundamental mode and that the influence of all the higher vibration modes is negligible. In such a case, the number of modes is $N = M = 1$ and Equations 3.71 and 3.73 can be written as:

$$M_1 = \Phi_1^T \, \mathbf{m} \, \Phi_1 = \sum_j \Phi_{1j}^2 m_j \tag{3.102}$$

$$K_1 = \Phi_1^T \mathbf{k} \, \Phi_1 \qquad (3.103)$$

Knowing M_1 and K_1, the exact value of the fundamental (first mode) frequency can be calculated as $\omega_1^2 = K_1/M_1$. A problem is that the equivalent SDOF system approach is typically used in order to avoid free vibration analysis; thus the mode shape Φ_1 needed for the determination of M_1 and K_1 is not known. The solution is to replace the fundamental mode shape Φ_1 with an approximation, Ψ, which is close to the actual shape. In such a case, by analogy with Equations 3.102 and 3.103, the equivalent mass m_{eq} and stiffness k_{eq} are obtained as:

$$m_{eq} = \Psi^T \mathbf{m} \Psi = \sum_j \Psi_j^2 m_j \qquad (3.104)$$

$$k_{eq} = \Psi^T \mathbf{k} \Psi = \sum_j \Psi_j^2 k_j \qquad (3.105)$$

and the approximate value of the fundamental natural frequency is:

$$\omega_1^2 \approx \frac{k_{eq}}{m_{eq}} \qquad (3.106)$$

The equivalent stiffness k_{eq} can be obtained also by an alternative approach. We take the displacements due to arbitrary lateral forces $\mathbf{f}$ as the approximate mode shape Ψ. Then the equation:

$$\mathbf{k} \Psi = \mathbf{f} \qquad (3.107)$$

applies. By multiplying both sides of Equation 3.107 from the left by Ψ^T, the alternative formula for k_{eq} can be written as:

$$k_{eq} = \Psi^T \mathbf{f} = \sum_j \Psi_j f_j \qquad (3.108)$$

Note that Equation 3.108 can be applied only in the special case of the assumed approximate mode shape Ψ, that is, for the displacements resulting from a static analysis. In this case, the absolute magnitude of displacements and not only the shape (i.e. the relative magnitude) should be used for Ψ, both in Equation 3.104 and in Equation 3.108. On the other hand, Equation 3.105 applies with any Ψ.

Using Equation 3.106, and denoting the lateral forces and the corresponding displacements in storey j as f_j and u_j, respectively, an approximation to the fundamental period can be obtained as:

$$T_1 = 2\pi \sqrt{\frac{\sum_j u_j^2 \, m_j}{\sum_j u_j \, f_j}} \qquad (3.109)$$

It is worth noting that this way of computing (approximately) the value of T_1 is usually referred to as the Rayleigh's method. The period T_1 from Equation 3.109 (and from similar formulas based on Equation 3.106) is always a little shorter than the exact value of the fundamental period. Accuracy depends on the quality of the approximation. However, the approach is robust; even for relatively poor approximations reasonable results are obtained. For instance, in buildings, just using lateral forces proportional to the masses of each storey as the arbitary loads to start the process, gives in most cases a very good approximation to T_1.

3.1.10 Modelling

In Eurocode 8 it is stated that, in the case of elastic analysis, 'the model of the building shall adequately represent the distribution of stiffness and mass in it so that all significant deformation shapes and inertia forces are properly accounted for under the seismic action considered'. It is difficult, however, to provide guidelines for the construction of mathematical models, which is a prime task of engineers.

Models of different levels of complexity can be used for the elastic analysis of buildings. At one end of the range, very sophisticated structural models with a very large number of degrees of freedom can be constructed by means of finite elements. Whereas columns and beams are typically modelled as one-dimensional (1D) elements, walls and slabs can be modelled by means of a large number of 2D or even three-dimensional (3D) finite elements. A number of computer programs are available for the elastic analysis of structures modelled with finite elements. However, taking into account the uncertainties related to the input data, especially to the characteristics of ground motion, even the most sophisticated structural models are able to predict only an approximation of the structural response to future earthquake ground motions. Moreover, it should be noted that ordinary buildings are expected to respond in the inelastic range during strong earthquakes, and that linear elastic analysis can, with the appropriate corrections, provide only rough estimates of the inelastic response. Finally, it is not easy to check the results of analyses obtained from sophisticated models. Thus, in seismic analyses it is reasonable to use simplified models, which represent an appropriate compromise between complexity and accuracy. These simplified models should take into account the most dominant characteristics that control the seismic response of typical building structures. The simplest possible model is an SDOF model, which may provide, in some cases, a reasonable approximation to the real behaviour.

In a typical building structure, a large proportion of the mass is concentrated at the levels of the floor diaphragms and at the roof. This means that it is appropriate to lump the masses at the floor levels. Horizontal concrete diaphragms are typically very stiff in a horizontal plane; so the assumption of infinitely rigid diaphragms is a reasonable one. Moreover, considering that the thickness of a typical diaphragm slab is much smaller than the cross-sectional dimensions of vertical elements, and therefore its flexural stiffness is much smaller than that of vertical elements, it is reasonable to assume that the diaphragms have no out-of-plane stiffness. These assumptions greatly simplify the model of the building structure.

In the majority of cases, there is no need to model structural walls with 2D finite elements. Since the height of a wall is typically much larger than its cross-sectional length, it is reasonable to model walls with 1D elements, possibly with shear deformation included.

Modelling of infills is not an easy task. According to Eurocode 8, 'infill walls which contribute significantly to the lateral stiffness and resistance of the building should be taken into account'. Infill walls can have an important effect especially in the case of frame structures, where they typically increase the initial stiffness and strength. However, the seismic response of infilled frames when subjected to strong ground motion is highly non-linear. After the failure of the infills, their influence disappears, whereas the basic frame structure continues to carry lateral loads. Quite frequently infills may even have detrimental effects. If not distributed in a regular way in plan and elevation of the building, they can cause a large torsional effect or a soft storey, respectively. Another possible adverse local effect is shear failure of columns due to the increased shear forces induced by the frame-infill interaction.

In linear elastic analysis, it is impossible to take into account all the effects of infills. A viable approach may be to use two models, one with infills and the other without them. A simple but effective model for infills is an equivalent diagonal strut which only carries compressive forces.

Soil-structure interaction may have either a beneficial or a detrimental effect on the behaviour of a structure. According to Eurocode 8, 'the deformability of the foundation shall be taken into account in the model, whenever it may have an adverse overall influence on the structural response'. The simplest way of modelling the influence of soil–structure interaction is the use of equivalent soil springs at the foundation (see Sections 6.1 and 7.2.2).

Actual building structures are three-dimensional, so that a spatial (three-dimensional, 3D) model is theoretically correct. Many building structures are not symmetrical in plan. In such a case, translational and torsional (about the vertical axis) vibrations are coupled, and a 3D model cannot be avoided. However, if the structure has two-way symmetry in plan, the vibrations in the two horizontal directions are uncoupled and the 3D model can be replaced by two 2D models, one in each horizontal direction.

Two-way symmetry is an idealised situation, which cannot be achieved in practice. For this reason seismic standards and codes have introduced the 'accidental eccentricity' of Section 3.1.8, to account for uncertainties in the location of masses and in the spatial variation of the seismic motion. As the nominal centre of mass at each floor is displaced from its nominal location in each direction by the accidental eccentricity, even a symmetric structure becomes asymmetric and, in principle, requires a 3D model. Nevertheless, standards and codes, including Eurocode 8, allow in some cases, as an approximation, the application of two 2D models instead of a 3D one. According to Eurocode 8, linear-elastic analysis may be performed using two planar models, one for each of the main horizontal directions, if the criteria for regularity in plan are satisfied. Depending on the importance of the building, two planar models can be used even if the criteria for regularity in plan are not satisfied, provided that a number of special regularity conditions are met (see Section 4.3.3.1). It should be noted, however, that it is rather impractical to check the in-plan regularity as required by Eurocode 8.

Not all of the degrees of freedom that are used in a static analysis need to be considered also as degrees of freedom in a dynamic analysis. Degrees of freedom can be separated in two groups: those with an assigned mass and those with zero mass. The degrees of freedom with an assigned mass can be called 'essential'; only these degrees of freedom have to remain in the model used for dynamic analysis. The other group of degrees of freedom, those without an assigned mass, can be eliminated by static condensation (Chopra 2007).

In a model with lumped masses at the floor levels and rigid floor diaphragms, the number of essential degrees of freedom is reduced to only three per floor diaphragm, corresponding to rigid-body motion in its (horizontal) plane: two horizontal translations and one (torsional) rotation (Figure 3.11). The total number of degrees of freedom of a model for dynamic analysis is thus equal to three times the number of storeys, irrespectively of the number of degrees of freedom of the model for static analysis. In the case of a planar model, the number of degrees of freedom is further reduced and is equal to the number of storeys: one displacement per storey (Figure 3.11).

An additional approximation, which allows simplification of the modelling for the majority of building structures, is the so-called *pseudo-3D model*. This is a spatial model of the whole structure, consisting of planar models of individual lateral load resisting systems (macro-elements or substructures, e.g. planar frames and walls) connected together by rigid diaphragms that are flexible (i.e. with zero stiffness) in their out-of-plane direction. So, a pseudo-3D model of a spatial frame is composed of separate planar frames in two directions

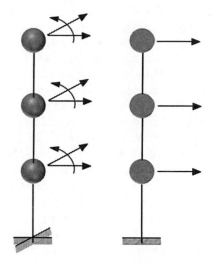

Figure 3.11 Essential degrees of freedom for dynamic analysis of a spatial (3D) and a planar (2D) model subjected to horizontal ground motion.

(Figure 3.12). In reality, two frames in two directions have common columns at the crossing line. In the pseudo-3D model, these columns are independently included in both planar frames. Compatibility of the axial deformations of these columns is not achievable. This is the approximation of the pseudo-3D model, which, in the majority of cases, does not have an important influence on the results.

The two sub-sections to follow address two particular aspects in linear elastic modelling that deserve special attention.

3.1.11 Elastic stiffness for linear analysis

The real force–deformation relation for reinforced concrete elements and structures is not linear, even in the case of a relatively small loading. The question is how the stiffness should

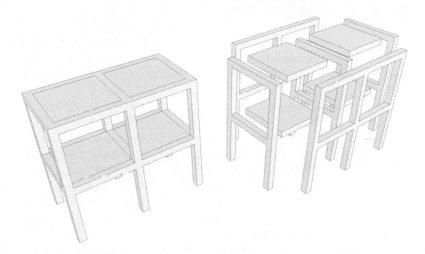

Figure 3.12 3D and pseudo-3D model composed of separate planar frames.

be determined for a linear elastic analysis. The upper bound for the stiffness of the elements is represented by the stiffness of the uncracked elements, whereas the lower bound is represented by the secant stiffness, which corresponds to the initiation of yielding of the reinforcement. The stiffness of the model influences both the seismic action and the displacements. A smaller stiffness means a longer period of vibration and a larger displacement, whereas the seismic action (acceleration, seismic loading) depends on the period and on the shape of the response spectrum. With the exception of structures whose fundamental period is in the very short-period range, the seismic loading typically decreases with decreasing stiffness, or does not change (this corresponds to the plateau of the Eurocode 8 acceleration spectrum). Since the expected behaviour of building structures subjected to strong earthquakes is non-linear, and since in the inelastic range deformations are more important than forces, it is required by Eurocode 8 that the effect of cracking on stiffness is taken into consideration, that is, a lower bound of stiffness which corresponds to the initiation of yielding of the reinforcement (see Section 3.3.3 and Equation 3.146). Details about the reinforcement, which are needed for the determination of the stiffness of cracked elements, are typically not known when the analysis starts. So, an iterative procedure is needed. According to Eurocode 8, such an impractical procedure may be avoided by assuming that the elastic flexural and shear stiffness properties of concrete elements are equal to one-half of the corresponding stiffness of the uncracked element.

3.1.12 Second-order effects in linear analysis

Seismic design codes require taking into account second-order (P-Δ) effects in buildings, whenever in the vertical members of any storey they exceed 10% of the total first-order ones. The criterion is the inter-storey drift sensitivity coefficient, θ, defined for storey i as:

$$\theta_i = \frac{N_{\text{tot},i}\Delta u_i}{V_{\text{tot},i}h_i} \tag{3.110}$$

where

- $N_{\text{tot},i}$ is the total gravity load in the seismic design situation at and above storey i.
- $V_{\text{tot},i}$ is the total seismic shear at storey i.
- h_i is the height of storey i.
- Δu_i is the inter-storey drift at storey i, that is, the difference of the lateral displacements at the top and bottom of the storey, u_i and u_{i-1}, at the floor's centre of mass. In Eurocode 8 it is the inelastic drift, estimated with the equal displacement rule according to Equation 3.116 in Section 3.2.2.2, via back-multiplying by the behaviour factor q the values of u_i, u_{i-1} from the linear analysis for the design spectrum.

Second-order effects may be neglected, if the value of θ_i does not exceed 0.1 at any storey. They should be taken into account for the entire structure, if at any storey θ_i exceeds 0.1. If θ_i does not exceed 0.2 at any storey, Eurocode 8 allows taking these effects into account without a full-fledged geometrically non-linear second-order analysis by multiplying by $1/(1-\theta_i)$ all first-order action effects from a linear elastic analysis for the seismic action. For concrete buildings, in the very uncommon case that θ_i exceeds 0.2 at any storey, an accurate second-order analysis is required by Eurocode 8.

Section 7.4.2 and Table 7.4 present the values of the inter-storey drift sensitivity coefficient of the example building of Chapter 7.

3.2 BEHAVIOUR FACTOR

3.2.1 Introduction

Experience has shown that the great majority of well-designed and constructed buildings survive strong ground motions, even if they were in fact designed for only a fraction of the forces that would develop if the structure behaved entirely as linearly elastic. As will be shown in the following sections, a reduction of seismic forces is possible thanks to the beneficial effects of energy dissipation in ductile structures and to inherent overstrength. This fact is taken into account in seismic design standards and codes, which use force reduction factors (e.g. the 'behaviour factor' q in Eurocode 8, or the 'response modification factor' R in US codes) to determine the seismic design loads. Such reduction factors are predominantly based on empirical observations of the behaviour of common structural systems during earthquakes. Consequently, on average they yield acceptable results. More recently, many numerical studies have also been performed aimed at determining appropriate values of reduction factors (e.g. FEMA 2009). Reduction factors are used in conjunction with linear analysis and, therefore, present a very simple and practical tool for seismic design. However, it is necessary to bear in mind that describing a complex phenomenon of response reduction for a particular structure, by means of a single average number, can be confusing and misleading. For this reason, the reduction factor approach, although it is very convenient for practical applications and has served the professional community well over decades, is able to provide only very rough answers to the problems encountered in seismic analysis and design. For a more realistic estimate of structural response during strong earthquakes, non-linear analysis is needed.

An illustration of the reduction of maximum acceleration in an inelastic SDOF system compared to its elastic counterpart with the same stiffness and mass ($T_n = 1$ s) is shown in Figure 3.13. The structural response in terms of absolute accelerations and relative displacements to the Ulcinj – Albatros N–S ground motion clearly demonstrates a substantial reduction in the maximum acceleration of the inelastic system compared to the elastic one, whereas the maximum displacements of both systems are approximately equal. Note that, at the end of the vibration, the whole input energy in the elastic system is dissipated by viscous damping, whereas in the case of the inelastic system both viscous damping and hysteretic behaviour contribute to the dissipation of energy.

3.2.2 The physical background of behaviour factors

Let us consider two idealised SDOF structural systems with the same mass and stiffness, that is, with the same natural period. One system shows an unlimited elastic behaviour, whereas the other one has a limited strength. The yielding point of the latter, inelastic system is defined by the yield strength f_y and the yield displacement u_y. The corresponding idealised force–displacement relationships are shown in Figure 3.14a.

Extensive research has shown that, for many systems with natural periods in the medium- and long-period range, the seismic demand in terms of displacements, u, is independent of the strength of the system and is approximately equal to the displacement demand, u_e, of an elastic system with the same natural period. This is the so-called *equal displacement rule*, which was stated by Veletsos and Newmark (1960), and has been used successfully for more than half a century. Many statistical studies have confirmed the applicability of the rule to structures on firm sites with fundamental periods in the medium- or long-period range, with relatively stable and full hysteretic loops. A discussion on the applicability of the equal displacement rule is provided, for example, in Fajfar (2000).

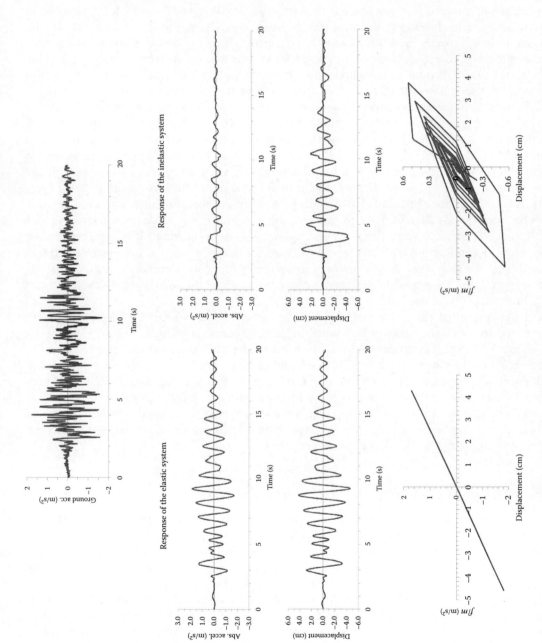

Figure 3.13 Response of elastic and inelastic SDOF system with $T_n = 1$ s to Ulcinj – Albatros, N–S ground motion (5% viscous damping).

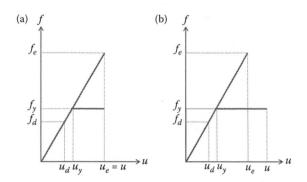

Figure 3.14 Idealised force–displacement relationships. (a) Equal displacement rule applies. (b) Equal displacement rule does not apply.

The system in Figure 3.14a can accommodate the imposed seismic demand either by large strength f_e (elastic system), or by a combination of smaller strength f_y and inelastic deformation capacity, defined by a ductility factor $\mu = u/u_y$ (yielding system). Note, however, that the reduction of strength may be conditioned not only by the available inelastic deformation capacity but also by the intent to limit damage in more frequent earthquakes (Section 1.3.2). The following relation applies:

$$q_\mu = \frac{f_e}{f_y} = \frac{u}{u_y} = \mu \qquad (3.111)$$

where q_μ is a reduction factor which determines the extent of possible reduction of the strength due to the inelastic deformation capacity. If the equal displacement rule is assumed to apply, it is equal to the ductility factor μ.

The problem can also be stated in a different way. Assuming that an inelastic deformation capacity defined by the ductility factor μ is provided (or, in the case of a serviceability limit state, an inelastic deformation is tolerated), the strength of the system should be equal at least to the required strength f_y, which represents the inelastic strength demand. This approach is actually used in design and can be written in the form:

$$f_y = \frac{f_e}{q_\mu} \qquad (3.112)$$

where f_e is the elastic strength demand, that is, the strength required for a structure which would remain in the elastic region during earthquake ground motion with a displacement demand u_e. The displacement demand and the related elastic strength demand can be obtained from the elastic acceleration spectrum as described in Section 3.1.2.2.

Expressions similar to Equation 3.112 can be found in various seismic standards and codes. However, an important difference should be noted between Equation 3.112 and the expressions in the standards and codes. In Equation 3.112, f_y represents the actual strength, whereas the seismic forces in standards and codes correspond to the design strength f_d which is, as a rule, lower than the actual strength. This difference reflects what is usually denoted as overstrength, which is an inherent property of properly designed, detailed, constructed and maintained highly redundant structures.

Taking into account the overstrength factor:

$$q_s = \frac{f_y}{f_d} \tag{3.113}$$

the following relation applies:

$$q = \frac{f_e}{f_d} = \frac{f_e}{f_y}\frac{f_y}{f_d} = q_\mu\, q_s \tag{3.114}$$

Thus, the total force reduction factor q, which is equal to the elastic strength demand f_e divided by the code prescribed seismic design action (force), f_d, can be defined as the product of the ductility-dependent factor q_μ and the overstrength factor q_s. The factors q and q_s are discussed in more detail in the next sections.

The seismic design force f_d can be obtained from the elastic strength demand as:

$$f_d = \frac{f_e}{q} \tag{3.115}$$

where q is the reduction factor defined in Equation 3.114.

In code procedures, including Eurocode 8, an elastic analysis is performed using the seismic design force, f_d. The resulting displacement is u_d (Figure 3.14). It should be emphasised that u_d is not the correct displacement to be used in design calculations. The actual displacement is $u = u_e$, which can be obtained as:

$$u = q\, u_d \tag{3.116}$$

The concept of reduction factors can be used also in the more general case when the equal displacement rule does not apply (Figure 3.14b), for example, for short-period structures. All the equations developed above still apply, except Equations 3.111 and 3.116. A relation between the elastic and inelastic displacement demand u_e and u, respectively, has to be known. Based on such a relation, a more general relation between the ductility factor μ and the reduction factor q_μ can be developed. Such a relation is typically dependent on the period T_n and is often called the $q_\mu - \mu - T$ relation. Several proposals based on statistical studies are available in the literature (see Section 3.2.3).

Using Figure 3.14b and the relations $u = \mu u_y$ and $f_e/f_y = q_\mu = u_e/u_y$, the inelastic displacement demand can be determined as:

$$u = \frac{\mu}{q_\mu}\, u_e \tag{3.117}$$

where u_e is the maximum relative displacement of the system with unlimited elastic behaviour subjected to the ground motion defined by the elastic acceleration spectrum A. An alternative form of Equation 3.117 is:

$$u = \mu\, q_s\, u_d \tag{3.118}$$

where u_d is the maximum relative displacement of the system obtained by linear analysis under the design loads f_d.

Equation 3.118 is a more general form of Equation 3.116. The equal displacement rule is a special case with $q_\mu = \mu$, resulting in $u = u_e = q\, u_d$.

3.2.3 The ductility-dependent factor q_μ

The ductility-dependent reduction factor q_μ has been the subject of extensive research. An overview of early proposals was presented by Miranda and Bertero (1994). Generally, the reduction factor q_μ is, in the medium-period (velocity-controlled) and long-period (displacement-controlled) regions, only slightly dependent on the period T_n, and is roughly equal to the prescribed target ductility μ (indicating the validity of the equal displacement rule). In the short-period (acceleration-controlled) region, however, the q_μ factor depends strongly on both T_n and μ. In the limit case of an infinitely rigid structure ($T_n = 0$), there is no reduction due to ductility ($q_\mu = 1$). Moderate influence of hysteretic behaviour and damping can be observed in the whole period region. The transition period from the period-dependent part to the, more or less, period-independent part of the q_μ spectrum is roughly equal to the transition period between the acceleration-controlled, short-period region and the velocity-controlled medium-period region T_C. This period is an important characteristic of the ground motion and is often referred to as the characteristic period or the 'predominant' period. It roughly corresponds to the period at which the largest amount of energy is imparted to the structure.

In the basic variant of the N2 method which has been adopted in Eurocode 8 as the method for non-linear pushover-based analysis, and is presented in Section 3.3, simple bilinear q_μ spectra, representing a simplified form of the relations proposed by Vidic et al. (1994), are used

$$q_\mu = (\mu - 1)\,\frac{T_n}{T_C} + 1 \qquad T_n \leq T_C \tag{3.119}$$

$$q_\mu = \mu \qquad T_n \geq T_C \tag{3.120}$$

According to Equations 3.119 and 3.120, in the medium- and long-period ranges, the equal displacement rule applies, that is, the displacement of the inelastic system is assumed to be equal to the displacement of the corresponding elastic system with the same period.

3.2.4 The overstrength factor q_s

Strength exceeding that required by codes (overstrength) is a major factor contributing to the seismic resistance of structures. The overstrength factor is defined at the level of the whole structure, as the ratio between the actual strength and the code-prescribed strength demands arising from the application of prescribed loads and forces. It results from the following groups of sources:

a. Redistribution of internal forces in the inelastic range in ductile, statically indeterminate (redundant) structures; difference between the design level and the required member strength (e.g. allowable vs. yield stresses, partial factors on resistance or material strengths); member oversize (due to discrete member sizes and/or desired uniformity of members for easier construction); minimum requirements according to code provisions regarding dimensioning and detailing; design for multiple combinations of actions (e.g. factored gravity loads); deformation constraints on system performance; architectural considerations.

    b. Conservatism in mathematical models; effects of structural elements that are not considered as a part of the lateral load resisting system; effects of non-structural elements.

    c. Higher material strength than the nominal one specified in design, strain hardening and strain rate effects.

The influence of the majority of (a) group factors can easily be at least approximately quantified by a non-linear pushover analysis; the (b) group sources are less reliable or require sophisticated mathematical modelling and may be neglected in practical design. The (c) group factors are uncertain and difficult to be quantified. They are typically not taken into account in deterministic analyses.

It is clear that overstrength may have its origin in a variety of sources, and that, in real structures, it varies widely, depending on the material and the type of the structural system, the structural configuration, the number of storeys, the detailing and the kind and date of the code to which the structure was designed.

3.2.5 Implementation in Eurocode 8

In order to avoid explicit inelastic structural analysis in design, the capacity of a structure to dissipate energy, through mainly ductile behaviour of its elements and/or other mechanisms, is taken into account by performing a linear elastic analysis based on a response spectrum which is reduced with respect to the elastic one, henceforth called a 'design spectrum'. This reduction is accomplished by introducing the behaviour factor q.

The code-suggested values of q-factors are essentially of an empirical origin. Thus, in addition to ductility, they generally automatically imply overstrength, although this is usually not explicitly realised. According to Eurocode 8, the behaviour factor q is a 'factor used for design purposes to reduce the forces obtained from a linear analysis, in order to account for the non-linear response of a structure, associated with the material, the structural system and the design procedures'. Furthermore, according to Eurocode 8, 'the behaviour factor q is an approximation of the ratio of the seismic forces that the structure would experience if its response was completely elastic with 5% viscous damping, to the seismic forces that may be used in the design, with a conventional elastic analysis model, still ensuring a satisfactory response of the structure'.

The q-factors in Eurocode 8 take into account both ductility and overstrength. In the majority of cases, a single value is prescribed, which includes both contributions to the reduction of the design forces. In some cases both contributions are taken into account explicitly, with the overstrength factor defined as α_u/α_1, where α_1 is the value by which the horizontal seismic design action is multiplied in order to form the first plastic hinge in the structure, while all the other design actions remain constant, and α_u is the value by which the horizontal seismic design action is multiplied, in order to form plastic hinges in a number of sections sufficient for the development of an overall plastic mechanism, while all other design actions remain constant (see Section 4.6.3 and Figure 4.13). Note that this definition is different from the definition of overstrength in Section 3.2.2 (see Figure 3.14), where the increase in the horizontal resistance is calculated with respect to the horizontal seismic design action (i.e. with $\alpha_1 = 1$), rather than with respect to the first plastic hinge in the structure.

The α_u/α_1 overstrength factor can be determined by non-linear static analysis (see Section 3.3). If such an analysis is not performed, conservative approximate values of α_u/α_1, provided in Eurocode 8, can be used (see Section 4.6.3).

The reduction of forces is realised by using a design acceleration spectrum S_d, which represents the elastic acceleration spectrum S_e for 5% damping, divided by the behaviour factor q. This reduction applies for periods longer than T_B. For short-period structures, the reduction due to ductility decreases. As stated before, for infinitely rigid structures ($T = 0$), there is no reduction due to ductility. However, it is assumed that an overstrength factor of at least 1.5 exists. Consequently, for $T = 0$, $S_d = S_e/1.5$. The design spectrum is linear between $T = 0$ and $T = T_B$.

For the horizontal components of the seismic action, the design spectrum, $S_d(T)$, is defined by the following expressions, using the Eurocode 8 notation for the natural period, T (whereas in the other parts of this book the natural period is denoted as T_n):

$$0 \leq T \leq T_B : \quad S_d(T) = a_g \cdot S \cdot \left[\frac{2}{3} + \frac{T}{T_B} \cdot \left(\frac{2.5}{q} - \frac{2}{3} \right) \right] \tag{3.121a}$$

$$T_B \leq T \leq T_C : \quad S_d(T) = a_g \cdot S \cdot \frac{2.5}{q} \tag{3.121b}$$

$$T_C \leq T \leq T_D : \quad S_d(T) \quad \begin{cases} = a_g \cdot S \cdot \dfrac{2.5}{q} \cdot \left[\dfrac{T_C}{T} \right] \\ \geq \beta \cdot a_g \end{cases} \tag{3.121c}$$

$$T_D \leq T : \quad S_d(T) \quad \begin{cases} = a_g \cdot S \cdot \dfrac{2.5}{q} \cdot \left[\dfrac{T_C T_D}{T^2} \right] \\ \geq \beta \cdot a_g \end{cases} \tag{3.121d}$$

where the parameters have the same meaning as in the case of the elastic spectrum defined in Section 3.1.3. Additionally, q is the behaviour factor and β is a lower-bound factor for the horizontal design spectrum, which is a nationally determined parameter (NDP) with a recommended value of 0.2.

Note that there is a discrepancy between Equation 3.121 and Equation 3.119, which indicates that the magnitude of reduction due to ductility starts decreasing at the period T_C (towards $T = 0$) rather than at T_B. On the other hand, the overstrength factor typically increases in the short-period region; this effect may counterbalance the smaller reduction due to ductility and justify the use of the full reduction also in the period range between T_B and T_C.

The values of the behaviour factor q are given in the relevant parts of Eurocode 8. They are in the range from 1.5 to 8 (6.75 in the case of reinforced concrete buildings), resulting in a factor of more than 5 between the design seismic action for two extreme cases of structures of the same Importance Class and at the same location. As a limiting case, for the design of structures classified as low-dissipative, no account is taken of any hysteretic energy dissipation; the smallest value $q = 1.5$, which is considered to account for overstrength, is used. For dissipative structures, the q-factors are larger, accounting for the hysteretic energy dissipation that mainly occurs in specifically designed zones, called dissipative zones. The q-values depend on the structural material, on the type and the regularity of the structural system and on the detailing. For example, since steel is a more ductile material than, say, masonry, the q-factors for steel structures are larger than for masonry structures. A statically determinate structure, for example, an inverted pendulum, has less overstrength than a statically indeterminate one, for example, a moment resisting frame; so, the former has smaller q-value than the latter. Obviously, a larger q-value corresponds to structures detailed for high ductility than that applying to those detailed for medium ductility (Ductility Class High, DC H, vs. Medium, DCM, see Sections 4.6.2 and 4.6.3).

For the vertical component of the seismic action, the design spectrum is given by Equations 3.121, with the design ground acceleration in the vertical direction, a_{vg}, replacing a_g, and S taken as equal to 1.0, and the other parameters as defined for the elastic vertical spectrum in Section 3.1.3. For the vertical component of the seismic action, a behaviour factor q up to 1.5 should generally be adopted for all materials and structural systems. The adoption of values for q greater than 1.5 in the vertical direction should be justified through an appropriate analysis.

For the calculation of displacements, the displacement determined by the linear elastic analysis based on design seismic action is multiplied by the displacement behaviour factor, q_d, which is assumed to be equal to q, unless otherwise specified. Thus, generally, Equation 3.116, based on the equal displacement rule, is applied. The fact that in the short-period range the equal displacement rule does not apply is recognised in a note in Eurocode 8, which states that: 'in general q_d is larger than q if the fundamental period of the structure is less than T_C'.

For the calculation and the magnitude of storey drifts and inter-storey drifts of the example building of Chapter 7, see Section 7.4.1, Table 7.3 and Figure 7.7.

3.2.6 Use of reduction factors for MDOF structures

The principle of the reduction of forces and the derivation of relevant equations, shown in the previous sections, is based on an SDOF system. Nevertheless, this approach has been widely used in standards and codes for any structure which is expected to deform in the inelastic range when subjected to strong ground motions, that is, also for multi-storey buildings modelled as MDOF systems. The application to MDOF systems raises some additional problems, as discussed below.

In the case of real structures, mostly MDOF models are used. The response spectrum approach, presented in previous chapters, is, by definition, not applicable to inelastic MDOF systems. However, the seismic behaviour of a large class of MDOF structural systems can be closely approximated by equivalent SDOF models. In such cases, all considerations of the previous sections of Chapter 3 can, with small modifications, be also applied to MDOF systems.

The starting point is a force–displacement relationship of the MDOF system obtained by a pushover analysis (i.e. a static analysis under monotonically increasing lateral loads). In the case of building structures, it is usually the base shear and the lateral displacement at the roof level which are, respectively, considered to represent the force and displacement. The force–displacement relationship of the equivalent SDOF system is obtained by a simple transformation of forces and displacements. From there onwards, all the equations derived for SDOF systems apply.

The relationship between the local and the global deformation quantities is very important for the behaviour of the structure, since the local quantities correspond to individual structural members and the global quantities to the structure as a whole. A suitable local deformation quantity is the chord rotation at a member end, defined as the angle between the normal to the member section at the member end and the chord connecting the two member ends (Fardis 2009). The most convenient global deformation quantity is the maximum displacement at the level of the roof of the building, u_r. The relationship between the local and global deformations depends significantly on the plastic mechanism. As an example, let us consider

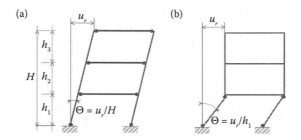

Figure 3.15 Idealised deformation shapes corresponding to (a) global and (b) local (storey) plastic mechanisms. Relations between the displacement at the roof and chord rotation in columns for different plastic mechanisms.

the three-storey frame structure of Figure 3.15. Depending on how the frame was designed, a favourable global plastic mechanism, with plastic hinges in all the beams and at the bottom of the ground storey columns (Figure 3.15a, see also Figure 2.9b and c), can form, or an unfavourable local storey mechanism, with plastic hinges at both ends of the columns in a single storey (Figures 3.15b and 2.9a). The idealised deformation shapes depicted in Figure 3.15, showing only plastic deformations, indicate that the same roof displacement u_r corresponds to different deformations of the individual members. In the favourable mechanism of Figure 3.15a, the chord rotations of all the beams and at the bottom of the columns in the ground storey amount, approximately, to u_r/H, where H is the total height of the frame. In the storey plastic mechanism (Figure 3.15b), however, the chord rotations at both ends of the columns amount, in the critical storey, to approximately u_r/h_j (i.e. to the storey drift), where h_j is the height of the storey. In the first case, the local chord rotations are about equal to the average drift ratio of the building, whereas in the second case they are much larger, depending on the number of storeys, according to the following approximate relationships between the local chord rotation ductility factor μ_l and the global ductility factor μ_g:

$$\mu_l = 1 + \frac{H}{h_i}(\mu_g - 1) \tag{3.122}$$

$$\mu_g = 1 + \frac{h_i}{H}(\mu_l - 1) \tag{3.123}$$

The local chord rotation ductility factor μ_l is defined as the maximum chord rotation divided by the chord rotation at yielding. The global ductility factor μ_g is defined as the maximum roof displacement divided by the roof displacement at the yield point of the idealised pushover curve. It is related to the reduction factor due to ductility (see, e.g. Equations 3.119 and 3.120). Equation 3.123 indicates that, in medium or high-rise buildings, the local ductility demand could be much larger than the global ductility demand, which is related to the reduction factor. For example, in a five-storey building with $h_j = 0.2H$ and a global ductility factor of $\mu_g = 4$, the local ductility factor from Equation 3.122 is as high as $\mu_l = 16$. Such a ductility capacity is difficult to attain, even with special detailing.

The discussion in this section demonstrates that, for MDOF systems, the approach with reduction factors usually represents a reasonable approximation for new buildings designed according to capacity design (see Section 4.5), which are expected to form a full-fledged, global plastic mechanism. By contrast, the q-factor approach is not appropriate for a storey plastic mechanism, which is typical of the majority of existing frame buildings. For this reason, in Eurocode 8, Part 3, which applies to existing buildings, the applicability of the q-factor approach is severely limited.

Example 3.3 at the end of this chapter illustrates the use of the concepts described in Sections 3.1 and 3.2 for the linear elastic analysis of a 3-storey pre-fabricated industrial building as per Eurocode 8.

3.3 NON-LINEAR ANALYSIS

Non-linear analysis is generally more complex than linear analysis. It is not mandatory in Eurocode 8. However, it is often the only reasonable choice when dealing with existing buildings in accordance with Eurocode 8, Part 3. Non-linear methods include time-history (also called response-history) analysis (Section 3.3.1) and pushover-based methods (Section 3.3.2). The use of non-linear analysis for practical applications is still evolving, and there are many areas where details of the implementation are open to judgment and alternative interpretations.

3.3.1 Equation of motion for non-linear structural systems and non-linear time-history analysis

The equation of motion for an MDOF system developed in Section 3.1.4 (Equation 3.55) is based on the assumption of linear elastic structural behaviour. When subjected to strong ground motion, most buildings are expected to deform into the inelastic range, where the relationship between restoring forces and deformations is non-linear. Equation 3.48, which was used for the determination of restoring forces in the case of linear elastic behaviour, is not valid in the inelastic range. It has to be replaced by a more general relationship between restoring forces and deformations, $\mathbf{f}_S(\mathbf{u})$ (hysteretic rules). Accordingly, Equation 3.55 is, in the case of an inelastic building, replaced by:

$$\mathbf{m}\ddot{\mathbf{u}} + \mathbf{c}\dot{\mathbf{u}} + \mathbf{f}_S = -\mathbf{m}\,\iota\,\ddot{u}_g \qquad (3.124)$$

where only the excitation due to the ground motion is considered.

Since the superposition rule does not apply in the non-linear range, the equation of motion, Equation 3.124 can only be solved by means of a numerical step-by-step integration method of differential equations (see Section 3.1.1.4). Such an analysis is called a *non-linear time-history analysis*. It is the most advanced analysis method and represents an approach which is perfectly correct from the standpoint of theory. However, due to its complexity, non-linear time-history analysis has, in practice, for the time being, only rarely been used. It is not only computationally demanding (a problem becoming less important with the development of advanced hardware and software), but also requires additional data, which are not needed in pushover-based non-linear analysis: a suite of accelerograms and data about the hysteretic behaviour of structural members (i.e. member response under large amplitude reversed loading). A consensus about a proper way to model viscous damping in the inelastic response of reinforced concrete structures has not yet been reached. Moreover, the complete analysis procedure is less transparent than in simpler methods. It is expected that, at some time in the future, non-linear time-history analysis will become the main analytical procedure in earthquake engineering. However, for all the reasons mentioned above, it is presently most rational to use simplified approximate procedures of non-linear analysis, for example, a pushover-based analysis, such as that described in Section 3.3.2.

3.3.2 Pushover-based methods

Pushover-based methods combine a non-linear static (i.e. pushover) analysis with the response spectrum approach. Seismic demand can be determined for an equivalent SDOF system from an inelastic response spectrum. A transformation of the MDOF system to an equivalent SDOF system is needed. This transformation represents the main limitation of the applicability of pushover-based methods. It is straightforward, if the structure vibrates in a single mode with a deformation shape that does not change over time (Section 3.1.9). These conditions are fulfilled only for a linear elastic structure with negligible influence of higher modes. Nevertheless, the assumption of a single time-invariant mode is used in pushover-based methods for inelastic structures, as an approximation.

Several variants of the pushover-based analysis have been proposed, and are available in the literature. In this book, the method implemented in Eurocode 8, that is, the *N2 method*, will be presented. The method was originally proposed in the late eighties (Fajfar and Fischinger 1987, 1989). Later, it was formulated in the acceleration–displacement (AD) format (Fajfar 1999, 2000) included in Eurocode 8. A further development of the N2 method is presented in Section 3.3.2.6. In the following sections, the steps of pushover-based analysis will be discussed with special consideration of the N2 method.

3.3.2.1 Pushover analysis

A non-linear static (pushover) analysis is performed by subjecting a structure to a monotonically increasing pattern of lateral forces, representing the inertial forces which would be experienced by the structure when subjected to ground shaking. Gravity loads are kept constant. Under incrementally increasing lateral loads, various structural elements yield sequentially. Consequently, at each event, the structure experiences a loss of stiffness.

Using a pushover analysis, a characteristic non-linear force–displacement relationship of the MDOF system can be determined. In the case of buildings, base shear and roof (top) displacement are usually chosen as representative forces and displacements, respectively.

The selection of an appropriate vertical distribution of lateral load is an important step in pushover analysis. A unique solution does not exist. Fortunately, the range of reasonable assumptions is usually relatively narrow and, within this range, different assumptions produce similar results. One practical possibility is to use two different displacement shapes (load patterns), and to envelope the results. According to Eurocode 8, the two load patterns are:

1. A 'uniform' pattern, where lateral forces are proportional to mass regardless of elevation.
2. A 'modal' pattern, consistent with the lateral force distribution determined in an elastic analysis.

The vector of lateral loads $\mathbf{f}$ is determined as:

$$\mathbf{f} = \alpha \mathbf{m} \Phi \qquad (3.125)$$

where $\mathbf{m}$ is the mass matrix. The magnitude of the lateral loads is controlled by the scale factor α. The distribution of lateral loads is related to the assumed displacement shape Φ, that is, it represents the displacement shape weighted by the masses. (Note that the displacement shape Φ is needed only for the transformation from the MDOF to the equivalent SDOF system, see Section 3.3.2.2). Consequently, the assumed load and displacement shapes are not mutually independent, as in some other approaches using pushover analysis. The procedure can start either by assuming the displacement shape Φ and determining the lateral load distribution according to Equation 3.125, or by assuming the lateral load distribution and determining the displacement shape Φ from Equation 3.125. Note that Equation 3.125 does not present any restriction regarding the distribution of lateral loads. In the derivation of formulas we use a planar structural model (see Section 3.1.5.4); the approach will later be extended to a 3D model. We further assume that the displacement shape Φ represents the fundamental vibration mode shape of the linear elastic structure. However, the developed expressions can be applied for any displacement shape and/or for any related distribution of lateral loads.

If the fundamental mode shape is used as the assumed displacement shape, and if it remains constant during ground shaking, that is, if the structural behaviour is linear elastic, then the distribution of lateral forces is the same as the distribution of 'seismic forces' that correspond to the fundamental mode (see Equation 3.84), so that Equation 3.125 is 'exact'. In the inelastic range, the displacement shape changes over time; Equation 3.125 represents an approximation of the 'seismic forces'. Nevertheless, by assuming lateral forces and displacements related according to Equation 3.125, the transformation from the MDOF to the equivalent SDOF system and vice-versa (Section 3.3.2.2) follows from simple mathematics, not only in the elastic but also in the inelastic range. No additional approximations are required, as in some other simplified procedures.

3.3.2.2 Transformation to an equivalent SDOF system

In the N2 method, seismic demand is determined by using response spectra. Inelastic behaviour is taken into account explicitly. Consequently, the structure should, in principle, be modelled as an SDOF system. Different procedures have been used to determine the characteristics of an equivalent SDOF system. One of them, used in the N2 method, is described below.

The starting point is the equation of motion for an MDOF system, Equation 3.124. For convenience, damping forces are not included. Damping will be taken into account later in the response spectrum. A planar MDOF model that explicitly includes only lateral (translational) degrees of freedom is used. With these assumptions, the equation of motion can be written as:

$$\mathbf{m}\ddot{\mathbf{u}} + \mathbf{f}_S = -\mathbf{m}\mathbf{1}\ddot{u}_g \qquad (3.126)$$

where $\mathbf{u}$ and $\mathbf{f}_S$ are vectors representing the displacements and the internal forces, $\ddot{u}_g$ is the ground acceleration as a function of time and $\mathbf{1}$ is a vector with all elements equal to 1, that is, it represents the influence vector ι for a planar building model.

We define the displacement vector $\mathbf{u}$ as:

$$\mathbf{u} = \Phi u_r \qquad (3.127)$$

where u_r is the time-dependent roof displacement and Φ is the displacement shape, normalised in such a way that the component at the roof is equal to 1.

By introducing Equation 3.127 into Equation 3.126, and by multiplying from the left-hand side with Φ^T, we obtain:

$$\Phi^T \mathbf{m}\Phi\ddot{u}_r + \Phi^T \mathbf{f}_S = -\Phi^T \mathbf{m}\mathbf{1}\ddot{u}_g \qquad (3.128)$$

By analogy with Equations 3.78 through 3.80 and Equations 3.91 and 3.92, we define the parameters M, L and Γ as:

$$M = \Phi^T \mathbf{m}\Phi = \sum_j \Phi_j^2 \, m_j \tag{3.129}$$

$$L = \Phi^T \mathbf{m}\mathbf{1} = \sum_j \Phi_j \, m_j \tag{3.130}$$

$$\Gamma = \frac{L}{M} = \frac{\sum_j \Phi_j \, m_j}{\sum_j \Phi_j^2 \, m_j} \tag{3.131}$$

Furthermore, we define:

$$f_S = \Phi^T \mathbf{f}_S \tag{3.132}$$

By taking into account Equations 3.129 through 3.132, Equation 3.128 becomes:

$$M\ddot{u}_r + f_S = -L\ddot{u}_g \tag{3.133}$$

which can be transformed into the equation of motion of the equivalent SDOF system:

$$L\ddot{u}_S + f_S = -L\ddot{u}_g \tag{3.134}$$

where the displacement of the equivalent SDOF system, u_S, and the displacement at the roof of the MDOF system, u_r, which is representative of the deformations in the MDOF system, are related by:

$$u_r = \Gamma u_S \tag{3.135}$$

A comparison of Equation 3.134 and Equation 3.2 shows that L represents the mass of the equivalent SDOF system, that is, $L \equiv m^*$ in the Eurocode 8 notation.

It will be shown that the same transformation with the Γ factor applies also to forces. The force in the equivalent SDOF system is f_S, whereas the base shear force V_b is representative of the forces in an MDOF system. In a static analysis, the external forces are equal to the internal forces (a pushover, i.e. a static analysis is being performed). Thus, the restoring forces $\mathbf{f}_S$ can be replaced by the lateral forces defined in Equation 3.125, resulting in:

$$f_S = \Phi^T \mathbf{f}_S = \Phi^T \alpha \mathbf{m}\Phi = \alpha M \tag{3.136}$$

The base shear force V_b can be computed as the sum of the lateral forces:

$$V_b = \sum f_i = \mathbf{f}^T \mathbf{1} = \alpha \Phi^T \mathbf{m}\mathbf{1} = \alpha L \tag{3.137}$$

By comparing Equations 3.136 and 3.137, the relation between the base shear in the MDOF system, V_b, and the force in the equivalent SDOF system, f_S, can be written as:

$$V_b = \Gamma f_S \tag{3.138}$$

Γ controls the transformation from the MDOF to the SDOF model and vice-versa. The same Γ value applies in the transformation of both displacements and forces, Equations 3.135 and 3.138. Thus, the force–displacement relationship determined for the MDOF system (the V_b–u_r diagram) also applies to the equivalent SDOF system (the f_S–u_S diagram), provided that both the forces and the displacements are divided by Γ. This can be visualised by changing the scale on both axes of the force–displacement diagram. The initial stiffness of the equivalent SDOF system remains the same as that defined by the base shear vs. roof displacement diagram of the MDOF system.

Equation 3.134 shows that L, determined from Equation 3.130, represents the mass of the equivalent SDOF system. Note a difference in the formulation from the equivalent mass in the case of linear elastic analysis in Section 3.1.

All equations presented in this section apply to a planar model for any assumed displacement shape Φ and, thus, for any related distribution of lateral loads. In a special case, the assumed displacement shape Φ represents the fundamental vibration mode shape of the linear elastic structure. This case corresponds to the 'modal' distribution of lateral forces in Eurocode 8. In such a case, L and M are the same as the corresponding values for the fundamental mode, developed in Section 3.1, and the transformation factor Γ represents the mode participation factor, Equation 3.78.

The same equations can also be used for a 3D building model (Section 3.1.10), with the only change that the influence vector $\mathbf{1}$ of the planar model is replaced by a general influence vector ι. Separate analyses are performed in each of the two horizontal directions. The procedure can be substantially simplified if the lateral loads, determined according to Equation 3.125, are applied in one direction only. This is a special case, which requires that the assumed displacement shape, too, has non-zero components in one direction only. In such a case, all the equations derived for the planar system can be directly used for the 3D system, by considering only the direction under investigation. Lateral loads are applied at the mass centres of different storeys, only in the investigated direction. Note that even in this special case of uncoupled assumed displacement shape, the displacements determined by pushover analysis of an asymmetric structure will be coupled, that is, they have components in three directions.

Static torsional effects are included. The dynamic torsional effects may, however, be quite different from the static ones. They can be estimated by performing a linear modal response spectrum analysis (see the extended N2 method in Section 3.3.2.6).

3.3.2.3 Idealisation of the pushover curve

Idealisation of the pushover curve can be performed either at the level of the MDOF system or at that of the SDOF system. In order to determine a simplified (elastic–perfectly plastic) force–displacement relationship, engineering judgement has to be used. In regulatory documents, some guidelines may be given. In Eurocode 8, the bilinear idealisation is based on the equal-energy principle and is performed at the SDOF level. The yield force f_{Sy}, which also represents the strength of the idealised equivalent SDOF system, is equal to the lateral force at the formation of the plastic mechanism. The initial stiffness of the idealised system is determined in such a way that the areas under the actual and the idealised force–deformation curves, up to the displacement at the formation of a plastic mechanism, are equal. Note that the displacement demand depends on the equivalent stiffness which, in the case of the equal-energy approach, depends on the target displacement. In principle, an iterative approach is needed. If the displacement at the formation of a plastic mechanism is used for the determination of the equivalent stiffness based on equal energy, as in Eurocode 8, a conservative estimate of displacement demand will, generally, be obtained. If the displacement demand is expected to be much lower than that corresponding to the plastic mechanism, it is reasonable to apply an iterative procedure (optional in Eurocode 8), and to base the equal energies on a smaller displacement, which leads to a higher equivalent stiffness. If, for a nearly elastic structure, the equivalent stiffness is based on the displacement corresponding to the formation of a plastic mechanism, the deformation quantities would be grossly overestimated.

The graphical procedure used in the basic N2 method requires a post-yield stiffness equal to zero. This is because the reduction factor q_μ is defined as the ratio of the required elastic strength to the yield strength. The influence of moderate strain hardening is incorporated in the demand spectra. It should be emphasised that moderate strain hardening does not have a significant influence on displacement demand, and that the proposed spectra apply approximately to systems with zero or small strain hardening.

The elastic period of the idealised bilinear system T^* can be determined as:

$$T^* = 2\,\pi \sqrt{\frac{L\,u_{Sy}}{f_{Sy}}} = 2\,\pi \sqrt{\frac{m^*\,u_{Sy}}{f_{Sy}}} \tag{3.139}$$

where f_{Sy} and u_{Sy} are the yield strength and displacement of the equivalent SDOF system, respectively, and L is the mass of the equivalent SDOF system. Note that Eurocode 8 uses a different notation: m^* instead of L. In the following text, the Eurocode 8 notation, m^*, is adopted for the mass of the equivalent SDOF system.

The so-called capacity diagram in AD format is obtained by dividing the forces in the force–deformation (f_S–u_S) diagram by the equivalent mass m^*, that is, as f_S/m^*. Note that f_S/m^* can be transformed into V_b/M^*, where M^* is, by analogy with Equation 3.86, the effective mass for the fundamental mode: $M^* = L^2/M$.

3.3.2.4 Seismic demand

Seismic demand is, in principle, represented by an inelastic response spectrum, which can be obtained from the elastic spectrum, if the appropriate q_μ–μ–T relation is known.

Starting from the usual acceleration spectrum (acceleration vs. period), inelastic spectra in acceleration–displacement (AD) format can be determined. For an elastic SDOF system, Equation 3.39 applies, repeated here for convenience:

$$D_e = \frac{T^{*2}}{4\,\pi^2}\,A_e \tag{3.140}$$

where A_e and D_e are the values in the elastic acceleration and displacement spectrum, respectively, at the period T^* for a fixed viscous damping ratio.

For an inelastic SDOF system with a bilinear force–deformation relationship, the acceleration spectrum (A_{in}) and the displacement spectrum (D_{in}) can be determined from Equations 3.112 and 3.117 by replacing forces with accelerations:

$$A_{in} = \frac{A_e}{q_\mu} \tag{3.141}$$

$$D_{in} = \frac{\mu}{q_\mu}\,D_e = \frac{\mu}{q_\mu}\,\frac{T^{*2}}{4\,\pi^2}\,A_e = \mu\,\frac{T^{*2}}{4\,\pi^2}\,A_{in} \tag{3.142}$$

where μ is the ductility factor defined as the ratio between the maximum displacement and the yield displacement and q_μ the reduction factor due to ductility, that is, due to the hysteretic energy dissipation of ductile structures. Note that q_μ is not the same as the reduction factor used in seismic codes. The code reduction factor, called in Eurocode 8 as behaviour factor q, takes into account both energy dissipation and overstrength (see Section 3.2).

Any inelastic spectrum can be employed in the analysis. In the basic version of the N2 method, implemented in Eurocode 8, a bilinear spectrum for the reduction factor q_μ is used, Equations 3.119 and 3.120; in the medium- and long-period ranges, this bilinear spectrum is based on the equal displacement rule, stating that the displacement of the inelastic system is equal to the displacement of the corresponding elastic system with the same period.

Starting from the elastic design spectrum, and using Equations 3.141, 3.142, 3.119 and 3.120, the demand spectra for the constant ductility factors μ in AD format can be obtained. The inelastic demand spectra corresponding to the Eurocode 8 elastic response spectrum for ground type B are shown in Figure 3.16. Note that construction of inelastic spectra is not, in fact, needed in the computational procedure. These spectra just help visualisation of the procedure.

The procedure for determining seismic demand for the equivalent SDOF system is illustrated in Figure 3.17. Figures 3.17a and 3.17b apply to short-period and to medium- or long-period structures, respectively. Both the demand spectra and the capacity diagram appear in the same graph. The intersection of the radial line corresponding to the elastic period of the idealised bilinear system, T^*, with the elastic demand spectrum in AD format defines the acceleration demand A_e, that is, the capacity required for elastic behaviour, and the corresponding elastic displacement demand, D_e. The yield acceleration represents both the acceleration demand, A_{in}, and the capacity of the inelastic system, f_S/m^*. The reduction factor q_μ can be determined as the ratio between the accelerations corresponding to the elastic and inelastic systems (Equation 3.111):

$$q_\mu = \frac{A_e(T^*)}{A_{in}} \qquad\qquad (3.143)$$

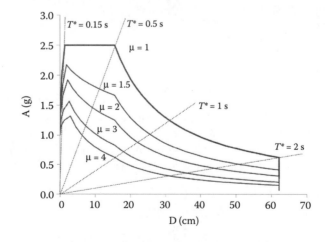

Figure 3.16 Inelastic demand spectra for constant ductility ratios in AD format normalised to 1.0 g peak ground acceleration, for elastic response spectrum of Type 1 as per Eurocode 8 for ground type B.

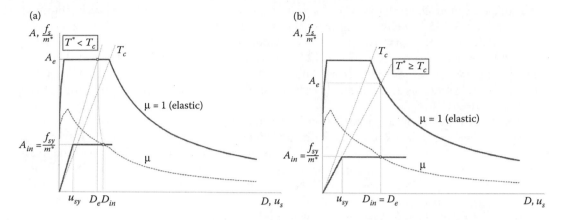

Figure 3.17 Determination of the seismic demand for an SDOF system with the period in the short- ($T^* < T_C$) (a) and medium/long-period range ($T^* \geq T_C$) (b).

If the period T^* is longer than or equal to the characteristic period of the ground motion T_C, the equal displacement rule, Equation 3.120, applies and the ductility demand is equal to the reduction factor due to ductility:

$$\mu = q_\mu \quad T^* \geq T_C \tag{3.144}$$

The inelastic displacement demand D_{in} is equal to the elastic displacement demand D_e (Equation 3.142 and Figure 3.17b).

If the period of the system is shorter than T_C, the ductility demand can be calculated from the rearranged Equation 3.119:

$$\mu = (q_\mu - 1)\frac{T_C}{T^*} + 1 \quad T^* < T_C \tag{3.145}$$

The inelastic displacement demand can be determined either from the definition of ductility or from Equations 3.142 and 3.145 as:

$$D_{in} = \mu u_{Sy} = \frac{D_e}{q_\mu}\left(1 + \left(q_\mu - 1\right)\frac{T_C}{T^*}\right) \tag{3.146}$$

In both cases (i.e. $T^* < T_C$ and $T^* \geq T_C$) the inelastic demand in terms of accelerations and displacements corresponds to the intersection point of the capacity diagram with the demand spectrum corresponding to the ductility demand μ. At this point, the ductility factor determined from the capacity diagram and the ductility factor associated with the intersecting demand spectrum are equal.

All the steps in the procedure can be performed numerically without using a graph. However, visualisation of the procedure may help in better understanding the relations between the basic quantities.

At this stage, the displacement demand can be modified if necessary, for example, to take into account larger displacements in the case of systems with narrow hysteresis loops or negative post-yield stiffness.

The displacement demand for MDOF systems (i.e. the target displacement), u_r, is obtained from Equation 3.135 by multiplying the displacement demand of the equivalent SDOF system, $D_{in} = u_s$, with the transformation factor Γ. Under monotonically increasing lateral loads with a fixed pattern as per Equation 3.124, the structure is pushed to the target displacement, u_r. It is assumed that the distribution of deformations throughout the structure in the static (pushover) analysis approximately corresponds to that which would be obtained in the dynamic analyses.

In the case of a 3D model, separate pushover analyses are performed in two horizontal directions. The relevant results (i.e. the displacements, storey drifts, joint rotations, and forces in brittle elements which should remain in elastic region), obtained by two independent pushover analyses in two orthogonal directions, are combined through the SRSS rule. In this way, torsional effects are included. Note, however, that these effects may be severely underestimated, especially in the case of torsionally flexible structures. For better estimation of torsional effects, the extended N2 method can be used (Section 3.3.2.6).

The target displacement u_r represents a mean value for the applied earthquake loading. There is a considerable scatter about that mean. Consequently, it is appropriate to investigate the likely building performance under extreme load conditions that exceed the design values, for example, to carry out the analysis to at least 150% of the calculated top displacement.

According to Eurocode 8, Part 3, the demands on both the 'ductile' and the 'brittle' components shall be those obtained from the non-linear analysis, using mean value properties of the materials.

3.3.2.5 Performance evaluation (damage analysis)

The expected performance can be assessed by comparing the seismic demands, determined in the previous section, with the capacities for the relevant performance level. Comparisons can be made both at the global and at the local level. In the case of inelastic behaviour, the relevant quantities are the roof displacement and the storey drifts, whereas at the local level a convenient quantity is member chord rotation. Forces and accelerations are relevant for brittle elements and for equipment which is sensitive to accelerations.

Collapse prevention is the main objective of any design. An adequate safety margin against collapse under the expected maximum seismic load needs to be assured. However, it is extremely difficult to predict a physical collapse which involves large deformations, significant second-order effects and complex material degradation due to localised phenomena. In spite of considerable research efforts, methods for the reliable assessment of collapse are not yet available. In practice, the near collapse (NC) limit state is often used as a conservative approximation of structural collapse. In Eurocode 8, Part 3, the NC limit state is defined as follows:

> 'The structure is heavily damaged, with low residual lateral strength and stiffness, although vertical elements are still capable of sustaining vertical loads. Most non-structural components have collapsed. Large permanent drifts are present. The structure is near collapse and would probably not survive another earthquake, even of moderate intensity.'

However, no guidance is provided as to how capacity at the NC limit state could be determined. The NC limit state of an individual structural element is usually defined as the point on its pushover curve at which the horizontal resistance drops by 20%, relative to the maximum previously attained. At the level of the structure, a commonly accepted quantitative definition of the NC limit state does not exist. An option is a similar definition as in the case of individual elements, for example, at a 20% drop of the lateral resistance of the structure. However, this definition, which seems to be the most appropriate, cannot be applied in non-linear dynamic analysis, or in pushover analyses with simplified models, for example, in the case of models without strength degradation. A more practical definition is based on the assumption that the NC limit state of the structure is reached when the first important vertical element (i.e. a column or a wall) reaches the NC limit state. Note, however, that this definition may be non-conservative in the case of a structure with significant second-order (P-Δ) effects.

The capacities of structural elements (beams, columns, walls) are empirically based. In Eurocode 8, information for the quantification of the capacity of components and/or mechanisms is provided in the relevant material-related Informative Annexes to its Part 3. Annex A applies to reinforced concrete structures. Expressions for the flexural deformation capacity and the deformation-dependent cyclic shear resistance given in Annex A are based on the results of statistical analyses using a very large database of test results (Biskinis et al. 2004; Biskinis and Fardis 2010a, 2010b).

In the case of ductile components and/or mechanisms, that is, beams, columns and walls under flexure with and without an axial force, the deformation capacity is defined in terms of the chord rotation θ, which is defined in Part 3 of Eurocode 8 as:

'the angle between the tangent to the axis at the yielding end and the chord connecting that end with the end of the shear span ($L_V = M/V$ = moment/shear at the end section), that is the inflection point. The chord rotation is also equal to the element drift ratio, that is, the deflection at the end of the shear span with respect to the tangent to the axis at the yielding end, divided by the shear span.'

Expressions are provided in Annex A for the chord rotation capacity at the component NC limit state, that is, the ultimate chord rotation capacity, for primary and secondary elements, corresponding to the mean-minus-sigma and the mean value, respectively, as fitted to the test results (Biskinis and Fardis 2010b). Expressions from (Biskinis and Fardis 2010a) are provided also for the chord rotation at yielding, which can be used, together with the ultimate chord rotation, to determine the chord rotation ductility capacity. In the case of 'brittle' mechanisms, that is, the shear mechanism of beams, columns or walls, the capacity is provided in terms of the cyclic shear resistance, which decreases with increasing plastic part of the ductility demand (Biskinis et al. 2004).

Note that the seismic demand to be compared with the capacity corresponding to the NC limit state, discussed in this section, is not the demand under the design seismic action as per Part 1 of Eurocode 8, whose recommended mean return period is 475 years (10% probability of being exceeded in 50 years, see Section 1.3). As pointed out in Part 3 of Eurocode 8, the limit state associated with the 'No Collapse' requirement of Part 1 of Eurocode 8 for the purposes of Life Safety is roughly equivalent to what is defined in Eurocode 8, Part 3, as limit state of Significant Damage. Instead, the demand corresponding to the NC limit state is typically based on a mean return period of 2475 years (2% probability of being exceeded in 50 years).

3.3.2.6 Influence of higher modes

The main assumption in basic pushover-based methods is that the structure vibrates predominantly in a single mode. This assumption is sometimes not fulfilled, especially in high-rise buildings and/or torsionally flexible, plan-asymmetric buildings. For such buildings, the contributions to the response from modes of vibration higher than the fundamental one in each principal direction should be taken into account.

At the time when Part 1 of Eurocode 8 was finalised, the extended version of the N2 method for plan-asymmetric buildings had not been fully developed yet. Nevertheless, based on the preliminary results, the clause 'Procedure for the estimation of torsional effects' was added, in which it is stated that:

'pushover analysis may significantly underestimate deformations at the stiff/strong side of a torsionally flexible structure'.

It is also stated that

'For such structures, displacements at the stiff/strong side shall be increased, compared to those in the corresponding torsionally balanced structure'

and that

'this requirement is deemed to be satisfied if the amplification factor to be applied to the displacements of the stiff/strong side is based on the results of an elastic modal analysis of the spatial model.'

Eurocode 8, Part 3, states that the approach in Part 1 (see previous paragraph) applies for the estimation of torsional effects. Furthermore, for buildings with a long fundamental period and for buildings irregular in elevation, it requires that:

'the contributions to the response from modes of vibration higher than the fundamental one in each principal direction should be taken into account'. Furthermore:

'this requirement may be satisfied … through special versions of the non-linear static analysis procedure that can capture the effects of higher modes on global measures of the response (such as interstorey drifts) to be translated then to estimates of local deformation demands (such as member hinge rotations). The National Annex may contain reference to complementary, non-contradictory information for such procedures.'

Such a procedure is the extended N2 method (Kreslin and Fajfar 2012), which combines two earlier approaches, taking into account higher mode effects in plan (Fajfar et al. 2005) and in elevation (Kreslin and Fajfar 2011), into a single procedure, enabling analysis of plan-asymmetric medium- and high-rise buildings. The extension is based on the assumption that the structure remains in the elastic range in higher modes. The seismic demand in terms of displacements and storey drifts can be obtained by combining the results of basic pushover analysis and those of elastic modal response spectrum analysis (RSA), which are both standard analyses, already present in Eurocode 8 and implemented in most commercial computer programs. Thus, the approach is conceptually relatively simple, straightforward and transparent.

In the elastic range, the vibration in different modes can be decoupled, with the analysis performed for each mode and seismic action component separately, according to the modal response spectrum analysis (RSA) of Section 3.1.5. The results obtained for different modes using design spectra are then combined through approximate combination rules, like the 'Square Root Sum of Squares' (SRSS) rule. This approach is widely accepted and used in practice, in spite of the approximations involved in the combination rules.

In the inelastic range, the superposition rule theoretically does not apply. However, the coupling between vibrations in different modes is usually weak (Chopra 2007); thus, for the majority of structures, some kind of superposition can be applied as an approximation in the inelastic range, too.

It has been observed that higher mode effects depend considerably on the magnitude of the plastic deformations. In general, higher mode effects in plan and in elevation decrease with increasing ground motion intensity. Thus, conservative estimates of amplification due to higher mode effects in plan and in elevation can usually be obtained by elastic analysis. The results of elastic analysis, properly normalised, mostly represent an upper bound to the results obtained for different intensities of ground motion in those parts of the structure where higher mode effects are important, that is, in the upper part of medium- or high-rise buildings, at the flexible sides of plan-asymmetric buildings and at the stiff sides of torsionally flexible plan-asymmetric buildings. One exception is the case of torsional de-amplification, which usually decreases with increasing plastic deformations.

The extended N2 method has been developed based on the above observations. It is assumed that an (in most cases conservative) estimate of the distribution of seismic demand throughout the structure can be obtained by combining (enveloping) the pushover results and the normalised results of elastic modal analysis. The target displacement may be determined as in the basic N2 method, or by any other procedure.

In principle, higher modes influence all quantities that are relevant for design. Torsional rotations affect displacements and, as a consequence, also affect storey drifts and local quantities. On the other hand, analyses have shown that, in elevation, the effect of higher modes is generally negligible for displacements, but should be taken into account when computing storey drifts and local quantities.

In the extended N2 method, it is assumed that the higher mode effects in the inelastic range are the same as in elastic range. Higher mode effects are determined by standard elastic modal response spectrum analysis in the form of correction factors, for the adjustment of results obtained by the usual pushover analysis. It is assumed that the structure remains in the elastic range when vibrating in higher modes, and that the seismic demands at different locations at the roof and at the mass centres along the height of the building can be estimated by combining the demands determined by a pushover analysis, which neglects higher mode effects, and the normalised demands from an elastic modal analysis which includes higher mode effects. Typically, the pushover analysis controls the response of those parts of the structure where the major plastic deformations occur; the elastic analysis determines the seismic demands at those parts in elevation where higher mode effects are important.

Higher mode effects in plan and in elevation can be considered simultaneously by two sets of correction factors. Possible de-amplification is not taken into account, thus the correction factors are not less than 1.0.

In order to predict the response of a building with a non-negligible effect of higher modes, the following procedure may be applied:

1. Perform the basic N2 analysis. In the case of a plan-asymmetric building, either two 2D (planar) models are used, one per horizontal direction, or a single model in 3D. Loading is applied at the mass centres (CM), independently in each of the two horizontal directions and with the + and − sign in each direction. The target displacement (displacement demand at the CM at roof level) is determined in each one of the two horizontal directions, as the larger of two values, for the + and − sign. It is assumed that the effect of higher modes on the target roof (top) displacement is negligible.

2. Perform the standard elastic modal response spectrum analysis of the 3D model independently, for excitation in two horizontal directions, considering all relevant modes (using, e.g. the CQC rule) and combine the results for both directions according to the SRSS rule. Determine the displacements and storey drifts at the CM of each storey. Determine the roof displacements for each frame or wall in plan. Normalise the results

in such a way that the roof displacement at the CM is equal to the target displacement (i.e. the roof displacement determined by the basic N2 method).

3. Determine the seismic demand using the results of steps 1 and 2. This can be achieved by applying two sets of correction factors, one in plan for displacements and another for storey drifts (in elevation). The set determined for displacements (in plan) also applies to the storey drifts. So, the resulting correction factor for the storey drift in a particular storey, and at a particular position in plan, is obtained as the product of two correction factors. These factors are defined for each horizontal direction separately and applied to the relevant results of the pushover analyses:

   a. The correction factor for displacements due to torsion is defined as the ratio of the normalised roof displacements from elastic modal analysis (step 2), to those from pushover analysis (step 1). The normalised roof displacement is the roof displacement at an arbitrary location divided by that at the CM. If the normalised roof displacement from elastic modal analysis is less than 1.0, then the value 1.0 is used, that is, no de-amplification due to torsion is taken into account. These correction factors depend on the location in plan.

   b. The correction factor for storey drifts due to higher mode effects in elevation is defined as the ratio between the normalised storey drifts from elastic modal analysis (step 2) and those from pushover analysis (step 1). As in the case of torsion, no de-amplification is taken into account, that is, if the ratio is less than 1.0, the value 1.0 is used. One correction factor is determined for each storey in the two horizontal directions.

The resulting correction factors for storey drifts (obtained as the product of two correction factors as described above) apply to all local deformation quantities (e.g. total joint rotations consisting of both elastic and plastic part). They also apply to the internal forces, provided that the resulting internal force does not exceed the force capacity of the structural member. If that capacity is exceeded, internal forces can be estimated from the deformations using the relevant force–deformation relationship. For more details on the determination of internal forces, see the procedure elaborated by Goel and Chopra (2005).

In the case of a planar (2D) structural model, the results obtained by the extended N2 method represent an envelope of the pushover results and the normalised RSA results. In a plan-asymmetric 3D model, the seismic demands at different locations at the roof and at the mass centres along the elevation, determined according to the procedure above, represent such an envelope. At other locations, they are mostly close to the envelope. If convenient from the computational point of view, the envelope of pushover results and RSA results, normalised to the target displacement of the mass centre at the roof, can simply be used in practice for all relevant quantities.

Two independent approximations determine the accuracy of the N2 method and other simplified pushover-based methods: the determination of the target displacement, and the distribution of seismic demand throughout the structure. The extended N2 method aims at providing an improved distribution of seismic demand, whereas the determination of the target displacement is the same as in the basic N2 method. Note, however, that the proposed distribution can be used in conjunction with any procedure for the determination of the target displacement.

3.3.2.7 Discussion of pushover-based methods

A pushover-based analysis represents a rational practice-oriented tool for the seismic analysis of structures. Compared to traditional elastic analyses, it provides a wealth of additional important information about the expected structural response, as well as insight into the structural aspects that control performance during severe earthquakes. Pushover-based analysis provides data on the strength and ductility of a structure, which cannot be obtained by elastic analysis. Furthermore, it exposes design weaknesses that could remain hidden in an elastic analysis, for example, in most cases it is able to detect the most critical parts of a structure.

Compared to non-linear response-history analysis, which usually provides the most reliable information on structural response (if performed correctly), pushover-based methods are a much simpler and transparent tool, requiring much simpler input data: an average spectrum is used, instead of a suite of accelerograms, and detailed data on the hysteretic behaviour of structural elements are not needed. There are no problems with the modelling of damping. The amount of computation time is only a fraction of that required by non-linear response-history analysis, and the use of the analysis results is straightforward. Of course, these advantages of pushover-based methods have to be weighed against their lower accuracy compared to non-linear response-history analysis.

For practical applications and educational purposes, graphical displays of the procedure are extremely important, even when all the results can be obtained numerically. Pushover-based methods achieved a breakthrough when the acceleration–displacement (AD) format was implemented, permitting visualisation of important demand and capacity parameters. A pushover-based analysis presented graphically in AD format helps to better understand the basic relations between seismic demand and capacity, and between the main structural parameters determining the seismic performance, that is, stiffness, strength, deformation and ductility. It permits visualisation of the response and its progression from low load levels to levels associated with the target displacement and beyond. It is a very useful tool for understanding the general seismic behaviour.

According to Eurocode 8, pushover-based analysis may be applied to verify the structural performance of newly designed or existing buildings for the following purposes:

1. The verification or revision of overstrength ratio values
2. The estimation of the expected plastic mechanisms and the distribution of damage
3. The assessment of structural performance of existing or retrofitted buildings
4. As an alternative to design based on linear-elastic analysis using the behaviour factor q

Like any approximate method, pushover-type methods are based on a number of assumptions. Their limitations should be observed. It cannot be expected that they will accurately predict the seismic demand for any structure and any ground motion. The basic pushover analysis is based on a very restrictive assumption, that is, a time-independent displacement shape. Thus, it is, in principle, inaccurate for structures where higher mode effects are significant, and it may not detect structural weaknesses that may be generated when the structure's dynamic characteristics change after formation of the first local plastic mechanism. Several different approaches have been proposed to improve the accuracy of pushover-based analyses in structures where the higher modes make important contributions. Some of them require quite complex analysis, defeating the purpose of using such methods.

The limitations of pushover-based methods have been discussed, for example, by Krawinkler and Seneviratna (1998), Fajfar (2000) and Krawinkler (2006).

Pushover-based methods are usually applied for the performance evaluation of a known structure, that is, an existing structure or a newly designed one. However, other types of analysis can also be applied and visualised in the AD format. Four quantities define the seismic performance: strength, displacement, ductility and stiffness. Design and/or performance evaluation begins by fixing one or two of them; the others are determined by calculations. Different approaches differ in the quantities that are chosen at the beginning of the design or evaluation. Let's assume that the approximate mass is known. In the case of seismic performance evaluation, stiffness (period) and strength of the structure have to be known; the displacement and ductility demands are calculated. In direct displacement-based design, the starting points are typically the target displacement and/or ductility demands. The quantities to be determined are stiffness and strength. The usual force-based design typically starts from the stiffness (which defines the period) and the approximate global ductility capacity. The seismic forces (defining the strength) are then determined, and finally displacement demand is calculated.

Note that, in all cases, the strength is the actual one and not the design base shear according to seismic codes, which is in all practical cases less than the actual strength. Note also that stiffness and strength are usually related quantities.

All these approaches can be easily visualised with the help of Figure 3.17.

3.3.3 Modelling

A model for inelastic analysis is, in principle, an extended model for linear elastic analysis, which additionally includes the strength of structural elements and their post-elastic behaviour.

Inelastic structural component models can be differentiated by the way in which plasticity is distributed through the cross sections of the members and along their lengths. The simplest models concentrate the inelastic deformations at the ends of the elements, by placing an inelastic spring there. The part of a member between the two inelastic springs remains fully elastic. All inelastic deformations are assumed to occur in these springs. The most complex models discretise the continuum along the member length and through the cross sections into small (micro) finite elements, with non-linear hysteretic constitutive properties and often numerous input parameters. Different models are used for the concrete and the reinforcement, and possibly also for bond. Somewhere in between these two extremes are the fibre models, which distribute plasticity by numerical integration through the member cross sections and along the member length. Details about different models are provided in Fardis (2009).

The most complex finite element models are, due to their severe computational requirements and numerous input parameters, appropriate only for studying details, for example, for the simulation of experiments on individual members or sub-assemblies. Even fibre models can be prohibitively complex for the simulation of whole realistic structures.

At present, the simplest model, that is, the one-component model with concentrated plasticity, proposed by Giberson (1967), seems to provide the best option for practical non-linear seismic response analysis. Several good reasons give support to this statement.

By concentrating the plasticity in zero-length springs with moment–rotation model parameters, such elements have a numerically efficient formulation. The model can work directly with chord rotations, so it can be directly related to experimental results obtained for RC members, which are typically given as force–deflection (or moment–chord rotation) relationships. Moreover, straightforward comparison of demand and capacity at the member level is possible. Inelastic member-end rotation depends solely on the moment acting at the end, so that any moment–rotation hysteretic model can be assigned to the spring, for example, an experimentally observed hysteretic behaviour. This decoupling of the inelastic behaviour between the two ends is possible if the inflection point stays steady after the first inelastic excursion of the member. In frame members (columns and beams), normally a skew-symmetric moment distribution along a member, with an inflection point at mid-span, is assumed. With one-component models, the modelling effort and the computational requirement are reasonable even in large 3D structures.

On the other hand, the simple one-component model with concentrated plasticity does have some shortcomings. In reality, the inelastic deformation of a reinforced concrete member is not concentrated at a critical location, but spreads along part of the member. Assuming a zero-length is thus an idealisation. The model does not take into account any coupling between the bending moments and the axial forces, nor between the two directions of bending. Fluctuation of the inflection point is also neglected.

In spite of these shortcomings, the performance of the one-component model with concentrated plasticity is usually good. Simulations of full-scale experiments have demonstrated that quite good agreement can be obtained with experimental results provided that the basic input data are appropriately chosen (e.g. Fajfar et al. 2006; Kosmopoulos and Fardis 2008; Dolšek 2010).

The one-component model with rigid plastic hinges or with inelastic springs is implemented in the majority of available computer programs that allow non-linear analysis. For the formation of the tangent flexibility and stiffness matrices of the model, see Fardis (2009).

For each hinge/spring model, it is necessary to determine the moment–rotation relationship. For non-linear dynamic analysis, the whole hysteretic behaviour has to be modelled, whereas for a pushover analysis only the cyclic envelope is needed. The relationship can be determined based on principles of mechanics and/or experimental data. As a minimum that would be sufficient as per Eurocode 8, the initial (elastic) stiffness, the strength and rotation capacity are needed, which determine a bilinear moment–rotation relation. A zero post-yield stiffness may be assumed. The corner point of the bilinear relation is the yield point of the member. The yield moment, M_y, can be determined from the principles of mechanics, based on the characteristics of the cross-section and the material characteristics of concrete and steel. Its value depends on the axial force N, which changes during the seismic response.

The axial forces due to gravity loads should be taken into account when determining moment–rotation relations for structural elements. Fluctuation of N, which usually does not have a large effect, cannot be considered in the one-component model. If the axial force varies considerably, as, for example, the axial force in coupled walls, a post-analysis check is suggested.

Determination of the yield chord rotation θ_y is tricky, since the actual non-linear force–deformation relation has to be replaced, in the case of a reinforced concrete member, by an equivalent linear relation. According to Eurocode 8, in reinforced concrete elements the elastic stiffness of the bilinear force–deformation relation should correspond to that of cracked sections and, indeed, to the initiation of yielding of the reinforcement. Unless a more accurate analysis of the cracked elements is performed, the elastic flexural and shear stiffness properties of concrete elements may be taken equal to one-half of the corresponding stiffness of the uncracked element. The relation between the yield rotation and the effective stiffness, EI_{eff}, is defined in Part 3 of Eurocode 8 as:

$$EI_{eff} = \frac{M_y L_V}{3\theta_y} \tag{3.147}$$

where the shear span L_V is the moment-to-shear ratio at the member end. A problem is that the secant stiffness to the yield point is usually much smaller than one-half of the stiffness of the uncracked elements. Thus, in a non-linear analysis of a structure, whose inelastic response is controlled by a single cross-section, for example, a column modelled as a cantilever beam, the deformation demand, which depends on the elastic stiffness, may be severely underestimated if one-half of the uncracked gross section stiffness is used. In Part 3 of Eurocode 8, expressions from Biskinis and Fardis (2010a) are provided for θ_y, and can be used in Equation 3.147 for the determination of a more realistic effective stiffness of members. On the other hand, in a more complex structure, for example, a frame, where some structural members yield, whereas others do not, a uniformly reduced stiffness to one-half of that corresponding to the uncracked gross sections may provide acceptable results at the global level. At the element level, however, the ductilities may be overestimated, due to underestimated yield rotations. This problem can be bypassed by evaluating performance in terms of the ultimate rotations rather than in terms of ductilities.

The end point of the bilinear moment–rotation diagram is the ultimate chord rotation, θ_u. In Eurocode 8, Part 3, empirical expressions, as per Biskinis and Fardis (2010b), are provided for θ_u. The ultimate chord rotation corresponds to a 20% drop in strength, and is intended to represent the NC limit state. It may be assumed that it represents the flexural deformation capacity of a member. Actually, a member has additional capacity beyond the NC limit state. So, in principle, it is possible to model the moment–rotation relation beyond θ_u. However, there is a lack of data on the descending branch of the moment–rotation curve. Moreover, simulating the behaviour beyond the NC limit state usually has only a very limited practical value.

Ductile flexural behaviour is possible only if the member shear strength exceeds its flexural strength (cf. Section 5.5). If this is not the case, brittle shear failure occurs before a plastic hinge can develop. Unless shear effects are included in the model by using a non-linear shear spring in series with the flexural springs, the shear force demand to capacity ratio has to be checked, to make sure that shear failure does not occur. If it does, the results of analysis beyond that point are not valid. Similarly, it is necessary to check the bond of the longitudinal bars.

In a deterministic inelastic analysis, it is reasonable to use a best estimate approach and to apply safety factors taking into account uncertainties only at the end. In such a case element properties should be based on mean values of the properties of the materials, as required by Eurocode 8. A safety factor is included in the Eurocode expressions for ultimate chord rotation and shear strength in the form of a factor γ_{el}, which addresses model uncertainty and depends on the standard deviation of test results.

Example 3.4, at the end of this chapter, illustrates non-linear modelling and pushover analysis for an idealised 4-storey frame.

EXAMPLE 3.1

To illustrate the concept and features of the vibration modes of MDOF systems, consider the oscillator with 3 degrees of freedom depicted in Figure 3.18. Assume that the mass matrix and stiffness matrix are given by:

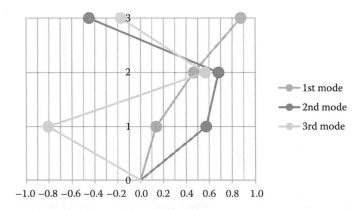

Figure 3.18 Oscillator of Example 3.1 with 3 horizontal degrees of freedom and corresponding mode shapes.

$$\mathbf{m} = \begin{bmatrix} 1 & 0 & 0 \\ 0 & 1 & 0 \\ 0 & 0 & 1 \end{bmatrix}$$

and

$$\mathbf{k} = \begin{bmatrix} 18,461.5 & -10,615.4 & 2769.2 \\ -10,615.4 & 10,153.8 & -3692.3 \\ 2769.2 & -3692.3 & 1615.4 \end{bmatrix}$$

For these matrices the three eigenvalues are:

$$\omega_1^2 = 85.5$$

$$\omega_2^2 = 3667$$

$$\omega_3^2 = 26,480$$

hence, the natural frequencies of the oscillator are:

$$\omega_1 = 9.25 \text{ rad/s}$$

$$f_n = \omega_n / 2\pi$$

$$\omega_3 = 163 \text{ rad/s}$$

With the values of ω_n^2 we can obtain (using Equation 3.59) the shape of each of the three natural modes Φ_n of the oscillator:

$$\Phi_1 = \begin{bmatrix} 0.1368 \\ 0.4650 \\ 0.8747 \end{bmatrix} \quad \Phi_2 = \begin{bmatrix} 0.5743 \\ 0.6822 \\ -0.4525 \end{bmatrix} \quad \Phi_3 = \begin{bmatrix} -0.8072 \\ 0.5642 \\ -0.1737 \end{bmatrix}$$

These shapes are depicted in Figure 3.18.

The orthogonality of the first and second mode shapes with respect to $\mathbf{m}$ is easily illustrated performing the product $\Phi_1^T \mathbf{m} \Phi_2$:

$$\begin{bmatrix} 0.1368 & 0.4650 & 0.8747 \end{bmatrix} \cdot \begin{bmatrix} 1 & 0 & 0 \\ 0 & 1 & 0 \\ 0 & 0 & 1 \end{bmatrix} \cdot \begin{bmatrix} 0.5743 \\ 0.6822 \\ -0.4525 \end{bmatrix}$$

$$= \begin{bmatrix} 0.1368 & 0.4650 & 0.8747 \end{bmatrix} \cdot \begin{bmatrix} 0.5743 \\ 0.6822 \\ -0.4525 \end{bmatrix} = 0$$

resulting in a zero value, as foreseen by Equation 3.61.

Likewise, performing the products around k would also illustrate the orthogonality of the mode shapes with respect to the stiffness matrix.

Additionally, performing the product $\Phi_1^T \mathbf{m} \Phi_1$:

$$[0.1368 \quad 0.4650 \quad 0.8747] \cdot \begin{bmatrix} 1 & 0 & 0 \\ 0 & 1 & 0 \\ 0 & 0 & 1 \end{bmatrix} \cdot \begin{bmatrix} 0.1368 \\ 0.4650 \\ 0.8747 \end{bmatrix}$$

$$= [0.1368 \quad 0.4650 \quad 0.8747] \cdot \begin{bmatrix} 0.1368 \\ 0.4650 \\ 0.8747 \end{bmatrix} = 1$$

we obtain a non-zero value. In fact, it results in a value equal to 1, indicating that the shape of the first mode is normalised with respect to the mass matrix, as is commonly done.

The reader may check that, for the other mode shapes, Equations 3.61 and 3.62 also hold. It is worth noticing that, with the mode shapes normalised with respect to the mass matrix, the products around the stiffness matrix result in: $\Phi_n^T \mathbf{k} \Phi_n = \omega_n^2$.

This example also serves to show that any deformed configuration $\mathbf{u}$ of the oscillator may be represented by a linear combination of the natural mode shapes. Consider, for instance, a deformed shape of the oscillator corresponding to a straight line. Such shape is described in the generalised coordinates of the system by:

$$\mathbf{u} = \begin{bmatrix} 0.333 \\ 0.667 \\ 1.000 \end{bmatrix}$$

It is easily verifiable that this configuration is a linear combination of the three mode shapes:

$$\mathbf{u} = q_1 \Phi_1 + q_2 \Phi_2 + q_3 \Phi_3$$

In fact, taking as weighting coefficients, $q_1 = 1.2304$, $q_2 = 0.1938$ and $q_3 = -0.0662$ and inserting the configurations of the three mode shapes Φ_n we obtain:

$$1.2304 \cdot \begin{bmatrix} 0.1368 \\ 0.4650 \\ 0.8747 \end{bmatrix} + 0.1938 \cdot \begin{bmatrix} 0.5743 \\ 0.6822 \\ -0.4525 \end{bmatrix} - 0.0662 \cdot \begin{bmatrix} -0.8072 \\ 0.5642 \\ -0.1737 \end{bmatrix} = \begin{bmatrix} 0.333 \\ 0.667 \\ 1.000 \end{bmatrix}$$

(the intended deformed configuration).

It is interesting to note that, by and large, the largest weighing coefficient is the one for the first mode ($q_1 = 1.2304$), as it would be expected, since the chosen straight deformed shape is mostly akin to the shape of the first mode.

EXAMPLE 3.2

For the oscillator with three degrees of freedom in Example 3.1, the *effective modal masses* of each mode are $M_1^* = 2.18$, $M_2^* = 0.646$ and $M_3^* = 0.174$, giving a total sum of 3, which is the mass of the oscillator, as expected. The *participating mass ratios* are 0.727, 0.215 and 0.058 with a total sum of 1. These values illustrate the dominant contribution of the first mode to the shear at the base of the oscillator, when subjected to an excitation at the base.

According to the two conditions of Eurocode 8 concerning the number of modes to be taken into account in the analysis, referred in Section 3.1.5.3, it can be observed that, with regard to the first condition, consideration of the two first modes in the evaluation of the seismic response would be sufficient (since the sum of the corresponding participating mass ratios exceeds 0.90). However, the second condition makes it necessary to also take into account the third mode (since its participating mass ratio exceeds 0.05).

EXAMPLE 3.3: ELASTIC ANALYSIS OF A THREE-STOREY STRUCTURE

In this example linear elastic analysis of a three-storey prefabricated industrial building as per Eurocode 8 is performed. As typical of prefabricated buildings, beam-to-column connections are considered as hinged. The building is symmetric in plan and all columns are the same; so, it is sufficient to analyse one column, modelled as a cantilever, with the corresponding floor masses and three degrees of freedom, namely the horizontal translations of the lumped masses. Storey height is 5 m at the first storey and 3.5 m at the upper two. The masses are 30 t, 28 t and 24 t, in the first, second and third storeys, respectively. The column is 0.8 m square, with axial forces corresponding to the masses. The Importance Class is II. The structure is located on type B ground (soil factor $S = 1.2$) at a location with reference peak ground acceleration $a_g = 0.25$ g. Thus, for an importance factor γ_I of 1.0, the design ground acceleration at the top of type B ground is 1.2×0.25 g $= 0.3$ g. Design takes place with a behaviour factor $q = 3$ (as in Ductility Class Medium as per Eurocode 8, see Section 4.6.3). The elastic and the design spectrum are shown in Figure 3.19. Concrete class is C 30/37, with Elastic Modulus $E_c = 33,000$ MPa.

Answer

The mass matrix **m** (in tons) is:

$$\mathbf{m} = \begin{bmatrix} 30 & 0 & 0 \\ 0 & 28 & 0 \\ 0 & 0 & 24 \end{bmatrix}$$

The easiest way to determine the stiffness matrix **k** is by inversion of the flexibility matrix **d**. An individual element of the flexibility matrix, d_{ij}, is the horizontal displacement of storey i due to a unit horizontal force at storey j. For cross-section constant along the height and neglecting shear deformations, d_{ij} can be calculated as:

$$d_{ij} = \frac{z_i^2}{6EI}(3z_j - z_i) \quad j \geq i$$

where the element stiffness is taken equal to one-half of the corresponding stiffness of the uncracked element: $EI = 0.5E_c bh^3/12 = 0.5 \times 33,000,000 \times 0.0341 = 563,200$ kNm². The distance from the base to storeys i and j is denoted as z_i and z_j, respectively. The flexibility matrix (in m/kN) is then:

$$\mathbf{d} = \begin{bmatrix} 0.7384 \cdot 10^{-4} & 0.1514 \cdot 10^{-3} & 0.2289 \cdot 10^{-3} \\ 0.1514 \cdot 10^{-3} & 0.3628 \cdot 10^{-3} & 0.5868 \cdot 10^{-3} \\ 0.2289 \cdot 10^{-3} & 0.5868 \cdot 10^{-3} & 0.1021 \cdot 10^{-2} \end{bmatrix}$$

The stiffness matrix (in kN/m) is:

$$\mathbf{k} = \mathbf{d}^{-1} = \begin{bmatrix} 0.1406 \cdot 10^6 & -0.1095 \cdot 10^6 & 0.3141 \cdot 10^5 \\ -0.1095 \cdot 10^6 & 0.1246 \cdot 10^6 & -0.4709 \cdot 10^5 \\ 0.3141 \cdot 10^5 & -0.4709 \cdot 10^5 & 0.2101 \cdot 10^5 \end{bmatrix}$$

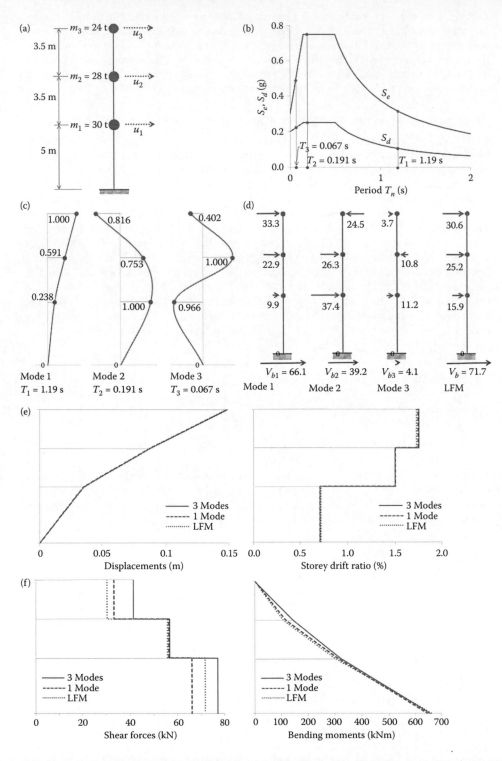

Figure 3.19 Example 3.3: (a) three degrees of freedom model; (b) elastic and design acceleration spectra; (c) natural modes and periods; (d) storey forces and base shears in the three modes and for lateral force method, LFM (kN); (e), (f) displacements, storey drift ratios, shear forces (kN), bending moments (kNm) from modal analysis with one or three modes and from the lateral force method.

Table 3.3 Results for the three modes of vibration

	Mode 1	Mode 2	Mode 3
ω_n(rad/s)	5.28	32.9	94.3
T_n (s)	1.19	0.191	0.067
L_n	47.69	31.50	−10.63
M_n	35.48	61.86	59.87
Γ_n	1.344	0.509	−0.178
$M_n^*(t)$	64.1	16.0	1.9
$M_n^*/\sum m_j$	0.78	0.20	0.02
$S_d(T_n) \equiv A_n(g)$	0.105	0.250	0.222

By solving the eigenvalue problem (Equation 3.59), the three natural frequencies, ω_n, and natural mode shapes, Φ_n, given in Table 3.3 and in part (c) of Figure 3.19 are obtained.

L_n and M_n are computed from Equations 3.91 and 3.92, respectively, Γ_n from Equation 3.78 and the effective modal mass M_n^* from Equation 3.86. These results are summarised in Table 3.3. Considering the effective modal masses, at least two vibration modes have to be taken into account, to comply with the Eurocode 8 requirements: sum of the effective modal masses of the modes included larger than 90% of the total mass, or all modes with effective modal masses greater than 5% of the total mass.

From the Eurocode 8 design acceleration spectrum in Equation 3.121, the spectral accelerations $S_d(T_n) \equiv A_n$ are obtained for the three vibration modes and are listed in the last row of Table 3.3. The seismic forces f_n are computed from Equation 3.94. They are shown in part (d) of Figure 3.19, alongside the corresponding base shears.

The modal displacements u_n are obtained as static displacements due to seismic forces, multiplied by the behaviour factor $q = 3$, see Equation 3.116. Practically the same displacements are obtained from Equation 3.93, with modal displacements, D_n, from the elastic spectrum. The floor displacements, the storey drift ratios (defined as the storey-relative displacement, i.e. inter-storey drift, divided by the storey height), the shear forces and the bending moments along the height of the column are presented in parts (e) and (f) of Figure 3.19: for the first mode separately and the SRSS combination values (Equation 3.87) of the three modes; the three periods are well-separated and the SRSS rule gives practically the same results as the CQC combination rule, Equation 3.88. These results show that higher modes have a substantial influence only on the shear forces in the upper and lowest storey and on the bending moment in the upper part.

It is allowed to apply the lateral force method, as the building is regular in elevation, and its fundamental period is less than 2.0 s. An approximation to the fundamental period can be obtained from Equation 3.109; selecting as lateral loads the gravity forces $(f_j = m_j g)$:

$$f_1 = 294 \text{ kN}, f_2 = 275 \text{ kN}, f_3 = 235 \text{ kN},$$

the following displacements are obtained by static analysis:

$$u_1 - 0.117 / \text{ m}, u_2 = 0.282 \text{ m}, u_3 = 0.469 \text{ m},$$

yielding $T_1 = 1.19$ s, that is, the same value (rounded to 2 decimal places) as obtained from the 'exact' eigenvalue analysis. The total mass is $M = 82$ t and the base shear from Equation 3.97 is 84.4 kN. With the Eurocode 8 reduction factor $\lambda = 0.85$, the design base shear is $V_b = 71.7$ kN. The seismic forces can be obtained from Equation 3.98, with z_1, z_2 and z_3 equal to 5 m, 8.5 m and 12 m, respectively, and are shown at the last diagram of

Part (d) of Figure 3.19. The resulting displacements, storey drift ratios, shear forces, and bending moments are compared in Parts (e) and (f) of Figure 3.19 to the modal analysis results (displacements and storey drifts obtained by static analysis are multiplied by $q = 3$). With the exception of the shear forces, very good agreement is observed. Note, however, that, in general, the agreement is not as good.

As pointed out in Section 1.3.2, the damage limitation requirement of Eurocode 8 is checked by comparing the demand in terms of the inter-storey drifts to a limit which depends on the type of non-structural elements. For Importance Class II, Eurocode 8 recommends taking the damage limitation earthquake as 50% of the design seismic action. A new analysis is not needed: the seismic demands (inter-storey drifts) for this earthquake are one-half of those for the design seismic action. The storey drift ratios resulting from modal analysis with three modes amount to 0.36%, 0.75% and 0.87% in the first, second and third storeys, respectively. As shown in Part (e) of Figure 3.19, these values differ very little from those obtained by other approaches. Most critical is the upper storey, where the drift ratio is less than the allowable value as per Eurocode 8 only when the non-structural elements do not interfere with the structure, or there are no non-structural elements (corresponding limit: 1%). The second storey respects the 0.75% limit for buildings with ductile non-structural elements attached to the structure, and the ground storey respects the 0.5% limit for brittle partitions.

As the building is flexible, second order (P-Δ) effects have to be checked through the inter-storey drift sensitivity coefficient θ (see Equation 3.110 in Section 3.1.12). It is the top (i.e. third) storey which is critical. Considering $N_3 = m_3g = 235$ kN and $h = 3.5$ m, and the values obtained by modal analysis with three modes: $d_3 = 0.061$ m and $V_3 = 41.4$ kN, the value of the coefficient θ at that storey is 0.097, that is, smaller than the threshold value of 0.1, above which second order effects have to be taken into account according to Eurocode 8.

EXAMPLE 3.4: NON-LINEAR ANALYSIS OF A FOUR-STOREY FRAME STRUCTURE

As an example of non-linear analysis, an idealised 4-storey frame representative of an existing structure built before the Eurocodes and depicted in Figures 3.20a and 3.20b is analysed in accordance with Eurocode 8. Smooth (plain) longitudinal bars are used, with a mean yield stress of 370 MPa. The mean value of concrete strength used in the analysis is 33 MPa and the modulus of elasticity is $E = 32,000$ MPa. According to Part 1 of Eurocode 8, a cracked element stiffness of one-half of the corresponding uncracked element stiffness is used in the mathematical model. The analyses are performed with the program ETABS (CSI 2002). Zero-length flexural plastic hinge elements are used at the ends of each elastic beam and column element, with bilinear moment–rotation relationships and zero hardening. The yield moments, M_y, of the end sections of the elements are determined by first principles. The M_y values of the beams are: $M_y = 163$ kNm, $M_y = 66$ kNm for moment inducing tension or compression at the top flange, respectively. Those of columns depend on the axial load due to gravity loading, hence on the position of the column. Assuming that the masses are uniformly distributed along the beams, the axial forces range from 70 kN in the exterior column of the top storey to 536 kN in the central column at the first storey; the corresponding mean axial stresses range from 0.58 to 4.47 MPa, and the resulting yield moments from $M_y = 61$ to 124 kNm. In the hinges, only plastic deformations occur. Before the yield rotation is attained, linear deformations take place only in the frame element. The yield rotation in the element, θ_y, is not part of the input data for the hinge, but determined automatically by the program as:

$$\theta_y = \frac{M_y L_V}{3EI}$$

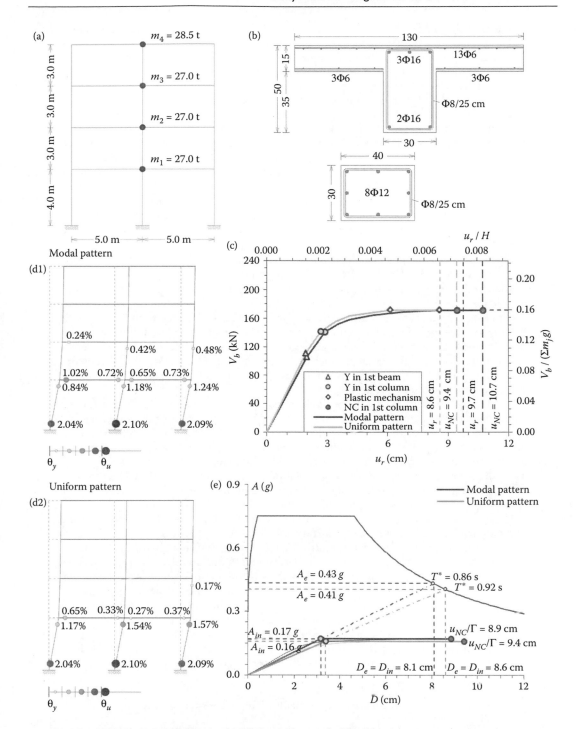

Figure 3.20 Example 3.4: (a) Geometry and lumped masses; (b) cross-sections and reinforcement of columns and beams; (c) pushover curves corresponding to two patterns of lateral loads; (d1 and d2) deformed configurations, plastic hinges and total rotations in plastic hinges, for two patterns of lateral loads, at the NC limit state; (e) determination of seismic demand and capacity–demand comparison for the equivalent SDOF system.

where M_y is the yield moment, L_V the shear span, E the modulus of elasticity of concrete and I the moment of inertia corresponding to a cracked section (taken as 50% of the value for the uncracked section). Plastic rotation beyond the yield point occurs in the hinge, in addition to the elastic rotation in the element. The deformation capacity of the columns and beams in terms of total (elastic plus inelastic part) chord rotation is determined from Equation A.1 in Part 3 of Eurocode 8. For the primary elements, it is divided by a conversion factor $\gamma_{el} = 1.5$ from mean to the mean–minus–sigma values of capacity. For a structure without detailing for earthquake resistance and with smooth reinforcement bars, the capacity from Equation A.1 in Part 3 of Eurocode 8 is reduced by a factor of $0.8/1.2 = 2/3$. The resulting capacities, θ_u, range from 1.98% to 2.28% in the columns (at the first storey, the values are 2.10% and 2.28% for the central and exterior column, respectively). In the beams, they are 2.19% for tension at the top and 1.86% for tension at the bottom. These capacities are used for the assessment of the seismic performance at the Near Collapse (NC) limit state as per Part 3 of Eurocode 8. For pushover analysis, unlimited ductility is assumed; the results are then valid only up to the failure of the first element.

The elastic structure (with cracked sections) has a fundamental period of $T_1 = 0.8$ s and a first vibration mode of: $\Phi_1^T = [0.466\ 0.719\ 0.899\ 1.000]$.

The effective modal mass (Equation 3.86) of the first mode amounts to 93.5% of the total mass.

For pushover analysis, two vectors of lateral loads are applied, as required by Eurocode 8, based on the 'modal' and the 'uniform' displacement shape. The 'modal' pattern uses the first mode shape from the elastic free vibration analysis. The lateral loads f are determined from Equation 3.125 as product of the assumed displacements Φ and the corresponding masses. Φ and normalised f are presented in Table 3.4. Second-order (P-Δ) effects are not taken into account.

Keeping the gravity loads constant and monotonically increasing the lateral forces, while maintaining a constant distribution of forces along the height of the structure, we obtain the deformed configurations of the frame in Figure 3.20c and the pushover curves in (d1) and (d2), representing the relationship between the base shear force V_b and the displacement at the roof u_r, for the two patterns of lateral loads. Some important events are marked: (a) yielding in the first beam and column, (b) formation of the plastic mechanism and (c) attainment of the NC limit state in the first column, namely in the central column of the first storey (assumed to represent also the NC limit state of the whole structure).

The difference between the pushover curves for the two lateral load patterns is small. The roof displacements at the NC limit state, u_{NC}, are 10.7 and 9.4 cm for the modal and the uniform lateral load patterns, respectively. The corresponding first-storey drift ratio is 1.9% in both cases. The presented results correspond to the lateral loads with a positive sign; a mirror image applies to the lateral loads with the opposite sign. An envelope of the results for both signs of the lateral loads should be taken into account. The results in Figure 3.20c demonstrate an unfavourable local mechanism in the first storey, typical of existing buildings: major damage is concentrated mostly in the columns of the first storey. This feature is more pronounced in the case of the uniform lateral load pattern.

The pushover analyses were performed assuming that shear failure of members does not occur. To confirm this assumption, the shear strength of all members should be larger

Table 3.4 Assumed displacement shapes and normalised lateral loads for the two load patterns in example 3.4

Pattern	Storey	1	2	3	4
Modal	Φ	0.466	0.719	0.899	1.000
Modal	f	0.44	0.68	0.85	1.00
Uniform	Φ	1.000	1.000	1.000	1.000
Uniform	f	0.95	0.95	0.95	1.00

Table 3.5 Properties of the equivalent SDOF system and seismic demand values

Pattern	Γ	T^* (s)	f_{sy} (kN)	D_y (cm)	$L \equiv m^*$ (t)	$D_e = D_{in}$(cm)	A_e (g)	A_{in} (g)	q_μ
Modal	1.209	0.86	142	3.2	84.4	8.1	0.43	0.17	2.5
Uniform	1.000	0.92	171	3.4	109.5	8.6	0.41	0.16	2.5

than the shear force corresponding to the yield moment. Indeed the shear strength from Equation A.12 in Part 3 of Eurocode 8 for primary elements is in all cases larger than the shear force corresponding to the respective yield moment.

For the NC limit state, Eurocode 8, Part 3 notes the use of a ground motion with a mean return period of 2475 years, corresponding to a probability of exceedance of 2% in 50 years. In this example, it is assumed that, for such a ground motion, the Eurocode 8 Type 1 acceleration spectrum for type B soil ($T_C = 0.5$ s) applies, with a peak ground acceleration of 0.3 g (Figure 3.20e). In order to determine the seismic demand for this structure as per Section 3.3.2.4, the MDOF system is transformed into an equivalent SDOF system (see Section 3.3.2.2) and the pushover curve is idealised as a bilinear force–deformation relationship (see Section 3.3.2.3). The results are plotted in Figure 3.20e; values of some of the quantities are summarised in Table 3.5. Note that $T^* > T_C$ and the equal displacement rule applies.

The seismic demand in terms of the roof displacement of the MDOF system (target displacement) u_r is obtained by multiplying the seismic demand of the equivalent SDOF system with the transformation factor Γ (Equation 3.135, where D is denoted as u_S). It is equal to 9.7 and 8.6 cm for the modal and the uniform lateral load patterns, respectively. The seismic demand for all other response quantities can be obtained from the results of the pushover analysis corresponding to a roof displacement equal to the target displacement. The seismic demand in terms of the first storey drift ratio is 1.7% for both lateral load patterns.

A comparison of demand and capacity at the level of the MDOF system (Figure 3.20c) and the SDOF system (Figure 3.20e) shows that the capacity is somewhat larger than the demand and the structure does not reach the NC limit state under the chosen ground motion. However, it should be pointed out that the structure is very near this limit state and that serious simplifications are involved in the analysis.

With a plot such as Figure 3.20e, it is easy to determine the ground motion at which the NC limit state would be attained, that is, the capacity of the structure in terms of elastic spectral accelerations. In such a case, the target displacement is equal to the displacement at the NC limit state. The seismic demand in terms of elastic spectral acceleration is represented by the crossing point of the vertical line through the NC displacement and the radial line representing the period of the structure. This crossing point represents a point on the elastic acceleration spectrum defining the ground motion at which the NC limit state is attained. It is estimated that this structure would attain the NC limit state under a ground motion with a peak ground acceleration of 0.33 g.

QUESTION 3.1

The lateral load-resisting system of a two-storey reinforced concrete building, 30×30 m in plan, is symmetric in both horizontal directions; it consists of a spatial frame with 36 0.4 m square columns (6 lines of 6 columns each) and four walls (0.2×2.5 m, two in the x-direction and two in the y-direction). To support the vertical loading, the beams are much stiffer than the columns. Both storey levels have rigid horizontal diaphragms. The height of both storeys is 3.5 m. The effective floor weight for the calculation of the mass is 8 kN/m².

Perform a modal response spectrum seismic analysis for a ground motion with design ground acceleration $a_g = 0.25$ g and Eurocode 8 type 1 spectrum for soil type B, and behaviour factor $q = 3.6$. Determine the seismic shear forces at the base of a wall and of a frame.

Concrete grade is C20/25, with an elastic modulus of $E = 30,000,000$ kPa.

Modelling and calculation procedure: Perform a planar analysis of a 2-DOF model (with the horizontal displacements of the two floors representing the two degrees of freedom), using half of the structural system consisting of one wall (modelled as a cantilever column) and three planar frames (with 6 columns each). You may assume that the beams are infinitely rigid and neglect the shear deformations of the wall.

QUESTION 3.2

An SDOF structural system has an elastic–perfectly plastic bilinear pushover curve (zero post-yield stiffness), total weight $W = 4000$ kN, lateral strength $F_y = 500$ kN, natural period $T_n = 1.0$ s and can accommodate a five-time larger displacement than the yield displacement (ductility factor $\mu = 5$). The seismic demand is defined by the Eurocode 8 Type 1 elastic spectrum on rock (ground type A), with $a_g = 0.25$ g and 5% damping. Is the system able to survive this ground motion? What is the maximum intensity of ground motion (expressed in terms of a_g, for the same spectral shape), which the system can survive? Assume that the equal displacement rule applies.

QUESTION 3.3

A 24×48 m industrial hall is covered with a space truss roof, 28×52 m in plan (Figure 3.21), supported along a perimeter of 24.6×48.6 m on 12 elastomeric bearings, at 12.3 m centres along the two short sides in plan, or at 12.15 m centres at the two long ones. The bearings are on top of an RC frame surrounding the hall; the frame has four 0.4 m-square

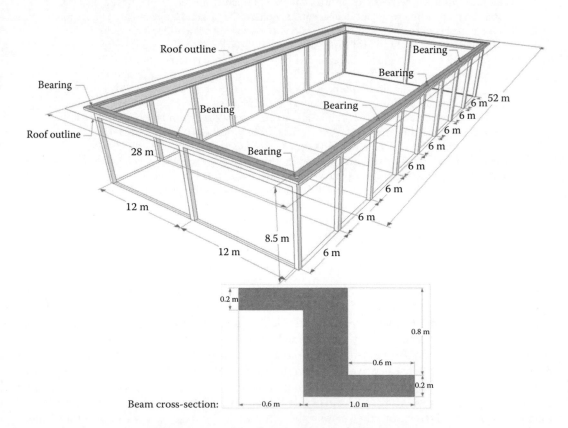

Figure 3.21 Question 3.3: Perimeter frame supporting the roof on bearings and cross-section of cap beam.

corner columns, seven intermediate ones on each long side in plan (eight spans of 6 m) and a single intermediate column on each short side (two spans of 12 m). The intermediate columns are 0.5 m-deep in the plane of the frame. The columns have a clear height of 7.5 m, are fixed at the base and have their tops connected along the perimeter by a 1 m-deep, 0.4 m-wide perimeter beam, which supports the roof on the bearings. The beam has a 0.2 m-thick top flange, protruding outwards from the web by 0.6 m. On the inside, the beam has a 0.6 m-wide, 0.2 m-thick bottom flange, which serves as the runway of an overhead crane or supports others types of equipment. The total permanent weight of this equipment (including the crane) is 200 kN. The quasi-permanent weight of the roof (including self-weight) is 1 kN/m² of plan area. The lateral stiffness of each bearing in seismic loading is 700 kN/m. Concrete grade is C20/25, with an Elastic modulus of 30,000,000 kPa.

The structure may be considered as a system with 6 DOFs:

  a. Three DOFs for the space frame roof (two horizontal displacements, u_{2X}, u_{2Y} in X and Y, and a rotation, θ_2, about the vertical).
  b. Similar DOFs, u_{1X}, u_{1Y}, θ_1 at the top of the perimeter RC frame (with the top and bottom flanges of the cap beam assumed to provide sufficient stiffness in a horizontal plane to consider the cap beam as a rigid diaphragm). The entire mass, M_1, of the perimeter beam, the upper-third of the columns and the 200 kN of equipment is lumped at the horizontal level at mid-depth of the cap beam.

The perimeter frames have in-plane lateral stiffness of:
$3(n_c + 1)(EI)_c(12k + 1)/[(3k + 1)H^3]$, where $k = (EI)_b/(EI)_c(H/L)$, with $(EI)_b$ denoting the effective rigidity of the beam, $(EI)_c$ that of an interior column for bending in the plane of the frame (strong axis), L, H, the theoretical bay length and the column height, respectively, and n_c the number of interior columns in the frame (as the outer columns of the frame have one-half the in-plane rigidity of the interior ones).

Set up the eigenvalue problem and solve it to determine the six periods, the corresponding mode shapes, the participation factors and the participating masses for excitation in directions X and Y. Note that DOFs of the roof and the top of the frame, which are in the same direction (i.e. the two translational ones in X, those in Y and the two rotations with respect to the vertical), are coupled, but, thanks to the two-way symmetry of the system, DOFs of different types are uncoupled. Calculate the modal displacements and seismic forces for the two DOFs of each horizontal direction due to an earthquake with a PGA on rock of 0.15 g, if the structure is supported on soil type C and the Type 1 Eurocode 8 spectrum applies. Combine modal results with the SRSS rule.

Chapter 4

Conceptual design of concrete buildings for earthquake resistance

4.1 PRINCIPLES OF SEISMIC DESIGN: INELASTIC RESPONSE AND DUCTILITY DEMAND

For the majority of buildings in seismic regions, the ground motion due to a nearby strong earthquake causes the most severe load among all loading conditions to which a building can possibly be subjected. On the other hand, the probability that such a ground motion will occur within the service life of the building is low. For example, we have already seen in Section 1.3 that, according to Eurocode 8, a building of ordinary importance is designed and constructed to withstand, without life-threatening local or global collapse, a 'design seismic action' associated with a recommended probability of exceedance of 10% in 50 years, which corresponds to a mean return period of 475 years. Since the probability is small, it is a common belief that for economic reasons it is not rational to build structures which would survive a strong earthquake without damage, that is, in the elastic range of behaviour.

The purpose of seismic design standards and codes, including Eurocode 8, is to ensure that in the event of a strong earthquake human lives are protected, meaning the structure does not suffer local, partial or overall collapse. It is not required that after a low-probability strong earthquake buildings remain undamaged and continue to perform their function immediately afterwards (with the exception of structures important for civil protection which should remain operational). In Eurocode 8, it is explicitly stated that the purpose of the standard is to limit (and not to prevent) damage.

However, as also pointed out in Section 1.3.2, Eurocode 8 (and other seismic design standards and codes) requires that, in case of an earthquake with larger probability of occurrence than the design seismic action, there should be no damage or associated limitations of use with costs disproportionately high compared to the overall cost of the building. This damage limitation requirement applies, according to Eurocode 8, to a seismic action with a recommended value of 10% probability of being exceeded in 10 years, which corresponds to a mean return period of 95 years.

Based on these seismic design requirements, it is expected that under the 'design seismic action' a building will deform in the inelastic range. In order to survive several inelastic deformation cycles, structures should have adequate capacity to dissipate energy without substantial reduction of the overall resistance against horizontal and vertical loading, also called ductility.

From the point of view of energy balance, input seismic energy is imparted to a structure and has to be dissipated by hysteretic behaviour and some non-yielding mechanisms, usually represented by viscous damping. Dissipation of energy by hysteretic behaviour is possible only in ductile structures, whereas it is very limited in brittle ones. Structures designed for earthquake resistance resist the seismic action thanks to a combination of strength and ductility. Assuming that the equal displacement rule applies, then, starting from a given seismic demand in terms of displacements u_{max}, it is possible to think of different structures, having the same mass and stiffness, but different idealised force–deformation relations representing different combinations of strength and ductility, as illustrated in Figure 4.1, but all of them able to accommodate the seismic demand, u_{max}. The extreme case is a structure with a very high strength, which is able to accommodate the imposed seismic demand in the elastic range. Such design is used for very important structures, like those in nuclear power plants, where damage related to inelastic deformations should be prevented even under a very strong ground motion. For more common structures, where damage is tolerated, the strength may be reduced. Nevertheless, the structure can only accommodate the seismic demand provided that it has adequate ductility capacity (i.e. capability to deform in the inelastic range).

The smaller the strength, the larger ductility is required. As elaborated in Section 4.6, Eurocode 8 leaves a certain choice of the ductility level to the designer. With increasing ductility capacity, various design requirements that have to be fulfilled increase in severity, but the required strength for a given level of seismic action decreases. It should be noted that structural damage is related to ductility. Therefore, design for high ductility may result in extensive structural damage under strong ground motions. Non-structural damage is related either to accelerations or to relative displacements, depending on the non-structural element. Since accelerations are related to forces, a design for high ductility is generally beneficial for acceleration-sensitive elements. As shown in Figure 4.1, displacements do not depend on the variant of design. It should be noted, however, that Figure 4.1 represents an idealised example. In practice, stiffness and strength are related to a certain extent: larger strength usually means larger stiffness and smaller displacements, which means less damage to deformation-sensitive non-structural elements.

Structures important for civil protection should remain operational after strong earthquakes too. In order to fulfil this requirement, only minor damage is allowed. So, they should be designed with an importance factor greater than 1.0, resulting in a larger strength than ordinary structures.

In seismic design standards and codes, the decrease of the seismic strength demand on account of inelastic action is achieved by using reduction factors (behaviour factor q in Eurocode 8, see Section 3.2) in conjunction with linear analysis. The use of reduction factors allows an approximate consideration of the inelastic behaviour of the structure in linear elastic analysis and has been widely adopted in seismic design standards and codes worldwide. However, for a more realistic estimate of the structural response during strong earthquakes, a non-linear analysis is needed.

The reader should be cautioned that the design seismic action is not intended to represent the strongest ground motion that can possibly occur at the site of the structure. In order to ensure

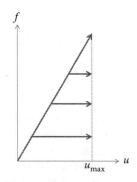

Figure 4.1 Structure resists seismic action through different combinations of strength and ductility.

that the structure does not collapse under a ground motion larger than the design earthquake, certain measures are taken to enhance its global ductility and eliminate pre-emptively the more dangerous collapse mechanisms. One of those measures is capacity design, described in detail in Section 4.5 and elaborated further in Sections 5.4.1, 5.5.1, 5.6.2.1 and 6.3.2.

4.2 GENERAL PRINCIPLES OF CONCEPTUAL SEISMIC DESIGN

4.2.1 The importance of conceptual design

It is very often said that in any human process the sooner an error is made, the greater is its potential for detrimental consequences. Translating this into a construction process, this means that an error made at the conceptual design phase, which precedes all other activities leading to the completion of the construction of a facility, may have severe future consequences. This is particularly true for seismic design. In fact, earthquakes being rare, they may spare a certain constructed facility for long; but when one strikes, it will unveil all defects that the structure may hide, particularly those due to inadequate seismic conceptual design. Such defects will be exposed in a dramatic fashion within a matter of a few seconds. This is why Eurocode 8 gives great importance to conceptual design aspects, as there is plenty of evidence from damage observation after earthquakes that simple and regular buildings tend to behave much better than irregular ones. Such favourable features should be incorporated at the earliest stages of design, when the interaction between the architect and the structural engineer is close. Only if they both understand the design requirements set out by the other, is it possible to achieve a solution that satisfies these requirements in a balanced and cost-effective manner. It is at this stage that a structural system appropriate for the specific conditions of the building and of the site needs to be chosen, so that architectural features are incorporated naturally into the design from the outset.

To support the conceptual design of buildings, Eurocode 8 lists a set of guiding principles that should be applied by the structural designer as a first step to achieve a building with good seismic response. The following subsections elaborate these principles.

4.2.2 Structural simplicity

Structural simplicity is characterised by the existence of clear and direct paths for the transmission of the inertial forces produced by the seismic excitation to the foundations of the building. It is an important objective, because the seismic response of simple structures is inherently less uncertain. Moreover, because the modelling, analysis, dimensioning and detailing of simple structures are subject to much less uncertainty, the prediction of the seismic behaviour thereof is more easily achievable. Hence the result is more reliable structures.

4.2.3 Uniformity, symmetry and redundancy

Uniformity in plan is characterised by an even distribution of the structural elements, allowing short and direct transmission of the inertial forces produced in the distributed masses of the building. In some cases, plan-wise uniformity may be realised by subdividing the entire building by seismic joints into dynamically independent units. Short and direct paths for the transmission of the inertial forces are achieved more easily if the distribution of stiffness and resistance closely resembles that of masses.

Another favourable feature is *symmetry*. If the building configuration is symmetrical or quasi-symmetrical, a symmetrical, well-distributed in-plan layout of structural elements is appropriate to achieve uniformity. Symmetry should be sought both in what concerns the plan-wise distribution of stiffness and strength, but also that of mass distribution, to avoid eccentricities that entail torsional response (twisting around a vertical axis), which tends to increase the horizontal displacements in some parts of the periphery of the building plan (the 'soft' or 'flexible' side) and hence is unfavourable.

The use of many evenly distributed structural elements increases the *redundancy* of the structure, that is, the structure becomes more reliable to resist the earthquake effects by accommodating the loss of some structural elements before becoming unstable. Redundancy also allows a more favourable distribution of the action effects in the non-linear range and widespread energy dissipation across the entire structure. Figure 4.2 illustrates schematically the concepts of uniformity, symmetry and redundancy of the in-plan structural layout.

Uniformity along the height of the building is possibly one of the most important features that should be pursued at the conceptual design stage. This is so because uniformity in height tends to eliminate transition zones up the building where concentrations of stresses or large ductility demands may prematurely cause collapse.

A frequent (and dangerous) situation of non-uniformity in height corresponds to the existence of a so-called *soft storey* at the ground floor of a building. This may occur in several cases:

- When the first floor height is significantly taller than those above, hence its stiffness is significantly smaller. This is particularly so in framed structures, as the lateral stiffness

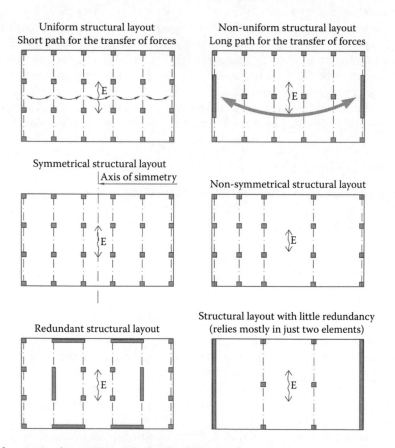

Figure 4.2 Uniformity in plan, symmetry and redundancy.

of columns varies with the cube of the clear height: for example, a 50% larger inter-storey height at the ground floor with regard to the storeys above results in that storey being three times more flexible than the others.

- When some of the vertical structural elements are discontinued from the second storey down, in view of obtaining an architecturally more open ground floor.
- Finally, although not strictly a structural feature, when stiff and strong *non-structural elements* (normally facade elements, but also partitions) are placed above an open ground floor. In view of the relevance of this situation, Eurocode 8 has some specific clauses intended to control possible negative effects of infills in masonry-infilled frames.

In all these cases, the horizontal stiffness (and normally the horizontal resistance as well) of the ground storey is much smaller than that of the storeys above, leading to the concentration of horizontal displacement in the ground storey (inter-storey drifts are very large in the first floor and very small in the ones above, see Figure 2.9a). As a result, collapse of the first storey is very likely (e.g. see Figures 2.10 and 2.11), due to the very high deformation demands and because the load-bearing capacity of the storey in such a deformed shape is very much decreased by second-order (or $P - \Delta$) effects in comparison with its original (undeformed) condition with vertical columns.

Note that a soft or weak storey may only exist if the lateral resisting system is mainly composed of a frame structure: structural walls, continuous along the full height of the building, normally preclude soft storey behaviour (see Figures 2.9d and 2.9e).

A soft storey at ground level is certainly the most severe situation, because therein the inter-storey seismic shear force is at its maximum and that storey supports the whole building. However, a soft storey at any other height of a building, for instance due to interruption of a shear wall at an intermediate height, may be extremely detrimental to the seismic response as well (see Figure 2.15 right).

4.2.4 Bi-directional resistance and stiffness

Horizontal seismic motion is by nature a bi-directional phenomenon and thus the building structure must be able to resist horizontal actions in any direction. Hence, the structural elements should be arranged in an in-plan orthogonal structural pattern, so that the building structure is able to direct the seismic action to these main structural directions. Having similar resistance and stiffness characteristics in both main directions is also a desirable feature, enabling essentially in-plane seismic response with little orthogonal response, that is, for a given direction of the seismic action, the response of the structure will be in that same direction; the structure responds in an uncoupled manner.

The choice of the stiffness characteristics of the structure should also aim at limiting the development of excessive displacements that might lead to instabilities due to second-order effects or lead to excessive damage of non-structural elements.

4.2.5 Torsional resistance and stiffness

In addition to lateral resistance and stiffness, building structures should possess adequate torsional resistance and stiffness, in order to limit the development of twisting motions around a vertical axis, which tend to stress the different structural elements in a non-uniform way. In this respect, arrangements in which the main elements resisting the seismic action are distributed close to the periphery of the building present clear advantages, since this increases the torsional stiffness of the structure (see Example 4.1).

4.2.6 Diaphragmatic behaviour at storey level

In buildings, floors play a very important role in the overall seismic behaviour of the structure. They act as horizontal diaphragms that collect and transmit the inertial forces to the vertical structural systems and ensure that those systems act together in resisting the horizontal seismic action.

The action of floors as diaphragms is especially relevant in cases of complex and non-uniform layouts of vertical structural systems, or when systems with different horizontal deformability characteristics are used together (e.g. in dual or mixed systems) in the same direction.

In-plane *stiffness* and in-plane *resistance* of floors are different issues, but both are important for the seismic response of the building.

Diaphragms should have sufficient in-plane stiffness for the distribution of horizontal inertial forces to the vertical structural systems, mobilising them with fairly similar displacements. If the in-plane stiffness of the floor system is small (in relation to the stiffness of the vertical systems), it will be unable to fulfil such an objective: similar structural systems placed at different positions in plan may suffer different horizontal displacements (see Figure 4.3). Horizontal displacements that differ across the building result in larger maximum displacements, which by default lead to a worse or less-controlled response.

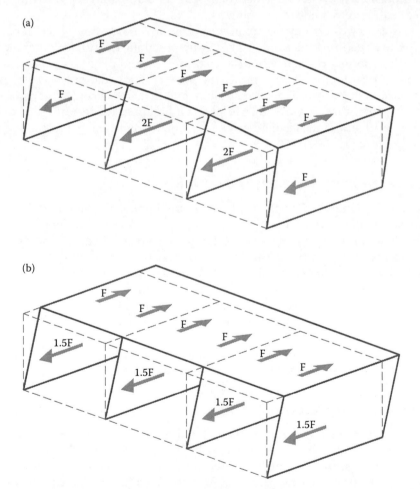

Figure 4.3 Effect of diaphragm stiffness. (a) Flexible diaphragm: uneven distribution of forces among the vertical elements. (b) Rigid diaphragm: uniform distribution of forces among the vertical elements.

In the past, when structural analysis capabilities were much more limited, the objective of having a so-called rigid diaphragm solution was sought, not only in view of its beneficial effects, but also and possibly primarily for the sake of simplifying the structural modelling and analysis. Nowadays, it is possible to model adequately the in-plane deformability of floor systems, hence the 'rigid diaphragm' assumption is not required anymore; yet it still is considered useful for simplifying the analysis. In any case, the key point is that the modelling assumptions in this respect should be adequate for the structural solution at hand, so that the analysis captures with accuracy the way in which the inertial forces are transferred to the vertical resisting systems. The need to include diaphragm deformability is particularly important in non-compact or very elongated in-plan shapes, or when there are significant changes in stiffness or offsets of vertical elements above and below the diaphragm (see Figure 4.4).

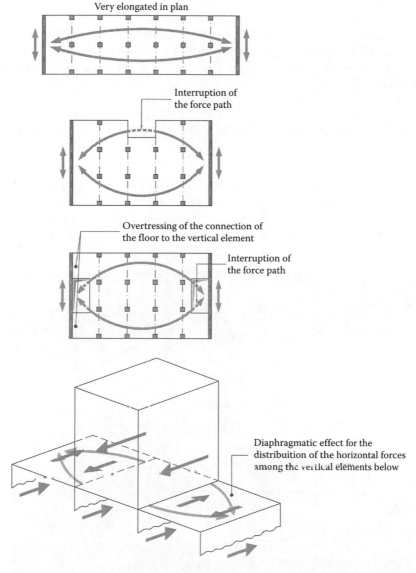

Figure 4.4 Irregular/unfavourable plan configurations for effective diaphragmatic action. Importance of diaphragmatic effect in case of vertical setbacks.

Another important aspect regarding diaphragmatic action is the in-plan resistance of floor systems, as well as the resistance of their connection to the vertical structural systems. In this respect, it must be emphasised that the resistance of diaphragms may be reduced significantly by large floor openings, especially if the latter are located in the vicinity of main vertical structural elements. This hinders the effective connection between the vertical and horizontal structure and limits the capability of transmitting the horizontal forces to the vertical elements, as intended (see Figure 4.4).

4.2.7 Adequate foundation

With regard to the seismic action, the design and construction of the foundation and of the connection to the superstructure should ensure that the whole building is subjected to a uniform seismic excitation. For structures composed of a discrete number of structural walls, likely to differ in width and stiffness, a rigid, box-type or cellular foundation, comprising a foundation slab and a cover slab, should generally be chosen (see Section 6.3.1). For buildings with individual foundation elements (footings or piles), Eurocode 8 requires to use a foundation slab or tie-beams between these elements in both main directions (Section 6.3.1 and Figures 4.11 and 6.12).

4.3 REGULARITY AND IRREGULARITY OF BUILDING STRUCTURES

4.3.1 Introduction

Despite the general principles of good conceptual design presented above, a precise definition of what is a regular or irregular structure in the context of the seismic response of buildings has eluded many attempts to achieve it. It is intuitive to classify the building shown in Figure 4.5 as irregular. However, there are so many variables and structural characteristics

Figure 4.5 Example of an irregular building.

that may (or should) be considered to establish, in abstract terms, the definition of regularity or irregularity of a building that, finally, the classification in each specific case has to rely mostly on the engineering experience and judgement of the structural designer.

In a certain way, Part 1 of Eurocode 8 recognises this difficulty and does not attempt to establish very strict rules for the distinction between regular and irregular buildings. It provides a relatively loose set of characteristics that a building should possess to be classified as regular. Such a classification essentially has the purpose of establishing some distinction in what concerns the more- or less-simplified structural model and the method of analysis to be used, as well as the value of the behaviour factor.

With this approach, Eurocode 8 does not forbid the design and construction of non-regular structures, but attempts to give incentive to choose regular structures, by making them easier to design and more economical, thanks to higher values of the behaviour factor. However, in most cases it is not possible to deter the architect from seeking original forms. It is in these cases that the interaction between the architect and the structural designer from the very beginning of the design process is essential. For this interaction to be successful, the structural engineer should be open-minded with regard to the architect's intention but, at the same time, be able to convey to the architect the structural implications of the intended irregular forms. Likewise, the architect should be able to understand these implications and be capable of accommodating into his/her architectural form the additional structural needs (e.g. the need for additional members or increased cross-sectional dimensions).

Out of the many possible irregularities in building structures, the final structural design should, in absolute terms, avoid the two most dangerous irregular situations: soft storeys and extremely flexible structures in torsion. Both cases correspond to extremely irregular stiffness distributions: the former regarding the distribution in height with a very weak storey and the latter regarding the distribution in plan, with the concentration of the stiffness at the centre or at a corner.

As in most other modern seismic design codes, in Part 1 of Eurocode 8 the concept of building regularity is presented separately for *regularity in plan* and *regularity in elevation*. Moreover, regularity in elevation is considered separately in the two main orthogonal directions in which the horizontal components of the seismic action are applied, meaning that the structural system may be characterised as regular in one of these two horizontal directions, but not in the other. Nonetheless, a building assumes a single characterisation (the most demanding) for regularity in plan, independent of direction.

In order to reduce stresses due to deformations associated with volumetric changes (thermal expansion and/or concrete shrinkage), buildings that are long in plan often have their structure divided by means of expansion joints into parts that can be considered as separate and structurally independent above the level of the foundation. The same practice is recommended in buildings with a plan shape consisting of several (close-to-) rectangular parts (L-, C-, H-, I- or X-shaped plans), for reasons of clarity and predictability of their seismic response (as well as for modelling and analysis simplification). Although this recommendation for compact in-plan shapes still holds nowadays, it must be recognised that its roots were laid in times when modelling and analysis capabilities were modest. At present, this objective should be somewhat balanced with the fact that too many independent units in the same building may be inconvenient, not only for reasons directly related to the seismic response (pounding between these units) but also for other reasons (maintenance of expansion joints and potential water leakage). The parts in which the structure is divided through such joints are considered as 'dynamically independent'. Structural regularity is defined and checked at the level of each individual 'dynamically independent' part of the building structure, regardless of whether these parts are analysed separately or together (the latter might be the case if they share a common foundation and are modelled together, or if the designer considers a single analysis as convenient for comparing the relative displacements of adjacent units to the width of the joint between them).

Eurocode 8 introduces primarily qualitative criteria for regularity, which can mostly be checked at the preliminary design stage by inspection, or through simple calculations, without doing the analysis of the building. This makes sense, as the main purpose of the regularity classification is to determine what type of modelling and linear analysis may be used for the design:

- 3D using a spatial model, or 2D using two separate planar models, depending on the regularity in plan
- Static, with the lateral force method, or modal response spectrum analysis, depending on the regularity in elevation

Moreover, regularity affects the value of the behaviour factor q that determines the design spectrum used in linear analysis.

It is generally more difficult to verify without analyses that a building is regular in plan than in elevation (unless this is clear by inspection). So, in case of doubt, the designer may very well presume that the building is irregular in plan and bears the very light penalties foreseen in Eurocode 8 (cf. Section 4.3.3), instead of carrying out the analyses necessary for the verification of regularity.

Irregularity in plan or elevation is the subject of Examples 4.2 to 4.4, where the implications for design and the suitability of the structural system for earthquake resistance are also discussed.

4.3.2 Criteria for irregularity or regularity in plan

For the structure of a building to be considered as *regular in plan*, six conditions have to be fulfilled at all storey levels:

CONDITION 1

The distribution in plan of the lateral stiffness and mass are *approximately symmetrical* with respect to two orthogonal horizontal axes. Normally, the horizontal components of the seismic action are applied along these two axes. As absolute symmetry is not required, it is up to the designer to judge whether this condition is met or not.

CONDITION 2

EN1998-1 states that 'the plan configuration shall be compact, that is, each floor shall be delimited by a polygonal convex line'. There is some tolerance with regard to this requirement: it is further stated that, if there are in-plan setbacks (re-entrant corners or edge recesses), we may still consider the structure as regular in plan under the following conditions:

- These setbacks do not affect the floor in-plane stiffness.
- For each setback, the area between the outline of the floor and a convex polygonal line enveloping the floor does not exceed 5% of the floor area.

In Figure 4.6, various plan configurations are presented illustrating the application of Condition 2 for in-plan regularity with regard to compactness; they show that edge recesses are more severely penalised by this condition in comparison with re-entrant corners. This is so because edge recesses disturb the horizontal force paths of the diaphragmatic effect more than re-entrant corners.

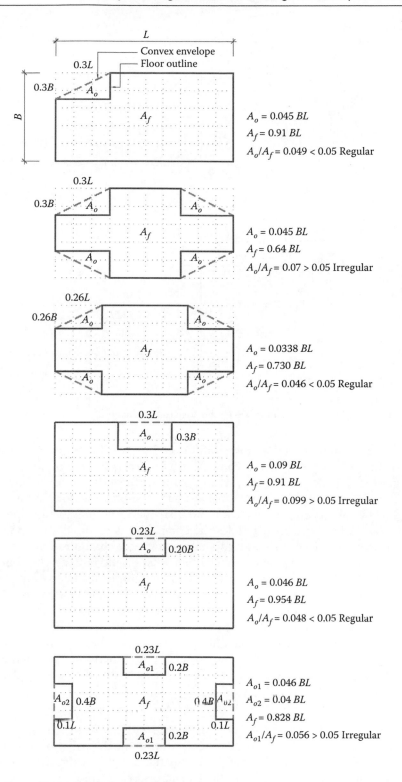

$A_o = 0.045\,BL$
$A_f = 0.91\,BL$
$A_o/A_f = 0.049 < 0.05$ Regular

$A_o = 0.045\,BL$
$A_f = 0.64\,BL$
$A_o/A_f = 0.07 > 0.05$ Irregular

$A_o = 0.0338\,BL$
$A_f = 0.730\,BL$
$A_o/A_f = 0.046 < 0.05$ Regular

$A_o = 0.09\,BL$
$A_f = 0.91\,BL$
$A_o/A_f = 0.099 > 0.05$ Irregular

$A_o = 0.046\,BL$
$A_f = 0.954\,BL$
$A_o/A_f = 0.048 < 0.05$ Regular

$A_{o1} = 0.046\,BL$
$A_{o2} = 0.04\,BL$
$A_f = 0.828\,BL$
$A_{o1}/A_f = 0.056 > 0.05$ Irregular

Figure 4.6 Regular and irregular plan configurations.

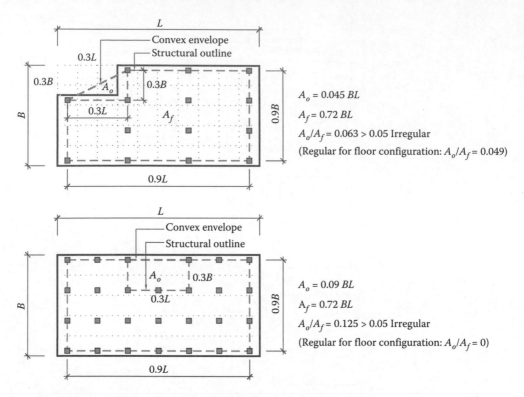

$A_o = 0.045 BL$

$A_f = 0.72 BL$

$A_o/A_f = 0.063 > 0.05$ Irregular

(Regular for floor configuration: $A_o/A_f = 0.049$)

$A_o = 0.09 BL$

$A_f = 0.72 BL$

$A_o/A_f = 0.125 > 0.05$ Irregular

(Regular for floor configuration: $A_o/A_f = 0$)

Figure 4.7 Irregular in-plan configurations due to the structural outline.

It should be also emphasised that this condition of Part 1 of Eurocode 8 includes two conditions to be checked, meaning that, even in cases where the 5% ratio of areas is fulfilled, the condition that the setbacks do not affect the in-plane stiffness may be controlling. An example of such a case is presented in Figure 4.7. The issue of in-plane stiffness is also addressed in Condition 3 for regularity in plan, related to floor stiffness and presented below.

Finally, it should be noted that, besides the compactness of the floor configuration, the outline of the structure in plan should also have a compact configuration. This means that the structure, as defined in plan by its vertical elements, should have an envelope (outline) of its exterior elements defining a convex polygonal line. Similar to the in-plane configuration, there is some tolerance with regard to this convexity requirement. So, it is acceptable that for each setback (of the structure) the area between the outline of the structure and the convex polygonal line enveloping does not exceed 5% of the outline area. The check for the structural outline compactness may be the conditioning factor for the regularity classification, as is illustrated in Figure 4.7, which presents two examples where the compactness check for the floor configuration is satisfied, whereas the structural outline does not satisfy it.

CONDITION 3

It should be possible to consider the floors as *rigid diaphragms*, in the sense that their in-plane stiffness is sufficiently large, so that the floor in-plane deformation due to the seismic action is negligible compared to the inter-storey drifts, and it has a minor effect on the distribution of seismic shears among the vertical structural elements.

Conventionally, a rigid diaphragm is defined as one in which, when it is modelled with its actual in-plane flexibility, its horizontal displacements due to the seismic action do not exceed anywhere by more than 10% the corresponding absolute horizontal displacements that would result from a rigid diaphragm assumption. Essentially, this conventional definition of a rigid diaphragm in the context of checking the in-plan regularity of building structures indicates that the use of a simpler modelling approach (i.e. a rigid diaphragm) is acceptable, provided that it does not induce differences in the distribution of the seismic load into the various lateral resisting systems larger than 10%. Nonetheless, it is neither required nor expected that fulfilment of this conventional definition is computationally checked, because this would require the analysis of the structure with the actual in-plane flexibility of the floors, making the rigid diaphragm assumption of no practical interest.

It is up to the designer to decide whether the rigid diaphragm assumption is justified, but it may be noted that, for instance, a solid reinforced concrete slab (or cast-in-place topping connected to a precast floor or roof through a clean, rough interface or shear connectors) may be considered as a rigid diaphragm, if its thickness and reinforcement (in both horizontal directions) are above the minimum thickness of 70 mm and the minimum slab reinforcement of Eurocode 2, as required in Part 1 of Eurocode 8 for concrete diaphragms.

It should also be emphasised that for a diaphragm to be considered rigid, it should also be free of large openings, especially in the vicinity of the main vertical structural elements. Anyway, if the designer does not feel confident about the rigid diaphragm assumption due to the large size of such openings and/or due to the small thickness of the concrete slab, then he/she may check it by applying the above conventional definition of a rigid diaphragm.

CONDITION 4

The aspect ratio or *slenderness of the floor plan*, $\lambda = L_{max}/L_{min}$, where L_{max} and L_{min} are, respectively, the larger and smaller in-plan dimension of the floor measured in any two orthogonal directions, should be no more than 4. This limit is complementary to Condition 3, which requires the in-plane rigidity of the diaphragm, and is intended to ensure that, independently of the result of Condition 3, in case of very slender floor plans, its deformability is explicitly considered in the structural model and conditions the distribution of the seismic forces among the vertical structural elements.

CONDITION 5

In approximately symmetrical buildings, according to Condition 1 above, in each of the two orthogonal horizontal directions, x and y, the 'static' eccentricity, e_0, between the floor centre of mass (C_M) and the storey centre of lateral stiffness (C_K), as illustrated in Figure 4.8, is not greater than 30% of the corresponding storey *torsional radius r*, that is:

$$e_{0x} \leq 0.3 \, r_x; \, e_{0y} \leq 0.3 \, r_y \tag{4.1}$$

The *torsional radius* r_x in Equation 4.1 is defined as the square root of the ratio of the torsional stiffness K_θ of the storey with respect to the centre of lateral stiffness, to the storey lateral stiffness K_y in direction y (orthogonal to x), as depicted in Figure 4.8; similarly for the torsional radius r_y:

$$r_x = \sqrt{K_\theta/K_y}, \quad r_y = \sqrt{K_\theta/K_x} \tag{4.2}$$

For single-storey buildings, Part 1 of Eurocode 8 allows to determine the centre of lateral stiffness and the torsional radius by considering the translational stiffness represented by

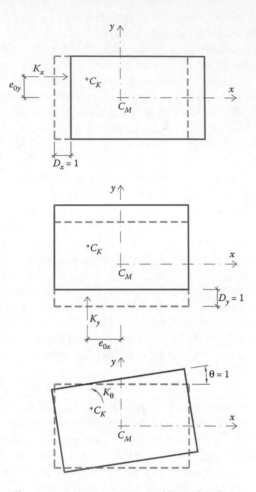

Figure 4.8 Centre of lateral stiffness, static eccentricities and lateral and torsional stiffness.

the moments of inertia of the cross-sections of the vertical elements, neglecting the effect of beams. Hence the position of the centre of lateral stiffness is given by:

$$x_{CK} = \frac{\sum x\,EI_y}{\sum EI_y} \quad y_{CK} = \frac{\sum y\,EI_x}{\sum EI_x} \tag{4.3}$$

with x, y defining the position of the various vertical elements, measured from the origin of any arbitrary plan reference (and x_{CK} and y_{CK} being also referred to such reference). In this context it is worth noting that if the origin of the plan reference is set at the centre of mass (C_M) of the floor, then the values computed with Equation 4.3 are the static eccentricities, e_{0x} and e_{0y}, referred to in Equation 4.1. It follows that the torsional radii in the two orthogonal directions are given by:

$$r_x = \sqrt{\frac{\sum((x - x_{CK})^2\,EI_y + (y - y_{CK})^2\,EI_x)}{\sum EI_y}}, \quad r_y = \sqrt{\frac{\sum((x - x_{CK})^2\,EI_y + (y - y_{CK})^2\,EI_x)}{\sum EI_x}}$$

$$\tag{4.4}$$

In Equations 4.3 and 4.4, EI_x and EI_y denote the section flexural rigidities for bending within a vertical plane parallel to horizontal directions x or y, respectively (i.e. about an axis parallel to axis y or x, respectively).

To illustrate the meaning of the condition expressed by Equation 4.1, in Figure 4.9 three schematic examples of the same floor configuration with different distributions of the lateral stiffness are presented, and its regularity is assessed in accordance to this condition. All cases are symmetric in relation to the x axis, and we observe only the situation regarding the asymmetry of stiffness in the y direction (i.e. only the ratio e_{0x}/r_x).

1. In the first case, the system at the extreme left has a lateral stiffness twice that of the other three systems. The eccentricity is $e_{0x} = 0.1L$ and the torsional radius $r_x = 0.528L$, leading to: $e_{0x}/r_x = 0.189 < 0.3$, which means that, in this respect, the *building is regular*.
2. The second case is similar to the first one, but the stiffness asymmetry in the y direction is larger, since the stiffness of the extreme left system is three times that of the others.

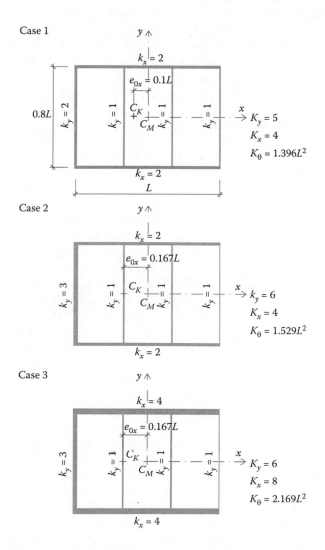

Figure 4.9 Three schematic distributions of stiffness and its effect on the ratio e_{0x}/r_x.

In this case we have a larger eccentricity, $e_{0x} = 0.167L$, and a slightly smaller torsional radius (essentially due to the increase of the translational stiffness): $r_x = 0.505L$, leading to $e_{0x}/r_x = 0.33 > 0.3$, which means that the building is *irregular in plan*.

3. Finally, the *third case* illustrates that it is possible to counteract this situation of irregularity in plan by increasing the stiffness of the systems in the orthogonal direction. In fact, doubling the stiffness of the orthogonal systems significantly increases the torsional stiffness, leading to a larger torsional radius $r_x = 0.601L$. Since the eccentricity remained unchanged at $e_{0x} = 0.167L$, the controlling ratio is $e_{0x}/r_x = 0.277 < 0.3$, which means that, in this respect, the *building is regular* again.

The examples presented refer to a one-storey building, where it is possible to define precisely the centre of lateral stiffness. In these cases, the centre of stiffness is the point where the application of a horizontal force produces only translation without any rotation around a vertical axis (see Figure 4.8). For multi-storey buildings, the centre of lateral stiffness is defined as the point in plan with the property that any set of horizontal forces applied at floor levels through that point produce only translation of the individual storeys, without any rotation with respect to the vertical axis. Conversely, any set of storey torques (i.e. of moments with respect to the vertical axis, z) produce only rotation of the floors about the vertical axis that passes through the centre of lateral stiffness, without horizontal displacement of that point in x and y at any storey. If such a point exists, the torsional radius, r, defined as the square root of the ratio of torsional stiffness with respect to the centre of lateral stiffness to the lateral stiffness in one horizontal direction, is unique and well-defined. It can be computed by applying to the building a set of storey torques, T_i, and, separately, a set of storey forces in the horizontal direction of interest but through the (unique) centre of lateral stiffness, with magnitudes proportional to those of the corresponding storey torques, $F_i = T_i/c$, with the lever arm, c, being arbitrary. The *torsional stiffness* is defined then as the ratio of: (a) the storey torsional moment (the torsional moment in this case being the sum of all storey torques applied above and at storey i) to (b) the corresponding storey twist (the twist being taken with respect to the base of the building). Similarly, the *lateral stiffness* is defined as the ratio of (a) the storey shear to (b) the corresponding horizontal displacement of the storey with respect to the base. Irrespective of the exact distribution of storey torques, T_i, and of storey forces, $F_i = T_i/c$, a unique value of r is computed.

Unfortunately, as already mentioned, the centre of lateral stiffness with this general definition, and with it the torsional radius, r, are unique and independent of the lateral loading only in single-storey buildings (because, by nature, in this case the lateral load is composed by just one force). In buildings of two storeys or more the result of such definition is not unique and depends on the distribution of lateral loading with height. This is especially so if the structural system consists of (planar) sub-systems, which develop different height-wise patterns of horizontal displacements under the same set of storey forces. In this context, it is appropriate to recall that moment frames exhibit a shear-beam type pattern of horizontal displacements, whereas walls behave more like vertical cantilevers. Part 1 of Eurocode 8 recognises this fact, stating that 'In multi-storey buildings only approximate definitions of the centre of stiffness and of the torsional radius are possible'. In spite of this, for buildings with all lateral-load-resisting systems running from the foundation to the top and having similar deformation patterns under lateral loads, it accepts approximate definitions of the centre of stiffness and torsional radius, r. Moreover, Eurocode 8 accepts that 'In frames and in systems of slender walls with prevailing flexural deformations, the position of the centres of stiffness and the torsional radius of all storeys may be calculated as those of the moments of inertia of the cross-sections of the vertical elements', meaning that, in such cases, Equations 4.3 and 4.4 may be applied independently at the different storeys.

For other cases, Eurocode 8 just indicates that the National Annex may provide guidance on the definition of the centre of stiffness and the torsional radius. For those cases, a procedure to determine these quantities which is fairly satisfactory is the following:

1. A set of storey horizontal forces, F_i, is selected as that of the (equivalent) lateral forces in the lateral force method of analysis (i.e. proportional to the product of storey mass, m_i, times its elevation from the base, z_i) and, as a *first step*, the building structure is analysed under a set of storey torques proportional to these forces: $F_i = T_i$ (with the lever arm taken equal to $c = 1$).

2. The centres of twist at each floor due to these storey torques are geometrically determined and the horizontal projection of the centre of twist at the elevation of 80% of the total height of the building H (i.e. $z = 0.8H$) may be considered as the centre of lateral stiffness of the whole building. In fact, the application of the set of horizontal forces, F_i, at this point at the different floor levels will produce translation of the individual storeys with minimum (in a least-squares-sense) rotation with respect to the vertical axis (twist).

3. Once this point is determined, a *second analysis* is performed, for each one of the two orthogonal horizontal directions, this time under the set of storey horizontal forces, $F_{x,i}$ (or $F_{y,i}$), numerically equal to T_i of the first analysis and applied through the centre of lateral stiffness determined for the building as a whole.

4. Then, for the calculation of the torsional radius r_x, the torsional stiffness and the lateral stiffness are computed as follows:
   a. Torsional stiffness: ratio of the total applied torsional moment, $T_{tot} = \Sigma_i T_i$ to the resulting rotation, $\theta_{0.8H}$, at $z = 0.8H$
   b. Lateral stiffness: ratio of the total applied shear (in direction y), $F_{y,tot} = \Sigma_i F_{y,i}$ to the resulting displacement $\delta_{y,0.8H}$, at $z = 0.8H$

As a matter of fact, with length units taken as those of the unit lever arm ($c = 1$), r_x is given by:

$$r_x = \sqrt{\delta_{y,0.8H}/\theta_{0.8H}} \tag{4.5}$$

Likewise, the other torsional radius r_y is calculated considering the displacement $\delta_{x,0.8H}$ resulting from the application of the horizontal forces in the x direction:

$$r_y = \sqrt{\delta_{x,0.8H}/\theta_{0.8H}} \tag{4.6}$$

It is worth noting that the results of the analysis performed for the seismic design of the structure cannot be used directly to determine the value of r_x (and r_y) according to the procedure outlined in the previous steps, because in the analyses for design, the horizontal forces, $F_{y,i}$ (and $F_{x,i}$), are applied at the storey centre of mass, whereas for the determination of $\delta_{y,0.8H}$ (and $\delta_{x,0.8H}$), the storey horizontal forces should be applied through the centre of lateral stiffness determined for the building as a whole. Therefore, two different sets of analysis (with forces applied at the centre of mass or forces applied through the stiffness centre) have to be performed, unless the stiffness centre coincides with the centre of mass at all storeys.

CONDITION 6

The *torsional radius* of the storey in each of the two orthogonal horizontal directions, x and y, of near-symmetry according to Condition 1 above is not less than the *radius of gyration* l_s of the floor mass:

$$r_x \geq l_s; r_y \geq l_s; \tag{4.7}$$

The *radius of gyration* of the floor mass in plan is defined as the square root of the ratio of the polar moment of inertia of the floor mass in plan with respect to the centre of mass of the floor to the floor mass. If the mass is uniformly distributed over a rectangular floor area with dimensions L and B (that include the floor area outside the outline of the vertical elements of the structural system), the radius of gyration l_s is:

$$l_s = \sqrt{(L^2 + B^2)/12} \qquad (4.8)$$

The value of l_s is mostly controlled by the longer dimension L of the floor shape. In particular, for rectangular shapes the value of l_s is:

- $l_s = 0.408\ L$ for $B/L = 1$ (square plan)
- $l_s = 0.370\ L$ for $B/L = 0.8$
- $l_s = 0.323\ L$ for $B/L = 0.5$
- $l_s = 0.298\ L$ for $B/L = 0.25$ (slenderness limit in Condition 4)

The factor on L does not change very much across quite different slenderness values (in fact for a rectangular shape, l_s tends asymptotically to $l_s = 0.289L$ as the slenderness increases).

The condition expressed by Equation 4.7 ensures that the fundamental frequency of the (primarily) torsional mode about the vertical axis z is higher than the fundamental (primarily) translational modes in each of the two horizontal directions, x and y (see Example 4.5 for a proof of this statement in an idealised, single-storey system) and prevents strong coupling of the torsional and translational response, which is considered uncontrollable and potentially very dangerous.

As a matter of fact, since the radius of gyration l_s is defined with respect to the centre of mass of the floor in plan, the 'torsional radii' r_x and r_y that should be used in Equation 4.7 to ensure this intended ranking of the frequencies of the three modes mentioned earlier are those defined with respect to the storey centre of mass, r_{mx} and r_{my}. These are related to the 'torsional radii' r_x and r_y defined with respect to the storey centre of lateral stiffness as:

$$r_{mx} = \sqrt{r_x^2 + e_{0x}^2}, \quad r_{my} = \sqrt{r_y^2 + e_{0y}^2} \qquad (4.9)$$

The greater the 'static' eccentricities e_{0x} and e_{0y} between the centres of mass C_M and stiffness C_K, the larger the margin provided by Equation 4.7 against a torsional mode becoming predominant.

Observing the value of $l_s = 0.37L$ for $B/L = 0.8$, which is the slenderness of the schematic cases presented in Figure 4.9, and correcting the values of the torsional radius according to Equation 4.9 we obtain:

1. First case: $r_{mx} = 0.538L$ leading to $r_{mx}/l_s = 1.45$
2. Second case: $r_{mx} = 0.532L$ leading to $r_{mx}/l_s = 1.44$
3. Third case: $r_{mx} = 0.624L$ leading to $r_{mx}/l_s = 1.69$

From these results it is clear that all those cases pass easily the condition expressed by Equation 4.7.

If the elements of the lateral-load-resisting system are distributed in plan as uniformly as the mass, then the condition of Equation 4.7 is satisfied (be it marginally) and does not need to be checked explicitly (see Example 4.6 for the proof), whereas if the main lateral-load-resisting elements, such as strong walls or frames, are concentrated near the plan centre, this condition may not be met and Equation 4.7 needs to be checked.

Finally, it is clear that if the natural frequencies of the structure are determined with a modal analysis, their values may be used directly to determine whether or not Condition 6 is satisfied for the building as a whole. In fact, if the frequency of the first (primarily) torsional mode of vibration is higher than the frequencies of the (primarily) translational modes in the two horizontal directions, x and y, then this condition for regularity in plan may be considered as satisfied.

In Example 4.7, the torsional radii of a frame building, the radius of gyration and the eccentricities of the centre of stiffness with respect to the mass centre are determined and the implications for the regularity in plan and for earthquake resistance are discussed. These properties are determined in Section 7.3.3 for the example building of Chapter 7, and used to characterise the regularity in plan and establish the q-factor value of that building.

4.3.3 Implications of irregularity in plan

4.3.3.1 Implications of regularity for the analysis model

For buildings regular in plan, the analysis for each one of the two horizontal components of the seismic action may be carried out using an independent 2D (planar) model for each one of the two horizontal directions of (near-) symmetry, x and y (see Section 3.1.5.4). The 2D model for each direction is geometrically similar to the 'pseudo-3D model' described in Section 3.1.10 with reference to Figure 3.12: the structure is considered to consist of a number of plane frames and/or walls (some of which may actually belong in a plane frame together with co-planar beams and columns). However, unlike the 'pseudo-3D model' of Section 3.1.10, where twisting of each rigid floor diaphragm around a vertical axis is considered, all parallel 2D frames or walls are constrained to have the same horizontal displacement at floor levels; that is, they are connected in series, with axially rigid links connected through hinges to the floors. The 2D model is analysed for the horizontal component of the seismic action parallel to it and gives internal forces and other seismic action effects only within vertical planes parallel to that of the analysis. The analysis does not give internal forces for beams or walls, which are in vertical planes orthogonal to the horizontal component of the seismic action considered. Bending in columns and walls is uniaxial, with axial force only due to the horizontal component of the seismic action, which is parallel to the plane of the analysis. The internal force results in columns that are common to two orthogonal plane frames are obtained from the two separate 2D analyses for the two horizontal components of the seismic action and are then combined via the 1:0.3 combination rule of Equation 3.100.

A 2D model according to the above does not take into account any eccentricity between the centres of mass and stiffness. However, even in buildings with zero eccentricities, the effects of the accidental eccentricities as per Section 3.1.8 should be taken into account. For buildings that are doubly symmetric in plan but are analysed with a 3D model, Eurocode 8 allows taking into account the accidental eccentricities of Section 3.1.8 in a very simple way. This is done by amplifying the results of the linear analysis for each translational component of the seismic action by $(1 + 0.6x/L)$, where x is the distance of the member in question to the mass centre in plan and L is the plan dimension, both are at right angles to the horizontal component of the seismic action. This factor is derived assuming that:

- Torsional effects are fully resisted by the stiffness of the structural elements in the direction of the horizontal component in question, without any contribution from any element stiffness in the orthogonal horizontal direction.
- The stiffness of the members resisting the torsional effects is uniformly distributed in plan.

The term $0.6/L$ is indeed equal to the storey torque due to the 'accidental eccentricity' of $0.05L$ acting on the storey seismic shear, V, divided by the moment of inertia of a uniform lateral stiffness, k_B, per unit floor area parallel to side B in plan, $k_B BL^3/12$, and further divided by the normalised storey shear, $V/(k_B BL)$. Normally there is also lateral stiffness, $k_L \approx k_B$, per unit floor area parallel to side L in plan; so, $k_L LB^3/12$ should be added to $k_B BL^3/12$ before dividing the storey torque $0.05LV$. The contribution of k_L is neglected and therefore the term $0.6x/L$ is safe-sided by an average factor of 2. Moreover, if it is analysed with a separate 2D analysis for each horizontal component according to the present section, because it meets the criteria in Section 4.3.3 for regularity in plan, the amplification factor of the simplified approach becomes $(1 + 1.2x/L)$ to cover the effect of the neglected static eccentricity, e, no matter whether there is actually one. Note that this eccentricity meets Condition 5 and Equation 4.1 in Section 4.3.2, that is, it cannot be large.

Eurocode 8 allows analysis with the lateral force method using two independent 2D models even in buildings that do not meet all conditions of Section 4.3.2 for regularity in plan, if they meet the following instead:

a. Partitions and claddings are well-distributed vertically and horizontally, so that any potential interaction with the structural system does not affect its regularity.
b. The height is less than 10 m.
c. In-plane stiffness of the floors is large enough to justify the rigid diaphragm assumption.
d. The storey centres of mass and stiffness lie approximately on (two) vertical lines.
e. The torsional radius r_x is at least equal to $\sqrt{l_s^2 + e_x^2}$ and r_y to $\sqrt{l_s^2 + e_y^2}$.

If conditions (a) to (c) are met, but not (d) and (e), then two separate 2D models may still be used, provided that, for design purposes, all seismic action effects from the 2D analyses are increased by 25%.

The aim of the above relaxation of the regularity conditions so as to use two independent 2D models is to help the designers of small buildings. For this reason, the extent of the application of this option is determined nationally: a note invites the National Annex to select the importance classes for which this relaxation applies; no recommendation is given for the selection.

Computer programs for linear analysis in 3D, static or modal, are presently ubiquitous in every-day seismic design practice. Therefore, the possibility of using two independent 2D models for linear analysis of regular buildings is of little practical interest. Nonetheless, this possibility is quite important for non-linear analysis, either static (pushover) or dynamic (response-history). Reliable, widely accepted and numerically stable non-linear constitutive models (including the associated failure criteria) are available only for members in uniaxial bending with (constant or little-varying) axial force; their extension to biaxial bending for widespread use in 3D analysis belongs to the future. So, in order to use non-linear analysis, the characterisation of a building structure as regular or irregular in plan is very important.

4.3.3.2 Implications of irregularity in plan for the behaviour factor q

As we will see in detail in Section 4.6.3, Eurocode 8 may reduce the behaviour factor of buildings designed for ductility owing to irregularity in plan. Moreover, we will see in the same section that if, at any floor, one or both conditions in Equation 4.7 is not met (i.e. if the radius of gyration of the floor mass exceeds the torsional radius in one or both of the two main directions of the building in plan), then the structural system is characterised as *torsionally flexible* and the behaviour factor q (apart from any reduction due to potential irregularity in elevation, as discussed below) is reduced to a relatively low value.

4.3.4 Irregularity and regularity in elevation

Eurocode 8 considers a building as *regular in elevation* if the building itself and its structure satisfy simultaneously the following five conditions:

CONDITION 1

The lateral-force-resisting systems (frames and walls) of the building are continuous from the foundation to the top of the building. Naturally, if there are in-plan setbacks along the height of the building, this rule requires that, in the zones of the building that do not extend to its total height, the lateral-force-resisting systems located therein should extend to the full height of such a zone. In fact, abrupt termination of lateral resisting systems at a certain level, particularly in the case of stiff walls, normally causes a lateral deformation pattern with a kink at that level. This deformation pattern deviates from the simple linear pattern as per Equation 3.98, which may be assumed for regular buildings. On the other hand, such a kinked deformation pattern produces stress concentrations and large ductility demands in adjacent elements. Moreover, at that level the transfer of forces from the different lateral resisting systems (due to the abrupt termination of one of them) may overstress the diaphragms close to that level (see Figure 4.4). This means that the modelling and dimensioning of the diaphragm at the design stage requires particular attention.

CONDITION 2

The storey mass and the lateral stiffness are constant along the height or decrease gradually and smoothly from the foundation (or ground level) to the top of the building. More precisely, what matters is that the ratio between storey mass and stiffness remains constant along the height. This condition is met if the mass and stiffness are constant along the height and also if the changes in mass and stiffness occur in a regular way for both quantities. This latter situation is normally found in buildings with structural systems well-distributed across the floor plan, because in those cases the progressive decrease of the storey mass associated with the decrease of the floor dimensions also corresponds to the progressive termination of those structural systems. By contrast, if the lateral resisting system is only composed of a small number of (stiff) elements, it may become more difficult to ensure that the progressive decrease of the floor dimensions (and hence its mass) does not entail abrupt changes of stiffness due to the termination, at some point, of those stiff elements.

CONDITION 3

In buildings with framed structures, there is no abrupt variation of the overstrength of the individual storeys relative to the design storey shear resulting from the analysis. For the verification of this condition, the contribution of masonry infills to the storey shear strength should be taken into account. For these purposes, the shear capacity of the storey can be computed as the sum, over all vertical elements, of the flexural resistance of each element (at the storey bottom) divided by the corresponding shear span (moment-to-shear ratio). In columns, the shear span may be considered as half the clear storey height, whereas in walls it may be considered as half the distance from the storey bottom to the top of the building.

Additionally, in case infill walls are used, their shear strengths (roughly equal to the minimum area of the horizontal section of the wall panel times the shear strength of bed joints) must be added to the storey shear capacity. It is the ratio of this actual shear force capacity of each storey to the storey shear force resulting from the analysis that should not have abrupt variations along the building height, for the building to be classified as regular in elevation.

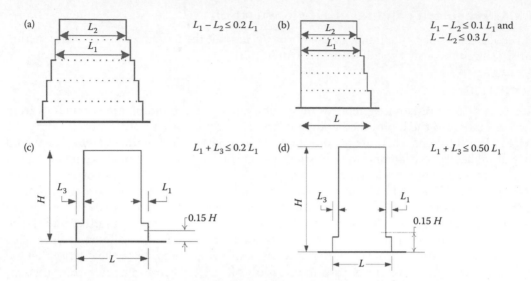

Figure 4.10 Criteria for regularity in elevation in buildings with setbacks: (a) symmetric setbacks; (b) asymmetric setbacks; (c) and (d) single setback at the lower part.

CONDITION 4

In buildings with *symmetrical setbacks* (along the height), the setback in each side and at any floor should not exceed 10% of the parallel dimension of the underlying storey. This condition applies both to buildings with gradual setbacks, as illustrated in Figure 4.10a, as well as to those with a tower and a podium as illustrated in Figure 4.10c. This condition is somewhat relaxed if the setback occurs within the bottom 15% of the total height of the building, H. In such a case it is accepted, still considering the building as regular, that this setback may reach up to 50% of the parallel dimension at the base of the building, as illustrated in Figure 4.10d. However, in this particular case there should be no undue reliance on the podium for transferring to the ground the seismic shears that develop in the tower. These shears should be transferred mainly through the vertical continuation to the ground of the structural systems of the tower; the podium should mainly transfer to the ground its own seismic shear. In other words, what is intended is that there is no need for the floors in the podium to transfer, through diaphragmatic action, an important part of the lateral forces coming from above to structural systems not placed within the vertical projection of the tower. The relevant clause of Eurocode 8 requires that the tower be designed for a seismic base shear at least equal to 75% of the base shear in a similar building without the podium.

CONDITION 5

In buildings with *asymmetrical setbacks*, any setback at any floor should not exceed 10% of the parallel dimension of the underlying storey. Additionally, the total setback at the top of each side of the building should not exceed 30% of the parallel dimension at the base of the building. These conditions are illustrated in Figure 4.10b.

4.3.5 Design implications of irregularity in elevation

4.3.5.1 Implications of regularity for the analysis method

Eurocode 8 allows different seismic analysis options or values of the behaviour factor q, depending on whether the building is regular or irregular in elevation.

If there is irregularity in elevation, it is unlikely to have a first mode shape that varies almost linearly from the base of the building to the roof. So, as such a mode shape underlies the lateral-load pattern of the lateral-force method of analysis (Equation 3.98), this method is not considered applicable to buildings characterised as irregular in elevation. The modal response spectrum method is, by contrast, capable of capturing well the effects of structural irregularity in elevation on the linear elastic response, but also, to a large extent, on the non-linear response as well. So, Eurocode 8 makes the application of modal response spectrum analysis mandatory for buildings which are irregular in elevation. This is not a penalty: in global terms, the results of modal response spectrum analysis are not more safe-sided over-all than those of the lateral force method. However, the modal response method estimates much better the peak dynamic response at the level of member internal forces and deformations, especially for structures irregular in elevation.

4.3.5.2 Implications of regularity in elevation for the behaviour factor q

It is more difficult to achieve a uniform distribution of inelastic deformations throughout the height of the structure as per Figures 2.9b to 2.9e in a height-wise irregular building. In fact, at the elevation(s) where the irregularity takes place, for example:

- At a large setback
- Where a lateral-force-resisting system is vertically discontinued
- When a storey has mass, lateral stiffness or overstrength higher than that in the storey below

it is likely to have a local concentration of inelasticity beyond the predictions of a linear analysis, even if it were a modal response spectrum analysis. Such a concentration will locally increase the deformation demands, above the building-average value that corresponds to the value of the q-factor used in the design. A possible way of tackling this problem would be to adopt stricter detailing in the regions likely to be affected by the structural irregularity, in order to enhance their ductility capacity to the level of the locally increased ductility demands. Instead, Eurocode 8 imposes a reduction of 20% to the value of the behaviour factor q used in the analysis, without relaxing the detailing requirements anywhere in the structure. The resulting 25% increase in the required resistance throughout the building is a serious disincentive to adopting a structural system which is irregular in elevation.

4.4 STRUCTURAL SYSTEMS OF CONCRETE BUILDINGS AND THEIR COMPONENTS

4.4.1 Introduction

The raison d'être of concrete buildings is to create horizontal surfaces for use/occupancy (floors) or protection (the roof). Most of the mass generating the inertial forces in an earthquake resides on these horizontal elements. Gravity loads are transferred from there to the ground via vertical elements, typically columns. Beams or girders span between columns, to facilitate the collection of gravity loads from the horizontal surfaces and facilitate their transfer to the columns (Figure 4.11a). Concrete walls are often used to resist horizontal forces and to brace the building laterally against second-order $(P - \Delta)$ effects (Figure 4.11b, see Section 5.2.3.4 and Equation 5.11 for the bracing role of walls under factored gravity loads).

Concrete walls can resist a horizontal earthquake very efficiently, working as vertical cantilevers. However, unlike in masonry buildings, it is not cost-effective to collect from

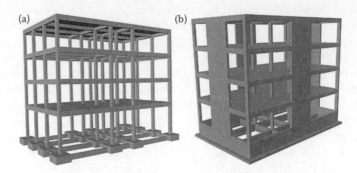

Figure 4.11 Structural systems: (a) frame resisting both gravity loads and lateral actions, on footings with two-way tie-beams; (b) wall-frame lateral-load-resisting system on two-way foundation beams.

the floors and transfer to the ground all gravity loads through what is called in Eurocode 8 'ductile' concrete walls. Note, though, that it may be cost-effective to use both for seismic and gravity actions only 'large, lightly reinforced walls' per Eurocode 8, highlighted in Section 4.4.2.1 but not covered in detail in this book. Ductile walls normally complement a combination of columns and floor beams, whose main role is to support the gravity loads acting on the horizontal surfaces of the building (Figure 4.11b). Normally, the beams are directly and rigidly connected to the columns. The resulting moment-resisting beam–column frame is also efficient in resisting horizontal or vertical earthquake forces within its plane. Therefore, besides their main role as a gravity-load-resisting system, frames of beams and columns double as earthquake-resisting systems; in fact, frames are the most common type of such a system in concrete buildings.

Inertial forces should find their way to the foundation via a smooth and continuous path in the structural system. From that point of view, cast-in-situ concrete is better for earthquake-resistant buildings than prefabricated elements of concrete, steel or timber that are assembled on site: the connections between such elements create discontinuities and potentially weak points in the flow of forces. So, cast-in-situ construction is the technique of choice for earthquake-resistant concrete buildings, at least in high seismicity regions.

4.4.2 Ductile walls and wall systems

4.4.2.1 Concrete walls as vertical cantilevers

A wall differs from a column in that, under lateral loading, it works as a vertical cantilever. A column, by contrast, needs to be combined with beams into a frame, in order to resist lateral loads efficiently: its moment resistance at the base is too small to make a meaningful contribution to the base shear of the building, if divided by the shear span (moment-to-shear ratio) of a vertical cantilever. Moreover, its lateral stiffness as a vertical cantilever is too low to be effective in reducing inter-storey drifts for damage limitation (see Section 1.3.2) or $P - \Delta$ (second-order) effects (see Section 3.1.12). To play its role as a vertical cantilever, the wall must be much stiffer than any beams it may be connected to at floor levels, so that these beams act only as parts of the horizontal diaphragm through which the wall receives the lateral forces from the floor, and not as horizontal elements of a frame encompassing both the wall and these beams. So, the wall's bending moment diagram under lateral loading looks like that of a vertical cantilever (see Figure 5.6 and Figures 7.17, 7.19, 7.24 and 7.25 in the example building of Chapter 7): the moment does not change sign within a storey (except possibly near the top of wall-frame systems); the moments decrease considerably from the wall base to the top, much more than the shears do. Besides, if beams frame into the wall at floor levels, the wall bending moment is normally larger right above a floor than right below it; as the same vertical bars cross these two sections and the increase in axial compression enhances the wall moment resistance, plastic hinges can form in the wall only above floor levels. Multiple plastic hinging may well develop up the height of the wall, if the wall moment resistance at floor levels and at the connection to the foundation is tailored to the elastic seismic moment demands. Even then, a soft-storey mechanism cannot form in the wall itself, as it requires plastic hinging in counter-flexure at two different locations up the height of the wall (cf. Figures 2.9a, 2.9d, 2.9e and 2.12).

To ensure that a wall plays the role of a stiff and strong vertical spine of the building and prevents a soft-storey mechanism, Eurocode 8 promotes localisation of the wall inelastic deformations at its base. A wall designed and detailed to dissipate energy in a single flexural plastic hinge at the base and remains elastic throughout the rest of its height is called in Eurocode 8 'ductile wall'. It is the main wall type addressed in Eurocode 8, but not the only one. An alternative is allowed, termed 'large lightly reinforced wall', where flexural overstrength over the seismic demands from the analysis is intentionally avoided anywhere up the height of the wall, in order to promote plastic hinging at several floor levels above the base and translate the global displacement demand into small rotational demands at several locations up the wall. The inelastic deformational demand at the base of the wall is thus reduced; it may even be eliminated, by allowing rocking of the wall's footing, instead of fixing the base of the wall against rotation – a prerequisite for plastic hinging at the base of a 'ductile wall'. In this way, the cumbersome and expensive detailing of the wall base region for ductility is avoided.

Large, lightly reinforced walls have certain advantages that ductile walls lack; for instance, rocking of a long footing and/or rotation of a long wall section about a neutral axis close to the compression edge of the wall raise the centroid of the wall section and, with it, the weights supported by the wall, cyclically (but temporarily) converting part of the vibration energy into recoverable and harmless potential energy of these weights, instead of inelastic deformation energy in plastic hinges, associated with permanent deformations and damage. Therefore, systems of large lightly reinforced walls designed according to their own special rules in Eurocode 8 may be more cost-effective under certain conditions than systems of ductile walls per Eurocode 8. However, as the use of large lightly reinforced walls is not common yet, this book covers only ductile walls.

4.4.2.2 What distinguishes a wall from a column?

Design codes define a wall as a vertical element with an elongated cross-section: a lower limit of 4 for the aspect ratio (long-to-short dimension) of a rectangular cross-section is used in Eurocode 2 for a vertical element to be considered as a wall. If the cross-section consists of rectangular parts, one of which has an aspect ratio greater than 4, the element is also classified as a wall. With this definition on the basis of the cross-sectional shape alone, a wall differs from a column in that it resists lateral forces mainly in one direction (parallel to the long side of the section) and can be designed for such a unidirectional resistance by assigning the flexural resistance to the two edges of the section ('flanges', or 'tension and compression chords') and the shear resistance to the 'web' between them, as in a beam. So, for the purposes of moment resistance and deformation capacity, the designer may concentrate the vertical reinforcement and provide concrete confinement only at the two edges of the section. Note that, if the cross-section is not elongated, the vertical element has to develop significant lateral-force resistance in both horizontal directions; so, it is meaningless to distinguish the 'flanges', where the vertical reinforcement is concentrated and the concrete confined, from the 'web', where they are not.

The above definition of a 'wall' is appropriate for dimensioning and detailing at the level of the cross-section, but meaningless for the intended role of a wall in the lateral-load-resisting system and for the usual practice to design, dimension and detail the wall as an entire element and not just at the cross-sectional level. Seismic design often relies on walls for the prevention of a storey-mechanism in the plane parallel to the wall's long direction, without checking if plastic hinges form in beams rather than in columns. However, walls can impose a beam-sway mechanism only if they act as vertical cantilevers (i.e. if their bending moment has the same sign throughout, at least in the lower storeys) and develop a plastic hinge only at the base. Whether a wall, as defined above, will indeed act as a vertical canti-lever and form a plastic hinge only at its base does not depend on the aspect ratio of its sec-tion, but on how stiff and strong the wall is relative to the beams it is connected to at storey levels; if these beams are almost as stiff and strong as the wall, then the wall works as a frame column rather than as a vertical cantilever. For a wall to play its intended role, the length dimension of its cross-section, l_w, should be large, not just relative to its thickness, b_w, but in absolute terms. To this end, and for the beam sizes commonly found in buildings, a value of at least 1.5 m for low-rise buildings or 2 m for medium- or high-rise ones is recommended for l_w. In fact, it can be shown (Fardis 2009) that the optimal value of l_w for moment and shear resistance, stiffness and ductility is about one-sixth of the total height of the wall, H_{tot}.

4.4.2.3 Conceptual design of wall systems

The walls of a wall system should be arranged in two orthogonal horizontal directions with as much two-way symmetry as possible. If the individual walls are all similar and symmetri-cally placed, they will be subjected at every storey to fairly uniform seismic force and defor-mation demands, minimising the uncertainty about the seismic response. In a system with (very) dissimilar walls, the stronger and stiffer ones will yield first, imposing on the rest their inelastic deflection pattern, notably one where storey drifts increase almost linearly to the top owing to the rotation of the plastic hinge at the base, while the walls that are still elastic tend to deflect as vertical cantilevers. In that case, besides the increased uncertainty of the post-elastic response, the floor diaphragms will be stressed hard to iron out the differences in height-wise deflection patterns between the stiffer walls, which have gone inelastic, and the more flexible ones, which remain elastic. Note, though, that the price of complete uniformity is poor redundancy: plastic hinges will develop almost simultaneously at all wall bases, and there will be little overstrength or redistribution of forces from certain walls to others.

Almost all our knowledge of the cyclic behaviour of concrete walls concerns walls with a two-way-symmetric rectangular or quasi-rectangular section (barbelled section, i.e. rectan-gular with each edge widened into a rectangular or square 'column' or compact flange – with an aspect ratio less than 4 – to enhance the moment resistance and prevent lateral instability of the compression zone). Such walls are modelled and dimensioned as prismatic elements having an axis through the centroid of the section. Lacking a better alternative, the same practice is applied when a rectangular wall runs into or crosses another wall at right angles, to create a wall with a composite cross-section of more than one rectangular parts – each part with an aspect ratio greater than 4 (L-, T-, U-, H-shaped walls, etc.). Such walls have high stiffness and strength in both horizontal directions, hence are subjected to biaxial bending and bi-directional shears during the earthquake. They are more cost-effective than the combination of their constituent parts as individual rectangular walls. However, pres-ent-day knowledge of their behaviour under cyclic biaxial bending and shear is very limited, and the rules used for their dimensioning and detailing still lack a sound basis. Moreover, their detailing for ductility is complex and difficult to implement on site. For this reason, it is recommended to make limited use of such walls in practical design. If non-rectangular walls are chosen, they should have a fairly simple section (e.g. one-way-symmetric U, or two-way-symmetric H).

Large openings should be avoided in ductile walls, especially near the base, where the plastic hinge forms. If they are necessary for functional reasons (doors or windows), they should not be staggered vertically, but should be arranged at every floor in a regular pattern, creating a coupled wall, with the lintels between the openings serving as coupling beams and designed as such. According to Eurocode 8, two walls are considered as coupled, if they are connected together (normally at each floor) through regularly spaced beams meeting special ductility conditions ('coupling beams') and this coupling reduces by at least 25% the sum of the bending moments at the base of the individual walls (the 'piers'), compared to that of the two 'piers' working independently.

4.4.2.4 Advantages and disadvantages of walls for earthquake resistance

Structural systems dominated by ductile walls have many advantages for earthquake resistance:

- The high lateral stiffness of walls reduces inter-storey drifts and structural or non-structural damage; it also overshadows the contribution of masonry infills to the lateral stiffness of the building and reduces the adverse effects: global ones, due to their potential irregularity in plan (eccentric placement) or elevation (open storey(s)), or local, notably shearing off weak columns, the creation of captive, squat columns, etc.
- Soft-storey mechanisms are precluded by the absence of wall counter-flexure within a storey.
- Rocking of the wall's footing or of the part of the wall above a plastic hinge raises the supported weights and is favourable for seismic performance.
- Overall, systems of walls are more cost-effective for earthquake-resistance than beam–column frames.

There also are drawbacks:

- Walls are inherently less ductile than beams or columns, more sensitive to shear and harder to detail for ductility.
- The small number of walls required for earthquake resistance leads to smaller redundancy and fewer alternative load paths.
- It is difficult to place several long walls without compromising the architectural function of the building, producing large eccentricities in plan, or creating a torsionally sensitive building (i.e. one with more lateral stiffness closer to the centre in plan than to the perimeter).
- It is not cost-effective to support the building's gravity loads with walls alone; certain beams and columns are needed anyway for that and may efficiently serve for earthquake resistance as well.
- It is hard to provide an effective foundation to a wall, especially with isolated footings. Because of the large bending moment and the relatively low vertical load of walls, the development of tensile forces in the foundation is often inevitable. A more favourable situation is to have the wall continue downwards from the ground floor into a basement (see Sections 4.4.5, 6.3.1 and 7.1). In such a case, the wall bending moment decreases within the basement from its maximum value at ground level (see Figures 7.17, 7.24 and 7.25), owing to the lateral restraint (horizontal forces) that the basement

floors provide; hence the moment applied at the foundation level may be substantially smaller, and vertical tension forces are avoided. The downside is that the sharp decrease of the wall's bending moment below the ground floor entails development of large shear forces in the wall (see Figures 7.17, 7.24 and 7.25).

- There is some uncertainty concerning certain features of the seismic response of walls and their systems: the cyclic behaviour and seismic performance of individual walls (which is more difficult to study experimentally or analytically than in the case of beams or columns); the rocking response and the associated lifting of the weight supported by the wall, the increase of wall shears after plastic hinging at the base (see Section 5.6.2.1), etc. Moreover, modelling for analysis, and dimensioning/detailing of walls is more challenging compared to frame columns (especially for non-rectangular walls).

4.4.3 Moment-resisting frames of beams and columns

4.4.3.1 Special features of the seismic behaviour of frames: The role of beam–column connections

In a lateral-load-resisting system comprising only uncoupled walls, the sum of the wall seismic shears at the base is equal to the resultant of the lateral seismic forces applied at storey levels (seismic 'base shear' of the building); the resultant moment of these storey lateral seismic forces with respect to the base (seismic 'overturning moment' at the base) is equal to the sum of bending moments at the base of the walls. So, walls resist the seismic overturning moments and shears directly, through bending moments and shears, respectively, in the walls themselves. In contrast, frames resist the seismic overturning moment not by the column moments, but through their axial forces (tensile at the windward side of the plan, compressive at the opposite, leeward one, see Figures 7.10, 7.13, 7.16, 7.20 and 7.23). The column bending moments resist indirectly the seismic storey shears: the algebraic difference of bending moments at the top and bottom of each column produces its contribution to the seismic shear of the storey. So, the seismic response of frame members is governed by flexure, or strictly speaking by normal action effects: bending moments and axial forces.

Elastic moment, shear and axial force diagrams due to the seismic action in the frames of the example building of Chapter 7 are depicted in Figures 7.8 to 7.16 and 7.20 to 7.23. Among other features, the seismic moment diagrams from the 'lateral force method' exhibit an abrupt change in the algebraic value of the seismic moment across any beam–column connection: the beam or column moment turns from large and positive at one face of a joint into large but negative at the opposite face. If the joint, being of finite dimensions, is considered as a part of the beam within the column, this abrupt change in the beam moment across the joint means that a large vertical shear force develops inside it, which is equal to the sum of the absolute values of beam moments at the joint faces divided by the column width in the plane of the frame (Figure 4.12a). By the same token, if the joint is considered as a part of the column between adjacent beam spans, the abrupt change in column moments across the joint implies a large horizontal shear force in it, which is equal to the sum of absolute values of column moments at the joint faces divided by the beam depth (Figure 4.12b). So, the core of the joint is subjected to very high shear stresses, equal to the sum of (the absolute values of) the beam or column seismic moments at opposite faces of the joint divided by the volume of this core (see Figure 2.21). Another repercussion of the rapid change in the algebraic value of seismic moments across any beam–column connection is that any beam or column longitudinal bars crossing the joint are under high tensile stresses on one side of the joint and under high compressive stresses on the other. This means that very high bond stresses develop all along the stretch of such bars within the joint; if plastic hinges form in the beam or the column at

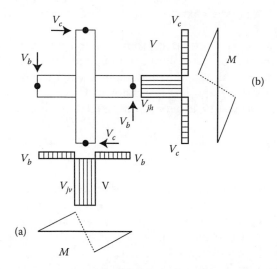

Figure 4.12 Seismic moments and shears in the beams and columns connected at a joint and seismic shears in the joint core: (a) joint considered as part of the beams; (b) joint considered as part of the column.

both sides of the joint, the value of these bond stresses may exceed the bar yield force divided by the lateral surface of the bar inside the joint (as a matter of fact, along the bottom bars, bond stresses may approach twice that value). Because the beams, rather than the columns, are expected to develop plastic hinges, to accommodate these bond stresses the column width should exceed a certain multiple of the beam bar diameter, as specified in Section 5.2.3.3 of Chapter 5. This often turns out to be a major constraint on the column size or the beam bar diameter. If the relevant rule in Eurocode 8 is not met, the bars may slip through the joint, thus increasing the apparent flexibility of the members framing into it and preventing them from plastic hinging next to it (Figure 2.22a). Although this will not have catastrophic consequences, it prevents the frame members connected to that joint from contributing to the strength, stiffness and energy dissipation capacity of the frame to their full potential.

4.4.3.2 Conceptual design of RC frames for earthquake resistance

The general layout and certain details of the geometry of an individual plane frame have a major impact on its seismic behaviour. Very important also is the overall layout of the frames in a frame structural system. The location of frames in plan and their span lengths are normally governed by architectural and functional considerations, while beam depths may be controlled by design for factored gravity loads (for the 'persistent and transient design situation' of EN 1990). Nevertheless, the structural designer is essentially free to choose the all-important geometric details of individual plane frames and has certain freedom concerning their overall geometry and location in plan.

Any single plane frame should run continuously from one side of the building plan to the other, without offsets, interruptions (i.e. missing beams between adjacent columns in a floor), or indirect supports of beams on other beams:

- If a beam does not continue straight from span to span, but its axis is offset at the column between them, there is no smooth flow of beam internal forces through a proper beam–column joint, neither continuity of the beam longitudinal bars across the column from one span to the next: these bars have to terminate there and be separately anchored at the joint.

- Even when the beam axis is not offset from span to span, the smooth flow of internal forces from the beam(s) to the column is impaired by a large eccentricity between the axis of the beam and the supporting column. The behaviour of strongly eccentric beam–column joints is, by-and-large, unknown. For that reason, Eurocode 8 sets an upper limit on the eccentricity, e, between the axis of the beam and the column at their connection:

$$e \le b_c/4 \qquad\qquad (4.10)$$

where b_c is the largest cross-sectional dimension of the column at right angles to the beam axis. Note that, if one lateral side of the beam is flush with one face of the column, this condition restricts the ratio of b_c to the beam width, b_w, not to be greater than 2.0. This is the case at the corner columns of the example building in Chapter 7 (see Figure 7.2).

- If a beam terminates at an (indirect) support on another beam, there is large uncertainty concerning its rotational restraint by the supporting beam via torsion in the latter. In approximation, an indirect support may be considered as a simple support; the indirectly supported beam is less effective in frame action than one connected to columns at both ends.

The ideal plane frame has:

1. Constant beam depth in all bays of a storey
2. Constant size of each column in all storeys
3. Approximately uniform spans
4. Interior columns of approximately the same size
5. Approximately the same height in all storeys

Note that, if points 1 to 4 above are met, and the exterior columns have one-half the stiffness of the interior ones, then, if the effect of column axial deformations is negligible, all beams in the storey will develop the same elastic seismic shears and bending moments (which will be equal at the two beam ends); all interior columns will also have the same elastic seismic shears and moments while their elastic seismic axial forces will be zero; the two exterior columns will develop half the seismic elastic shears and moments of interior ones and will resist the full seismic overturning moment, via seismic axial forces equal to the seismic overturning moment at storey mid-height divided by the distance between the axes of the two exterior columns. If all members of such a frame are dimensioned to resist exactly the elastic seismic moments, all beam ends in a storey will be subjected to (about) the same inelastic chord rotation demands; all columns, interior or exterior, will also develop (about) the same inelastic rotation demands at storey bottoms; the same at column tops. Such uniformity reduces uncertainty concerning the distribution of seismic action effects among frame members. If the two exterior columns have more than one-half the stiffness of interior ones, their share of storey elastic seismic shears will increase (alongside their elastic moments, as well as those at the two outer beam ends), but less than proportionally; seismic axial forces in interior columns will be non-zero, but small.

Beams with long span may have their top reinforcement at the supports governed by factored gravity loads (the 'persistent and transient design situation' as per EN 1990), rather than by the 'seismic design situation'. This will result in beam overstrength, $M_{Rd,b}$, relative to the moment demand, $M_{Ed,b}$, in the 'seismic design situation'. The overstrength is carried over to the capacity design of columns around joints (per Equation 5.31 in Section 5.4.1 of Chapter 5) and to the capacity design shears of beams and columns (per Equations 5.42 and 5.44, respectively, in Section 5.5.1), penalising them and creating some uncertainty whether plastic hinges will form in the beams or the columns. Besides, the large hogging moments due to quasi-permanent gravity loads at the ends of long span beams may prevent reversal of yielding in sagging flexure at any plastic hinges that may form there. As a result, inelastic elongations accumulate in the top reinforcement and the beam gradually grows longer, pushing out the supporting columns and possibly forcing exterior ones to separate from the exterior beams which are at right angles to the elongating one(s).

Beams with low span-to-depth ratio have to be dimensioned for high shear forces, from the seismic analysis or from capacity design in shear (see Equations 5.42 in Section 5.5.1). At the ends of such beams the shear due to quasi-permanent gravity loads is small (see last term in Equations 5.42) and a reversal of the seismic action will also cause an almost full reversal of the sign of the acting shear (cf. Equation 5.43 in Section 5.5.1), exhausting the shear capacity of the beam in both diagonal directions or causing sliding shear failure along through-depth cracks at the end section(s) of the beam. To resist such effects, diagonal reinforcement or stirrups at ±45° are needed at the ends of short beams (see Equations 5.42 and 5.43 in Section 5.5.1 and Equations 5.49 to 5.51 in Section 5.5.3). Moreover, unless diagonally reinforced, short beams have low deformation capacity and poor ductility.

For the reasons detailed above, short beam spans should be avoided, while spans of 4 to 5 m should be preferred over longer ones, at least for the storey heights and gravity loads commonly encountered in buildings.

In frame systems (with the frames preferably having an individual geometry according to the above), frames should be arranged in two orthogonal horizontal directions in a way that maximises two-way symmetry and minimises irregularities in plan of the type highlighted in Section 4.3.2. If such frames are all similar and symmetrically placed, they will be subjected at every storey to fairly uniform seismic force and deformation demands, without undue concentration in a single frame, member or location thereof and risk of early failure. Whatever has been said in the first paragraph of Section 4.4.2.3 concerning dissimilar walls in a wall system applies by analogy to systems of frames with very different strength and stiffness: the stronger and stiffer ones will yield earlier during the response, imposing on the rest a deflection pattern where storey drifts increase almost linearly to the top, instead of following the storey shear force pattern. The floors will be subjected to larger in-plane forces, to bridge the differences between the drift patterns of the stiffer and already inelastic frames and those of the more flexible ones, which remain elastic. Complete uniformity will again result, though, in a less progressive formation of the overall plastic mechanism, and plastic hinges will develop almost simultaneously in the various frames, be it where expected. Moreover, the storeys have little overstrength after the first plastic hinge formation and cannot redistribute forces from certain locations to others.

4.4.3.3 Advantages and drawbacks of frames for earthquake resistance

The advantages of RC frames for earthquake resistance may be summarised as follows:

- Frames place few constraints on a building's architectural design, including the façade.
- Frames may be cost-effective for earthquake resistance, because beams and columns are placed anyway for gravity loads; so, they may also provide earthquake resistance in both horizontal directions, if their columns are large.

- Two-way frame systems, comprising several multi-bay plane frames per horizontal direction, are highly redundant, offering multiple load paths.
- Thanks to their geometry (notably their slenderness), beams and columns are inherently ductile, less prone to (brittle) shear failure than walls.
- Frames with concentric connections and regular geometry have well-known and understood seismic performance, thanks to the numerous experimental and analytical studies carried out in the past; moreover, they are rather easy to model and analyse for design purposes.
- It is easier to design an earthquake-resistant foundation element for a smaller vertical member than for a larger one (i.e. for a column in comparison to a wall).

There also are disadvantages:

- Frames are inherently flexible; the cross-section of their members may be governed by the inter-storey drift limitation under the moderate earthquake for which limitation of damage to structural and non-structural elements is desired (see Section 1.3.2).
- Column counter-flexure in the same storey allows soft-storey mechanisms (Figure 2.9a), which lead to collapse.
- Earthquake-resistance requirements on frames lead to large columns.
- The reinforcement detailing of frames for ductility requires workmanship of high level for its execution and good supervision on site (especially to fix the dense reinforcement and place/compact the concrete through the beam–column joints in two-way frames).
- Sizing and detailing of beam–column joints for bond and anchorage of beam bars crossing them is quite challenging. Difficulties increase with the use of higher strength materials, as the size of joints made of higher concrete strength is smaller, while higher steel strength implies higher bond stresses.
- There is still some uncertainty concerning the seismic response and performance of frames:
  - The effects of eccentric connections or strongly irregular layouts in 3D.
  - The size of the effective slab width in tension (see Figure 2.22b) and the extent to which slab bars in it and parallel to the beam increase its flexural capacity for hogging moment, $M_{Rd,b}^-$, hence the beam capacity design shears per Equations 5.42 in Section 5.5.1 of Chapter 5 and the likelihood of plastic hinging in the columns, despite meeting the capacity design rule around joints per Equation 5.31 in Section 5.4.1.
  - The behaviour of columns of two-way frames under cyclic biaxial bending with varying axial force, which may even cause plastic hinging in columns which meet the capacity design rule of Equation 5.31 in separate uniaxial bending per horizontal direction.

4.4.4 Dual systems of frames and walls

4.4.4.1 Behaviour and classification per Eurocode 8

Walls and frame systems each have their advantages and disadvantages as lateral-load-resisting systems. Walls seem to have a better balance of advantages against drawbacks; nevertheless, a concrete building always has beams and columns to support the gravity loads; it is a waste not to use them for earthquake resistance. Therefore, frames and walls may well be cost-effectively combined in a single lateral-load-resisting system.

Eurocode 8 uses the fraction of the elastic seismic base shear taken by all the system's frames according to linear analysis for the seismic action, to distinguish whether frames or walls dominate the lateral-load-resisting system:

- When frames or walls take at least 65% of the seismic base shear, we have a 'frame system' or a 'wall system', respectively.
- When the percentage of the seismic base shear taken by frames or walls is between 35% and 65%, the frame-wall system is called 'dual'; if the fraction of elastic base shear taken by the walls is from 50% to 65%, the system is a 'wall-equivalent dual'; if it is between 35% and 50%, it is a 'frame-equivalent' one.

Eurocode 8 considers a wall system as a 'coupled wall system', if coupled walls, as defined at the end of Section 4.4.2.3, provide more than 50% of the total wall resistance.

The building in Chapter 7 is classified as a 'wall-equivalent dual' in the X-direction and as a 'wall system' one in Y (see Section 7.3.1).

Dual systems combine the strength, stiffness and immunity to soft-storey effects of wall systems with the ductility, deformation capacity and redundancy of frames. The walls prevent non-structural damage in frequent, moderate earthquakes, helping the building meet the inter-storey drift limits of Eurocode 8 under the damage limitation earthquake (Section 1.3.2). The frames serve as a second line of defence in strong earthquakes, in case the deformation capacity of the less ductile walls is exhausted and some walls lose part of their strength and stiffness.

The way frames and walls share the horizontal seismic action comes out of their different horizontal deflection pattern under lateral loading:

- Frames have a shear-beam-type of lateral displacement pattern, in which inter-storey drifts follow the height-wise pattern of the storey seismic shears: they decrease from the base to the top.
- Walls fixed at the base deflect like vertical cantilevers: their inter-storey drifts increase from the base to the roof.

If frames and walls are combined in the same structural system, the floor diaphragms impose on them common floor displacements. As a result, the walls restrain the frames at lower floors, taking the full inertia loads of these floors, while near the top the frame is called upon to resist the full floor inertia loads and, in addition, to hold back the walls, which – if alone – would have developed a large deflection at the top. So, in rough approximation, the walls of dual systems may be considered to be subjected to:

- The full inertia loads of all floors
- A concentrated force at roof level, in the reverse direction with respect to the peak seismic response and the floor inertia loads

The concentrated force at the top exceeds the resultant inertia loads in the upper floors, that is, the storey seismic shear there. So, the walls are often in reverse bending and shear in the upper storeys with respect to the storeys below (see Figures 7.17, 7.24 and 7.25). If the frame is considered to be subjected to just the concentrated force at the top, equal and opposite to the one it applies there to the wall(s) and in the same sense as the floor inertia loads, then it has in all storeys a roughly constant seismic shear and about the same bending moments (see Figures 7.8, 7.9, 7.11, 7.12, 7.14, 7.15, 7.21 and 7.22). Thus, even when the cross-sectional dimensions of frame members are kept the same in all storeys, their reinforcement requirements for the seismic action do not decrease from the base to the top. As a matter of fact, the reinforcement required in the columns may even decrease in the lower storeys, thanks to the favourable effect of the higher axial load on flexural strength. Therefore, in dual systems column size may never decrease in the upper storeys.

4.4.4.2 Conceptual design of dual systems

Dual systems have a more complicated seismic response than pure frame or wall ones. The resulting larger uncertainty concerning their seismic behaviour and performance may be their only drawback as a system. They are a quintessential example of systems of dissimilar subsystems; hence, whatever has been said in the first paragraph of Section 4.4.2.3 and the last one of Section 4.4.3.2 applies to them as well. Their conceptual design should aim to reduce the uncertainties arising from this feature. For instance, floor diaphragms should be thicker and stronger within their plane than what is required in pure frame systems. Another uncertainty arises from any rocking of the walls at the base, which will shift part of the storey shear from the walls to the frames. Rocking of wall footings with uplift is an intrinsically complex phenomenon, not reliably modelled in the context of seismic design practice. Its underestimation will lead to unsafe design of the frames, while its overestimation is unsafe for the walls.

Note that in a system consisting only of walls, the distribution of seismic shear between them will be practically unaffected by the rotation of the walls at the foundation level: rotations will mainly increase the absolute magnitude of storey drifts. The effect of footing rotation is even smaller in pure frame systems, practically affecting the seismic action effects only in the ground storey; moreover, such rotation is much smaller than in wall footings, because the higher axial load of the column resists uplift; more importantly, the smaller the cross-section of a vertical element compared to the plan dimensions of its footing, the smaller its rotation. So, it is dual systems that suffer from the increased uncertainty due to the rotations of footings with respect to the ground.

A prudent design of a dual system would reduce differential rocking. Ideally, this could be achieved by providing full fixity of walls and columns at the foundation level. However, full fixity is unfeasible, except at the top of a rigid basement (as in the example building of Chapter 7). In all other cases, great attention should be paid in the analysis phase to the modelling of soil compliance under the foundation elements, especially those of walls. Moreover, sensitivity studies should be carried out concerning the assumptions made and the values of properties used in the analysis.

Tall buildings often have a strong wall near the centre in plan, for example, around a service core housing elevators, stairways, vertical piping, etc., and stiff and strong perimeter frames. In such systems outrigger beams may be used to advantage, increasing the global lateral stiffness and strength and mobilising the perimeter frames against the seismic overturning moment.

4.4.5 Foundations and foundation systems for buildings

Foundations are used to transfer the gravity loads from the structure to the ground. During earthquakes they should also be capable of transferring horizontal loads and overturning moments developed by the inertial forces acting on the building masses. In addition, foundations may be subjected to differential movements imposed by the soil: for example, along the height of deep embedded foundations, across wide raft foundations or between foundations belonging to the same building unit, etc. During earthquakes, foundations have a tendency to settle, slide or possibly uplift.

Different types of foundation systems may be encountered in buildings: shallow isolated or spread footings, box-type foundations, rafts, caissons and piles.

Caissons and piles are typically used when either the surface soil layers have poor mechanical characteristics and the bearing resistance should be sought deeper, or when large tensile forces are developed and the foundations must be 'tied-down' to the ground. This type of deep foundations systems is not covered in this book.

- An isolated footing is defined as a single concrete element placed underneath the vertical element to transfer directly its axial force due to gravity loads to the ground. It may be square, rectangular or even circular in plan.
- A spread footing is defined as an isolated footing with a large plan dimension, extending under several columns. It is especially appropriate under closely spaced columns or a long concrete wall.
- A box-type foundation is a box extending throughout the building footprint area and comprising a wall all around the perimeter, working as a deep spread footing, plus two rigid horizontal diaphragms: one at the top level of the perimeter wall and another at the bottom. Such a system is very convenient for buildings with a basement.
- A raft foundation is a slab extending across the whole building footprint and supporting all its columns or walls.

As a general rule, deformations of shallow foundations during earthquake must remain limited, because they take place below the ground surface and are therefore difficult to inspect and repair after an earthquake. Furthermore, inelastic deformations of soils and foundations are hard to predict accurately, although it is recognised that they may provide a significant source of energy dissipation. Foundations should, therefore, be stiff enough to ensure a uniform transmission to the ground of the actions from the superstructure; to this end, special attention should be paid to the effects of horizontal differential displacements between vertical elements. To avoid such displacements, all individual footings are usually placed at the same level and interconnected through tie-beams. As the seismic action acts in both horizontal directions, a two-way tie-beam system is necessary. The main role of tie-beams is to reduce the magnitude and impact of differential settlements and/or horizontal movements between adjacent footings, due to large unbalanced vertical loads and/or variations in the underlying soil. By having all footings at the same level, one avoids attracting forces to one of them located deeper, which may act as a skirt.

If the contact pressures are too large compared to the foundation bearing capacity or, if despite of the tie-beams, large differential settlements cannot be ruled out, isolated footings are often replaced by a raft, which acts as a single footing under the entire building, transferring vertical loads to the ground throughout its plan area. As for individual footings, it is strongly recommended to have the raft over a horizontal surface, instead of different levels.

The conceptual design of shallow foundation systems for buildings is revisited in more detail in Section 6.3.1.

The choice of a foundation system and of a structural system of the superstructure that suits the layout of the building and the foundation conditions is the subject of Examples 4.8 and 4.9.

4.5 THE CAPACITY DESIGN CONCEPT

4.5.1 The rationale

The fundamental period of concrete buildings, T, is normally in the constant spectral pseudovelocity part of the response spectrum or beyond that part: $T \geq T_C$. As pointed out in Sections 3.2.2 and 3.2.3, in that range inelastic seismic displacements are roughly equal to the elastic ones ('equal displacement rule'). A prime target of seismic design is to apportion the given total seismic displacement demand to the various elements of the building, entrusting inelastic deformations only to those elements that can reliably withstand them, while keeping in the elastic range those which cannot. The tool for such a control of the inelastic seismic response is 'capacity design'. This tool establishes a strength hierarchy among the individual elements which ensures that, along the full load path of the inertial forces to the foundation ground, the strength of the structural system is governed by ductile elements, not brittle ones. Although capacity design is implemented during the detailed design phase, its effectiveness depends strongly on the structural layout and member sizes chosen during conceptual design.

The elements to which the global displacement demands are channelled via capacity design are chosen using the following criteria:

1. The inherent 'ductility' of the element: its capacity to sustain large inelastic deformations and dissipate energy in cyclic loading, without material loss of force-resistance, or the lack thereof, i.e., the inherent brittleness of the element.
2. Importance for the stability of other elements and the integrity of the whole: vertical elements are more important than horizontal; the foundation is the most important part of the system; so, they must be shielded from inelastic deformations which may jeopardise their integrity.
3. Accessibility and convenience to inspect and repair.

On the basis of these criteria, a hierarchy of elements is established, which determines if and in which order they may enter the inelastic range during the seismic response. 'Capacity design' is the tool to enforce this hierarchy. It works as follows:

The elements higher in the hierarchy are identified; their required design resistance is then determined not from the analysis, but via 'capacity design', that is, using only equilibrium and the force capacities of those elements which are ranked as less important, more accessible or inherently more 'ductile' (hence the term 'capacity design'), so that these latter elements exhaust their force resistance (yield) before the former do and indeed shield them from yielding.

4.5.2 The role of a stiff and strong vertical spine in the building

A prime aim of 'capacity design' is to prevent a 'storey-sway' plastic mechanism, in which inelastic deformations concentrate in a single storey (Figure 2.9a) and may lead to failure and collapse of its vertical elements, triggering overall collapse. As, for given fundamental period T, the global inelastic displacement demand at roof level is roughly given ('equal displacement rule'), it should be uniformly spread to all storeys, instead of a single one. For this to be kinematically possible, the beam–column nodes along any vertical element should stay on the same line during the seismic response. To this end, vertical elements should (see Figure 2.9b to 2.9e):

- Stay in the elastic range up their full height
- Rotate about their base, either at a flexural 'plastic hinge' they form just above their connection to the foundation (Figures 2.9b and 2.9d), or by rigid-body rotation of their individual footings relative to the ground (Figures 2.9c and 2.9e)

Such a side-sway plastic mechanism is kinematically possible, only if plastic hinges also form at both ends of every single beam of the system ('beam-sway' mechanisms). This produces the widest possible spreading of the global displacement demand through the structural system and minimises the local deformation demands on individual members or locations.

If the intended distributed plastic hinging in Figures 2.9b to 2.9e takes place simultaneously throughout the structure, beam ends and the bases of vertical elements will develop a chord rotation (angle between the normal to the member section at a member end and the chord connecting the two member ends) approximately equal to the roof displacement, δ, divided by the total building height, H_{tot} (i.e. to the average drift ratio of the building, δ/H_{tot}). Besides, the chord rotation ductility factor demand at member ends (peak chord rotation demand during the response, divided by the chord rotation at yielding of that end) is roughly equal to the demand value of the top displacement ductility factor, μ_δ. Under the design seismic action, μ_δ is about equal to q_μ (see Equation 3.120), that is, well within the capacity of concrete members with end regions detailed for ductility per Section 5.7. So, in the context of protecting life and fulfilling the no-collapse requirement, the 'beam-sway' mechanisms of Figures 2.9b to 2.9e allow to achieve, relatively easily and economically, fairly high q-factor values.

In the 'storey-sway' mechanism of Figure 2.9a, all inelastic deformations take place in the single 'soft-storey', with plastic hinging at both ends of all vertical elements in the storey in counter-flexure. The chord rotation demands at the ends of these vertical elements approach the ratio of the roof displacement, δ, to the soft-storey height, h_i. So, they are H_{tot}/h_i times larger than those of a 'beam-sway' mechanism. The chord rotation ductility factor is about equal to H_{tot}/h_i times the global displacement ductility factor, μ_δ, derived from the q_μ-factor via Equations 3.119 and 3.120 (cf. Equations 3.122 and 3.123 and Figure 3.15 in Section 3.26). No mid- or high-rise building can withstand such demands in its columns.

To spread the global inelastic deformation demands to the entire structural system and prevent a 'soft-storey', the building needs a strong and stiff spine of vertical elements, which by virtue of their geometry and/or design will stay elastic above their base under any earthquake. This is achieved by overdesigning them (except at the base section) relative to the horizontal ones and/or the action effects from the analysis. Sections 5.4.1 and 5.6.1.1 present in detail how this is pursued through 'capacity design' of columns or walls, respectively.

In addition to their vital role in spreading the total deformation and energy dissipation demands to the entire structural system, vertical elements also meet prioritisation criteria 1 and 2 of Section 4.5.1 for choosing which elements to capacity-design; compared to beams, they are:

- Inherently less 'ductile', because axial compression adversely affects ductility
- More important for the stability and integrity of the whole structure

However, concerning criterion 3, columns are easier to repair than beams, as they are accessible from all sides.

Eurocode 8 promotes beam-sway mechanisms through multiple means, direct or indirect:

- In frame- or frame-equivalent dual systems: by capacity design of the columns to be stronger in flexure than the beams and, therefore, be spared from plastic hinging (see Section 5.4.1)
- In wall- and wall-equivalent dual systems: by overdesigning them above the base, to remain elastic in flexure (see Section 5.6.1.1) and by entitling them to q-factor values comparable to those of frame- or frame-equivalent dual systems (see Section 4.6), despite their poorer redundancy and the inherently lower ductility of walls
- Through the Eurocode 8 limits on inter-storey drifts (computed for elastic response to the damage limitation seismic action, using the cracked stiffness of concrete members, see Section 1.3.2): these limits cannot be met without walls or good-size columns

4.5.3 Capacity design in the context of detailed design for earthquake resistance

Capacity design is applied as follows in the context of detailed seismic design using linear analysis with the q-factor (cf. Section 5.1.1):

1. Detailed design starts with dimensioning for the ultimate limit state (ULS) in flexure (for the bending moment and axial force pairs from the analysis for all applicable ULS design situations) and detailing of the longitudinal reinforcement at those locations which are considered as appropriate/convenient to detail for cyclic ductility and energy dissipation and where flexural plastic hinges are foreseen/allowed in a 'beam-sway' plastic mechanism (called 'dissipative zones' in Eurocode 8):
   a. All beam ends connected to vertical elements (see Sections 5.3.1, 5.3.2, 5.7.1, 5.7.3 and 5.7.4).
   b. The base section of all vertical elements (at the connection to the foundation, see Sections 5.4.2 for columns, 5.6.1 for walls).
   c. The top and bottom regions of those columns which Eurocode 8 exempts from capacity design (see Sections 5.4.1 and 5.4.2).
2. All elements in shear and the regions of vertical elements outside 'dissipative zones' in flexure are dimensioned to stay elastic after flexural yielding of the 'dissipative zones'. To this end, they are overdesigned with respect to the relevant action effects from the analysis, normally through 'capacity design', employing equilibrium and the overstrength flexural capacities, $\gamma_{Rd}M_{Rd}$, of the already dimensioned 'dissipative zones'.
3. 'Dissipative zones' are detailed to provide ductility capacity according to the deformation demands imposed on them by force-based design with the chosen q-factor.
4. The ground, and normally the foundation elements themselves, are normally capacity-designed to stay elastic when the 'dissipative zones' in the superstructure reach their overstrength flexural capacities (Section 6.3.2); Eurocode 8 allows also the option to dimension and detail foundation elements for ductility, as in the superstructure, despite the difficulty of repairing them.

4.6 DUCTILITY CLASSIFICATION

4.6.1 Ductility as an alternative to strength

According to Equations 3.119 and 3.120, the design seismic forces are approximately inversely proportional to the global displacement ductility factor, μ_δ. Therefore, increasing the available global ductility reduces the internal forces for the dimensioning of structural members, hence possibly their cost (see also Section 4.1 and Figure 4.1). Apart from any cost benefits, ductility has several advantages as a substitute for strength:

- A high q-factor makes it feasible, or easier, to verify the foundation soil, which is normally done on the basis of strength, not of deformation capacity.
- Reduced strength serves as a physical upper limit on the inertial forces and the response accelerations that can develop in the structure, 'isolating' from them, hence protecting, any contents of the building and non-structural parts that are sensitive to acceleration.
- An ample ductility supply enhances robustness and resilience of the building to earthquakes stronger than the design seismic action and its sensitivity to the uncertain details of the ground motion.

However, high lateral force resistance, in lieu of enhanced ductility, offers other advantages:

- By helping the structure to stay elastic under more frequent, moderate earthquakes, higher strength reduces structural damage and improves usability after the event. Structural damage is also reduced under the design seismic action.
- Detailing of members just for strength, instead of ductility, is easier and simpler. It can be done more reliably, especially when the technical level of workmanship is not very high.
- Force-based design against non-seismic actions (including wind) provides certain lateral strength for free, to be used for earthquake resistance as well, without costly and demanding detailing of members for ductility.
- If the structural layout is unusually complex and irregular, outside the scope of seismic design standards addressing mainly ordinary layouts, the designer may feel more confident for his/her design by narrowing the gap between the results of linear analysis used to dimension the members and the non-linear seismic response to the design seismic action, through a lower q value.

In view of the different advantages of both possible design choices (ductility vs. strength), it is up to the designer to decide what is the best option for each specific situation at hand. In this context, and as explained in detail in the next section, Eurocode 8 introduces three different 'Ductility Classes', leaving the choice to the designer. However, national authorities may set some limitations to such a choice.

4.6.2 Ductility Classes in Eurocode 8

Eurocode 8 allows trading ductility for strength by providing rules for three alternative ductility classes (DCs):

1. Ductility Class Low (DC L)
2. Ductility Class Medium (DC M)
3. Ductility High (DC H)

4.6.2.1 Ductility Class L (low): Use and behaviour factor

Buildings of DC L are not designed for ductility; only for strength. Except certain minimum conditions for the ductility of reinforcing steel (see Table 5.6 in Section 5.7.2), they have to follow just the dimensioning and detailing rules specified in Eurocode 2 for non-seismic actions, for example, wind. Although they are expected to stay elastic under the combination of the design seismic action and the concurrent gravity loads (the 'seismic design situation'), they can use a behaviour factor value of $q = 1.5$ instead of $q = 1.0$, thanks to member overstrength due to (cf. Section 3.2.4):

- The difference between the mean strength of steel and in-situ concrete and the design values (5%-fractile strengths divided by partial material factors, see Section 5.1.2)
- The possible control of the amount of reinforcement in some critical sections by the requirements for non-seismic actions or by minimum reinforcement
- The use of the same reinforcement at the cross-sections of a beam or column across a joint, determined by the most demanding of these two sections; rounding-up of the number and/or diameter of bars, etc.

DC L buildings are not cost-effective for moderate or high seismicity. Moreover, lacking engineered ductility, they may also lack a reliable safety margin against earthquakes stronger than the design seismic action. So, they are not considered suitable for moderate or high seismicity regions. Eurocode 8 recommends using DC L only for 'low seismicity cases', but leaves this decision to the National Annex, along with the definition of what is a 'low seismicity case': its recommendation is to consider it as such, if the design ground acceleration on rock, a_g (including the importance factor, γ_I), does not exceed 0.08 g, or the design acceleration on the ground, a_gS, is not more than 0.1 g (see Section 3.1.3 for a_g and S).

Eurocode 8 allows also to use DC L if the seismic design base shear at the level of the foundation or the top of a rigid basement for $q = 1.5$ is less than the base shear due to the design wind, or any other lateral action for which design is based on linear analysis.

4.6.2.2 Ductility Classes M (medium) and H (high) and their use

Seismic design for lateral strength alone without engineered ductility is an extreme, for use only in the special cases highlighted in Section 4.6.2.1. In the prime case of seismic design, that is, based on ductility and energy dissipation, Eurocode 8 gives the option to design for more strength and less ductility or vice-versa, by choosing between Ductility Class M or H.

Buildings of DC M or H have q-factor values higher than the default value of 1.5 used for DC L and considered as due to overstrength alone. DC H buildings enjoy higher values of q than DC M ones; in return, they are subject to stricter detailing rules (see Tables 5.1 to 5.5) and have higher safety margins in capacity design against shear (see Sections 5.5 and 5.6). However, unlike DC L, DC M does not systematically require more steel than DC H: the total quantities of materials are essentially the same; in DC H, transverse reinforcement and vertical members have a larger share of the total quantity of steel than in DC M.

DC M and H are expected to achieve about the same performance under the design seismic action, but DC M is slightly easier to design and implement and may give better performance in moderate earthquakes. DC H may provide larger safety margins than M against collapse under earthquakes (much) stronger than the design seismic action and may be more economic for high seismicity, especially if there is a strong local tradition and expertise in seismic design and on-site implementation of complex detailing.

Eurocode 8 does not relate the choice between DC M and H to seismicity or the importance of the structure, nor puts limits to their application. Countries are free to choose for the various parts of their territory and types of construction. They would better, though, leave this choice to the designer, depending on the specifics of the project.

4.6.3 Behaviour factor of DC M and H buildings

In Eurocode 8, the value of the behaviour factor, q, of DC M and H buildings depends on:

- The Ductility Class
- The type of lateral-force-resisting-system
- The regularity or lack thereof of the structural system in elevation

The value of the q-factor is linked, indirectly (through the ductility classification) or directly (see Section 5.7.3), to the local ductility and detailing requirements for members.

Table 4.1 lists the values of the q-factor for buildings which are regular in elevation per the Eurocode 8 criteria in Section 4.3. These values are called basic values, q_o, of the q-factor and are the ones linked to local ductility demands and member detailing (see Section 5.7.3). The value of q used for the calculation of the seismic action effects from linear analysis is reduced with respect to q_o:

Table 4.1 Basic value, q_o, of behaviour factor per EC8 for height-wise regular buildings

Lateral-load-resisting structural system:	DC M	DC H
1 Inverted pendulum	1.5	2
2 Torsionally flexible	2	3
3 Uncoupled wall system, not in one of the two categories above	3	$4\alpha_u/\alpha_1$
4 Any structural system other than the above	$3\alpha_u/\alpha_1$	$4.5\alpha_u/\alpha_1$

1. In buildings irregular in elevation per Eurocode 8 (see Sections 4.3.4 and 4.3.5): to $q = 0.8q_o$.
2. In wall, wall-equivalent dual or 'torsionally flexible' systems, to $(1 + \alpha_o)q/3 \geq 0.5q$, where q may be reduced per 1 above if there is irregularity in elevation, and α_o (≤ 2) is the mean aspect ratio of the walls in the system (sum of wall heights, h_{wi}, divided by the sum of wall cross-sectional lengths, l_{wi}); this last reduction reflects the adverse effect of low shear span ratio on wall ductility for $\alpha_o < 2$ (a value corresponding to a mean shear span ratio of the walls in the system less than about 1.65, which are non-ductile).

The above reductions of q notwithstanding, DC M and H buildings are entitled to a final q-factor value of 1.5, considered to be always available thanks to overstrength alone.

An 'inverted pendulum system' is, per Eurocode 8, a building with at least 50% of the mass in the top third of its height, or with energy dissipation possible only at the base of one element (Eurocode 8 excludes from this category one-storey frame systems having all columns connected at the top through beams in both horizontal directions and a maximum value of normalised axial load, ν_d, among all combinations of the design seismic action with the concurrent gravity loads, which is less or equal to 0.3). The low q-factors of 'inverted pendulum system' in row 1 are due to poor redundancy and sensitivity to $P - \Delta$ effects or overturning moments.

According to Eurocode 8, a system is 'torsionally flexible' if, at any floor, the radius of gyration of the floor mass exceeds the torsional radius in one or both of the two main directions in plan. As pointed out in Section 4.3.2, it is also considered in Eurocode 8 as planwise irregular. Its low q-factor value in row 2 of Table 4.1 reflects the increased likelihood of twisting about the vertical, to which the perimeter elements of the building are sensitive.

The types of system in rows 3 and 4 of Table 4.1 have been defined in Section 4.4.4.1. Except for uncoupled wall systems of DC M, their q-factor includes explicitly an overstrength factor α_u/α_1 due to redundancy of the structural system. This is in addition to the factor of 1.5 due to overstrength of materials and elements (as in DC L), which is hidden in the DC M or H q-factor values. α_u/α_1 is the ratio of: a) the seismic action that turns the building into a full side-sway plastic mechanism, to b) the seismic action at formation of the first plastic hinge in the system (with the quasi-permanent gravity loads acting together with both these seismic actions); α_1 is the lowest value of $(M_{Rd} - M_V)/M_E$ among all members (M_{Rd} is the design value of moment resistance at the member end and M_E, M_V the bending moments there from elastic analysis for the design seismic action and the quasi-permanent gravity loads, respectively); α_u may be computed as the ratio of:

1. The seismic base shear causing a full plastic mechanism according to non-linear static ('pushover') analysis per Section 3.3.2, to
2. The base shear due to the design seismic action

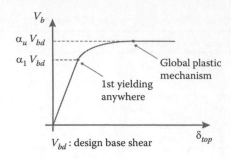

Figure 4.13 Definition of factors α_u and α_1 on the basis of a base shear vs. top displacement diagram from pushover analysis.

For consistency with α_1, pushover analysis should use the design values, M_{Rd}, of moment resistance at member ends (Figure 4.13).

A practitioner is unlikely to carry out iterations of: (a) pushover analyses and (b) design based on elastic analysis, just to compute α_u/α_1 for the q-factor. So, Eurocode 8 gives default values of α_u/α_1. For buildings regular in plan, the default values increase with the redundancy of the system, as follows:

- $\alpha_u/\alpha_1 = 1.0$ for wall systems with only two uncoupled walls per horizontal direction
- $\alpha_u/\alpha_1 = 1.1$ for
  - One-storey frame systems or frame-equivalent dual ones
  - Wall systems with two or more uncoupled walls in the horizontal direction considered
- $\alpha_u/\alpha_1 = 1.2$ for
  - One-bay multi-storey frame systems and frame-equivalent dual ones
  - Wall-equivalent dual systems
  - Coupled wall systems
- $\alpha_u/\alpha_1 = 1.3$ for multi-storey multi-bay frames or frame-equivalent dual systems

In a building which is irregular in plan per Eurocode 8 (see Section 4.3.2), the default value of α_u/α_1 is the average of:

- 1.0
- The default value given above for buildings regular in plan

Values higher than the default may be used for α_u/α_1, but up to a maximum of 1.5, provided that the value used is confirmed by pushover analysis, after design with the resulting q-factor.

Buildings in rows 3 and 4 of Table 4.1 may use different q-factors in the two main horizontal directions, depending on the structural system and its vertical regularity or not in these two directions, but not by virtue of Ductility Class, which is the same for the entire building.

The relative magnitude of the values of q highlighted in the present section reflects the position of Eurocode 8 on the effects of the type and regularity of the lateral-force-resisting-system on its earthquake resistance. This is an aspect to keep in mind during conceptual design.

Examples 4.10 to 4.12 at the very end of this chapter illustrate some implications of the choice of Ductility Class, and of the corresponding value of the behaviour factor, for the design.

4.7 THE OPTION OF 'SECONDARY SEISMIC ELEMENTS'

Eurocode 8, like other seismic codes, distinguishes the structural members that have a secondary role and contribution to earthquake resistance from the rest, calling them 'secondary seismic' and 'primary seismic' members, respectively (henceforth called 'secondary' and 'primary' members). The contribution of 'secondary' members to the lateral stiffness and earthquake resistance of the building is not taken into account in the analysis for the seismic action. The building structure is considered to rely for its earthquake resistance only on 'primary' members: 'secondary' members are not considered as part of the lateral-load-resisting system.

Only 'primary' members are designed and detailed for earthquake resistance following the rules of Eurocode 8. By contrast, 'secondary' members follow the rules of Eurocode 2 and are fully considered and designed only for the non-seismic combinations of actions. The only requirement of Eurocode 8 on them is to maintain support of gravity loads under the most adverse displacements and deformations imposed on them in the seismic design situation, that is, by the design seismic action and the concurrent gravity loads (see Section 5.9).

The designer is free to choose which members, if any, he/she may consider as 'secondary', subject to two restrictions introduced in Eurocode 8:

1. The total contribution to lateral stiffness of all 'secondary' members may not exceed 15% of that of all 'primary' ones.
2. The characterisation of some of members as 'secondary' may not change the classification of the structure from irregular to regular.
   So:
   a. If a frame, a column or a wall does not continue through the full height of the relevant part of the building, it cannot be classified as 'secondary'.
   b. If there is an abrupt change in the storey stiffness or (in infilled frame buildings) in the storey overstrength, this variation cannot be smoothened out by classifying some vertical elements as 'secondary'.
   c. The eccentricity between any storey's centres of mass and stiffness may not be reduced from over 30% of the storey's torsional radius to less, and the torsional radius in any direction may not increase from less than the radius of gyration of the masses to more, by classifying some vertical elements as 'secondary', etc.

The main reason to consider as 'secondary' some of the members of a building designed for DC M or H is if they do not fall within the scope of Eurocode 8 for seismic design based on energy dissipation and ductility: flat slab frames and post-tensioned girders are prime examples. So, if the designer wants to use these types of concrete elements in a DC M or DC H building, he/she may have to rely for the seismic action only on walls or strong frames (usually along the perimeter), designating flat slabs, post-tensioned girders and their supporting columns as 'secondary' members. As a matter of fact, in frame or frame-equivalent dual systems, columns supporting post tensioned girders had better be taken as 'secondary' anyway: normally the large size of prestressed girders makes it unfeasible to satisfy the strong-column/weak-beam capacity design rule, Equation 5.31; moreover, such columns should have a cross-section sufficient for the support of gravity loads, but otherwise as small as feasible, in order to reduce the 'parasitic' shears developing in these columns upon post-tensioning at the expense of the axial force in the girder.

The designer may also want to consider as 'secondary' those members which – owing to architectural constraints – do not conform to the rules for geometry, dimensioning or detailing for energy dissipation and ductility, for example, beams which:

- Are connected to columns at an eccentricity violating Equation 4.10
- Are supported on columns which are not large enough to satisfy the Eurocode 8 rule for bond and maximum diameter of the top bars of the beam within the joint (see Section 5.2.3.3); or
- Connect closely-spaced columns and hence develop a high seismic shear force (e.g. a large capacity-design shear from Equations 5.42, owing to the short clear span, l_{cl}) that cannot be verified for the ULS in shear.

Unlike the cases which are outside the scope of Eurocode 8's design rules for energy dissipation and ductility, those cases mentioned earlier should preferably be accommodated through proper selection of the local structural layout, instead of resorting to 'secondary' members. There are two good reasons for doing so:

1. The earthquake 'perceives' the structure as built, neither 'knowing' much nor 'caring' about the considerations and assumptions made in its design calculations. So, the 'primary' members may perform well thanks to their ductility, but the 'secondary' ones may suffer serious damage.
2. A structural system that cannot be utilised in its entirety for the engineered earthquake resistance of the building is a waste of resources. This is more so, given the conservatism of the special design requirements for 'secondary members' (see Section 5.9).

That said, the option of designing the entire structural system for strength instead of ductility (see Section 4.6.2.1) may be worth considering. In the framework of Eurocode 8, this means selecting DC L (Low) and $q = 1.5$. Then it is not necessary to make a distinction between 'secondary' and 'primary' members, as all members can be designed and detailed according to Eurocode 2, both for seismic and for non-seismic actions, without any regard to the special detailing and dimensioning rules of Eurocode 8 for energy dissipation and ductility.

EXAMPLE 4.1

The building shown in Figure 4.14, 20×35 m in plan, has columns on a 5×5 m grid and shear walls (with dimensions shown in m, 250 mm in thickness) in three alternative arrangements, (a), (b), (c), all with the same total cross-sectional area of the shear walls. Compare the three alternatives, taking into account the restraint of floor shrinkage, the lateral stiffness and the torsional one with respect to the vertical axis, the vertical reinforcement required for the same total flexural capacity at the base, the static eccentricity, the system's redundancy, etc.

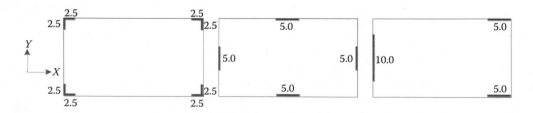

Figure 4.14 Example 4.1.

Answer

The volume of concrete is the same in all three options. At first sight, option (a) seems to make better use of it, because all four walls have biaxial strength and stiffness and are well placed to maximise the overall torsional stiffness with respect to the vertical axis. However, the walls of the two other options provide larger total lateral stiffness to both horizontal directions, as well as torsional stiffness with respect to the vertical. For the same vertical reinforcement ratio, they also give larger flexural resistance than those in option (a) thanks to their geometry and, secondarily, their larger axial load (due to their larger tributary floor area). Moreover, in option (a) the walls restrain shrinkage of the floors and may lead to cracking. It is also difficult to provide an effective foundation to a wall at a corner in plan, as in option (a). Compared to (b), option (c) provides larger total lateral stiffness and flexural resistance in horizontal direction Y, as well as torsional stiffness with respect to the vertical axis. It has very large eccentricity of the centre of mass with respect to those of stiffness and resistance (which are almost at the centre of the 10 m long wall); this large eccentricity is less of a problem than it seems at first sight, because it is partly resisted by the contribution to torsion about the vertical axis of the two walls in X (similarly to case 3 in Figure 4.9). The main handicap of option (c) is its lack of redundancy in direction Y and the lack of a load path other than through the 10 m long wall. For these reasons, the ideally balanced option (b) seems better. However, its two walls per direction still provide poor redundancy.

EXAMPLE 4.2

In the structural systems sketched in elevation as (a) and (b) (Figure 4.15), cross-hatched regions denote walls and vertical lines are columns. Compare the two systems with regard to: (i) regularity in elevation and (ii) suitability for earthquake resistance.

Answer

Regularity in elevation: System (a) is irregular in elevation, because the wall, which is its main source of lateral force resistance, does not continue to the top. If the criterion for irregularity in elevation is storey lateral stiffness and resistance, system (b) may nominally be less irregular than (a), because these properties are nominally not so much affected by the offset in the wall at floor 4, as by the termination of the wall there in case (a).

Suitability for earthquake resistance: System (b) has a very severe discontinuity in the load path at floor 4, which will lead to more adverse and uncertain response than the termination of the wall at that floor in system (a). In principle, system (a) can be designed and detailed for the concentration of inelastic deformation demands at the bottom of the

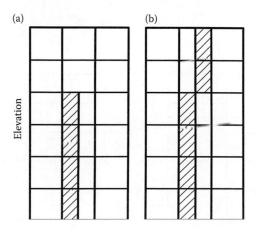

(a)   (b)

Elevation

Figure 4.15 (a–b) Example 4.2.

fifth storey columns and can be capacity-designed against a soft-storey mechanism at that storey. System (b) cannot be reliably designed for predictable seismic response; it is absolutely unsuitable for earthquake resistance.

EXAMPLE 4.3

Compare the two systems (a) and (b) (Figure 4.16) concerning earthquake resistance.

Answer

Both systems are irregular in elevation, owing to the drastic change of the horizontal dimension at floor 2. However, system (b) is much more adverse for earthquake resistance for many reasons: (1) The outer columns do not continue to the ground; at the second floor their action effects need to be transferred to the central columns, which continue to the ground, via the horizontal elements and the floor diaphragm at that level; (2) above floor 2, only the central part of the frame is engaged in inelastic action for earthquake resistance; the outer ones follow its displacements, staying in the elastic regime; (3) the central part of the frame, which provides almost all of the earthquake resistance, has less redundancy and a smaller number of possible load paths; and (4) the resultant of lateral forces is applied higher up, while the width of the base (distance between the outer columns) is much smaller; this combination increases very much the seismic axial forces at the base of the outer columns and the footings underneath, making the verification of these columns at the ULS in flexure with axial load very difficult, as well as that of their footings for the corresponding seismic action effects.

EXAMPLE 4.4

Comment on the layout of the framing plan shown in Figure 4.17 concerning earthquake resistance in the two horizontal directions X or Y (dots are columns, lines depict beams).

Answer

The building is characterised by perfect symmetry and uniformity in plan. At each corner, the area between the outline of the floor and the convex polygonal line enveloping the floor is about 2% of the floor area, well below the 5% limit set in Eurocode 8 for regularity in plan. In direction X, all the frames are continuous from one side to the opposite. However, in Y, all interior frames are one-bay; there is no continuous frame from one side to the other, except for the two 3-bay exterior ones. So, the building suffers in that direction from lower redundancy and multiplicity of load paths, fewer plastic hinges in beams (56 per storey in direction X and 36 per storey in direction Y) and less cost-effective use of the concrete in the frames.

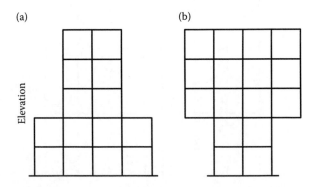

Figure 4.16 (a–b) Example 4.3.

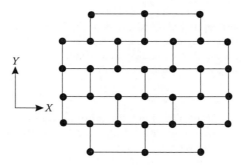

Figure 4.17 Example 4.3.

EXAMPLE 4.5

A system, whose centres of mass and lateral stiffness coincide in plan, has three uncoupled DOFs: the translations in the two orthogonal horizontal directions, X and Y, and twisting about the vertical axis, Z. Show that the torsional rigidity conditions of Eurocode 8 (i.e. torsional radii greater than the radius of gyration of the mass) imply that the period of the twisting mode, T_θ, is shorter than those of the translational ones in X, T_X, and Y, T_Y.

Answer

$T_X = 2\pi\sqrt{(M/K_X)}$, $T_Y = 2\pi\sqrt{(M/K_Y)}$, $T_\theta = 2\pi\sqrt{(I_\theta/K_\theta)}$, where: K_X, K_Y, K_θ: lateral stiffness in X, Y, torsional stiffness about a vertical axis through the centre of mass and stiffness, M, I_θ: mass and rotary moment of inertia about vertical axis through centre of mass and stiffness.

$$T_X > T_\theta \rightarrow M/K_X > I_\theta/K_\theta \rightarrow K_\theta/K_X > I_\theta/M \rightarrow r_Y = \sqrt{(K_\theta/K_X)} > l_s = \sqrt{(I_\theta/M)}$$

$$T_Y > T_\theta \rightarrow M/K_Y > I_\theta/K_\theta \rightarrow K_\theta/K_Y > I_\theta/M \rightarrow r_X = \sqrt{(K_\theta/K_Y)} > l_s = \sqrt{(I_\theta/M)}$$

EXAMPLE 4.6

A building has storey masses uniformly distributed over the floor area and a structural system consisting of several regularly spaced and similar plane frames in each one of the two orthogonal horizontal directions, X and Y, except for the two exterior frames in each direction, which have half the lateral stiffness of an individual interior frame of the same direction. Show that such a building cannot fulfill the torsional rigidity conditions in Eurocode 8 ($r_X \geq l_s$, $r_Y \geq l_s$), except as equalities and, indeed, only in the special case where the total lateral stiffness is the same in the two directions X and Y.

Answer

Let us denote by k_X, k_Y, m the lateral stiffness in X, Y, and the mass per unit floor area; they all have a constant value over the plan. Moreover, because of the uniformity of k_X, k_Y, m over the plan, the centres of mass and of lateral stiffness coincide. Let us introduce $a = k_Y/k_X$. The total lateral stiffness in X, Y, and the torsional stiffness about a vertical axis through the centre of stiffness are:

$K_X = \int_A k_X dA = k_X A$, $K_Y = \int_A k_Y dA = a k_X A$, $K_\theta = \int_A (y^2 k_X + x^2 k_Y) dA = k_X \int_A (y^2 + ax^2) dA$ $= k_X (I_X + a I_Y)$, where A, I_X, I_Y are the area and the moments of inertia with respect to the centroidal axes X and Y of the floor plan.

The torsional radii are: $r_Y = \sqrt{(K_\theta/K_X)} = \sqrt{[(I_X + a I_Y)/A]}$, $r_X = \sqrt{(K_\theta/K_Y)} = \sqrt{[(I_X + a I_Y)/(aA)]}$.
The radius of gyration of the mass is: $l_s = \sqrt{(I_\theta/M)} = \sqrt{[\int_A (y^2 m + x^2 m) dA]/[\int_A m dA]} = \sqrt{[(I_X + I_Y)/A]}$.

$r_X \geq l_s \rightarrow (I_X + a I_Y)/(aA) \geq (I_X + I_Y)/A \rightarrow , 1 \geq a$; $r_Y \geq l_s \rightarrow (I_X + a I_Y)/A \geq (I_X + I_Y)/A \rightarrow , a \geq 1$.

Therefore: $a = 1$, $r_X = l_s$, $r_Y = l_s$.

EXAMPLE 4.7

Discuss the suitability for earthquake resistance of the three-storey building depicted in Figure 4.18 (cross-sectional dimensions in cm), establish the eccentricity of the centre of mass (as centroid of floor plan) to the centre of stiffness (from the moments of inertia of the columns) and compare with the torsional radii.

Answer

Judging from the cross-sectional size alone, the columns, unless much more heavily reinforced than the beams (cross section height of beams is larger than the corresponding depth of columns), are weaker than the beams at all interior or exterior joints, with the exception of the connection of C8 and B10. So, the building is prone to a soft-storey mechanism. Beam B3 is indirectly supported on B9, and B7 is indirectly supported on B4; so, B3 does not form a proper moment resisting frame with C4, nor B7 with C3. Beams B5, B6 are offset; so, their connection to column C8 is doubly eccentric, and the behaviour of that beam–column joint for bending around global axis X is uncertain; the same can be said for the 2-bay frame these beams form with C7, C8, C9. There are only three frames, which are continuous from one side in plan to the opposite without offsets: that of B1, B2 along direction X, those of B9, B10 and B11, B12 in Y. There is a two-way eccentricity of the centre of mass with respect to the centre of stiffness, estimated below in a coordinate system X–Y with origin at the exterior corner of column C1:

Floor area = $9.825 \times 10.25 + 3.25 \times 0.5 = 102.33$ m^2
Co-ordinates, X_{CM}, Y_{CM}, of Centre of Mass, as centroid of the floor plan:

* $X_{CM} = (9.825^2 \times 10.25 + 3.25^2 \times 0.5)/(2 \times 102.33) = 4.86$ m.
* $Y_{CM} = (9.825 \times 10.25^2/2 + 3.25 \times 0.5 \times 10.5)/102.33 = 5.21$ m.

Moments of inertia of the floor plan with respect to its centroid:

* $I_X = (9.825^3 \times 10.25 + 3.25^3 \times 0.5)/3 - 102.33 \times 4.86^2 = 829.11$ m^4.
* $I_Y = (3.25 \times 10.75^3 + 6.575 \times 10.25^3)/3 - 102.33 \times 5.21^2 = 928.35$ m^4.

Radius of gyration of the floor plan area with respect to its centroid:

$l_s = \sqrt{[(829.11 + 928.35)/102.33]} = 4.144$ m

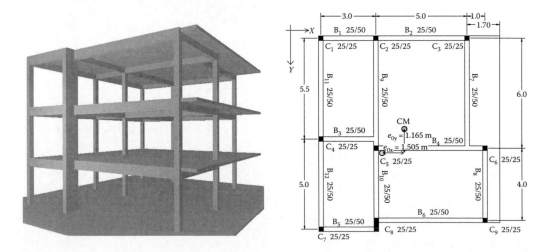

Figure 4.18 Example 4.7.

Co-ordinates, X_{CK}, Y_{CK}, of centre of stiffness, as centroid of the moments of inertia of the columns (the moment of inertia of a 0.25 m square column is symbolised by I):

- For response parallel to X: $\sum I_X = 3I$ (for C8) + $8I$ (for C1 to C7 and C9) = $11I$.
- For response parallel to Y: $\sum I_Y = 27I$ (for C8) + $8I$ (for C1 to C7 and C9) = $35I$.
- $X_{CK} = (0.125 \times 3I + 3.125 \times 29I + 8.125 \times I + 9.125 \times 2I)/(35I) = 3.355$ m,
- $Y_{CK} = (0.125 \times 3I + 5.625 \times I + 6.125 \times 2I + 10.125 \times I + 10.625 \times I + 10.375 \times 3I)/(11I) = 6.375$ m.

Torsional stiffness: $(0.125 - 3.355)^2 \times 3I + (3.125 - 3.355)^2 \times 29I + (8.125 - 3.355)^2 \times I + (9.125 - 3.355)^2 \times 2I + (0.125 - 6.375)^2 \times 3I + (5.625 - 6.375)^2 \times I + (6.125 - 6.375)^2 \times 2I + (10.125 - 6.375)^2 \times I + (10.625 - 6.375)^2 \times I + (10.375 - 6.375)^2 \times 3I = 320.2(m^2)I$

Torsional radii with respect to centre of stiffness and comparison with radius of gyration of floor plan:

- $r_X = \sqrt{[320.2I/35I]} = 3.025$ m $< l_s = 4.144$ m,
- $r_Y = \sqrt{[320.2I/11I]} = 5.395$ m $> l_s = 4.144$ m.

The building is torsionally flexible in direction y.
Eccentricities, e_{oX}, e_{oY}, of the centre of mass with respect to the centre of stiffness:

- $e_{oX} = X_{CM} - X_{CK} = 4.86 - 3.355 = 1.505$ m, $|e_{oX}| > 0.3r_X = 0.908$ m,
- $e_{oY} = Y_{CM} - Y_{CK} = 5.21 - 6.375 = -1.165$ m, $|e_{oY}| < 0.3r_Y = 1.62$ m.

The eccentricity in x is large enough to consider the building as irregular in plan.

EXAMPLE 4.8

A multi-storey building, with a quadrilateral plan as shown in Figure 4.19, has interior columns in an irregular pattern in plan that serves architectural and functional considerations. Partition walls and interior beams supporting the slab have different layout in different storeys. However, there is no constraint to the type, location and size of the lateral force resisting components and sub-systems on the perimeter. Proposals are made and justified for the choice of the lateral-load-resisting system and its foundation.

Answer

The irregular pattern of interior columns in plan and the varying layout of interior beams at different storeys prohibit the use of continuous in plan and elevation clear frames inside the building. So, the seismic action should be fully resisted by strong frames around the perimeter, preferably combined with a wall at about mid-length of each side. Interior beams should serve the support of slabs, as well as the pattern and the constraints due to architectural/functional considerations, with the minimum possible cross-section, to minimise the share of the seismic base shears resisted by the interior frames, at the expense of the contribution of the exterior lateral-load-resisting system;

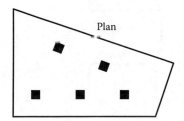

Figure 4.19 Example 4.8.

flat slabs, directly supported on the columns without beams, may be used at the interior. Only the lateral-load-resisting system on the perimeter counts then towards earthquake resistance ('primary seismic element' per Eurocode 8); the interior system does not ('secondary seismic elements' in Eurocode 8, taken into account only against gravity loads). A (nearly-basement-high) box foundation system is most appropriate, comprising a deep foundation beam on the perimeter for the lateral-load-resisting elements, footings for the interior columns, a top slab and a grid of tie-beams or a concrete slab at the bottom, connecting the footings with the base of the perimeter foundation beam, as convenient.

EXAMPLE 4.9

A three- to four-storey building is built on a slope (Figure 4.20). Wing ABCD (in plan) has three storeys and a frame structural system. Wing EFGH has a concrete core at the centre for an elevator shaft and staircase. Propose a foundation system for the two wings of the building and a structural system for the superstructure.

Answer

As there is no basement under wing ABCD, a general excavation to achieve the same foundation level, or to rigidly connect the foundation of the two wings, for them to have the same horizontal displacements, is not cost-effective. Moreover, the T-shape of the building in plan and the eccentric position of the elevator-cum-staircase shaft make the building irregular and introduce considerable uncertainty concerning its seismic response. Besides, if part ABCD does not have a basement anyway, it is not sensible to construct one just to provide a footing for the central core and a box foundation to the whole building.

The best option is to separate ABCD and EFGH into two statically independent, planwise regular wings, founded at different levels. Stiff lateral-load-resisting elements are

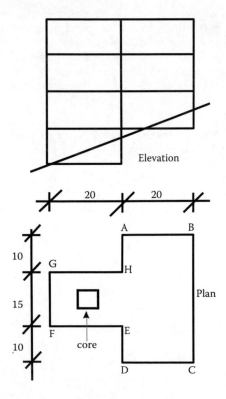

Figure 4.20 Example 4.9.

needed on the perimeter of EFGH, to increase the torsional stiffness of that part and balance the effect of the central core for the shaft.

EXAMPLE 4.10

A large building in a moderate seismicity region has three to five storeys over different parts of its plan and continuous concrete walls over most of the perimeter, with irregularly placed openings of various sizes. Choose the best option for its seismic design: low strength and high ductility or the opposite?

Answer

A large number of concrete walls, be it with openings, can provide a low-to-mid-rise building with sufficient strength to resist the design seismic action in a moderate seismicity region elastically (i.e. with $q = 1.5$), even with little reinforcement. Moreover, Eurocode 8 design and detailing rules for ductile walls of DC M or H are not meant for long walls with irregular openings. It is preferable to design such walls for nearly elastic response. Last but not least, linear analysis with a q-factor significantly larger than 1.5 cannot predict with any confidence the inelastic response of a system of geometrically complex walls to the design earthquake. Therefore, the prudent and, in all likelihood, most cost-effective choice for the seismic design of the building is for high strength and low ductility.

EXAMPLE 4.11

A cooling tower, with circular horizontal section and concrete shell thickness of 120 mm, is designed for wind with an average design value (including the partial safety factor) of $p = 2$ kN m^2 of projected vertical surface area. The thin tower shell, with its double curvature, is fairly stiff: its dynamic response is like that of a rigid body on flexible supports (a series of diagonal concrete columns), with uniform response acceleration up the tower and fundamental period in the constant pseudo-acceleration spectral range. Estimate the design ground acceleration at the site (including the importance factor), Sa_g, above which seismic design for DC L (i.e. with $q = 1.5$) governs over design for wind actions.

Answer

If H denotes the total height of the tower's shell and R_m the mean value of its diameter up the height, the design value of the lateral wind force is $2R_mHp$ and that of the seismic base shear (lateral seismic load resultant) is $0.85 \times (2\pi(R_mHt)\varepsilon S_{a,d})$, where $\varepsilon = 25$ kN m^3 is the unit weight of reinforced concrete and $S_{a,d} = 2.5Sa_g/q$ the design spectral acceleration (Sa_g in g's). For $2R_mHp > 0.85 \times 2\pi(R_mHt)\varepsilon S_{a,d} = 0.85 \times 2\pi(R_mH\varepsilon t)(2.5Sa_g/q)$, we need: $0.4qp/(0.85\pi\varepsilon t) > Sa_g$, that is, $0.4 \times 1.5 \times 2/(0.85 \times \pi \times 25 \times 0.12) > Sa_g$, that is, $Sa_g < 0.15$ g for wind to govern over seismic design with $q = 1.5$.

EXAMPLE 4.12

A concrete building has an aspect ratio ('slenderness') in elevation (: ratio of height from the foundation, H, to width of the base, B, parallel to the seismic action) of 4. The design ground acceleration at the site is $Sa_g = 0.3$ g, the corner period of the spectrum is $T_C = 0.6$ s and the fundamental period of the structure is $T = 0.8$ s. The building is designed with Eurocode 8's default values of q for plan-wise irregular, height-wise regular multi-storey, multi-bay frame- or frame-equivalent dual systems: $q = 4.5 \times 1.15 = 5.175$ for DC H, $q = 3 \times 1.15 = 3.45$ for DC M, and $q = 1.5$ for DC L. Determine the appropriate DC for the design of the building, if the design requirement is to have the resultant of the seismic lateral force (acting at 2/3 of the building's height from the foundation, H) and of the total weight of the building passing through: (a) the edge of the foundation in plan (nominal overturning, failure of the ground under the toe of the foundation);

(b) one-third of the base width, B, from the centre in plan (: safety factor of 1.5 against overturning); (c) one-sixth of B from the centre in plan (: uplift starts, for linear distribution of soil pressures).

Answer

The total design lateral force, V, equal to the building weight, W, times $0.85 \times (2.5Sa_g/q)$ (T_C/T) (Sa_g in g's), acting at 2/3 of the building's height from the foundation, H, produces an overturning moment at the base:

$$M_o = 0.85 \times (2.5Sa_g/q)(T_C/T)(2H/3)W$$

a. For the resultant of V and W to pass through the edge of the base in plan (nominal overturning):

$M_o = 0.85 \times (2.5Sa_g/q)(T_C/T)(2H/3)W \leq WB/2$, that is, $q \geq 2.55$ (requiring DC H or DC M).

b. For the resultant of V and W to pass from a point at one-third of the base width, B, from the centre (safety factor of 1.5 against overturning):

$M_o = 0.85 \times (2.5Sa_g/q)(T_C/T)(2H/3)W \leq WB/3$, that is, $q \geq 3.825$ (requiring DC H).

c. For the resultant of V and W to pass from a point $B/6$ from the centre (uplift starting):

$M_o = 0.85 \times (2.5Sa_g/q)(T_C/T)(2H/3)W \leq WB/6$, that is, $q \geq 7.65$ (not possible).

Witness how the choice of ductility class affects the design of the foundation. Even for a tall building in a high seismicity area, it is easy to prevent nominal overturning or failure of the ground under the toe of the foundation, if design is for DC M or H. If the goal is to retain a safety factor of 1.5 against nominal overturning (a common conventional goal in foundation design), design can only be for DC H. However, it is unfeasible to prevent uplift of the foundation under these conditions.

QUESTION 4.1

The building shown in Figure 4.21 consists of several structurally independent units separated by wide joints. All elements shown in light grey are of structural concrete. Do the criteria of Eurocode 8 for regularity in plan and elevation seem overall to be met? Which structural features of the building seem favourable for its earthquake resistance and which ones adverse? Does the building give an overall impression of being deficient in terms of seismic resistance?

Figure 4.21 Question 4.1.

QUESTION 4.2

The six-storey building in Figure 4.22 has an open ground floor, except for the 200 mm-thick solid masonry infills (shown cross-hatched) along the property lines between walls T2, T6 and T1. There are similar infills in the five storeys above, supplemented with 200 mm-thick infills with many openings on the street sides between walls T1, T2 and T5, T6 (shown in elevation), and 100 mm-thick masonry partitions at the interior, solid or with openings. Columns (denoted by K) and walls (denoted by T) are shown in solid dark. The complex core of walls at the centre houses an elevator and stairs. Ground storey beams are shown with the width of their web. Which features of the structural system and of the layout of the infills may adversely affect the earthquake resistance of the building? What may have contributed to the full failure/disintegration of all intermediate columns K1 to K3 and K12 to K14 of the façades at the ground floor in a past earthquake? What may have kept the beams supported on these columns from collapsing upon losing their intermediate supports and before propping?

QUESTION 4.3

A four-storey hotel building (Figure 4.23) has an open ground floor for the restaurant. Storeys 2 to 4 have one row of rooms along each long side in plan, separated by a corridor. The two short sides of the perimeter are fully infilled in all storeys, except for certain openings at the ends of the corridor at storeys 2 to 4 and along the right-hand side of the ground floor. There is a staircase near the upper left-hand corner, with straight flights between landings at floor levels and in-between floors. Interior and exterior walls are of 0.1 m- or 0.2 m-thick brick masonry, respectively. Columns, denoted by C.., are shown with their rectangular or L-shaped section; beams, denoted by B.., are shown with the width of their web; cross-section dimensions are written next to the member no. in meters (e.g. 0.2/0.7 next to a beam means web width 0.2 m and cross-sectional depth 0.7 cm). Comment on the features of the structural design and of the layout of infills which are important for earthquake resistance and seismic performance. How do they relate to the almost full collapse of this building (the extreme left-hand bay with the staircase survived, as well as one long-side façade and the frame along the right-hand side in plan)?

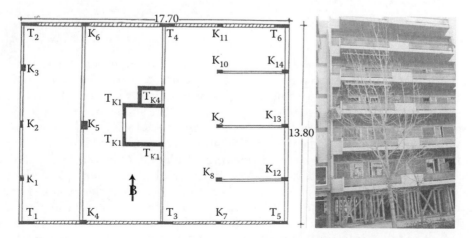

Figure 4.22 Question 4.2.

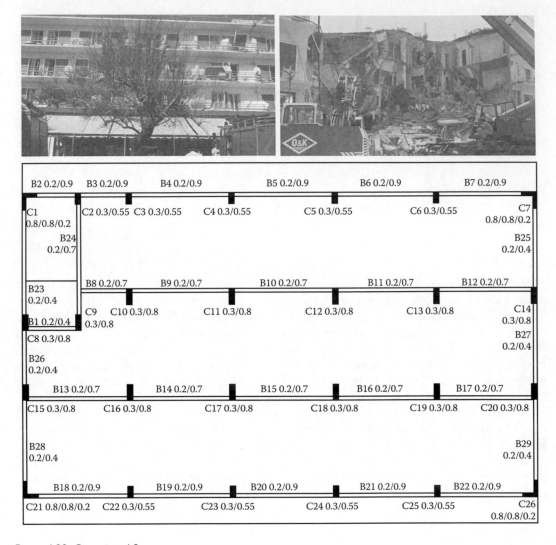

Figure 4.23 Question 4.3.

QUESTION 4.4

In a six-storey building (Figure 4.24) the ground floor is open, except for the 200 mm-thick solid masonry infills (shown hatched) along the property line on the left-hand-side K1-T1-K10 and the three bays around the staircase/elevator shaft area K8-K12-T5-T4. There are similar infills at the five overlying stories, but are supplemented with 200 mm-thick solid infills along the property line at the top side K15-K16-K17, 200 mm-thick infills with many openings along the rest of the perimeter and 100 mm-thick masonry partitions at the interior, solid or with openings. Columns (denoted by K) and walls (denoted by T) are shown solid dark, while beams with the projection of their web. Where do the centres of mass and stiffness seem to be located at the ground floor level (considering also the effect of the infills)? Does the building seem regular in plan according to the criteria of Eurocode 8? Comment on those features of the structural system and of the layout of infills in plan and elevation that adversely affect the seismic resistance

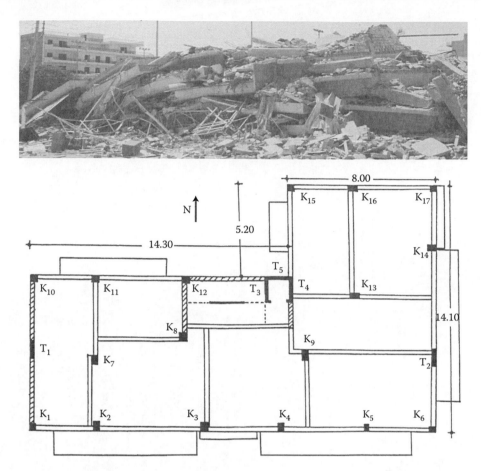

Figure 4.24 Question 4.4.

and performance. Which side of the building seems more likely to have kick-started its collapse in a past earthquake?

QUESTION 4.5

A seven-storey building has the ground storey open. Floors 6 and 7 are set back along the façade; floor 6 along the left-hand side too. The framing plan of storeys 2 to 5 is shown in Figure 4.25a and the foundation plan in Figure 4.25b. Five columns of the façade in storeys 2 to 5 (at the bottom side in Figure 4.25a) are supported at the tip of cantilevering beams without continuing to the foundation. Three concrete walls – shown in light grey in Figures 4.25a and 4.25b – are added at the only feasible locations on the perimeter for the purposes of seismic strengthening; the new footings for the added walls are shown with a dashed outline in Figure 4.25b. The axonometric view in Figure 4.25c shows the as-built configuration; that in Figure 4.25d has the three added walls.

Calculate the coordinates of the centres of mass and stiffness at ground storey (from the outline of the plan and the moments of inertia of vertical elements, respectively), the eccentricities, the torsional radii and the radius of gyration of the as-built and the retrofitted building, and characterise both of them as regular or not in plan and elevation,

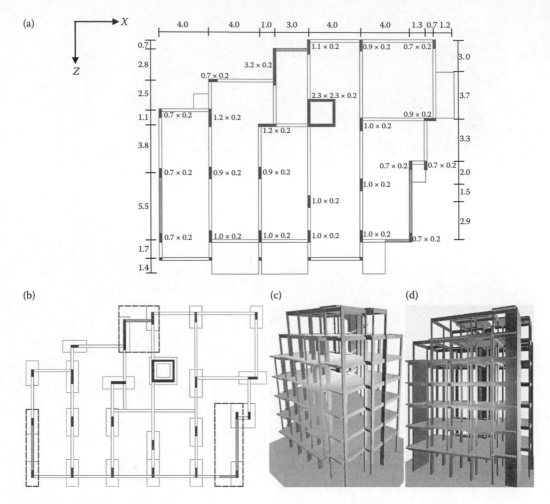

Figure 4.25 (a–d) Question 4.5.

according to all Eurocode 8 criteria. Comment on the effectiveness of the retrofitting concerning regularity and on features of the structural layout of both the as-built and the retrofitted building which are important for its earthquake resistance, stressing the ones you consider adverse.

QUESTION 4.6

For the depicted two-storey building (Figure 4.26), locate the centres of mass and stiffness at the ground storey from the outline of the plan and the moments of inertia of vertical elements (estimating their cross-sectional size from the other dimensions in plan, including a beam width of 0.3 m). Determine the eccentricities, the torsional radii and the radius of gyration. Characterise the building as regular or not in plan and elevation according to all Eurocode 8 criteria. Comment on the features of the structural layout which are important for the earthquake resistance of the building, pointing out the ones you consider unfavourable.

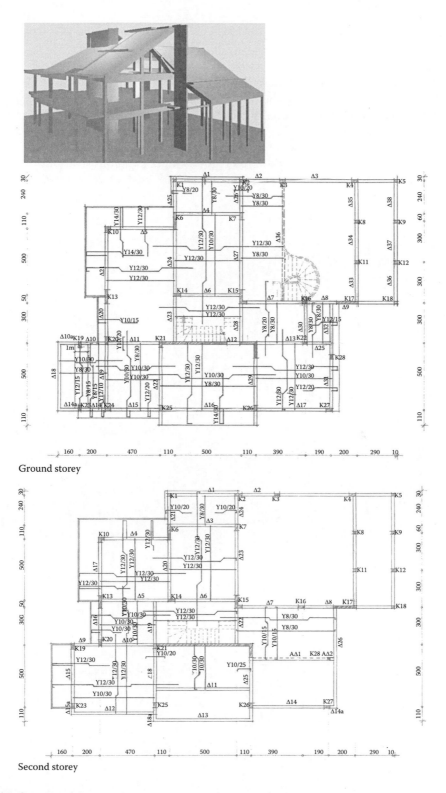

Ground storey

Second storey

Figure 4.26 Question 4.6.

QUESTION 4.7

For the building of Question 3.3:

1. Calculate the torsional radii and the radius of gyration for:
   a.   The space truss roof itself, on bearings, or
   b.   The perimeter frame, with the roof mass considered fixed to the cap beam.
   Check the condition for torsional flexibility, Equation 4.7; what is the conclusion of this check concerning regularity in plan? How does it compare with the conclusion from the natural periods calculated in Question 3.3?
2. Would you characterise the building as regular in plan and/or elevation?
3. Propose an appropriate Ductility Class and behaviour factor value for the design.

Chapter 5

Detailed seismic design of concrete buildings

5.1 INTRODUCTION

5.1.1 Sequence of operations in the detailed design for earthquake resistance

The subject of this chapter, detailed design, is the third stage in the overall design process. The second stage is the analysis for the design actions; Chapter 3 focussed specifically on the analysis for the seismic action. The analysis stage is preceded by conceptual design, the subject of Chapter 4. A sub-stage of conceptual design is 'sizing of members', that is, the selection of their cross-sectional dimensions, which in turn determine the member elastic stiffness, a necessary input to the analysis of the structural system for any action – including the seismic one. Therefore, 'sizing' should take place before any analysis and, as such, is part of conceptual design. However, it is addressed in this chapter, because it relates closely to detailed design rules and requirements for them, which are dealt with in other sections of this chapter.

Capacity design introduces strong interdependence across various phases of detailed design of a building per Eurocode 8, in the same member as well as between different ones, especially in frames:

- If the columns are capacity designed around joints to be stronger in flexure than the beams (see Section 4.5.2 and Equation 5.31 in Section 5.4.1), the longitudinal reinforcement of the beams should be known beforehand; to this end, the beams are the first members to be dimensioned; in fact, beam ends and the base section of walls are normally the only places whose detailed design is based exclusively on analysis results – in this case on their bending moments.
- Dimensioning of a column or a beam in shear depends on the longitudinal reinforcement of the column/beam itself and that of the members framing into it at either end, so that it is carried out after the amount and layout of the beam and column reinforcement have been determined (see Equations 5.42, 5.44 in Section 5.5.1).
- Dimensioning of any storey of a Ductility Class High (DC H) wall in shear depends on the vertical reinforcement at the base of the bottom storey (see first bullet point above and Equation 5.54 in Section 5.6.2.1); it should be undertaken after the amount and layout of the latter is determined.
- The design of isolated footings and of their tie-beams and the verification of the soil underneath depend on the longitudinal reinforcement of the vertical elements they support (see Section 6.3.2); so, it should take place afterwards.

The operations in detailed design should follow the sequence highlighted above, so that all the information needed at every step is available beforehand. Integrated computer programs for detailed design as per Eurocode 8 should be structured to perform these operations in the above sequence; even when the columns do not need to be capacity designed in order to be stronger than the beams as per Equation 5.31 in Section 5.4.1, it is convenient to follow the same general sequence, but using, in that case, for the dimensioning of the column in flexure its bending moments from the analysis, $M_{Ed,c}$, instead of the beam moment resistances, $M_{Rd,b}$.

The sequence suggested above is followed in the detailed description of the dimensioning steps across this chapter and in Section 6.3 of Chapter 6, as well as in the full example of Chapter 7. Throughout these chapters the abbreviation DC H, DC M and DC L is used for Ductility Class High, Medium and Low, respectively.

5.1.2 Material partial factors in ultimate limit state dimensioning of members

Eurocode 8 adopts the Eurocode 2 approach for ultimate limit state (ULS) design, where the general ULS verification, Equation 1.1, uses a design value of force or moment resistance, R_d, calculated from the design values of material strengths; the latter are obtained by dividing the nominal or characteristic values by the corresponding material partial factors:

- $f_{cd} = f_{ck}/\gamma_c$, for concrete
- $f_{yd} = f_{yk}/\gamma_s$, for steel

Being safety elements, the partial factors γ_c and γ_s are nationally determined parameters (NDPs). Eurocode 8 does not recommend values for them, but mentions the options of:

1. Using the values $\gamma_c = 1.5$, $\gamma_s = 1.15$, recommended in Eurocode 2 for the ULS design against non-seismic actions, or
2. Setting $\gamma_c = 1.0$, $\gamma_s = 1.0$, which are the recommended values for design against accidental actions

Option 1 is very convenient for the designer, as he/she may then dimension the elements to provide a design value of force resistance, R_d, at least equal to the largest design value of the action effect due to the non-seismic or the seismic combinations of actions. With option 2, elements have to be dimensioned once for the action effect due to the non-seismic combinations and then for that due to the seismic ones, each time using different values of γ_c and γ_s for R_d.

5.2 SIZING OF FRAME MEMBERS

5.2.1 Introduction

If member sizes are not judiciously selected from the outset, the designer will encounter problems in the detailed design phase after the analysis. For example:

- Failure of undersized members to meet the ULS verification in shear or (more rarely) in flexure, no matter the amount of their reinforcement.
- Extreme congestion of reinforcement in undersized members.
- Poor utilisation of materials in oversized members (which may thus have the minimum longitudinal reinforcement only), with undesirable distribution of overstrengths over the structure, leading to a concentration of inelasticity in members which are not oversized and so forth.

Such problems, especially of the first two types, require revising the member sizes and repeating the analysis. Besides, even if these two types of problem do not arise, a poor choice of member sizes may lead to substandard overall seismic performance and low cost-effectiveness of the building.

The following paragraphs give guidance on how to size beams and columns, in order to help meet the rules of Eurocodes 8 and 2 during the detailed design phase. Except for the procedure in Section 5.2.3.4, which, strictly speaking, requires knowledge of beam longitudinal reinforcement, the rest may be used during conceptual design before any analysis.

5.2.2 Sizing of beams

To facilitate continuity of the beam top and bottom bars across a column, the cross-section of beams should be the same in all bays of a plane frame.

Beam depth is often controlled by gravity loads, or (in frame buildings without shear walls) by drift control under the damage limitation earthquake (see Section 1.3.2); it is normally chosen around one-tenth of the average bay length in the frame.

The importance of a sizeable web width is sometimes overlooked; instead, the designer often tries to accommodate the beam within the thickness of a nonstructural wall under the beam. The web should be sufficiently wide:

1. To avoid undue congestion of longitudinal bars (preferably placed in one layer),
2. To provide at least the minimum concrete cover of stirrups at the sides of the beam per Part 1-1 of Eurocode 2, and
3. To provide at least the minimum mean axial distance of longitudinal bars to the concrete surface per Part 1-2 of Eurocode 2.

Note that, depending on the environmental exposure class and the specified fire rating, requirements under 2 and 3, respectively, may be quite restrictive. Concerning point 1: at the supports on columns, most of the top beam bars should be placed within the beam stirrups, but some may be outside these stirrups in a top slab; if h_f is the thickness of that slab, top bars may be placed within an effective flange width in tension (cf. Figure 2.22b) extending per Eurocode 8 on each side of the beam to the face of the column parallel to it, and even beyond, by:

- $4h_f$, if the column is interior in the direction of the beam and a similarly deep beam frames into the column in the transverse direction;
- $2h_f$, if the column is exterior in the direction of the beam and supports a similar transverse beam, or is interior but without a transverse beam.

Note also that if the column cross-sectional depth in the direction of the beam is small, the onerous restriction of the beam bar size as per Eurocode 8 (highlighted in Section 5.2.3.3) may result in a large number of small diameter bars, aggravating bar congestion at the supports on columns and requiring even wider beam webs. Examples at beam ends supported by relatively thin perimeter walls or columns may be found in Chapter 7 (see Figures 7.34 to 7.39).

The ideal connection of a beam with a column is concentric, with the column being wider than the web of the beam on each side by at least 50 mm, to allow the beam longitudinal bars to pass through the confined core of the column section between its outermost bars. If a fully concentric connection is not feasible, the eccentricity between the beam and the column centroidal axes is limited by Eurocode 8 according to Equation 4.10. To meet this condition, perimeter beams having the exterior side flush with that of the exterior columns should have a web wider than one-half of the largest cross-sectional dimension of the column at right angles to the beam axis. An example of this may be seen at the corner columns of the building in Chapter 7 (see Figure 7.2).

5.2.3 Sizing the columns

5.2.3.1 Introduction

Storey seismic shears and column axial forces decrease from the base to the roof; so, one may be tempted to reduce the column section in the upper storeys. However, field experience and tests provide strong evidence that such a reduction is detrimental for the seismic performance of columns at intermediate or upper storeys, especially in medium- to high-rise buildings. Besides, when the column section changes from one storey to the next, it is difficult to detail the transition of column bars through the joint. Moreover, for the same reinforcement ratio (often the minimum of 1% per Eurocode 8) and cross section, the column moment resistance decreases in the upper storeys owing to the reduction in column axial compression. Recall in this respect from Section 4.4.4.1 (and witness in Figures 7.9, 7.12, 7.15, 7.22) that the column seismic moments in dual (frame–wall) systems are often smaller in the lower storeys, while those due to gravity loads are invariably larger at the top storey (see Figures 7.27, 7.28, 7.30); so, if the column is smaller in the upper storeys, it may require more vertical reinforcement there. Therefore, except for serious architectural reasons, the size of a column should be kept constant in all storeys, as determined from the most critical one.

The most cost-effective option, which also serves the requirement to have a clear structural system, is to have as uniform a size of columns in the building as feasible: experience from past earthquakes shows that larger columns in the system are more likely to fail than the smaller ones, even when they have higher vertical steel ratio.

Eurocode 8 sets a minimum length of 200 or 250 mm, for a side of a DC M or H column, respectively. In addition, if the sensitivity coefficient for second-order effects, $\theta = P\delta/Vh$ (see Section 3.1.12) exceeds 0.1, the column sides should be at least equal to 5% of the distance of the inflection point to the column end further away, for bending within a plane parallel to the side. It should be pointed out, though, that these minimum values may govern only the narrow sides of sections composed of more than one rectangular part (T-, L-, C-, etc.). For all other column sections, the Eurocode 8 or 2 rules highlighted in Sections 5.2.3.3 to 5.2.3.4 are normally far more restrictive.

5.2.3.2 Upper limit on normalised axial load in columns

To ensure a minimum flexural ductility of the column, Eurocode 8 sets upper limits on its axial load ratio:

- For DC M:

$$v_d \leq 0.65 \tag{5.1a}$$

- For DC H:

$$v_d \leq 0.55 \tag{5.1b}$$

where $v_d = N_d/(A_c f_{cd})$, with N_d denoting the column axial load in the seismic design situation and A_c the column cross-sectional area. In order to choose A_c from the outset, so that

Equations 5.1 and other restrictions listed below are met, a value of N_d should be estimated before the analysis:

- If beams, parallel to the horizontal seismic action component considered, frame into the column from both sides (as they normally do in interior columns), a back-of-the-envelope calculation may give N_d as the total column tributary plan area in all floors (column tributary plan area in a typical floor times the number of overlying storeys) times the estimated quasi-permanent gravity load per unit floor area in the seismic design situation (typically from 7 to 9 kN/m²).
- If beams frame into the column along the horizontal direction of the seismic action only from one side (e.g. as in a column which is exterior in that direction of the earthquake), this value of N_d due to quasi-permanent gravity loads should be increased at each storey by the maximum possible beam shear, taken as the sum of the hogging moment resistance at the beam end framing into the column, plus the sagging one at the opposite end, divided by the clear beam span; if the minimum value of N_d in the seismic design situation is sought (e.g. for use in Equations 5.2), the beam shear is computed from the sum of the sagging moment resistance at the end framing into the column plus the hogging one at the opposite end and subtracted from the value of N_d due to quasi-permanent gravity loads. This calculation cannot be practically done before dimensioning the beams, so:
  - If the column in question is exterior, the total seismic axial force in the row of exterior columns of rectangular-in-plan buildings may be taken to be equal to the total seismic overturning moment at storey mid-height, divided by the plan dimension parallel to the horizontal direction of the seismic action: if the total building height is H_{tot} and all storeys have approximately the same mass, the total seismic overturning moment at mid-height of the ground storey – with storey height H_{st} – may be taken as the base shear times $(2/3)H_{tot} - H_{st}/6$; exterior columns share this force in proportion to their cross-sectional area.
  - If the column is interior in plan, its seismic axial force may be neglected (but this is an approximation questionable for columns not connected to beams on both sides).

5.2.3.3 Column size for anchorage of beam bars in beam–column joints

Eurocode 8 sets a very restrictive, albeit fully warranted, lower limit to the column depth, h_c, parallel to a beam framing into the column, to accommodate the very high bond stresses along the length of a beam bar inside an interior beam–column joint, or the anchorage of beam bars terminating in a joint, either exterior or not (cf. Figure 2.22a). If the bar diameter is d_{bL}, then the limit is:

- In an interior beam–column joint:

$$\frac{d_{bL}}{h_c} \leq \frac{7.5f_{ctm}}{\gamma_{Rd}f_{yd}} \cdot \frac{1 + 0.8\nu_d}{1 + k\dfrac{\rho_2}{\rho_{1,max}}} \tag{5.2a}$$

- In a beam–column joint that is exterior in the direction of the beam:

$$\frac{d_{bL}}{h_c} \leq \frac{7.5f_{ctm}}{\gamma_{Rd}f_{yd}} \cdot (1 + 0.8\nu_d) \tag{5.2b}$$

where:

- For DC M, $\gamma_{Rd} = 1.0$, $k = 0.5$.
- For DC H, $\gamma_{Rd} = 1.2$, $k = 0.75$.
- The value of $v_d = N_{Ed}/f_{cd}A_c$ is the minimum in all combinations of the design seismic action with the quasi-permanent gravity loads from the analysis or the rough estimation outlined in the previous sub-section ($v_d = 0$ for net axial tension, as may occur in exterior columns of medium- or high-rise buildings).
- $\rho_{1,\text{max}}$ is the maximum value of beam top reinforcement ratio and ρ_2 is the bottom steel ratio (taken as $\rho_2 = 0.5 \rho_{1,\text{max}}$, if the beam reinforcement is not known yet and only the combination of its maximum allowed diameter and the column depth is being sought).

The most critical issue for Equations 5.2 appears to be the top storey, but as the dependence on v_d is rather weak, those actually controlling are normally the storeys which require the largest amount of beam reinforcement at the support, to be accommodated with the minimum possible number of (larger) bars. For common axial load ratio values (e.g. $v_d \sim 0.2$) and steel grades (500 MPa) and for a low concrete grade ($f_{ck} = 20$ MPa), a column depth h_c of about 40 d_{bL} is required for DC H! The required size is relaxed to about 30 d_{bL} for medium-high axial loads and higher concrete grades. If DC M is chosen, the required column size is reduced by about 25%.

For a sample application of this section to the beams and columns of the seven-storey example building, see Section 7.6.2.1.

5.2.3.4 Sizing of columns to meet the slenderness limits in Eurocode 2

The Eurocode 2 rules concerning second-order effects in the analysis and verifications for gravity loads pose strong demands on the size of columns. Buildings designed for earthquake resistance do not necessarily meet by default the complex local and global lateral stiffness rules of Eurocode 2 against second-order effects. According to them, if such effects are important, the ULS resistance of members should be verified for internal forces from an analysis satisfying equilibrium in the deformed state and accounting for all effects that increase local and global deformations: cracking, material nonlinearities, creep, biaxial bending, soil flexibility, soil–structure interaction, postulated deviations of vertical members from the vertical, etc. This applies to the combination of actions taken into account at the ULS, that is, the 'persistent and transient' design situation in EN1990, where permanent and variable actions enter multiplied with partial load factors and deviations of vertical members from the vertical are considered. The seismic design situation is excluded, as it is covered by the specific provisions of Eurocode 8 for second-order effects (see Section 3.1.12). Eurocodes 2 and 8 both allow ignoring these effects, if they are less than 10% of the first-order ones and give simplified criteria to check whether they are. Given that the type of analysis per Eurocode 2, which accounts for second-order effects, is onerous, the designer should avoid it by meeting these criteria. Being deemed-to-satisfy rules, these simplified criteria are safe-sided and give larger member sizes than required by a rigorous analysis with second-order effects.

Eurocode 2 gives a simplified criterion for the slenderness ratio of isolated columns, λ:

$$\lambda \equiv \frac{l_0}{i_c} \leq \lambda_{\text{min}} = 20 \frac{ABC}{\sqrt{n}} \tag{5.3}$$

where l_0 is the column effective length, i_c the radius of gyration of the uncracked column section, and $n = N_{Ed}/A_c f_{cd}$, with N_{Ed} the column axial force in the 'persistent and transient' design situation in EN1990 (i.e. with permanent and imposed actions multiplied with their partial load factors). If necessary, N_{Ed} may be estimated before the analysis as the product of the total column tributary plan area in all floors (column tributary plan in a typical floor times the number of overlying storeys) times the sum of factored permanent load per unit floor area (estimated between 10 and 12 kN/m² in typical buildings) and the factored specified imposed load per unit floor area. Eurocode 2 gives default values for A, B and C:

- $A = 0.7$ for A, corresponding to an 'effective' creep coefficient of 2.1.
- $B = 1.1$ for $B = \sqrt{(1 + 2\, \rho_{tot} f_{yd}/f_{cd})}$; the default value corresponds to a column steel ratio, ρ_{tot}, slightly over the minimum of 0.4% recommended in Eurocode 2; $B = 1.2$ suits better the 1% minimum steel ratio in Eurocode 8.
- $C = 0.7$ for $C = 1 - M_{01}/M_{02}$, where M_{01}, M_{02} are the first-order end moments of the column, with $|M_{02}| \geq |M_{01}|$; $C = 0.7$ is recommended if the building is not braced by walls per Equation 5.11 below, or if the column's first-order moments are mainly due to lateral loads or postulated deviations from the vertical.

The effective length of the column is derived from its clear height, H_{cl}, as follows:

- For buildings not braced by walls with total moment of inertia meeting Equation 5.11:

$$l_0 = H_{cl} \cdot \max\left\{ \sqrt{1 + 10\frac{k_1 k_2}{k_1 + k_2}}; \left(1 + \frac{k_1}{1 + k_1}\right)\left(1 + \frac{k_2}{1 + k_2}\right) \right\} \tag{5.4a}$$

- For buildings braced by walls per Equation 5.11:

$$l_0 = 0.5 H_{cl} \cdot \sqrt{\left(1 + \frac{k_1}{0.45 + k_1}\right) \cdot \left(1 + \frac{k_2}{0.45 + k_2}\right)} \tag{5.4b}$$

In Equations 5.4, k_i is the column rotational stiffness at end node i (=1, 2) relative to the total restraining stiffness (moment, M_i, applied to the node divided by the resulting rotation, θ_i) of the members framing in node i in the plane of column bending considered:

$$k_i = \frac{\theta_i}{M_i} \sum \frac{EI_{c,eff}}{H_{cl}} = \frac{\sum EI_{c,eff}/H_{cl}}{4\sum EI_{c,eff}/H_{cl} + 4\sum EI_{b,eff}/L_{cl}} \tag{5.5}$$

The sums of column or beam stiffness values in Equation 5.5 are taken around node i. L_{cl} is the clear length of a beam framing into node i within the plane of column bending considered, and $EI_{b,eff}$ is this beam's cracked flexural rigidity, taking into account creep, which is computed with the design concrete modulus, $E_{cd} = E_{cm}/1.2$, divided further by $(1 + \varphi_{eff})$, where φ_{eff} is the final creep coefficient times the fraction of the total bending moment in the combination of actions due to quasi-permanent loads. If d is the beam effective depth, b the width of a flange that is in compression over its whole thickness, t, and b_w the thickness of the web, $EI_{b,eff}$ may be taken as

$$EI_{b,eff} = E_s bd^3 \left\{ \begin{array}{l} \dfrac{1}{a}\left[\dfrac{\xi^2}{2}\left(\dfrac{1+\delta}{2} - \dfrac{\xi}{3} \right)\dfrac{b_w}{b} + \left(1 - \dfrac{b_w}{b} \right)\left(\xi - \dfrac{t}{2d} \right)\left(1 - \dfrac{t}{2d} \right)\dfrac{t}{2d} \right] + \\[4mm] \dfrac{(1-\delta)}{2}\left[(1-\xi)\rho_1 + (\xi - \delta)\rho_2 + \dfrac{\rho_V}{6}(1-\delta) \right] \end{array} \right\} \tag{5.6}$$

where $\alpha = (1 + \varphi_{eff})E_s/E_{cd}$ is the ratio of effective elastic moduli (steel-to-concrete) and the neutral axis depth (normalised to d) is computed as

$$\xi = \sqrt{\alpha^2 A^2 + 2\alpha B} - \alpha A \tag{5.7}$$

where

$$A = \frac{b}{b_w}(\rho_1 + \rho_2 + \rho_v) + \frac{1}{\alpha}\frac{t}{d}\left(\frac{b}{b_w} - 1 \right),$$

$$B = \frac{b}{b_w}(\rho_1 + \rho_2\delta + 0.5\rho_v(1+\delta)) + \frac{1}{2\alpha}\left(\frac{t}{d} \right)^2\left(\frac{b}{b_w} - 1 \right) \tag{5.8}$$

In Equations 5.6 through 5.8 ρ_1, ρ_2 are the ratios of tension and compression reinforcement, ρ_v the ratio of longitudinal reinforcement at the sides of the web between the tension and compression steel (all ratios normalised to bd) and $\delta = d_1/d$ the centroidal distance of compression bars from the extreme compression fibres, normalised to d.

If there is no distinct compression flange, Equations 5.6 through 5.8 are applied with $b = b_w$. If there is one, but Equations 5.7, 5.8 give a neutral axis depth, ξd, less than the compression flange thickness, t, then Equations 5.6 through 5.8 are (re-)applied in a simplified form, with b_w taken equal to b.

If two beams parallel to the plane of column bending frame into opposite faces of node i, they should be considered in turn with the top flange of one beam taken in tension and that of the other in compression. Note that, although strictly speaking, the effective width of the slab should be included in b when the compression flange includes the slab, it makes little difference in the outcome of Equation 5.6 if the web width is taken as b, provided that this value is also used when ρ_1, ρ_2, ρ_v are normalised to bd.

Concerning the cracked flexural rigidity of a column, $EI_{c,eff}$, Eurocode 2 gives an approximation:

$$EI_{c,eff} = E_s I_s + E_{cd}\sqrt{\frac{f_{ck}(\text{MPa})}{20}}\frac{K_2 I_c}{1 + \varphi_{eff}} \tag{5.9}$$

where E_s and I_s are the elastic modulus and the moment of inertia of the section's reinforcement (which, if unknown, may be obtained at this stage from the minimum steel ratio of 1% in Eurocode 8) with respect to the centroid of the section; I_c is the moment of inertia of the uncracked gross concrete section, and K_2 is taken as

$$K_2 = \frac{n\lambda}{170} = \frac{1}{170}\frac{N_{Ed}}{A_c f_{cd}}\frac{l_0}{i_c} \le 0.20 \tag{5.10}$$

A minimum value $k_i = 0.1$ is recommended in Eurocode 2 at a column end where the column is fixed against rotation (here at the base of the ground storey). Note that the column one storey above has its lower end less restrained and hence may be more critical, despite its lower axial load. So, the minimum column size meeting Equation 5.3 throughout all storeys should be sought in the two lowest storeys.

As both the unknown effective length of the column, l_0, and the size of its section (through i_c and A_c) enter in Equations 5.3 through 5.5 and 5.9, 5.10 in an implicit nonlinear way, they have to be found by iterations, after dimensioning the top beam reinforcement at the supports to determine $EI_{b,eff}$.

Eurocode 2 allows to consider the building as braced in a given horizontal direction and to apply Equation 5.4b in lieu of Equation 5.4a, if it has walls with a total moment of inertia of the uncracked gross section in that horizontal direction, ΣI_w, which meets the following condition at the top of the foundation or of a rigid basement:

$$F_{V,Ed} \leq 0.31 \frac{n_{st}}{n_{st} + 1.6} \frac{E_{cd} \sum I_w}{H_{tot}^2} \tag{5.11}$$

where $F_{V,Ed}$ is the total vertical load acting on all n_{st} overlying storeys of the building throughout the plan area and H_{tot} is the height of the walls above the top of the foundation or of a rigid basement.

Equation 5.11 presumes that the bracing walls are cracked. If it can be shown that they stay uncracked while performing their bracing function in the ULS combination of actions considered (i.e. for the 'persistent and transient' design situation, with factored permanent and imposed loads and geometric imperfections), then the right-hand side of Equation 5.11 is multiplied by 2, reducing by 50% the minimum required value of ΣI_w. The bracing walls should be dimensioned at the ULS to resist the full lateral force on the building due to the deviation from the vertical postulated in Eurocode 2.

If the building is laterally braced by walls meeting the criterion of the above paragraph in a horizontal direction, then Equation 5.3 can be met with a column depth of reasonable magnitude in that direction (as in the example building of Chapter 7); otherwise they may come out quite large (Fardis et al. 2012). Walls that are collectively sufficient to laterally brace the building as per the previous paragraph, normally take a large enough fraction of the elastic base shear for the lateral force resisting system to qualify as dual (be it frame-equivalent). So, it is the columns of frame systems that are more severely penalised by the Eurocode 2 rules on second-order effects. On the positive side, the resulting large columns of laterally unbraced buildings, as well as the large walls necessary in braced ones, impart significant lateral force resistance and stiffness, thanks to which a building may perform well in an earthquake it has not been designed for (Fardis et al. 2012).

As pointed out at the closing of Section 5.2.1, in order to apply Equations 5.6 through 5.8 the beam longitudinal reinforcement should be known. This is possible only when the procedure in the present sub-section is applied in the context of detailed design after the beams are fully dimensioned and the designer wants to check if second-order effects may indeed be neglected. In that phase the value of N_{Ed} to be used in Equations 5.3, 5.10 is the one from the analysis in EN1990's 'persistent and transient' design situation. To size the columns at the conceptual design phase, the procedure may be applied with the beam longitudinal reinforcement estimated in the two lowest storeys for the purposes of Equations 5.6 through 5.8, for example, with $\rho_v = 0$, and the top and bottom reinforcement taken from the corresponding maximum and minimum steel ratios, respectively, as per Eurocode 8.

For a sample application of this section to one of the columns of the seven-storey example building, see Section 7.6.2.2.

5.3 DETAILED DESIGN OF BEAMS IN FLEXURE

5.3.1 Dimensioning of the beam longitudinal reinforcement for the ULS in flexure

The top and bottom bars at the two ends of each beam are dimensioned for the ULS in flexure with no axial force for the envelope of bending moments resulting from the analysis under:

a. The combination of factored gravity loads ('persistent and transient design situation' per EN1990), and
b. The combination of quasi-permanent gravity loads, $G + \psi_2 Q$, with plus and minus the design seismic action

The beam seismic moments in (b) are the final outcome of the combination of the moments due to the horizontal components of the design seismic action, E_X, E_Y, per Equations 3.99 or 3.100 in Section 3.1.7 and include the effect of the accidental eccentricities of these components (see Section 3.1.8).

The cross-sectional area of the top reinforcement, A_{s1}, of each end region is dimensioned as the tension reinforcement required for an acting moment, M_{Ed}, equal to the maximum hogging moment at the column face; normally in this dimensioning, combination (b), with the beam seismic moments taken as hogging, controls over (a). The cross-sectional area of the bottom reinforcement, A_{s2}, is dimensioned as the tension reinforcement for an acting moment, M_{Ed}, equal to that at the column face, or at a nearby section where the sagging moment attains its maximum; in this case M_{Ed} is obtained from combination (b), but with the beam seismic moment taken as sagging. Besides, the main bottom bars of the beam are dimensioned from a section around mid-span, normally where the sagging moment from combination (a) attains its maximum value within the span.

The cross-sectional area, A_s, of the tension reinforcement may be conveniently dimensioned from the extreme value of the pertinent acting moment M_{Ed} (i.e. the extreme sagging moment for the bottom reinforcement or the extreme hogging one for the top bars), by taking the internal lever arm of the beam (between its tension and compression chords), z, as equal to the distance between the tension and compression bars, $d - d_2$, where d is the effective depth of the section and d_2 is the distance of the centroid of the compression bars from the extreme compression fibres:

$$A_s = |M_{Ed}| / (f_{yd}(d - d_2)) \tag{5.12}$$

where f_{yd} is the design yield stress of steel. Note that the absolute value of M_{Ed} is used; its sign determines the side of the section (top or bottom) where the tension fibres are and the tension reinforcement area, A_s, is placed.

Alternatively to Equation 5.12, the reinforcement may be dimensioned with stricter adherence to the assumptions in Eurocode 2 for the ULS in flexure without axial force for concrete grade, f_{ck}, up to 50 MPa. This alternative employs the dimensionless acting moment:

$$\mu_d = |M_{Ed}| / (b_{eff} d^2 f_{cd}) \tag{5.13}$$

where b_{eff} is the effective width of the compression flange and f_{cd} is the design strength of concrete. From the value of μ_d, the mechanical ratio of tension reinforcement, defined as

$$\omega = A_s/(b_{eff}d) \cdot (f_{yd}/f_{cd}) \tag{5.14a}$$

is computed as

$$\omega = 0.973\left(1 - \sqrt{1 - \frac{2\mu_d}{0.973}}\right) \tag{5.15a}$$

or

$$\omega = 1 - \sqrt{1 - 2\mu_d} \tag{5.15b}$$

Equation 5.15a is obtained if the standard parabolic–rectangular σ–ε law of concrete is adopted for ULS design; Equation 5.15b results, instead, from the rectangular stress block in the extreme 80% of the compression zone. Neither of these expressions accounts for the presence of longitudinal bars in the compression zone. It is necessary to account for it, if the dimensionless acting moment, μ_d, is so large that the (normalised to d) neutral axis depth, ξ, reaches a value beyond which the tension reinforcement is not even in the yielding state when the extreme compression fibres exhaust the ultimate strain of concrete in ULS design for bending, ε_{cu2} (=0.35% for $f_{ck} \leq 50$ MPa), that is, when ξ reaches the value:

$$\xi_{\lim} = \frac{\varepsilon_{cu2}}{\varepsilon_{cu2} + f_{yd}/E_s} \tag{5.16}$$

If the standard parabolic–rectangular σ–ε law of concrete is adopted for the ULS design, the limit value of μ_d corresponding to the value of ξ from Equation (5.16) is

$$\mu_{d,\lim} = 0.81\,\xi_{\lim}\left(1 - 0.416\,\xi_{\lim}\right) \tag{5.17a}$$

whereas, if the rectangular stress block in the extreme 80% of the compression zone is adopted:

$$\mu_{d,\lim} = 0.8\,\xi_{\lim}\left(1 - 0.4\,\xi_{\lim}\right) \tag{5.17b}$$

The part of μ_d that exceeds the value of $\mu_{d,\lim}$ from Equations 5.17 is assigned to a resisting moment produced by compression reinforcement with cross-sectional area A_s' and a mechanical ratio defined as

$$\omega' = A_s'/(b_{eff}d) \cdot (f_{yd}/f_{cd}) \tag{5.14b}$$

and computed from

$$\omega' = \frac{\mu_d - \mu_{d,\lim}}{1 - d_2/d} \tag{5.18}$$

Then the tension reinforcement is obtained as

$$\omega = \omega_{\lim} + \omega' \tag{5.19}$$

where $\omega_{\lim}$ is the value of ω given by Equations 5.15 for $\mu_d = \mu_{d,\lim}$. The second term is the part of the tension reinforcement needed to act as a couple with the compression reinforcement from Equation 5.18.

The required cross-sectional areas of the top and bottom reinforcement, A_{s1}, A_{s2}, are determined as above, using as acting moment, M_{Ed}, the extreme hogging moment for A_{s1} and the extreme sagging one for A_{s2}. The final value of A_{s2} may not be taken less than the compression reinforcement area, A_s', obtained from the extreme hogging moment via Equations 5.14b, 5.13 and 5.15 through 5.18 (if Equations 5.13 through 5.19 are used in lieu of Equation 5.12).

Eurocode 8 allows counting in A_{s1} the cross-sectional area, $\Delta A_{s,\text{slab}}$, of all slab bars that are simultaneously:

- Parallel to the beam
- Within the effective flange width in tension per Eurocode 8, which extends on each side of the web beyond the face of the column parallel to the beam by the widths given in Section 5.2.2
- Well anchored within the joint or beyond

However, the design of the beams in flexure is normally a separate procedure from the design of the slabs; therefore, $\Delta A_{s,\text{slab}}$ is not available at this phase of detailed design. So, most often the designer fails to profit from this allowance, to reduce the amount of real beam top reinforcement by $\Delta A_{s,\text{slab}}$, presuming this convenient omission to be safe-sided. Indeed it is so for the ULS in flexure of the beam, but it is unsafe wherever the beam moment resistance is used as a demand in 'capacity-design' calculations (see Sections 5.3.4 and 5.4.1).

All the above apply to beams in pure bending, without axial load. As a matter of fact, the values of beam axial forces which may come out of the analysis depend heavily on the modelling of the floor diaphragms and/or the way the external lateral loads are applied to the floors. So, normally they are fictitious and would better be neglected in dimensioning the beams. At any rate, the way to consider a (real) axial force in dimensioning a beam for the ULS in flexure is presented in Section 6.3.8, on the occasion of a postulated axial force for the design of tie-beams between footings.

5.3.2 Detailing of beam longitudinal reinforcement

In translating A_{s1} and A_{s2} into a combination of bar diameters and numbers, the designer should respect the detailing rules of Eurocode 8 summarised in Table 5.1. The rule at the fourth row concerning the maximum ratio of tension reinforcement, $\rho_{\max}$, is the only one of these rules which is not prescriptive; at the same time it is the most restrictive: as the value of A_{s1} to be accommodated within a given beam width, b, cannot be reduced below what is necessary to resist the acting moment, M_{Ed}, the best way to meet the rule for $\rho_{\max}$ is by increasing the ratio of compression reinforcement, ρ', at the end section.

The bar diameters chosen should also respect the maximum allowed by Equations 5.2 for a given section depth, h_c, of the column where these bars are anchored (at exterior columns) or pass through (at interior ones).

The bars chosen on the basis of the two end sections and the one around mid-span where the span bottom bars are determined, are terminated according to the positive and negative moment envelopes; they extend beyond the point where they are not needed according to the envelope by the 'tension shift' length $z = 0.9d - d_2$ in Eurocode 2.

Table 5.1 EC8 detailing of the longitudinal bars in primary beams (in secondary ones as in DC L)

	DC H	DC M	DC L
'critical region' length at member end	$1.5h$		h
$\rho_{min} = A_{s,min}/bd$ at the tension side		$0.5f_{ctm}/f_{yk}^{a}$	$0.26f_{ctm}/f_{yk}^{a}$, $0.13\%^{b}$
$\rho_{max} = A_{s,max}/bd$ in critical regions[b]		$\rho' + 0.0018f_{cd}/(\mu_\varphi\varepsilon_{yd}f_{yd})^{c}$	0.04
$A_{s,min}$, top and bottom bars		$2\Phi14$ (308 mm²)	–
$A_{s,min}$, top bars in the span		$0.25A_{s,top\text{-}supports}$	–
$A_{s,min}$, bottom bars in critical regions		$0.5A_{s,top}^{d}$	–
$A_{s,min}$, bottom bars at supports		$0.25A_{s,bottom\text{-}span}^{b}$	
Anchorage length for diameter d_{bL}^{e}		$l_{bd} = a_{tr}[1-0.15(c_d/d_{bL}-1)](d_{bL}/4)f_{yd}/(2.25f_{ctd}a_{poor})^{f,g,h,i}$	

[a] f_{ctm} (MPa) $= 0.3(f_{ck}(MPa))^{2/3}$: 28-day, mean tensile strength of concrete; f_{yk} (MPa): nominal yield stress of longitudinal steel.
[b] NDP (nationally determined parameter) per EC2; the value recommended in EC2 is given here.
[c] ρ': Steel ratio at the opposite side of the section; μ_φ: curvature ductility factor corresponding via Equations 5.64 to the basic value of the behaviour factor, q_o, applicable to the design; $\varepsilon_{yd} = f_{yd}/E_s$.
[d] This $A_{s,min}$ is additional to the compression steel from the ULS verification of the end section in flexure under the extreme hogging moment from the analysis for the seismic design situation.
[e] Anchorage length in tension is reduced by 30% if the bar end extends by $\geq 5d_{bL}$ beyond a bend $\geq 90°$.
[f] c_d: Concrete cover of anchored bar, or one-half the clear spacing to the nearest parallel anchored bar, whichever is smaller.
[g] $a_{tr} = 1 - k(n_wA_{sw} - A_{s,tmin})/A_s \geq 0.7$, with A_{sw}: Cross-sectional area of tie leg within the cover of the anchored bar; n_w: number of such tie legs over the length l_{bd}; $k = 0.1$ if the bar is at a corner of a hoop or tie, $k = 0.05$ otherwise; $A_s = \pi d_{bL}^2/4$ and $A_{s,tmin}$ is specified in EC2 as equal to $0.25A_s$.
[h] $f_{ctd} = f_{ctk,0.05}/\gamma_c = 0.7f_{ctm}/\gamma_c = 0.21f_{ck}^{2/3}/\gamma_c$: Design value of 5%-fractile tensile strength of concrete.
[i] $a_{poor} = 1.0$ if the bar is in the bottom 0.25 m of the beam depth, or (in beams deeper than 0.6 m) ≥ 0.3 m from the beam top; otherwise, $a_{poor} = 0.7$.

Eurocode 2 considers the bar stress to drop off linearly along the anchorage length, l_{bd} (given at the last row of Table 5.1), from f_{yd} to zero. So, if the inclination of the moment envelope (i.e. the shear force, V) exceeds the bar yield force, $f_{yd}\pi d_{bL}^2/4$, times the ratio of the internal lever arm, z, to l_{bd} (i.e. if $V > (2.25\pi f_{ctd}a_{poor})d_{bL}z/\{a_{tr}[1-0.15(c_d/d_{bL}-1)]\}$ with the notation of Table 5.1), then the bar has to be further extended by l_{bd} beyond the point it is not needed according to the moment envelope. Otherwise, the bar fully contributes along l_{bd} in resisting the moment.

The full string of beams ('continuous beam') in a frame should be designed in bending all together, combining reinforcement requirements to the right and left of interior joints. It is also recommended to combine different top or bottom bars into continuous ones, if their ends come close or overlap. To this end, few bar sizes (even a single size) should be used all along each string of beams.

The rules in Table 5.1 for DC L do not apply to deep beams, defined in Eurocode 2 as those with depth, h, less than one-third of their span. Eurocode 2 also requires skin reinforcement at the lateral sides of 1 m deep beams or deeper. For the purposes of detailing the beam longitudinal reinforcement, beams may be defined as deep if their depth is more than the smaller of 1 m, or one-third of the span. Eurocode 2 prescribes an orthogonal reinforcement mesh per lateral side of a deep beam, with maximum bar spacing which is the lesser of 300 mm or twice the web thickness and cross-sectional area per side and direction not less than 150 mm²/m or 0.05% of the concrete area (i.e. 0.1% total for both sides). Much more demanding are the requirements of Eurocode 2 for skin reinforcement placed to control cracking at the web of 1 m deep beams or deeper (see the next section). Note that the minimum ratio of tension steel, ρ_{min}, in row 2 of Table 5.1 is also to control potential cracking throughout the tension zone. However, if it is concentrated just at the tension chord, its effectiveness in that role is reduced at points further away. So, if the beam is deeper than 1 m, Eurocode 2 assigns that role to skin reinforcement distributed over the entire tension zone. In that case, placing at the tension chord a quantity, $A_{s,\min}$, of minimum reinforcement equal to $\rho_{\min}$ times the full effective cross-sectional area, bd, would be not just duplication, but a waste. To avoid it, it is recommended here for deep beams:

- To determine the minimum reinforcement concentrated at the tension chord, $A_{s,\min}$, as $\rho_{\min}$ from row 2 of Table 5.1, times the product of b and a depth of 1 m
- To distribute over the depth of the section horizontal skin reinforcement at a steel ratio of $\rho_{\min}$; that reinforcement should also be dimensioned for crack control in the web, according to the next section

5.3.3 Serviceability requirements in Eurocode 2: Impact on beam longitudinal reinforcement

5.3.3.1 Introduction

Eurocode 2 includes important serviceability limit state (SLS) requirements concerning the level of stresses in steel or concrete and the crack width under service loads, as well as the amount and form of reinforcement necessary to control cracking due to non-quantified imposed deformations and other ill-defined, often random, causes. These requirements are relevant to beams, but have very little to do with seismic design; moreover, they are normally met by default in ordinary beams designed and detailed for earthquake resistance. So, strictly speaking, they are outside the scope of this book. However, they often control the longitudinal reinforcement in oversized beams, such as deep foundation beams, especially those that double as perimeter walls of basements. As a matter of fact, in Chapter 7 they are applied to such elements of the example building and found to control their longitudinal reinforcement. For all these reasons, and because there is still a gap for this topic in literature concerning the application of Eurocode 2, these SLS requirements are highlighted here, alongside guidance on how to apply them to beams. They are relevant to those regions of a beam where tension may build up under service conditions: normally the top flange at the end sections of beams in the superstructure and the bottom one in the span; the reverse in foundation beams.

For a sample application of Sections 5.3.2 and 5.3.3 to the beams of the seven-storey example building, see Section 7.6.2.1.

5.3.3.2 Stress limitation SLS

The SLS of stress limitation imposes stress limits on concrete and steel under service conditions. The limits are nationally determined parameters (NDPs) in Eurocode 2 with recommended values:

- Under the 'characteristic' gravity loads, $G + Q$
  - Concrete stress, $\sigma_{c,G+Q} \leq 0.6f_{ck}$
  - Steel stress, $\sigma_{s1,G+Q} \leq 0.8f_{yk}$
- Under the quasi-permanent gravity loads, $G + \psi_2 Q$
  - Concrete stress, $\sigma_{c,G+\psi2Q} \leq 0.45f_{ck}$

Once the amount of tension, compression and web reinforcement in the beam section is determined on the basis of Sections 5.3.1 and 5.3.2 and so forth, the above limits are checked as follows:

$$\sigma_{s1,G+Q} = E_s \frac{M_{G+Q}}{EI_{b,eff}}(1 - \xi)d \leq 0.8f_{yk} \qquad (5.20)$$

$$\sigma_{c,G+Q} = E_{cm} \frac{M_{G+Q}}{EI_{b,eff}} \xi d \leq 0.6 f_{ck} \tag{5.21a}$$

$$\sigma_{c,G+\psi 2Q} = E_{cm} \frac{M_{G+\psi 2Q}}{EI_{b,eff}} \xi d \leq 0.45 f_{ck} \tag{5.21b}$$

where E_s and E_{cm} are the elastic moduli of steel and concrete (mean values) and $EI_{b,eff}$, ξ are determined from Equations 5.6 through 5.8. As the actions causing these stresses are (almost fully) long term, it makes sense to use in Equations 5.6 through 5.8 the value $\alpha = (1 + \varphi_\infty) E_s / E_{cm}$, with φ_∞ the final value of the creep coefficient. This is safe-sided for σ_{s1}, but reduces the estimate of σ_c.

5.3.3.3 Crack width SLS

The characteristic value of crack width, w_k, under the quasi-permanent gravity loads, $G + \psi_2 Q$, is checked against an upper limit value, w_{max}, which is an NDP, with a recommended value of 0.3 mm for the common environmental exposure classes in buildings. According to Eurocode 2, for long-term loading (such as the quasi-permanent gravity loads), w_k may be computed as

$$w_k = 1.7 \left(2(c_{nom} + d_{bw}) + 0.1 \frac{d_{bL,mean}}{\rho_{eff}} \right) \frac{\max \left(\left(\sigma_{s1,G+\psi 2Q} - 0.4 f_{ctm} \frac{1 + \alpha \rho_{eff}}{\rho_{eff}} \right); 0.6 \sigma_{s1,G+\psi 2Q} \right)}{E_s} \tag{5.22}$$

Coefficient 1.7 in Equation 5.22 converts the mean estimate of the crack width to a characteristic value; the term that follows is a semi-empirical best estimate of the crack spacing; the last term is an estimate of the difference in mean tensile strains of steel and concrete between adjacent cracks. Concerning symbols, c_{nom} is the nominal concrete cover of the stirrup (minimum required for durability, plus a tolerance of 10 mm), d_{bw} is the diameter of the stirrup, $d_{bL,mean}$ is the mean longitudinal bar diameter in the tension zone and $\sigma_{s1,G+\psi 2Q}$ is the steel stress due to the quasi-permanent gravity loads, computed from Equation 5.20, using $M_{G+\psi 2Q}$ instead of M_{G+Q}. The tension steel ratio:

$$\rho_{eff} = A_{s1} / A_{c,eff} \tag{5.23}$$

refers to the effective area of concrete in tension surrounding the tension reinforcement, A_{s1}. For sections in bending, with rectangular tension zone having width b_w, Eurocode 2 defines $A_{c,eff}$ as

$$A_{c,eff} = \min \left(2.5 d_1; \frac{h - \xi d}{3} \right) b_w \tag{5.24a}$$

where d_1 is the centroidal distance of A_{s1} from the extreme tension fibres and $h - \xi d$ the depth of the tension zone. If A_{s1} is spread in a well-defined T- or L-shaped tension zone with flange width b and thickness t, and if b_w denotes the width of the web, then (Figure 5.1):

$$A_{c,eff} = \min \left(2.5 d_1; \frac{h - \xi d}{3} \right) b_w + \min \left(t; 2.5 d_1; \frac{h - \xi d}{3} \right) (b - b_w) \tag{5.24b}$$

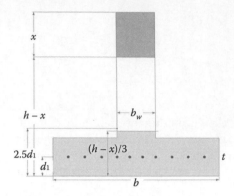

Figure 5.1 Effective concrete area in tension for crack control.

Note that, strictly speaking, the case of Equation 5.24a is not that of the effective flange width in tension introduced in Section 5.2.2, which contributes to the hogging moment resistance of the beam's end section with its slab bars. That concept is introduced in Eurocode 8, not in Eurocode 2, and refers to the ULS, not the SLS. By the same token, the slab bars in that effective width in tension are not included in A_{s1} for the purposes of Equation 5.23. A representative case where Equation 5.24b does apply is the strip footing of the deep foundation beam of Figure 7.42, with all eight of its longitudinal bars included in A_{s1}.

Eurocode 2 defines α for use in Equation 5.22 as $\alpha = E_s/E_{cm}$. However, as the crack width is computed for the quasi-permanent loads, it makes more sense to use the value $\alpha = (1 + \varphi_\infty) E_s/E_{cm}$, as for Equations 5.20, 5.21.

Eurocode 2 differentiates the estimation of crack width in the web of deep beams where skin reinforcement is placed (see the last parts of Section 5.3.2 and 5.3.3.4): the coefficient in front of $d_{bL,\mathrm{mean}}/\rho_{eff}$ in the term representing the mean crack spacing is 0.2, instead of 0.1; moreover, a mean value of $\sigma_{s1,G+\psi2Q}$ over the web is used, equal to one-half the maximum steel stress computed over the section.

5.3.3.4 Minimum steel for crack control

Should cracking occur due to non-quantified imposed deformations or another ill-defined cause, possibly often random, the steel crossing the crack should be able to keep the crack width below the applicable limit value, w_{max}. To this end, its cross-sectional area in the tension zone, $A_{s,\min}$, should be sufficient to resist the tensile force released when the part of the so-far uncracked section that is in tension cracks, and, indeed, developing a steel stress, σ_s, which is low enough to keep the resulting crack width below w_{max}. According to Eurocode 2, the usual limit value $w_{\mathrm{max}} = 0.3$ mm is achieved, if $A_{s,\min}$, develops a stress, σ_s, which depends on the mean bar diameter, $d_{bL,\mathrm{mean}}$, as

- if 8 mm $< d_{bL,\mathrm{mean}} \leq 12$ mm:

$$\sigma_s \, (\mathrm{MPa}) = 280 + 20 \times (12 - d_{bL,\mathrm{mean}}) \tag{5.25a}$$

- if 12 mm $< d_{bL,\mathrm{mean}} \leq 16$ mm:

$$\sigma_s \, (\mathrm{MPa}) = 240 + 10 \times (16 - d_{bL,\mathrm{mean}}) \tag{5.25b}$$

- if 16 mm $< d_{bL,\text{mean}} \leq 25$ mm:

$$\sigma_s \text{ (MPa)} = 200 + (40/9) \times (25 - d_{bL,\text{mean}}) \qquad (5.25c)$$

- if 25 mm $< d_{bL,\text{mean}} \leq 32$ mm:

$$\sigma_s \text{ (MPa)} = 160 + (40/7) \times (32 - d_{bL,\text{mean}}) \qquad (5.25d)$$

If the tension zone in the uncracked section is rectangular, with width that of the web, b_w, and depth, $y_{cg,t}$, equal to the distance of the centroid of the uncracked section to the extreme tension fibres, then, the minimum reinforcement of beams in flexure per Eurocode 2 is

$$A_{s,\text{min}} = 0.4 k_h b_w y_{cg,t} \frac{f_{ctm}}{\sigma_s} \qquad (5.26a)$$

where k_h reflects the reduction of the net tensile force in deep sections due to non-uniform self-equilibrating stresses:

- if $h \leq 0.3$ m:

$$k_h = 1.0 \qquad (5.27a)$$

- if 0.3 m $< h \leq 0.8$ m:

$$k_h = 1.21 - 0.7 \, h(\text{m}) \qquad (5.27b)$$

- if $0.8 \, m < h$:

$$k_h = 0.65 \qquad (5.27c)$$

If the tension zone in the uncracked beam has a T- or L-shape, and b and t denote the width and the thickness of the tension flange, while b_w still stands for the width of the web, then

$$A_{s,\text{min}} = \left[0.4 k_h b_w y_{cg,t} + \max\left(0.5; 0.9 k_b \left(1 - \frac{t}{2 y_{cg,t}} \right) \right) (b - b_w) t \right] \frac{f_{ctm}}{\sigma_s} \qquad (5.26b)$$

where k_b is the counterpart of k_h for a wide tension flange:

- if $(b - b_w) \leq 0.3$ m:

$$k_b = 1.0 \qquad (5.28a)$$

- if $0.3\text{ m} < (b - b_w) \leq 0.8\text{ m}$:

$$k_b = 1.21 - 0.7(b - b_w)\text{ (m)} \tag{5.28b}$$

- if $0.8\text{ m} < (b - b_w)$:

$$k_b = 0.65 \tag{5.28c}$$

The rules in Eurocode 2 concerning the minimum skin reinforcement for crack control in deep beams allow taking $\sigma_s = f_{yk}$ and $k_b = 0.5$, giving a minimum ratio of horizontal web reinforcement:

$$\rho_{h,\min} = 0.5 k_c \frac{f_{ctm}}{f_{yk}} \tag{5.29}$$

where k_c reflects the distribution of stresses within the tributary area of the skin reinforcement – it is the counterpart of 0.4 in Equations 5.26a and of $0.9k_b(1-0.5t/y_{cg,t})$ in Equation 5.26b. The most adverse condition is a uniform stress distribution, as in pure tension; then $k_c = 1.0$. This gives the same minimum steel ratio as listed at the second row of Table 5.1 for DC M or H beams, but this time distributed over the sides of the web. Eurocode 2 points out that, if the target is to control the crack width in the web to $w_{\max} = 0.3$ mm, the value of σ_s that corresponds to the diameter of the skin reinforcement according to Equations 5.25 should be used in Equation 5.29, instead of f_{yk}.

5.3.4 Beam moment resistance at the end sections

After dimensioning and detailing the beam longitudinal bars at the two end sections, the design values of beam moment resistance at these sections are computed from the final cross-sectional areas of its reinforcement. If there is only top and bottom reinforcement in the section, A_{s1} and A_{s2}, the design values of moment resistance in hogging or sagging bending, respectively, may be estimated as

$$M_{Rd,b}^- = \min(A_{s1}, A_{s2})f_{yd}(h - d_1 - d_2) + \max\left[0, (A_{s1} - A_{s2})\right] \cdot$$
$$f_{yd}\left[h - d_1 - 0.5(A_{s1} - A_{s2})f_{yd}/(b_w f_{cd})\right] \tag{5.30a}$$

$$M_{Rd,b}^+ = A_{s2}f_{yd}\max\left[\left(h - d_2 - 0.5A_{s2}f_{yd}/(b_{eff}f_{cd})\right), (h - d_1 - d_2)\right] \tag{5.30b}$$

where:

- d_1, d_2 are the centroidal distances of A_{s1}, A_{s2}, from the top or bottom of the beam section, respectively.
- b_w, b_{eff} are the effective widths in compression of the bottom flange (normally that of the web) and the top flange, respectively.

Often, a simpler option is considered to provide sufficient accuracy:

$$M_{Rd,b}^- = zA_{s1}f_{yd}; \quad M_{Rd,b}^+ = zA_{s2}f_{yd} \tag{5.30c}$$

where the internal lever arm, z, may be taken equal to $0.9d$.

The values from Equations 5.30 are used in the 'capacity design' of columns in flexure, Equation 5.31, and for the 'capacity design' shears in the beam itself and the columns connected to it (see Equations 5.42 and 5.44, respectively). Eurocode 8 stresses that, whenever Equation 5.30a is used for these 'capacity design' purposes, the area, $\Delta A_{s,slab}$, of all slab bars which are: (a) parallel to the beam, (b) placed on each side of it within the effective flange width in tension per Eurocode 8 (given in Section 5.2.2) and (c) well anchored within the joint or beyond, should be included in A_{s1}, no matter whether they are relied upon to provide the tension reinforcement area, A_s, required for the ULS in flexure under the extreme hogging moment according to Section 5.3.1 (see also second paragraph from the end of that section).

The moment resistance of deep beams, having uniformly distributed reinforcement between its top and bottom ones, A_{s1} and A_{s2}, may be determined from Section 5.4.3, applicable to asymmetrically reinforced column sections with uniformly distributed reinforcement along the lateral sides, by setting the axial load equal to zero.

5.4 DETAILED DESIGN OF COLUMNS IN FLEXURE

5.4.1 Strong column–weak beam capacity design

To pursue the desired global ductility, Eurocode 8 promotes beam-sway mechanisms and takes measures to prevent a soft storey (cf. Sections 2.2, 4.5.2, 5.4.1). A soft-storey mechanism (Figure 2.9a) develops in a frame system when the top and bottom ends of (all) the columns in a storey yield in opposite bending and start undergoing unrestrained flexural rotations there, without a notable increase of their bending moments beyond the corresponding moment resistance, M_{Rc} (this is, in fact, how a flexural 'plastic hinge' is defined). The way to prevent soft storeys in frames is by forcing flexural plastic hinges out of the columns and into the beams, so that a beam-sway mechanism develops (Figure 2.9b and c). To this end, within any vertical plane in which a soft storey is to be prevented, the two columns framing into a beam–column joint from above and below are dimensioned to be jointly stronger by 30% than the (one, two or more) beams connected to the same joint from any side (Figure 5.2):

$$\sum M_{Rd,c} \geq 1.3 \sum M_{Rd,b} \tag{5.31}$$

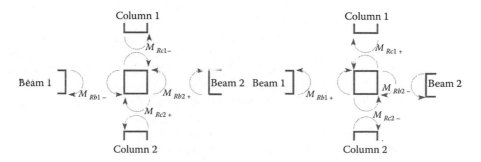

Figure 5.2 Direction of action of column and beam moment resistances around a joint in the capacity design check of the column for both directions of the response to the seismic action.

where:

- $M_{Rd,c}$: design value of column moment resistance at the face of the joint, in the vertical plane of bending in which a soft storey is to be prevented (i.e. with moment vector at right angles to that plane), with the sum referring to the column sections above and below the connection.
- $M_{Rd,b}$: design value of beam moment resistance at the face of the joint, with the sum extending to all beam ends connected to the joint; beams that are not in the vertical plane in which a soft storey is to be prevented but at an angle α to it, enter Equation 5.31 with their $M_{Rd,b}$-value multiplied by cos α.

Normally Equation 5.31 is checked within two orthogonal vertical planes. For the usual columns with section composed of rectangular parts (including L- or T-sections etc.), these vertical planes are chosen parallel to the column sides, facilitating the calculation of $M_{Rd,c}$. In the most common case where the beams connected to the column at the joint are parallel to the column sides, they have $\alpha = 0$ in one of the two horizontal directions in which Equation 5.31 is checked and $\alpha = 90°$ in the orthogonal one.

The check of Equation 5.31 takes place twice in each of the two vertical planes considered: first with both column moments, $M_{Rd,c}$, acting clockwise on the joint in the direction about the normal to that plane and then counterclockwise (Figure 5.2). Beam moment resistances, $M_{Rd,b}$, are taken to act on the joint in the opposite sense with respect to those of the columns. The values of $M_{Rd,b}$ may be calculated from Equations 5.30; the beams connected to one side of the joint with respect to the normal of the vertical plane are hogging and their $M_{Rd,b}$-value is computed from Equation 5.30a; those connected to the opposite are sagging and Equation 5.30b is used for them.

For the application of Equation 5.31, see Examples 5.1 and 5.2 at the end of this chapter.

The calculation of column moment resistance, $M_{Rd,c}$, for known column reinforcement and given axial force, N, is addressed in Section 5.4.3. The value of N to be used in this calculation should be the most safe sided for the fulfilment of Equation 5.31, notably the minimum compressive or maximum tensile force in the range of values derived from the analysis for the 'seismic design situation'. In general, this extreme value of N is obtained by subtracting from the axial load due to the quasi-permanent gravity loads, $G + \psi_2 Q$, the value of N due to the design seismic action. However, the application of this general rule should be physically consistent with the sense of action (clockwise or counterclockwise) of $\Sigma M_{Rd,b}$ in Equation 5.31, and hence its value. Section 5.8.2 deals in more detail with the value of N in capacity design calculations.

Equation 5.31 is called 'capacity design' of columns in flexure, because the demand for the required (design value of) column moment resistance, $M_{Rd,c}$, is not an action effect from the analysis, but the (design values of the) 'capacities', $M_{Rd,b}$, at the locations where plastic hinges are allowed (even promoted), in this case at the beam ends. This design rule employs only equilibrium (of moments) and is independent of the magnitude of the design seismic action; so, it achieves its goal for any earthquake, no matter how strong it is. Note at this point that, although the equilibrium of moments is meant to refer to the 'centre' of the joint, where the beam and column axes theoretically intersect, the transfer of $M_{Rd,c}$, $M_{Rd,b}$, from the faces to the centre of the joint is omitted in Equation 5.31 for convenience. This is safe-sided, provided that $1.3 h_b / h_c > H_{cl} / L_{cl}$, with h_b, h_c denoting the cross-sectional depths of the beam and columns, respectively, and L_{cl}, H_{cl} the average clear span of the beams on either side of the joint, or the average clear storey height above and below it, respectively, all in the vertical plane in which Equation 5.31 is checked (see Fardis 2009).

Eurocode 8 exempts the following cases of columns from the enforcement of Equation 5.31:

1. In the horizontal direction(s) where walls take at least 50% of the elastic base shear, that is, the system qualifies as a wall system or a wall-equivalent dual one; the reason is that a wall (be it with the minimum length-to-thickness ratio of 4 per Eurocode 2) is very unlikely to yield in counter-flexure at both the top and bottom sections in a storey (Figure 2.9d,f); so, if there are plenty of them in a horizontal direction, they prevent soft-storey mechanisms.
2. Around the joints of the top floor, no matter the structural system; one reason is that it makes little difference for the plastic mechanism whether the plastic hinge forms at the top of a top storey column, or at the ends of the beams connected to it; another reason is the good ductility of the top storey columns thanks to their low axial load. Note also that it is hard to meet Equation 5.31 with only one column section at the left-hand side.
3. In two-storey buildings of any structural system, provided that none of the ground storey columns has axial load ratio, $v_d = N_d/(A_c f_{cd})$, above 0.3, for the maximum column axial load, N_d, in any combination of the seismic design situation (design seismic action plus concurrent gravity loads, $G + \psi_2 Q$); such columns have sufficient ductility to withstand concentration of the entire deformation demand in one storey instead of two, with consequent doubling of the ductility demand in ground storey columns.
4. In one-out-of-four columns per plane frame with columns of similar size, in a horizontal direction not exempted from Equation 5.31 on the basis of 1 above; it is worth profiting from this exemption at interior joints rather than at exterior ones, where a beam frames from one side only and Equation 5.31 is easily met.

Eurocode 8 presumes that a plastic hinge will form at any column end where Equation 5.31 is not checked by virtue of the exemptions above and requires to detail these plastic hinge regions so that they can develop significant inelastic deformations after plastic hinging. In fact, the same detailing rules apply in these regions as those applied at the base of the column, where a plastic hinge is allowed anyway.

5.4.2 Dimensioning of column vertical reinforcement for action effects from the analysis

The base section at the bottom storey of a column (the connection to the foundation), as well as all columns exempted from the capacity-design rule, Equation 5.31, are dimensioned for the ULS in biaxial flexure with axial force, using triplets M_y–M_z–N from the analyses for the combination of the design seismic action with the quasi-permanent gravity loads, $G + \psi_2 Q$. This is combination (b) in Section 5.3.1 for the seismic design situation; combination (a) is normally not critical for the dimensioning of primary columns and may be ignored.

The column sections right above and right below a beam–column joint are served by the same vertical bars. Besides, as pointed out in Section 5.2.3.1, it is good practice to avoid changing the column section from one storey to the next. So, these two sections are dimensioned as a single one, for all M_y–M_z–N triplets that the analysis gives for them in the seismic design situation, each triplet being the single triplet due to the quasi-permanent gravity loads $G + \psi_2 Q$ plus a seismic one (see Section 5.8 for the number and composition of the seismic triplets, depending on the analysis method and the use of Equation 3.99 or 3.100). Most critical of all triplets is the one giving the largest amount of reinforcement in one of the two sections; however, it is not easy to screen out non-critical ones. Generally, for the usual range of values of the dimensionless axial load, $N_d/A_c f_{cd}$, most critical among triplets with similar biaxial moments is the one having the lowest axial compression.

There are several iterative algorithms for the ULS verification of sections with any shape and amount and layout of reinforcement for a combination M_y–M_z–N. They employ section analysis and the σ–ε laws used for design (elastic–perfectly plastic for steel, normally parabolic–rectangular for concrete) to find the strain distribution which satisfies equilibrium. It is checked then whether, in that strain distribution, the conventional ultimate strain of concrete, $\varepsilon_{cu,2}$, is exceeded at the corners of the section. However, there is no general algorithm for the direct calculation of the section reinforcement for a given M_y–M_z–N triplet. The traditional manual approach with design charts is not practical for the large number of columns of a real building; it is also very restrictive for the bar layout and the steel grade. A practical, yet approximate, step-by-step computational procedure is proposed in the following paragraphs for the direct dimensioning of symmetrically reinforced rectangular sections under a set of M_y–M_z–N triplets.

1. The mechanical reinforcement ratio, $\omega_{1d} = A_{s1}/(bd)\cdot(f_{yd}/f_{cd})$, of the steel bars placed along each one of two opposite sides of the section of length b, is estimated under uniaxial moment, M, with axial force, N, neglecting the orthogonal moment component; d is the effective depth at right angles to the vector of M (cf. Equations 5.13); each layer of bars with cross-sectional area A_{s1} is at centroidal distance d_1 from the nearest side of the section of length b. M, N and d_1 are normalised as

$$\mu_d = M/(bd^2 f_{cd}), \quad \nu_d = N/(bd f_{cd}), \quad \delta_1 = d_1/d \tag{5.32}$$

Section analysis is used, with the material σ–ε laws and criteria adopted in Eurocode 2 for the ULS design:
a. Elastic–perfectly plastic steel, with a yield stress of f_{yd} and unlimited strain capacity
b. Parabolic–rectangular σ–ε law for concrete, with design strength f_{cd} at strain ε_{c2}, with ultimate strain ε_{cu2} (for $f_{ck} \le 50$ MPa, $\varepsilon_{c2} = 0.002$, $\varepsilon_{cu2} = 0.0035$)

Depending on the value of the dimensionless axial load, ν_d, there are three possible cases:
i. The most usual case is to have yielding of the tension and the compression reinforcement; this happens if:

$$\delta_1 \frac{\varepsilon_{cu2} - \varepsilon_{c2}/3}{\varepsilon_{cu2} - \varepsilon_{yd}} \equiv \nu_2 \le \nu_d < \nu_1 \equiv \frac{\varepsilon_{cu2} - \varepsilon_{c2}/3}{\varepsilon_{cu2} + \varepsilon_{yd}} \tag{5.33a}$$

where $\varepsilon_{yd} = f_{yd}/E_s$. Then, the neutral axis depth, x, normalised to d as $\xi = x/d$, is

$$\xi = \frac{\nu_d}{1 - (\varepsilon_{c2}/3\varepsilon_{cu2})} \tag{5.34a}$$

to be substituted in terms of ν_d in the following equation, to be solved directly for ω_{1d}:

$$(1 - \delta_1)\omega_{1d} = \mu_d - \xi\left[\frac{1-\xi}{2} - \frac{\varepsilon_{c2}}{3\varepsilon_{cu2}}\left(\frac{1}{2} - \xi + \frac{\varepsilon_{c2}}{4\varepsilon_{cu2}}\xi\right)\right] \tag{5.35a}$$

ii. The second commonest case is to have the tension bars yielding, but the compression ones elastic; this happens if v_d is less than v_2, as given at the left-hand side of Equation 5.33a:

$$v_d \leq \delta_1 \frac{\varepsilon_{cu2} - \varepsilon_{c2}/3}{\varepsilon_{cu2} - \varepsilon_{yd}} \equiv v_2 \tag{5.33b}$$

Then ξ and ω_1 are related to the dimensionless axial force and moment through:

$$\left[1 - \frac{\varepsilon_{c2}}{3\varepsilon_{cu2}}\right]\xi^2 - \left[v_d + \omega_{1d}\left(1 - \frac{\varepsilon_{cu2}}{\varepsilon_{yd}}\right)\right]\xi - \omega_{1d}\frac{\varepsilon_{cu2}\delta_1}{\varepsilon_{yd}} = 0 \tag{5.34b}$$

$$\omega_{1d}\frac{(1 - \delta_1)}{2}\left(1 + \frac{\xi - \delta_1}{\xi}\frac{\varepsilon_{cu2}}{\varepsilon_{yd}}\right) = \mu_d - \xi\left[\frac{1 - \xi}{2} - \frac{\varepsilon_{c2}}{3\varepsilon_{cu2}}\left(\frac{1}{2} - \xi + \frac{\varepsilon_{c2}}{4\varepsilon_{cu2}}\xi\right)\right]$$

$$\tag{5.35b}$$

By replacing ω_{1d} from Equation 5.35b into Equation 5.34b, a strongly nonlinear equation is obtained for ξ, to be solved numerically-iteratively; ω_{1d} is then determined from Equation 5.35b.

iii. The most rare (and undesirable) case is to have yielding compression bars and the tension ones elastic; this happens if v_d exceeds v_1, given by the right-hand side of Equation 5.33a:

$$v_1 \equiv \frac{\varepsilon_{cu2} - \varepsilon_{c2}/3}{\varepsilon_{cu2} + \varepsilon_{yd}} \leq v_d \tag{5.33c}$$

Then ξ and ω_{1d} are related to each other and to v_d, μ_d, through:

$$\left[1 - \frac{\varepsilon_{c2}}{3\varepsilon_{cu2}}\right]\xi^2 - \left[v_d - \omega_{1d}\left(1 + \frac{\varepsilon_{cu2}}{\varepsilon_{yd}}\right)\right]\xi - \omega_{1d}\frac{\varepsilon_{cu2}}{\varepsilon_{yd}} = 0 \tag{5.34c}$$

$$\omega_{1d}\frac{(1 - \delta_1)}{2}\left(1 + \frac{1 - \xi}{\xi}\frac{\varepsilon_{cu2}}{\varepsilon_{yd}}\right) = \mu_d - \xi\left[\frac{1 - \xi}{2} - \frac{\varepsilon_{c2}}{3\varepsilon_{cu2}}\left(\frac{1}{2} - \xi + \frac{\varepsilon_{c2}}{4\varepsilon_{cu2}}\xi\right)\right] \tag{5.35c}$$

Substituting for ω_{1d} in Equation 5.35b the expression from 5.35c, a highly nonlinear equation results for ξ, to be solved numerically-iteratively; ω_{1d} is then determined from Equation 5.35c.

2. The procedure in step 1 above is applied first with all M_y–N pairs in the set of M_y–M_z–N combinations, with b the side length parallel to the vector of M_y and dimensions d, d_1 at right angles to it. The most critical pair gives the total area of reinforcement, A_{sy}, along each side parallel to the M_y-vector. This is repeated with all M_z–N pairs and the roles reversed, to find the total area of reinforcement, A_{sz}, along each one of the two other sides – those parallel to the M_z-vector. As the M_y–N pair from which A_{sy} is derived most likely does not belong in the same M_y–M_z–N combination as the pair M_z–N giving A_{sz}, these reinforcement requirements are superimposed on the section

Table 5.2 EC8 detailing rules for vertical bars in primary columns (in secondary ones: as in DC L)

	DC H	DC M	DC L
$\rho_{min} = A_{s,min}/A_c$		1%	$0.1N_d/A_cf_{yd}, 0.2\%$[a]
$\rho_{max} = A_{s,max}/A_c$		4%	4%[a]
Diameter, d_{bL}		≥8 mm	
Number of bars per side		≥3	≥2
Spacing along the perimeter of bars restrained by a tie corner or hook	≤150 mm	≤200 mm	–
Distance along perimeter of unrestrained bar to nearest restrained one		≤150 mm	
Lap splice length[b]		$l_0 = 1.5[1-0.15(c_d/d_{bL}-1)]a_{tr}(d_{bL}/4)f_{yd}/(2.25f_{ctd})$[c,d,e]	

[a] NDP (nationally determined parameter) per EC2; the value recommended in EC2 is given here.
[b] Anchorage length in tension is reduced by 30% if the bar end extends by ≥ $5d_{bL}$ beyond a bend ≥ 90°.
[c] c_d: Minimum of: concrete cover of lapped bar and 50% of clear spacing to adjacent lap splice.
[d] $a_{tr} = 1 - k(2n_wA_{sw} - A_{s,min})/A_s$, with $k = 0.1$ if the bar is at a corner of a hoop or tie, $k = 0.05$ otherwise; A_{sw}: cross-sectional area of a column tie; n_w: number of ties in the cover of the lapped bar over the outer third of the length l_0; $A_s = \pi d_{bL}^2/4$ and $A_{s,min}$ is specified in EC2 as equal to A_s.
[e] $f_{ctd} = f_{ctk,0.05}/\gamma_c = 0.7f_{ctm}/\gamma_c = 0.21f_{ck}^{2/3}/\gamma_c$: design value of 5%-fractile tensile strength of concrete.

and translated into a bar layout meeting the Eurocode 8 detailing rules in Table 5.2 for column vertical bars, with the corner ones counting to both sides (Figure 5.3).

3 If available, an iterative algorithm may be used in the end to verify that the section with the selected layout of reinforcement satisfies the ultimate strain of concrete, ε_{cu2}, under any one of the M_y–M_z–N triplets. If it does not, one bar may be added to each side, till the section meets the verification criteria.

The procedure above can be applied to sections composed of more than one rectangular parts, orthogonal to each other (L, T, etc.). In Step 1, such a section is replaced by an equivalent rectangular, having cross-sectional area the same as the actual one and the same effective depth at right angles to the vector of the uniaxial bending moment considered. The reinforcement areas, A_{sy} and A_{sz}, coming out of this exercise are distributed along the corresponding extreme tension and compression fibres of the section, while meeting the detailing rules in Table 5.2. If Step 3 is carried out, it should be done for the actual cross-sectional shape and bar layout.

'Capacity design' through Equation 5.31 normally governs over the ULS verification in biaxial bending and axial force using the action effects from the analysis for the seismic design situation. So, if Equation 5.31 should be met at a joint, it makes sense to use it from the outset as the basis for dimensioning the column vertical reinforcement, instead of the analysis results. To this end, in Step 1 above, each uniaxial moment from the analysis is

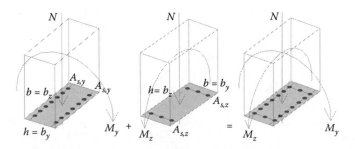

Figure 5.3 Dimensioning of reinforcement in column section for biaxial moments with axial force.

replaced by half the maximum value of the right hand side of Equation 5.31 for clockwise or counterclockwise beam moments on the joint in the vertical plane of bending of interest, that is, at right angles to the pair of column sides where the so-computed reinforcement is arranged. Step 1 is repeated with half the maximum value of the right-hand side of Equation 5.31 in a vertical plane of bending orthogonal to the first one, to determine the reinforcement along the other pair of sides of the section. As pointed out in Section 5.4.1, the value of the column axial force to be used is the minimum compressive or the maximum tensile one over the combinations in the seismic design situation which produce N-values consistent with the sense of action of the $\Sigma M_{Rd,b}$ on the joint – clockwise or counterclockwise; it is normally different for the two directions of bending.

To avoid the onerous ULS design/verification of sections in biaxial bending with axial force, Eurocode 8 allows to replace it with separate uniaxial verifications, but under pairs $(M_y/0.7) - N$ and $(M_z/0.7) - N$, that is, with the moments increased by 43%. Unless $M_y \approx M_z$, this simplification is too conservative (especially if seismic action effects are obtained from Equation 3.100 in Section 3.1.7). So, if the computational capability of a truly biaxial verification is available, there is no need to resort to this uniaxial approximation. Note also that, normally, the detailing rules in Table 5.1 and/or rounding up of the reinforcement required to meet the analysis results for the beams produce a value of $\Sigma M_{Rd,b}$ in Equation 5.31, which exceeds by more than 10% the beam capacity strictly necessary according to the analysis. So, if 'capacity design', Equation 5.31, is met for the most safe-sided (minimum compressive or maximum tensile) column axial force, the simplified uniaxial ULS verification per Eurocode 8 is also met automatically; there is no reason to do both.

If 'capacity design', through Equation 5.31, applies at the top section of a column in the bottom storey, it may give more vertical reinforcement there than at the base section of the same column, which is not subject to 'capacity design' and is dimensioned only for the ULS in bending with axial force under the action effects from the analysis for the seismic design situation. If so, it is a good practice to place at the base section the same reinforcement as at the top. Indeed, this is required in Eurocode 8 for DC H buildings. This ensures that, after plastic hinging at the base of the column, the moment at the top will not increase to much larger values than at the bottom.

5.4.3 Calculation of the column moment resistance for given reinforcement and axial load

This section is about the design values of the ULS moment resistance of a column, $M_{Rd,c}$, to be used in Equation 5.31 and in the other 'capacity design' calculations of Section 5.5.

In rectangular sections $M_{Rd,c}$ refers to centroidal axes parallel to the sides. The assumptions made in the previous section apply. The same for the notation introduced there, including Equations 5.32. Additional notation is introduced here for the reinforcement (the same as the one used in Section 5.2.3.4 for Equation 5.6):

- The mechanical reinforcement ratio $\omega_{1d} = A_{s1}/(bd) \cdot (f_{yd}/f_{cd})$ refers to the tension flange only; for generality, the compression flange may have different reinforcement, A_{s2}, with mechanical ratio $\omega_{2d} = A_{s2}/(bd) \cdot (f_{yd}/f_{cd})$; its centroidal distance from the extreme compression fibres is still d_1.
- There are intermediate bars between the tension and compression reinforcement, uniformly distributed along the length $(h - 2d_1)$ of the cross-sectional depth h; their total cross-sectional area, A_{sv}, is taken smeared along that length, with mechanical reinforcement ratio:

$$\omega_{vd} = A_{sv}/(bd) \cdot (f_{yd}/f_{cd}) \tag{5.36}$$

Only one-half of each corner bar is included in A_{s1} or A_{s2}; the other half counts as part of A_{sv}.

There are three cases (i)–(iii), analogous to those in Section 5.4.2, but generalised to accommodate the more general reinforcement layout:

i. Tension and compression reinforcement yield, if the normalised axial load is in the range:

$$\omega_{2d} - \omega_{1d} + \frac{\omega_{vd}}{1 - \delta_1}\left(\delta_1 \frac{\varepsilon_{cu2} + \varepsilon_{yd}}{\varepsilon_{cu2} - \varepsilon_{yd}} - 1\right) + \delta_1 \frac{\varepsilon_{cu2} - \varepsilon_{c2}/3}{\varepsilon_{cu2} - \varepsilon_{yd}} \equiv \nu_2 \leq \nu_d <$$

$$\nu_1 \equiv \omega_{2d} - \omega_{1d} + \frac{\omega_{vd}}{1 - \delta_1}\left(\frac{\varepsilon_{cu2} - \varepsilon_{yd}}{\varepsilon_{cu2} + \varepsilon_{yd}} - \delta_1\right) + \frac{\varepsilon_{cu2} - \varepsilon_{c2}/3}{\varepsilon_{cu2} + \varepsilon_{yd}}$$

(5.37a)

The design value of moment resistance is obtained from:

$$\frac{M_{Rd,c}}{bd^2 f_{cd}} = \xi\left[\frac{1 - \xi}{2} - \frac{\varepsilon_{c2}}{3\varepsilon_{cu2}}\left(\frac{1}{2} - \xi + \frac{\varepsilon_{c2}}{4\varepsilon_{cu2}}\xi\right)\right]$$

$$+ \frac{(1 - \delta_1)(\omega_{1d} + \omega_{2d})}{2} + \frac{\omega_{vd}}{1 - \delta_1}\left[(\xi - \delta_1)(1 - \xi) - \frac{1}{3}\left(\xi \frac{\varepsilon_{yd}}{\varepsilon_{cu2}}\right)^2\right]$$

(5.38a)

with the normalised neutral axis depth computed from:

$$\xi = \frac{(1 - \delta_1)(\nu_d + \omega_{1d} - \omega_{2d}) + (1 + \delta_1)\omega_{vd}}{(1 - \delta_1)\left(1 - \frac{\varepsilon_{c2}}{3\varepsilon_{cu2}}\right) + 2\omega_{vd}}$$

(5.39a)

ii. The tension bars yield, the compression ones are elastic, if ν_d is less than ν_2 from Equation 5.37a:

$$\nu_d \leq \omega_{2d} - \omega_{1d} + \frac{\omega_{vd}}{1 - \delta_1}\left(\delta_1 \frac{\varepsilon_{cu2} + \varepsilon_{yd}}{\varepsilon_{cu2} - \varepsilon_{yd}} - 1\right) + \delta_1 \frac{\varepsilon_{cu2} - \varepsilon_{c2}/3}{\varepsilon_{cu2} - \varepsilon_{yd}} \equiv \nu_2$$

(5.37b)

Then:

$$\frac{M_{Rd,c}}{bd^2 f_{cd}} = \xi\left[\frac{1 - \xi}{2} - \frac{\varepsilon_{c2}}{3\varepsilon_{cu2}}\left(\frac{1}{2} - \xi + \frac{\varepsilon_{c2}}{4\varepsilon_{cu2}}\xi\right)\right] + \frac{(1 - \delta_1)}{2}\left(\omega_{1d} + \omega_{2d}\frac{\xi - \delta_1}{\xi}\frac{\varepsilon_{cu2}}{\varepsilon_{yd}}\right)$$

$$+ \frac{\omega_{vd}}{4(1 - \delta_1)}\left[\xi\left(1 + \frac{\varepsilon_{yd}}{\varepsilon_{cu2}}\right) - \delta_1\right]\left[1 + \frac{\varepsilon_{cu2}}{\varepsilon_{yd}}\left(\frac{\xi - \delta_1}{\xi}\right)\right]\left[1 - \frac{\delta_1}{3} - \frac{2}{3}\xi\left(1 + \frac{\varepsilon_{yd}}{\varepsilon_{cu2}}\right)\right]$$

(5.38b)

with ξ the positive root of the equation:

$$\left[1 - \frac{\varepsilon_{c2}}{3\varepsilon_{cu2}} + \frac{\omega_{vd}}{2(1-\delta_1)}\frac{\left(\varepsilon_{cu2} + \varepsilon_{yd}\right)^2}{\varepsilon_{cu2}\varepsilon_{yd}}\right]\xi^2 - \left[v_d + \omega_{1d} - \omega_{2d}\frac{\varepsilon_{cu2}}{\varepsilon_{yd}} + \frac{\omega_{vd}}{1-\delta_1}\left(1 + \frac{\varepsilon_{cu2}}{\varepsilon_{yd}}\delta_1\right)\right]\xi$$

$$- \left[\omega_{2d} - \frac{\omega_{vd}\delta_1}{2(1-\delta_1)}\right]\frac{\varepsilon_{cu2}}{\varepsilon_{yd}}\delta_1 = 0 \tag{5.39b}$$

iii. The third case is to have the compression bars yielding and the tension ones elastic; this happens if v_d exceeds v_1 at the right-hand side of Equation 5.37a:

$$v_1 \equiv \omega_{2d} - \omega_{1d} + \frac{\omega_{vd}}{1-\delta_1}\left(\frac{\varepsilon_{cu2} - \varepsilon_{yd}}{\varepsilon_{cu2} + \varepsilon_{yd}} - \delta_1\right) + \frac{\varepsilon_{cu2} - \varepsilon_{c2}/3}{\varepsilon_{cu2} + \varepsilon_{yd}} \leq v_d \tag{5.37c}$$

Then:

$$\frac{M_{Rd,c}}{bd^2 f_{cd}} = \xi\left[\frac{1-\xi}{2} - \frac{\varepsilon_{c2}}{3\varepsilon_{cu2}}\left(\frac{1}{2} - \xi + \frac{\varepsilon_{c2}}{4\varepsilon_{cu2}}\xi\right)\right] + \frac{(1-\delta_1)}{2}\left(\omega_{1d}\frac{1-\xi}{\xi}\frac{\varepsilon_{cu2}}{\varepsilon_{yd}} + \omega_{2d}\right)$$

$$+ \frac{\omega_{vd}}{4(1-\delta_1)}\left[1 - \xi\left(1 - \frac{\varepsilon_{yd}}{\varepsilon_{cu2}}\right)\right]\left[1 + \frac{\varepsilon_{cu2}}{\varepsilon_{yd}}\left(\frac{1-\xi}{\xi}\right)\right]\left[\frac{1}{3} - \delta_1 + \frac{2}{3}\xi\left(1 - \frac{\varepsilon_{yd}}{\varepsilon_{cu2}}\right)\right] \tag{5.38c}$$

with ξ the positive root of the equation:

$$\left[1 - \frac{\varepsilon_{c2}}{3\varepsilon_{cu2}} - \frac{\omega_{vd}}{2(1-\delta_1)}\frac{\left(\varepsilon_{cu2} - \varepsilon_{yd}\right)^2}{\varepsilon_{cu2}\varepsilon_{yd}}\right]\xi^2 + \left[\omega_{2d} + \omega_{1d}\frac{\varepsilon_{cu2}}{\varepsilon_{yd}} - v_d + \frac{\omega_{vd}}{1-\delta_1}\left(\frac{\varepsilon_{cu2}}{\varepsilon_{yd}} - \delta_1\right)\right]\xi$$

$$- \left[\omega_{1d} + \frac{\omega_{vd}}{2(1-\delta_1)}\right]\frac{\varepsilon_{cu2}}{\varepsilon_{yd}} = 0 \tag{5.39c}$$

Note that v_1 at the right-hand side of Equation 5.37a and the left-hand side of 5.37c is the dimensionless 'balance load'. For that value of v_d, Equations 5.38a and 5.38c give the maximum possible value of $M_{Rd,c}$ that the section can develop. As we will see in Section 5.5, this moment resistance at 'balance load' is of interest for the capacity design shears of beams and columns.

For an application of the above to a rectangular column section, see Example 5.3 at the end of this chapter.

If the cross-section consists of more than one rectangular part in two orthogonal directions (L-, T- or C-sections, etc.), it is convenient to compute the moment resistance, $M_{Rd,c}$, with respect to centroidal axes parallel to the two orthogonal directions of the sides, since the beams connected to the column are parallel or normal to the sides of the rectangular parts. If an iterative algorithm of the type mentioned in Section 5.4.2 (e.g. at Step 3) is available for the ULS verification of sections with any shape and reinforcement layout under any

M_y–M_z–N combination, it can be used for the calculation of $M_{Rd,c}$, by setting the strain all-along the extreme fibres of the compression flange equal to ε_{cu2} and searching for a neutral axis depth that equilibrates the axial load N. If such an algorithm is not available, $M_{Rd,c}$ may be estimated considering the section as rectangular, with width b being that of the compression flange. This is acceptable, if the width of the compression zone is constant between the neutral axis and the extreme compression fibres (i.e. the compression zone lies within a single one of the rectangular parts of the section).

5.5 DETAILED DESIGN OF BEAMS AND COLUMNS IN SHEAR

5.5.1 Capacity design shears in beams or columns

The monotonic or cyclic behaviour of concrete members in flexure is fairly ductile: after flexural yielding they can sustain significant inelastic deformations (i.e. rotations), with little loss of moment resistance. Their inherent flexural ductility can easily be improved further, by using dense, closed stirrups, or similar types of transverse reinforcement, which laterally confine the concrete and prevent the longitudinal bars from buckling. By contrast, concrete members are inherently brittle in shear, whether monotonic or cyclic: if they reach their shear resistance before yielding in flexure, they suffer a drastic, and often sudden, drop in resistance right after its peak. For this reason, shear failure of members before flexural yielding should be prevented by all means. Eurocode 8 accomplishes this goal by enforcing 'capacity design' of all members in shear.

Capacity design of beams or columns in shear is a more straightforward and effective application of the 'capacity design' concept than that in Section 5.4.1 for columns in flexure. This section will show that, once plastic hinges are presumed to form at the relevant member ends, equilibrium of moments suffices to establish the maximum possible shear force in the member which is physically permitted by the 'capacities', M_{Rd}, of the plastic hinges. By designing against this 'capacity design' shear, instead of the shear force from the analysis for the seismic design situation, we preclude shear failure of the beams or the column not only before flexural yielding, but also afterwards; indeed indefinitely, for any level of earthquake.

With the internal forces at beam ends in Figure 5.4a taken as positive, equilibrium of moments about one end gives the shear force at the other:

$$V_1 = V_{g+\psi q,1} + \frac{M_1 + M_2}{l_{cl}} \tag{5.40a}$$

$$V_2 = V_{g+\psi q,2} - \frac{M_1 + M_2}{l_{cl}} \tag{5.40b}$$

where $V_{g+\psi q,1}$ and $V_{g+\psi q,2}$ are the moments of the transverse load acting between the two ends with respect to end 2 or 1, respectively, divided by the clear span of the beam, l_{cl} (i.e. the reactions to this load when the beam is simply supported). The maximum value of V_1 develops when M_1 and M_2 in the sum $M_1 + M_2$ both attain their maximum possible positive values; when M_1 and M_2 attain their algebraically minimum negative values, V_2 reaches its minimum possible value.

If the beam is connected at both ends to stronger columns, which satisfy Equation 5.31 without the 1.3 factor, the maximum possible positive values of M_1 and M_2 are

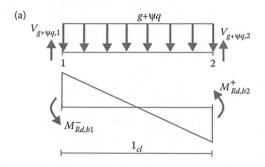

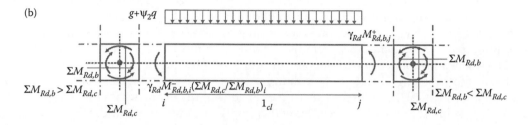

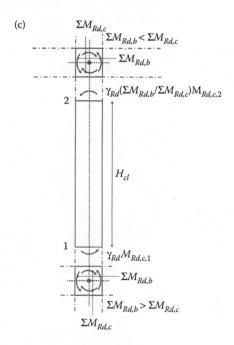

Figure 5.4 End moments considered for the capacity design in shear of: (a) an isolated beam; (b) a beam connected to columns with plastic hinging around the joint at the beam or the column ends; (c) a column connected to beams with plastic hinging around the joint at the column or the beam ends.

the corresponding moment resistances, taken for convenience as equal to their design values, M_{Rd}, times an overstrength factor, $\gamma_{Rd} \geq 1.0$. Accordingly, in Equation 5.40a we take

$$M_1 = \gamma_{Rd} M_{Rd,b1}^{-}, \ M_2 = \gamma_{Rd} M_{Rd,b2}^{+} \tag{5.41a}$$

and in Equation 5.40b

$$M_1 = -\gamma_{Rd}M^+_{Rd,b1}, \quad M_2 = -\gamma_{Rd}M^-_{Rd,b2} \tag{5.41b}$$

Using in Equations 5.40a, 5.40b the values of M_1, M_2 from Equations 5.41a, 5.41b, respectively, we obtain the maximum possible ('capacity design') shears at ends 1 and 2, respectively, of a beam that is weaker than the columns it is connected to.

Beams connected to weaker columns (i.e. not satisfying Equation 5.31 without the 1.3 factor) will most likely not develop plastic hinges at their ends before the columns. Assuming that at end i (1, 2) of the beam, the beam moment is negative and that the sum of the beam design moment resistances around the joint exceeds that of the columns in the sense associated with negative beam moment at that end, $M^-_{Rd,bi}$ in Equation 5.41a should be replaced by the beam moment at column hinging above and below the joint at end i. Assuming that the moment input from the yielding columns to the elastic beams is shared by the two beams connected to the joint in proportion to their own moment resistance, the beam moment at end i after the columns yield can be taken equal to $M^-_{Rd,bi}[\Sigma M_{Rd,c}/ \Sigma M_{Rd,b}]_i < M^-_{Rd,bi}$, with $\Sigma M_{Rd,b}$ referring to the sections of the beam across the joint at end i and $\Sigma M_{Rd,c}$ being the sum of moment resistances of the column above and below the joint, for bending in the vertical plane of the beam (for columns with sides at an angle ψ to that plane, the $M_{Rd,c}$-values with respect to centroidal axes parallel to the sides enters $\Sigma M_{Rd,c}$ multiplied by $\sin\psi$). This is also similar for the positive sense of bending of the beam at end i. So, a rational generalisation of Equations 5.40, 5.41 for the design value of the maximum shear at a section x in the part of the beam closer to end i (with j denoting the other end of the beam, see Figure 5.4b) is

$$\max V_{i,d}(x) = \frac{\gamma_{Rd}\left[M^-_{Rd,bi}\min\left(1;\dfrac{\sum M_{Rd,c}}{\sum M_{Rd,b}}\right)_i + M^+_{Rd,bj}\min\left(1;\dfrac{\sum M_{Rd,c}}{\sum M_{Rd,b}}\right)_j\right]}{l_{cl}} + V_{g+\psi q,o}(x) \tag{5.42a}$$

where

- All moments and shears enter as positive.
- If $M_{Rd,bi}$ acts on the joint in the clockwise direction, so does $(\Sigma M_{Rd,b})_i$, but $(\Sigma M_{Rd,c})_i$ acts on the joint in the counterclockwise direction.
- $V_{g+\psi q,o}(x)$ is the shear force at cross-section x due to the quasi-permanent gravity loads, with the beam simply supported (index: o); if the gravity load is not uniformly distributed along the beam, it may be conveniently computed from the results of the analysis of the full structure for these gravity loads alone: then $V_{g+\psi q,o}(x)$ may be taken as the shear force $V_{g+\psi q}(x)$ at cross-section x in the full structure, corrected for the shear force $(M_{g+\psi q,1} - M_{g+\psi q,2})/l_{cl}$ due to the bending moments $M_{g+\psi q,1}$ and $M_{g+\psi q,2}$ at the beam end sections 1 and 2 in the full structure.
- $\gamma_{Rd} = 1.2$ for beams of DC H and $\gamma_{Rd} = 1$ for DC M.

With $V_{g+\psi q,o}(x)$ taken positive at sections x in the part of the beam closer to end i, the minimum shear in that section is

$$\min V_{i,d}(x) = -\frac{\gamma_{Rd}\left[M^+_{Rd,bi}\min\left(1;\dfrac{\sum M_{Rd,c}}{\sum M_{Rd,b}}\right)_i + M^-_{Rd,bj}\min\left(1;\dfrac{\sum M_{Rd,c}}{\sum M_{Rd,b}}\right)_j\right]}{l_{cl}} + V_{g+\psi q,o}(x)$$

(5.42b)

The moments and shears at the right-hand side of Equation 5.42b being positive, its outcome may be positive or negative. If it is negative, the shear at section x may change sense of action during the seismic response (from downwards to upwards, or vice versa). We will see in Section 5.5.3 that, in the dimensioning of the transverse reinforcement of beams in DC H buildings, Eurocode 8 uses the ratio:

$$\zeta_i = \frac{\min V_{i,d}(x_i)}{\max V_{i,d}(x_i)}$$

(5.43)

as a measure of the reversal of shear at end i (similarly at end j).

Equation 5.42a gives safe-sided results even in a situation when the positive plastic hinge develops not at the very end j of the beam but at a point nearby, where the available moment resistance in positive bending is exhausted for the first time by the demand moment under the combination of quasi-permanent gravity loads and the seismic action that causes beam or column yielding – whichever occurs first – around the joint at end i.

Owing to the transverse gravity loads on the beam and the, in general, different longitudinal reinforcement of its two ends, the value of the capacity design shear from Equation 5.42a varies along the beam. The absolutely maximum value of shear at a certain cross-section x, $\max V_{i,d}(x)$ from Equation 5.42a, is in the same direction (down- or upwards) as the shear at x due to the quasi-permanent gravity loads in the simply supported beam, $V_{g+\psi q,o}(x)$; the absolutely minimum, $\min V_{i,d}(x)$ from Equation 5.42b, is in the opposite direction; $\max V_{i,d}(x)$ and $\min V_{i,d}(x)$ take place when the beam exhausts at end i its moment resistance in hogging or sagging bending, respectively.

If a beam is not connected at end i to a beam–column joint, but is ('indirectly') supported on another beam or girder, it is not expected to develop sizeable seismic moments there, let alone its moment resistance under seismic loading. The capacity design shear along the beam may then be estimated by replacing $M^+_{Rd,bi}$ or $M^-_{Rd,bi}$ in Equations 5.42 with the moment at end i under the quasi-permanent gravity loads alone, $M_{g+\psi q,i}$ (taken positive if it induces tension to the same flange of the beam as the moment resistance it replaces, or negative otherwise).

The picture is much simpler in columns, as there is no transverse load between the two ends; so the capacity design shear is constant all along the column. The design shear force parallel to a set of sides of a column, having clear height H_{cl} within the plane of bending (in general, equal to the distance of the top of the beam or slab at the base of the column to the soffit of the beam at the top), symmetric cross-section and reinforcement (so that $M_{Rd,c}$ is the same clockwise or counterclockwise) and ends indexed by 1 and 2, is given by a parallel to Equation 5.42a:

$$V_{CD,c} = \frac{\gamma_{Rd}\left[M_{Rd,c1}\min\left(1;\dfrac{\sum M_{Rd,b}}{\sum M_{Rd,c}}\right)_1 + M_{Rd,c2}\min\left(1;\dfrac{\sum M_{Rd,b}}{\sum M_{Rd,c}}\right)_2\right]}{H_{cl}}$$

(5.44)

$M_{Rd,c1}$ and $M_{Rd,c2}$ are moment resistances with respect to centroidal axes at right angle to the shear force being computed. The possibility of having plastic hinges at the end(s) of the column itself or in the beams connected to it is taken into account (Figure 5.4c); $\Sigma M_{Rd,c}$ refers to the sections of the column above and below the joint and $\Sigma M_{Rd,b}$ to the beam sections on opposite sides of it (for a beam at an angle α to the column shear force being calculated, $M_{Rd,b}$ enters $\Sigma M_{Rd,b}$ multiplied by $\cos \alpha$); the sense of action of $\Sigma M_{Rd,c}$ on the joint is the same as that of $M_{Rd,ci}$, while that of $\Sigma M_{Rd,b}$ is opposite.

Eurocode 8 specifies $\gamma_{Rd} = 1.3$ for columns in buildings of DC H and $\gamma_{Rd} = 1.1$ for those of DC M.

To obtain the largest absolute value of the beam shear force from Equation 5.42a and the algebraically minimum value of ζ in Equation 5.43, the values of $\Sigma M_{Rd,c}$ to be used in Equations 5.42 should be the maximum ones within the range of fluctuation of the column axial load from the analysis for all combinations of the design seismic action with the quasi-permanent gravity loads. The maximum moment resistance, $M_{Rd,c}$, is normally obtained from the maximum compressive force in that range of N, that is, the value of N due to the quasi-permanent gravity loads, $G + \psi_2 Q$, plus the value due to the design seismic action. However, if that sum exceeds the 'balanced load' $\nu_1 bdf_{cd}$ (see the right-hand side of Equation 5.37a or the left-hand side of Equation 5.37c in Section 5.4.3 for ν_1), $M_{Rd,c}$ is taken equal to the moment resistance at 'balance load' mentioned after Equation 5.39c.

Concerning the capacity design shear of columns, $V_{CD,c}$ from Equation 5.44: as we will see in Section 5.5.4, the shear capacity of a column as per Eurocode 2 increases with increasing axial compression. To find which one is the most critical shear verification of the column, we may have to consider more than one possible axial force values for $M_{Rd,ci}$ ($i = 1, 2$) in Equation 5.44, namely:

1. The minimum compression, which normally minimises the demand, $V_{CD,c}$, and the capacity, $V_{Rd,c}$.
2. The maximum compression, which maximises $V_{Rd,c}$ and most often $V_{CD,c}$ as well (except in case 3 below).
3. The 'balanced load' mentioned in the previous paragraph, if it is less than the maximum compression in 2 above; this 'balanced load' maximises $V_{CD,c}$ and gives an intermediate value of $V_{Rd,c}$.

More detailed guidance concerning the extreme values of N due to the design seismic action is given in Section 5.8.2.

For an application of this subsection to a beam, see Example 5.4 at the end of this chapter. For sample applications of the sub-section and the rest of Section 5.5 to the beams and columns of the seven-storey example building, see Sections 7.6.2.2 and 7.6.2.3.

5.5.2 Dimensioning of beams for the ULS in shear

Eurocode 2 uses for the ULS resistance in shear the variable strut inclination truss model: a model with angle of inclination, θ, of the compression stress field in the web with respect to the member axis which varies in the range:

$$0.4 \leq \tan \theta \leq 1 \ (22° \leq \theta \leq 45°) \tag{5.45}$$

According to this model:

1. Transverse reinforcement with design value of yield stress f_{ywd} and geometric ratio $\rho_w = A_{sh}/b_w s_h$ (where A_{sh} is the total area of transverse reinforcement with spacing s_h along the beam) contributes a shear resistance equal to

$$V_{Rd,s} = \rho_w b_w z f_{ywd} \cot\theta \qquad (5.46)$$

2. The shear resistance cannot exceed the following limit value, without failure of the web in diagonal compression:

$$V_{Rd,max} = 0.3 b_w z \left(1 - \frac{f_{ck}(\text{MPa})}{250} \right) f_{cd} \sin 2\theta \qquad (5.47)$$

The design shear force at section x along the beam, $V_{Ed}(x)$, is the maximum of the two values:

- From capacity design, Equation 5.42a
- From the analysis for the gravity loads in the 'persistent and transient design situation'

The general procedure for dimensioning in shear a section x of a beam is the following:

1. $V_{Ed}(x)$ is set equal to $V_{Rd,max}$ and Equation 5.47 is inverted for a value of θ.
2. In the very unlikely case that $V_{Rd,max}$ is less than $V_{Ed}(x)$ even for $\theta = 45°$, the width of the web is increased so that $\theta \leq 45°$.
3. In the very usual case when the condition $V_{Ed}(x) = V_{Rd,max}$ gives a θ-value below the lower limit in Equation 5.45, θ is set equal to that limit.
4. The shear reinforcement is dimensioned by setting: $V_{Ed}(x) = V_{Rd,s}$ for the final value of θ.
5. Dimensioning of the shear reinforcement starts at a section at a distance d from the face of a supporting column; the so-dimensioned shear reinforcement at the section is maintained to the face of the column.
6. Apart from point 5 above, a reverse 'shift rule' applies to the shear reinforcement determined at section x: it can be maintained constant over a distance $z \cot \theta$ in the direction of increasing shears, that is, toward the nearest support.

The above apply both to the 'seismic' design situation and the 'persistent and transient' one.

The shear reinforcement chosen should respect the detailing rules prescribed in Eurocodes 2 and 8, summarised in Table 5.3.

Apart from the special dimensioning rules for DC H beams highlighted in Section 5.5.3, the only difference that design against seismic actions as per Eurocode 8 or for non-seismic ones as per Eurocode 2 makes for beams in shear is the special detailing prescribed in Eurocode 8 for the stirrups in the end regions where plastic hinges are likely to form. These are termed 'critical regions', and a conventional length is specified for them. The prescribed maximum stirrup spacing as a multiple of the longitudinal bar diameter aims at preventing buckling of these bars (which is much more likely for bars subjected to alternate tension and compression, as is the case during the earthquake action).

The stirrup diameter and spacing are constant within each 'critical region', obeying the relevant detailing rules in Table 5.3. They are determined from the condition $V_{Ed}(x) = V_{Rd,s}$

Table 5.3 EC8 detailing rules for the transverse reinforcement of primary beams

	DC H	DC M	DC L
	Outside critical regions		
Spacing, $s_h \leq$		0.75d	
$\rho_w = A_{sh}/b_w s_h \geq$		$(0.08\sqrt{f_{ck}(MPa)})/f_{yk}(MPa)$[a]	
	In critical regions		
Diameter, $d_{bw} \geq$		6 mm	
Spacing, $s_h \leq$	$6d_{bL}^b$, $h/4$, $24d_{bw}$, 175 mm	$8d_{bL}^b$, $h/4$, $24d_{bw}$, 225 mm	–

[a] NDP (nationally determined parameter) per EC2; the value recommended in EC2 is given here.
[b] d_{bL}: minimum diameter of all top and bottom longitudinal bars within the critical region.

at a distance $x = d$ from the column face. A practical implication of the different detailing of 'critical regions' is that shear reinforcement in the rest of the beam is dimensioned from the condition $V_{Ed}(x) = V_{Rd,s}$ at a distance, x, from the column face equal to the 'critical region' length plus $z \cot \theta$. It is normally kept constant between the 'critical regions', as controlled by the most demanding section beyond a distance of $z \cot \theta$ from their ends.

5.5.3 Special rules for seismic design of critical regions in DC H beams for the ULS in shear

In DC H beams, additional Eurocode 8 rules further differentiate the dimensioning of 'critical regions' in shear from the rest of the beam. For the dimensioning of these regions in the 'seismic design situation', Eurocode 8 sets in Equations 5.46, 5.47 the strut inclination, θ, equal to 45°. This choice (i.e. $\tan \theta = 1$) gives the minimum value of $V_{Rd,s}$ in the range of θ per Eurocode 2, Equation 5.45. It amounts to a classical Mörsch-Ritter 45°-truss for the design in shear without a concrete contribution term. The reason of this choice is that in plastic hinges the shear resistance due to the transverse reinforcement decreases with increasing inelastic cyclic deformations (Biskinis et al. 2004); the magnitude of these deformations is significant in beams of DC H. Despite this apparently large penalty on $V_{Rd,s}$, the density of beam stirrups in the 'critical regions' of DC H beams is usually controlled by the detailing requirements at the last row of Table 5.3.

Another aspect where shear design in 'critical regions' of DC H beams deviates from the Eurocode 2 rules in Section 5.5.2 is the use of inclined bars against shear sliding at the end section of a beam at an instant in the response when the end section is cracked through its depth and the shear force is high. This may happen if the shear force has large reversals and a high peak value. Because a through-cracked section is not crossed by stirrups, Eurocode 8 requires for it inclined bars against sliding shear, if, with ζ from Equation 5.43, both of the following criteria are met:

$$-1 \leq \zeta < -0.5 \tag{5.48}$$

$$\max V_{i,d} > (2 + \zeta)f_{ctd}b_w d \tag{5.49}$$

where max $V_{i,d}$ is the maximum design shear force from Equation 5.42a at the end section of the beam 'critical region' at end i; the design value of the 5%-fractile of the tensile strength of concrete is $f_{ctd} = f_{ctk,0.05}/\gamma_c = 0.7f_{ctm}/\gamma_c = 0.21f_{ck}^{2/3}/\gamma_c$ (MPa). The limit shear at the right-hand side of Equation 5.49 varies (depending on the magnitude of the shear reversal) from one-third to one-half the value of $V_{Rd,max}$ for $\theta = 45°$.

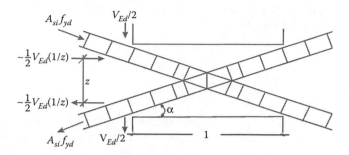

Figure 5.5 Coupling beam with diagonal reinforcement per Eurocode 8.

If both Equations 5.48 and 5.49 are met, the end section should be crossed by inclined bars at an angle $\pm\alpha$ to the beam axis. These bars, with total cross-sectional area A_s, should resist with the vertical components $A_s f_{yd} \sin\alpha$ of their yield force – in tension and compression – one-half of max $V_{i,d}$ from Equation 5.42a. The other half should be resisted by stirrups, according to the recommendation in Eurocode 2 to take at least one-half of the design shear force with shear links:

$$A_s f_{yd} \sin\alpha \geq 0.5 \max V_{i,d} \qquad (5.50)$$

If the beam is short, inclined bars are conveniently placed along its two diagonals in elevation (see coupling beam of Figure 5.5); then $\tan\alpha \approx (d - d_1)/L_{cl}$. If it is not short, the inclination of its diagonals to the beam axis is small and inclined bars placed along them are not efficient; in such cases it is more cost-effective to place two sets of shear links: one at an angle $\alpha = 45°$ to the beam axis, the other at $\alpha = -45°$. However, constructability and reinforcement congestion hamper this option. Normally, there is neither risk from sliding shear nor a need for inclined reinforcement, if we avoid beams that are short and not loaded by significant gravity loads (in such beams, the first term in the right-hand side of Equation 5.42b is large and the second one is small).

5.5.4 Dimensioning of columns for the ULS in shear

In columns designed to Eurocode 8, plastic hinging under the design seismic action is the exception. If it does take place, it leads to lower ductility demands than in DC H beams, and hence, leads to a smaller reduction of shear resistance. So, Eurocode 8 neglects this reduction for columns.

Columns are subjected to almost full shear reversals, while their capacity design shears from Equation 5.44 normally exceed the limit at the right-hand side of Equation 5.49 for $\zeta = -1$. Nevertheless, Eurocode 8 does not require for them inclined bars against shear sliding, trusting their axial force to close through-cracks of the end section against the low plastic strains that may build up in the vertical bars. Sliding is also resisted by clamping and dowel action of the large diameter intermediate bars between the corners, which remain elastic when the peak shear and moment occur in the column. So, the dimensioning of columns in shear takes place according to the Eurocode 2 alone, that is, taking into account the effect of axial load on shear resistance as follows:

1. A compressive axial force, N_d, increases the shear resistance, $V_{Rd,s}$, due to the transverse reinforcement by the transverse component of the strut which carries N from the

compression zone at the top section of the column to that of the bottom at an inclination of z/H_{cl} to the column axis:

$$V_{Rd,s} = \frac{z}{H_{cl}}N_d + \rho_w b_w z f_{ywd} \cot\theta \qquad (5.46a)$$

2. Eurocode 2 introduces in $V_{Rd,\max}$ an empirical multiplicative factor, which is a function of $v_d = N_d/A_c f_{cd}$ and takes into account: (a) the contribution of N_d to shear resistance, at the same time as (b) the burden placed on the inclined compression field accompanying the tension in the transverse reinforcement by the normal stress component in the strut due to N for $v_d > 0.5$:

$$V_{Rd,\max} = 0.3\min(1.25; 1 + v_d; 2.5(1 - v_d))b_w z\left(1 - \frac{f_{ck}(\text{MPa})}{250}\right)f_{cd}\sin 2\theta \qquad (5.47a)$$

With these modifications in the shear resistance formulas, steps 1 to 4 of the general procedure in Section 5.5.2 for dimensioning beams in shear are also applicable to columns, using all along the column $V_{CD,c}$ from Equation 5.44, instead of $V_{Ed}(x)$. This procedure is followed separately in the two transverse directions of the column, using the corresponding values of $V_{CD,c}$ as design shears. In rectangular columns, the side length at right angles to the plane of bending is used as b_w in Equations 5.46a, 5.47a and 90% of the effective depth, d, in the other direction as z.

If the section comprises more than one rectangular parts along two orthogonal directions, it is simpler and safe-sided to assign the design shear of each transverse direction only to the longest part of the section in that direction (i.e. to one leg per direction in a T- or L-section). That part plays the role of the web; only the stirrup legs in it which are parallel to the design shear contribute to the area of transverse reinforcement per unit height of the column, A_{sh}/s_h, in the direction considered.

Column sides longer than about 250 mm in DC H or 300 mm in DC M should have intermediate vertical bars engaged at a corner of a stirrup or by the hook of a cross-tie (see relevant rule in Table 5.2, row 3 from the bottom). The legs of these intermediate stirrups or cross-ties contribute to the shear resistance per Equation 5.46a at right angles to the column side; their cross-sectional area enters in $\rho_w = A_{sh}/b_w s_h$, multiplied by $\cos\alpha$, where α is the angle between the leg and the direction of the shear force. Although the cross-sectional area and/or spacing of intermediate stirrups or cross-ties may well differ from those of the perimeter hoops, they are usually chosen the same, for simplicity.

The transverse reinforcement should respect the detailing rules in Table 5.4. Except those concerning the effective mechanical ratio $a\omega_{wd}$ of stirrups, which have a fundamental basis explained in Sections 5.7.3 to 5.7.5, these rules are empirical. As in beams, the rule prescribing the maximum stirrup spacing in 'critical regions' as a multiple of the diameter of longitudinal bars aims at preventing buckling.

If the stirrup diameter and/or spacing are not controlled by the design shear, $V_{CD,c}$, which is constant along the column, but by the detailing rules, which are different in 'critical regions' and outside, the transverse reinforcement may be chosen different in each 'critical region' in a storey and in-between these regions. For simplicity, the transverse reinforcement is often chosen the same throughout the storey, as controlled by the most demanding of the two 'critical regions'.

Table 5.4 EC8 detailing rules for transverse reinforcement in primary columns

	DC H	DC M	DC L
Critical region length[a] ≥	$1.5h_c$, $1.5b_c$, 0.6 m, $H_{cl}/5$	h_c, b_c, 0.45 m, $H_{cl}/6$	h_c, b_c,
	Outside the critical regions		
Diameter, $d_{bw} \geq$	6 mm, $d_{bL}/4$		
Spacing, $s_w \leq$	$20d_{bL}$, h_c, b_c, 400 mm		
At lap splices of bars with $d_{bL} > 14$ mm, $s_w \leq$	$12d_{bL}$, $0.6h_c$, $0.6b_c$, 240 mm		
	In critical regions[b]		
Diameter, $d_{bw} \geq$[c]	6 mm, $0.4\sqrt{(f_{yd}/f_{ywd})}d_{bL}$	6 mm, $d_{bL}/4$	
Spacing, $s_w \leq$[c,d]	$6d_{bL}$, $b_o/3$, 125 mm	$8d_{bL}$, $b_o/2$, 175 mm	As outside critical regions
Mechanical ratio $\omega_{wd} \geq$[e]	0.08		–
Effective mechanical ratio $a\omega_{wd} \geq$[d,e,f,g]	$30 \mu_\varphi^* v_d \varepsilon_{yd} b_c/b_o - 0.035$		–
	In the critical region at the base of the column (at the connection to the foundation)		
Mechanical ratio $\omega_{wd} \geq$	0.12	0.08	–
Effective mechanical ratio $a\omega_{wd} \geq$[d,e,f,h,i]	$30 \mu_\varphi v_d \varepsilon_{yd} b_c/b_o - 0.035$		–

[a] h_c, b_c, H_{cl}: column sides and clear length.

[b] For DC M: If a value of $q \leq 2$ is used for the design, the transverse reinforcement in critical regions of columns with an axial load ratio $v_d \leq 0.2$ may follow only the rules for DC L columns.

[c] For DC H: In the two lower storeys of the building, the requirements on d_{bw}, s_w apply over a distance from the end section not less than 1.5 times the critical region length.

[d] Index c denotes the full concrete section; index o the confined core to the centreline of the perimeter hoop; b_o is the smaller side of this core.

[e] ω_{wd}: volume ratio of confining hoops to confined core (to centreline of perimeter hoop) times f_{ywd}/f_{cd}.

[f] $a = (1 - s/2b_o)(1 - s/2h_o)(1 - \{b_o/[(n_h - 1)h_o] + h_o/[(n_b - 1)b_o]\}/3)$: confinement effectiveness factor of rectangular hoops at spacing s, with n_b legs parallel to the side of the core with length b_o and n_h legs parallel to the side of length h_o.

[g] For DC H: at column ends protected from plastic hinging through the capacity design check at beam–column joints, μ_φ^* is the value of the curvature ductility factor that corresponds per Equations 5.64 to 2/3 of the basic value, q_o, of the behaviour factor applicable to the design; at the ends of columns where plastic hinging is not prevented, because of the exemptions from the application of Equation 5.31, μ_φ^* is taken equal to μ_φ defined in footnote h (see also footnote i); $\varepsilon_{yd} = f_{yd}/E_s$.

[h] μ_φ: curvature ductility factor corresponding per Equations 5.64 to the basic value, q_o, of the behaviour factor applicable to the design.

[i] For DC H: The requirement applies also in the critical regions at the ends of columns where plastic hinging is not prevented, because of the exemptions from the application of Equation 5.31.

5.6 DETAILED DESIGN OF DUCTILE WALLS IN FLEXURE AND SHEAR

5.6.1 Design of ductile walls in flexure

5.6.1.1 Design moments of ductile walls

To ensure that flexural plastic hinging is limited to the wall base and the wall stays elastic above it, despite higher mode response after the plastic hinge develops at the base, Eurocode 8 requires designing in flexure the rest of the wall height for a linear envelope of the positive and negative wall moments derived from the analysis for the design seismic action. The linear envelope is shown schematically in Figure 5.6, for simplicity without the tension shift; real examples of wall moment diagrams from the analysis are depicted in the upper half of Figure 7.44 in Chapter 7, alongside the design envelopes fitted to them per Eurocode 8 and shifted upwards by the tension shift. Thanks to the resulting flexural overstrength, the rest of the wall does not need to be specially detailed for flexural ductility, nor to be designed in

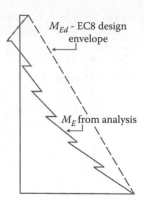

Figure 5.6 Bending moment diagram of a wall from the analysis and moment-envelope per Eurocode 8 for the design of a ductile wall in flexure (for simplicity the envelope does not include the 'tension-shift').

shear accounting for the cyclic decay of shear resistance in plastic hinges; so, its design and construction are much simpler and possibly less costly.

A wall flange longer than 4-times its thickness qualifies itself as a wall in the orthogonal direction (walls with T-, L-, H- or C-section). Then its design moments in that direction are obtained from a linear envelope as depicted in Figure 5.6 and not directly from the analysis.

5.6.1.2 *Dimensioning and detailing of vertical reinforcement in ductile walls*

The detailed design of a wall starts with dimensioning its vertical reinforcement at the base section for the normal action effects (moment(s) and axial force) derived from the analysis for the seismic design situation, per the Eurocode 2 criteria and rules for the ULS in flexure with axial force. The present section describes the dimensioning procedure, after an introduction about the distribution of vertical reinforcement over a wall section.

A wall differs from a frame column in the shape of its seismic moments diagram from the analysis. It differs from an isolated column, cantilevering from the foundation without connection to any floor beam, only in the shape of the cross-section, which, in a wall, comprises one or more elongated rectangular parts – conventionally per Eurocodes 2 and 8 with ratio of sides above 4.0. If it consists of a single elongated rectangular part, the wall develops essentially uniaxial moments and shears (in a vertical plane of bending in the long direction of the section), even when the seismic response is equally strong in the two horizontal directions. The main impact of the section geometry on the wall design, though, even for sections with two or more elongated rectangular parts (L-, T-, I-, C-sections, etc.), is the clear separation of the two ends of the section in the long direction. These end regions provide most of the moment resistance through vertical stresses – tensile at one end, compressive at the other – and play the prime role for flexural ductility: only they are enclosed in steel hoops for concrete confinement and anti-buckling restraint of vertical bars. In that respect, they resemble the top and bottom 'flanges' of a beam. Another common point with beams is that the part of the section between the longitudinally (and heavily) reinforced 'flanges' resists the shear, acting as a 'web'. A third commonality is, of course, the (essentially) uniaxial bending, parallel to the 'web'. By contrast, a column (even a big one behaving as a vertical cantilever) works in both transverse directions and requires vertical bars and confinement all around the section.

Like a deep beam, a wall has longitudinal reinforcement in the web as well, to control the width of flexural or shear cracks in that part of the wall too. This reinforcement is placed in

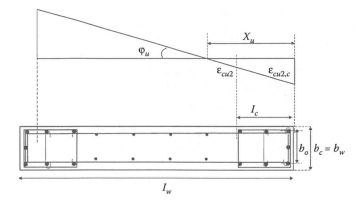

Figure 5.7 Schematic arrangement of vertical reinforcement in a ductile wall section and determination of boundary element length.

two curtains (one near each face of the web, see Figure 5.7 and examples in Figures 7.45 to 7.47 of Chapter 7) and is normally chosen on the basis of the prescriptive minimum requirements of Eurocodes 2 and 8, listed in Table 5.5 under 'Web' and 'vertical bars'. Suppose that the detailing rules concerning the minimum steel ratio and bar diameter and the maximum bar spacing in each curtain give a ratio of vertical web reinforcement $\rho_v = A_{sv}/bs_v$, where A_{sv} is the cross-sectional area of two web bars (one per curtain), s_v is the bar spacing along the length, l_w, of the wall section and b is the width of the compression flange (this ρ_v is normalised to the compression flange width, b, whereas the minimum and maximum web reinforcement ratios in Table 5.5 are normalised to the actual thickness of the web, $b_{wo} \le b$). The corresponding mechanical ratio is $\omega_{vd} = \rho_v f_{yd}/f_{cd}$.

The vertical reinforcement that is concentrated near the tension edge is considered for the present purposes as lumped at the centroid of its cross-sectional area, A_{s1}, at an effective depth d; we define its mechanical ratio as $\omega_{1d} = A_{s1}/(bd) \cdot (f_{yd}/f_{cd})$. We assume that the vertical reinforcement concentrated near the compression edge has the same cross-sectional area as the tension one, $A_{s2} = A_{s1}$, and it is assumed to be lumped at its centroid at a distance d_1 from the extreme compressive fibres; its mechanical ratio is $\omega_{2d} = \omega_{1d}$.

Using as a basis Equations 5.37 through 5.39 in Section 5.4.3, which take into account the web reinforcement with a uniform mechanical ratio, ω_{vd}, between the tension and the compression reinforcement, we can modify the procedure proposed in Section 5.4.2 for rectangular columns with symmetric tension and compression reinforcement only, to calculate $\omega_{1d} = \omega_{2d}$, for known ω_{vd} and given dimensionless parameters as defined by Equation 5.32 in Section 5.4.2. Note that walls have a low axial load ratio; in fact for walls Eurocode 8 sets an upper limit of 0.35 for DC H and of 0.4 for DC M to the ratio of the maximum axial load from the analysis for the seismic design situation to $A_c f_{cd}$, which exceeds ν_d as defined in Equation 5.32. We then have only cases (i) and (ii), as follows:

i. The tension and the compression reinforcement both yield, if ν_d is in the range:

$$\frac{\omega_v}{1-\delta_1}\left(\delta_1 \frac{\varepsilon_{cu2}+\varepsilon_{yd}}{\varepsilon_{cu2}-\varepsilon_{yd}} - 1\right) + \delta_1 \frac{\varepsilon_{cu2}-\varepsilon_{c2}/3}{\varepsilon_{cu2}-\varepsilon_{yd}} \equiv \nu_2 \le \nu_d < \nu_1$$

$$\equiv \frac{\omega_v}{1-\delta_1}\left(\frac{\varepsilon_{cu2}-\varepsilon_{yd}}{\varepsilon_{cu2}+\varepsilon_{yd}} - \delta_1\right) + \frac{\varepsilon_{cu2}-\varepsilon_{c2}/3}{\varepsilon_{cu2}+\varepsilon_{yd}} \qquad (5.51a)$$

Table 5.5 EC8 detailing rules for ductile walls

	DC H	DC M	DC L
Critical region height, h_{cr}	$\geq \max(l_w, H_w/6)$[b] $\leq \min(2l_w, h_{storey})$ if wall ≤ 6 storeys $\leq \min(2l_w, 2h_{storey})$ if wall > 6 storeys		–
	Boundary elements		
a) In critical height region:			
- Length l_c from wall edge $\geq$	$0.15l_w$, $1.5b_w$, part of the section where ε_c >0.0035		–
- Thickness b_w over $l_c \geq$	0.2 m; $h_{st}/15$ if $l_c \leq \max(2b_w, l_w/5)$, $h_{st}/10$ otherwise		–
- Vertical reinforcement:			
ρ_{min} over $A_c = l_c b_w$	0.5%		0.2%[a]
ρ_{max} over A_c		4%[a]	
Spacing along perimeter of bars restrained by tie corner or cross-tie hook	≤ 150 mm	≤ 200 mm	–
- Confining hoops (index w)[c]:			
Diameter, $d_{bw} \geq$	6 mm, $0.4\sqrt{(f_{yd}/f_{ywd})}d_{bL}$	6 mm,	wherever $\rho_L > 2\%$ in section: as over rest of the wall (see case b below)
Spacing, $s_w \leq$[d]	$6d_{bL}$, $b_o/3$, 125 mm	$8d_{bL}$, $b_o/2$, 175 mm	
$\omega_{wd} \geq$[c]	0.12	0.08	–
$\alpha\omega_{wd} \geq$[d,e]	$30\,\mu_\varphi(\nu_d + \omega_\nu)\varepsilon_{yd}b_w/b_o - 0.035$		–
b) Over the rest of the wall height:	Wherever in the section $\varepsilon_c > 0.2\%$: $\rho_{\nu,min} = 0.5\%$; elsewhere: 0.2% In parts of the section where $\rho_L > 2\%$: distance of unrestrained bar in compression zone to nearest restrained bar ≤ 150 mm; hoops with $d_{bw} \geq \max(6 \text{ mm}, d_{bL}/4)$, spacing $s_w \leq \min(12d_{bL}, 0.6b_{wo}, 240 \text{ mm})$[a] till distance $4b_w$ above or below floor slab/beam; $s_w \leq \min(20d_{bL}, b_{wo}, 400 \text{ mm})$[a] beyond that distance		
	Web		
Thickness, $b_{wo} \geq$	$\max(150 \text{ mm}, h_{storey}/20)$		–
Vertical bars (index: v):			
$\rho_v = A_{sv}/b_{wo}s_v \geq$	0.2%, but 0.5% wherever in the section $\varepsilon_c > 0.002$		0.2%[a]
$\rho_v = A_{sv}/b_{wo}s_v \leq$		4%	
$d_{bv} \geq$	8 mm		–
$d_{bv} \leq$	$b_{wo}/8$		–
Spacing, $s_v \leq$	$\min(25d_{bv}, 250 \text{ mm})$		$\min(3b_{wo}, 400 \text{ mm})$
Horizontal bars (index: h):			
$\rho_{h,min}$	0.2%		$\max(0.1\%, 0.25\rho_v)$[a]
$d_{bh} \geq$	8 mm		–
$d_{bh} \leq$	$b_{wo}/8$		–

continued

Table 5.5 (continued) EC8 detailing rules for ductile walls

	DC H	DC M	DC L
Spacing, $s_h \leq$	min($25d_{bh}$, 250 mm)		400 mm
$\rho_{v,min}$ at construction joints[f]	$\max\left(0.25\%;\ \dfrac{1.3f_{ctd} - N_{Ed}/A_c}{f_{yd} + 1.5\sqrt{f_{cd}f_{yd}}}\right)$		–

[a] NDP (Nationally Determined Parameter) per EC2; the value recommended in EC2 is given here.
[b] l_w: long side of rectangular wall section or rectangular part thereof; H_w: total height of wall; h_{storey}: storey height.
[c] (In DC M only) The DC L rules apply to the confining reinforcement of boundary elements, if: under the maximum axial force in the wall from the analysis for the seismic design situation, the wall axial load ratio $v_d = N_{Ed}/A_c f_{cd}$ is ≤ 0.15; or, if $v_d \leq 0.2$ but the q-value used in the design is $\leq 85\%$ of the q-value allowed when the DC M confining reinforcement is used in boundary elements.
[d] Footnotes d, e, f of Table 5.4 apply for the confined core of boundary elements.
[e] μ_φ: value of the curvature ductility factor corresponding through Equations 5.64 to the product of the basic value q_o of the behaviour factor times the ratio $M_{Ed,o}/M_{Rd,o}$ of the moment at the wall base from the analysis for the seismic design situation to the design value of moment resistance at the wall base for the axial force from the same analysis; $\varepsilon_{yd} = f_{yd}/E_s$; ω_{vd}: mechanical ratio of vertical web reinforcement.
[f] N_{Ed}: minimum axial load from the analysis for the seismic design situation (positive for compression); $f_{ctd} = f_{ctk,0.05}/\gamma_c = 0.7f_{ctm}/\gamma_c = 0.21f_{ck}^{2/3}/\gamma_c$: design value of 5%-fractile tensile strength of concrete.

Then, with a normalised neutral axis depth computed as

$$\xi = \frac{(1 - \delta_1)v_d + (1 + \delta_1)\omega_v}{(1 - \delta_1)(1 - \varepsilon_{c2}/3\varepsilon_{cu2}) + 2\omega_v} \tag{5.52a}$$

we find the symmetric edge reinforcement, $\omega_{1d} = \omega_{2d}$, from

$$(1 - \delta_1)\omega_{1d} = \mu_d - \xi\left[\frac{1 - \xi}{2} - \frac{\varepsilon_{c2}}{3\varepsilon_{cu2}}\left(\frac{1}{2} - \xi + \frac{\varepsilon_{c2}}{4\varepsilon_{cu2}}\xi\right)\right]$$

$$- \frac{\omega_{vd}}{1 - \delta_1}\left[(\xi - \delta_1)(1 - \xi) - \frac{1}{3}\left(\frac{\xi\varepsilon_{yd}}{\varepsilon_{cu2}}\right)^2\right] \tag{5.53a}$$

ii. The tension steel yields but the compression one is elastic; v_d is less than v_2 from Equation 5.51a:

$$v_d \leq \frac{\omega_{vd}}{1 - \delta_1}\left(\delta_1\frac{\varepsilon_{cu2} + \varepsilon_{yd}}{\varepsilon_{cu2} - \varepsilon_{yd}} - 1\right) + \delta_1\frac{\varepsilon_{cu2} - \varepsilon_{c2}/3}{\varepsilon_{cu2} - \varepsilon_{yd}} \equiv v_2 \tag{5.51b}$$

Then ξ is the positive root of the equation:

$$\left[1 - \frac{\varepsilon_{c2}}{3\varepsilon_{cu2}} + \frac{\omega_{vd}}{2(1 - \delta_1)}\frac{(\varepsilon_{cu2} + \varepsilon_{yd})^2}{\varepsilon_{cu2}\varepsilon_{yd}}\right]\xi^2 - \left[v_d + \omega_{1d}\left(1 - \frac{\varepsilon_{cu2}}{\varepsilon_{yd}}\right) + \frac{\omega_{vd}}{1 - \delta_1}\left(1 + \frac{\varepsilon_{cu2}\delta_1}{\varepsilon_{yd}}\right)\right]\xi$$

$$- \left[\omega_{1d} - \frac{\omega_{vd}\delta_1}{2(1 - \delta_1)}\right]\frac{\varepsilon_{cu2}\delta_1}{\varepsilon_{yd}} = 0 \tag{5.52b}$$

We can replace ω_{1d} in Equation 5.52b in terms of ξ and the dimensionless moment from:

$$\omega_{1d}\frac{(1-\delta_1)}{2}\left(1+\frac{\xi-\delta_1}{\xi}\frac{\varepsilon_{cu2}}{\varepsilon_{yd}}\right)=\mu_d-\xi\left[\frac{1-\xi}{2}-\frac{\varepsilon_{c2}}{3\varepsilon_{cu2}}\left(\frac{1}{2}-\xi+\frac{\varepsilon_{c2}}{4\varepsilon_{cu2}}\xi\right)\right]$$

$$-\frac{\omega_{vd}}{4(1-\delta_1)}\left[\xi\left(1+\frac{\varepsilon_{yd}}{\varepsilon_{cu2}}\right)-\delta_1\right]\left[1+\frac{\varepsilon_{cu2}}{\varepsilon_{yd}}\left(\frac{\xi-\delta_1}{\xi}\right)\right]\left[1-\frac{\delta_1}{3}-\frac{2}{3}\xi\left(1+\frac{\varepsilon_{yd}}{\varepsilon_{cu2}}\right)\right] \qquad (5.53b)$$

The resulting very nonlinear equation is solved numerically for ξ; ω_{1d} is then found from Equation 5.53b.

The edge-reinforcement area, $A_{s1}=\omega_{1d}(bd)\cdot(f_{cd}/f_{yd})$, from Equations 5.53 is implemented as a number of bars near the edge of the section, normally spread over a certain distance, l_c, from it, for example, along a 'boundary element' (see Figure 5.7 and examples in Figures 7.45 to 7.47 of Chapter 7). The minimum l_c value specified by Eurocode 8 within the critical region at the base of the wall is given at the top of the 'boundary elements' part of Table 5.5. The distance d_1 of this reinforcement from the section edge refers to the centroid of these bars. Note that, because ω_{vd} is considered uniform between the centroids of ω_{2d} and ω_{1d}, a fraction $(l_c/d-\delta_1)/(1-\delta_1)$ of the total web reinforcement area, $\rho_v bd$, falls within the distance l_c over which the edge reinforcement is spread and should be added to $A_{s1}=\omega_{1d}(bd)\cdot(f_{cd}/f_{yd})$ before translating it into an edge-reinforcement area.

The minimum web reinforcement continues to the top of the wall. There are two ways to decide at which levels the edge reinforcement placed at the base section is curtailed:

1. We take away from each edge region one pair of bars at a time (on opposite long faces of the wall), or even two such pairs or more. As long as the distance from the wall base is less than the critical region height, h_{cr}, given at the top of Table 5.5, the length l_c of the 'boundary element' still applies; the bars removed are chosen from the unrestrained ones along the perimeter of the 'boundary element', preferably far from the extreme fibres. Above the critical region height, the pair of bars removed is the one further away from the extreme fibres; the size of the 'boundary element' shrinks accordingly, below the minimum specified by Eurocode 8 for the critical region; the minimum web reinforcement extends over the freed space in the section. The remaining moment resistance of the section is computed using Equations 5.37 through 5.39 from Section 5.4.3 and compared to the linear M-envelope per Eurocode 8 (Figure 5.6), in order to find the level where the reduced edge reinforcement suffices. Note that this level should be consistent with the value of the wall axial force, N, used in Equations 5.37 through 5.39 with the reduced amount of reinforcement. The process continues with further pairs of bars removed from each edge, to the top of the wall.

2. The second approach lends itself better to systematic dimensioning within an integrated computational environment. It presumes that bars start at floor level of each storey and serve the bottom section of the storey above; at floor level of that storey the bars are lap-spliced with some of the edge bars starting there, or are continued for anchorage if they are not needed anymore. The dimensioning procedure described above for the wall base (Equations 5.51 through 5.53) is repeated at the bottom section of each storey, with the values of the moment and axial force applying there, to dimension the edge reinforcement, which should come from the storey below, in order to cover the requirements in flexure with axial force at the bottom section of the storey. In all storeys whose base falls within the critical region height, h_{cr}, in Table 5.5,

the layout of the bars placed near each edge follows that in the 'boundary element' of the base, as far as the outline and the location of restrained bars along the perimeter of the 'boundary element' are concerned. Above the critical region, little care is taken to follow the same pattern as in the critical region or to place the bars very close to those coming from the storey below; the overriding consideration is to spread the bars over a distance l_c from the extreme fibres, so that the maximum steel ratio is not violated within the area $A_c = l_c b_w$.

If the wall section comprises two or more elongated rectangular parts at right angles to each other (as in T-, L-, C- or H-sections), it should be designed in flexure as a whole, for the M_y–M_z–N triplet of the entire section, assuming that it remains plane. The three-step procedure proposed in Section 5.4.2 for dimensioning the vertical reinforcement of columns, rectangular or not, under M_y–M_z–N triplets, may be adapted to Equations 5.51 through 5.52, to account for web bars distributed between the two edges in the direction considered. The so-modified procedure normally gives a safe-sided estimate of the vertical reinforcement near the corners of the non-rectangular section. The full vertical reinforcement placed over the section should also meet the detailing rules in Table 5.5 for boundary elements, web minimum reinforcement and so forth. Note that the size of any boundary elements needed around the non-rectangular section may be estimated from the strain profile(s) obtained in the course of Step 3 of the procedure, namely through the iterative algorithm for the ULS verification of sections with any shape and layout of reinforcement for any combination M_y–M_z–N.

Strictly speaking, even a rectangular wall is subjected to biaxial bending with axial force, M_y–M_z–N. So, although this is rarely done for rectangular walls, after the vertical reinforcement is estimated and placed according to the pertinent detailing rules, the base section of each storey may be verified for the ULS in bending with axial force for all M_y–M_z–N combinations from the analysis for the seismic design situation. The moment in the strong direction of the wall, let us say M_y, is obtained from the linear M-envelope in Figure 5.6; the value of M_z is that from the analysis.

The edge bars curtailed according to procedure 1 or 2 above – or any alternative – should extend vertically above the level where they are not needed anymore for the ULS in bending with axial force by a length equal to $z \cot \theta / 2$, according to the 'shift rule' as per EC2, where θ is the value of the strut inclination used at that level in the design of the wall in shear (see Section 5.6.2). They are extended by their anchorage length, only if the inclination of the moment envelope to the vertical (which is constant up the wall, see Figure 5.6) exceeds the bar yield force, $f_{yd} \pi d_{bL}^2 / 4$, times the ratio of the internal lever arm, z, to l_{bd} (i.e. $(2.25 \pi f_{ctd}) d_{bL} z / \{a_{tr}[1 - 0.15(c_d / d_{bL} - 1)]\}$ with the notation of Table 5.2).

5.6.2 Design of ductile walls in shear

5.6.2.1 Design shears in ductile walls

'Ductile walls', designed to develop a flexural plastic hinge only at the base, are protected by Eurocode 8 provisions from shear failure throughout their height. The design value of moment resistance at the wall's base section, $M_{Rd,o}$, and equilibrium alone do not suffice for the derivation of the maximum seismic shears that can develop at various levels of the wall, because, unlike in the cases of Figure 5.4, the forces applied on the wall at intermediate levels are unknown and vary during the seismic response. It is reasonable to assume that, if $M_{Rd,o}$ exceeds the bending moment at the base from the elastic analysis for the design seismic action, $M_{Ed,o}$, the seismic shears at any level of the wall exceed those from the same elastic analysis in proportion to $M_{Rd,o}/M_{Ed,o}$. This amounts to multiplying the shear forces from the elastic analysis for the design seismic action, V'_{Ed}, by a capacity-design magnification factor ε:

$$\varepsilon = \frac{V_{Ed}}{V'_{Ed}} = \gamma_{Rd} \left(\frac{M_{Rd,o}}{M_{Ed,o}} \right) \leq q \qquad \left(\text{if } h_w/l_w \leq 2 \right) \tag{5.54a}$$

where γ_{Rd} covers overstrength at the base, for example, due to steel strain hardening. If the wall is rectangular, $M_{Rd,o}$ can be estimated according to Section 5.4.3, Equations 5.37 through 5.39. The last paragraph of Section 5.4.3 applies to non-rectangular walls.

Eurocode 8 adopts Equation 5.54a with $\gamma_{Rd} = 1.2$, for DC H walls with ratio of wall height to horizontal dimension, $h_w/l_w \leq 2$ ('squat'). If a DC H wall has $h_w/l_w > 2$ ('slender'), Eurocode 8 amplifies further the shear forces from the elastic analysis, V'_{Ed}, to account for an increase of shears after plastic hinging at the base due to higher modes. It follows in this respect the approach in Eibl and Keintzel (1988) and Keintzel (1990). That approach essentially presumes that:

1. $M_{Ed,o}$ and V'_{Ed} are computed via 'lateral force' elastic analysis, with a first mode period T_1.
2. The behaviour factor, q, should be applied only to the first mode results; higher mode response is elastic – at least as far as the wall shears are concerned.
3. Higher mode periods lie in the constant–spectral–acceleration plateau of the elastic spectrum; their spectral acceleration is equal to $S_a(T_C)$, where T_C is the upper corner period of the plateau.
4. The ratio of the sum-of-the-squares of higher mode participation factors to the square of the participation factor of the first mode is equal to 0.1, that is, a very safe-sided estimate.

These considerations lead to an increase of ε in DC H walls with $h_w/l_w > 2$ per Eurocode 8:

$$\varepsilon = \frac{V_{Ed}}{V'_{Ed}} = \sqrt{ \left(\gamma_{Rd} \frac{M_{Rd,o}}{M_{Ed,o}} \right)^2 + 0.1 \left(q \frac{S_a(T_C)}{S_a(T_1)} \right)^2 } \leq q \; \left(\text{if } h_w/l_w > 2 \right) \tag{5.54b}$$

where T_1 is the period of the first mode in the horizontal direction closest to that of the wall shear force.

Equation 5.54b gives very safe-sided (i.e. high) values, especially if $M_{Ed,o}$ and V'_{Ed} are computed via a 'modal response spectrum' elastic analysis (Antoniou et al. 2014).

The value of $M_{Rd,o}/M_{Ed,o}$, and hence of ε in Equations 5.54, may well exceed 1.0 if:

- The wall base is oversized with respect to the seismic demand, $M_{Ed,o}$, and has the minimum vertical reinforcement in its web and – sizeable – boundary elements. To reduce this type of overstrength, the wall should not be so thick as to have the minimum requirements per Table 5.5 control its vertical reinforcement.
- The analysis for the design seismic action produces a high axial force at the wall base. The vertical reinforcement at the wall base is governed by the sign of the design seismic action giving – alongside the moment, $M_{Ed,o}$ – the minimum axial compression from the analysis for the seismic design situation. When the sign of the design seismic action reverses, we have the maximum axial compression but the same moment, $M_{Ed,o}$, producing a large overstrength: $M_{Rd,o} \gg M_{Ed,o}$. Such an overstrength is acute in walls placed near the corners in plan of high-rise buildings, in piers of coupled walls and so on.

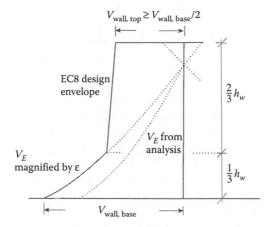

Figure 5.8 Design shear forces per Eurocode 8 up a ductile wall in a dual (wall–frame) system.

As a shield to excessive values of ε from Equations 5.54 due to the above reasons, Eurocode 8 sets for DC H walls the ceiling of q to its value, so that the final design shear, V_{Ed}, does not exceed the value qV'_{Ed} corresponding to a fully elastic response.

To avoid the intrinsic complexity and conservatism of Equations 5.54, Eurocode 8 allows to DC M walls the following simplification:

$$\varepsilon = \frac{V_{Ed}}{V'_{Ed}} = 1.5 \tag{5.55}$$

Note that, unlike Equations 5.54 which may be over-conservative, Equation 5.55 gives a very low safety margin (if any) against flexural overstrength at the base or inelastic higher mode effects.

Higher mode effects on inelastic shears are larger at the upper storeys of the wall, especially in dual (frame–wall) systems. In such systems the frames restrain the walls at the upper storeys, to the extent that the wall shears in the top storey or the one below from the 'lateral force' analysis reverse sign and are opposite to the total seismic shear applied. In general, the 'lateral force' analysis gives very small wall shears in the upper storeys, which will not come anywhere close to the relatively high values that may develop there owing to higher modes, even after multiplication by the factor ε of Equations 5.54 and 5.55 (see Figure 5.8, where dotted curves represent the shear force from the analysis and its value multiplied by ε). To deal with this problem, Eurocode 8 sets a minimum for the design shear force at the top of the ductile walls in dual (frame–wall) systems, equal to half the magnified shear at the base, increasing linearly to the magnified value of the shear, $\varepsilon V'_{Ed}$, at one third of the wall height from the base (Figure 5.8).

For a sample application of the part of this sub-section and of the rest of Section 5.6 referring to DC M walls, see Section 7.6.2.4 for the walls of the seven-storey example building.

5.6.2.2 *Verification of ductile walls in shear: Special rules for critical regions of DCH walls*

The general Eurocode 2 rules highlighted in Section 5.5.2 for beams and 5.5.4 for columns apply to the verification in shear of DC M walls throughout their height and of DC H ones outside their critical height. The contribution of the axial load to $V_{Rd,s}$ is small; so, Equation 5.46 may be applied. The internal lever arm, z, may be taken equal to 80% of the length, l_w, of the wall section. The web reinforcement should also meet the prescriptive detailing rules in Table 5.5.

Three types of special rules apply in the critical region of DC H walls:

1. The design value of shear resistance, as controlled by diagonal compression in the web, $V_{Rd,max}$, is taken as 40% of the value given by Equation 5.47 per Eurocode 2. A value of θ is not fixed in the range of Equation 5.45, but, with such a drastic reduction in $V_{Rd,max}$, it makes sense to take $\theta = 45°$. This large reduction is fully supported by the available cyclic test data (Biskinis et al. 2004; Fardis 2009). It has not been extended to DC M walls as well, because, if applied alongside the magnification of design shears per Section 5.6.2.1 (Equation 5.55), it might be prohibitive to use ductile walls for earthquake resistance in Eurocode 8. However, caution should be exercised in exhausting the liberal $V_{Rd,max}$-value of Equation 5.47a in ductile walls of DC M.

2. Unlike columns (but like DC H beams), ductile walls of DC H should be explicitly verified against sliding, because their axial load level is too low to mobilise sufficient friction, and the web bars are of smaller diameter and more sparse than in columns. The base of every storey within the critical height of the wall should be verified in sliding shear. The design value of the resisting shear against horizontal sliding along the base section in a storey is given by Eurocode 8 as

$$V_{Rd,SLS} = V_{fd} + V_{dd} + V_{id} \qquad (5.56)$$

that is, as the sum of:

- The friction resistance, V_{fd}, of the compression zone:

$$V_{fd} = \min\left(\mu_f\left[\left(\sum A_{sj}f_{yd} + N_{Ed}\right)\xi + \frac{M_{Ed}}{z}\right]; 0.3\left(1 - \frac{f_{ck}(\text{MPa})}{250}\right)f_{cd}b_{wo}\left(\xi d\right)\right) \qquad (5.57)$$

with:
- μ_f: friction coefficient, equal to 0.7 for rough interfaces or equal to 0.6 for smooth ones
- $\sum A_{sj}$: total area of vertical bars in the web and of those placed in the boundary elements specifically against shear sliding without counting in the ULS for bending
- M_{Ed}, N_{Ed}: values from the analysis for the seismic design situation
- ξ: normalised neutral axis depth, from Equations 5.39 in Section 5.4.3
- A design value of the dowel resistance, V_{dd}:

$$V_{dd} = \sum A_{sj} \min\left(1.3\sqrt{f_{cd}f_{yd}}; 0.25f_{yd}\right) \qquad (5.58)$$

For concrete class above C20/25, the second term governs in Equation 5.58; it reflects yielding of the dowel in pure shear without axial force, with a safety margin of about 2.3;

- The horizontal projection, V_{id}, of the design resistance of any inclined bars, with a total area (in both directions) $\sum A_{si}$, placed at an angle $\pm \varphi$ to the base of the wall:

$$V_{id} = \sum A_{si}f_{yd}\cos\varphi \qquad (5.59)$$

Inclined bars should preferably cross the base section at about mid-length, to avoid affecting – through the couple of vertical components of their tension and compression forces – the moment resistance, $M_{Rd,o}$, used in Equations 5.54 for the design shear, V_{Ed}, or the location of the plastic hinge; they should extend at least to a level of $l_w/2$ above the base section, making an inclination $\varphi = 45°$ not only convenient, but also very cost-effective. Although inclined bars are needed only if $V_{fd} + V_{dd}$ is less than the design shear, V_{Ed}, Eurocode 8 requires placing them always at the base of 'squat' walls (i.e. those with $h_w/l_w \leq 2$) of DC H, at a quantity sufficient to resist, through V_{id}, at least $0.5V_{Ed}$. In such walls, Eurocode 8 requires inclined bars at the base of all storeys, at a quantity sufficient to resist at least 25% of the storey design shear.

Note that the minimum clamping reinforcement required across construction joints (i.e. at the base section of each storey) in a DC H wall according to the last row in Table 5.5 counts also against shear sliding. The second, non-prescriptive term in this requirement comes from the condition that cohesion, plus friction, plus dowel action at such a joint exceeds the shear stress at shear cracking of the concrete cast right above the joint.

3. A special rule applies for dimensioning the web reinforcement ratios, horizontal ρ_h, and vertical ρ_v, in those storeys of DC H walls where the maximum shear span ratio, $\alpha_s = M_{Ed}/(V_{Ed}l_w)$ (normally at the base of the storey) is less than 2. This rule is a modification of the Eurocode 2 rule for the calculation of shear reinforcement in members with $0.5 < \alpha_s < 2$:

$$V_{Rd,s} = V_{Rd,c} + \rho_h b_{wo}(0.75 l_w \alpha_s) f_{yhd} = V_{Rd,c} + \rho_h b_{wo}\left(0.75\frac{M_{Ed}}{V_{Ed}}\right)f_{yhd} \tag{5.60}$$

where

- ρ_h is the ratio of the horizontal reinforcement, normalised to the web width, b_{wo}, and f_{yhd} its design yield strength;
- $V_{Rd,c}$ is the design shear resistance of members without shear reinforcement per Eurocode 2 (in kN):

$$V_{Rd,c} = \left\{ \min\left[\frac{180}{\gamma_c}(100\rho_L)^{1/3},\ 35\sqrt{1+\sqrt{\frac{0.2}{d}}}\,f_{ck}^{1/6}\right]\left(1+\sqrt{\frac{0.2}{d}}\right)f_{ck}^{1/3} \right.$$
$$\left. + 0.15\min\left(\frac{N_{Ed}}{A_c},\ 0.2\frac{f_{ck}}{\gamma_c}\right)\right\}b_{wo}d \tag{5.61}$$

where ρ_L is the tension reinforcement ratio, γ_c is the partial factor for concrete, f_{ck} is in MPa, b_{wo} and d are in meters, the wall gross cross-sectional area, A_c, is in m² and N_{Ed} is in kN (in the critical region of the wall, $V_{Rd,c} = 0$ if N_{Ed} is tensile).

The ratio of vertical web reinforcement, ρ_v, is then dimensioned to provide a 45° inclination of the concrete compression field in the web, together with the horizontal reinforcement and the vertical compression in the web due to the minimum axial force in the seismic design situation, $\min N_{Ed}$:

$$\rho_h f_{ydh} b_{wo} z \leq \rho_v f_{ydv} b_{wo} z + \min N_{Ed} \tag{5.62}$$

Note that M_{Ed} in α_s comes from the M-envelope of Figure 5.6, which does not exhibit inflection points in any storey, and V_{Ed} comes from Equations 5.54 – and the envelope in Figure 5.8, if the wall belongs to a dual (frame–wall) system. So, α_s may be less than 2 at upper storeys of walls with large l_w, giving unduly large web reinforcement. Judgment should be used in such cases, as these expressions are meant for the base region of 'squat' walls (those with $h_w/l_w \leq 2$); besides, owing to the very limited knowledge and data at the time on the cyclic behaviour and shear failure of squat walls, Equations 5.60, 5.62 are conservative (safe-sided).

The special rules for the critical region of DC H walls (especially no. 1 above concerning $V_{Rd,max}$), in conjunction with the magnification of shear forces from the analysis per Equations 5.54 in Section 5.6.2.1, make it difficult to verify DC H walls in shear. It is normally not very effective to increase the web thickness, b_{wo}, in order to meet this verification: this will increase proportionally the value of $V_{Rd,max}$, but will also increase, albeit less than proportionally, the seismic shear force from the analysis. It is much more effective to keep the ratio $M_{Rd,o}/M_{Ed,o}$ at the wall base as low as possible, preferably close to 1.0 (see discussion after Equation 5.54b in Section 5.6.2.1).

5.7 DETAILING FOR DUCTILITY

5.7.1 'Critical regions' in ductile members

Of the two constituents of reinforced concrete, steel is ductile in tension but not in compression, as bars may buckle, shedding their force and risking fracture. Concrete is brittle, unless its lateral expansion is well restrained by confinement. So, the only way to build a RC member which is ductile and can reliably dissipate energy during inelastic seismic response is by combining:

- Reinforcing bars in the direction where tensile principal stresses are expected to develop; and
- Concrete and reinforcement in the direction of compressive principal stresses, with dense ties to laterally confine the concrete and restrain the bars against buckling.

This is feasible wherever principal stresses and strains develop during the seismic response invariably in the directions where reinforcement can be conveniently placed. In one-dimensional RC members (beams, columns, slender walls), it is convenient to place the reinforcement in the longitudinal and transverse directions. Cyclic flexure indeed produces at the extreme fibres of a RC member principal stresses and strains in the longitudinal direction and allows effective use of reinforcement, both to take up directly the tension and to restrain the concrete and the compression bars transverse to their compressive stresses. Flexure is the only mechanism of force transfer in such a member, which allows using to advantage and reliably the ductility of tension reinforcement and effectively enhancing the ductility of concrete and of compression bars through lateral restraint. The regions of the member dominated by flexure under seismic loading are its ends, where the seismic moments take their maximum value. After flexural yielding of the end section, a flexural plastic hinge develops there, dissipating energy in alternate positive and negative bending. Eurocode 8 calls this region 'critical region', which has a more conventional connotation than the term 'dissipative zone', used also in Eurocode 8 for the part of a member or connection of any material where energy dissipation takes place by design.

A 'critical region' is a conventionally defined part of a primary RC member, up to a certain distance from:

1. The base section of a ductile wall, that is, at the connection to the foundation or the top of a rigid basement.
2. An end of a column or beam connected to a beam or a vertical element, respectively, no matter whether the relative magnitude of the moment resistances of the members around the connection show that a plastic hinge at that end is likely. Cantilevering beams not designed for a vertical seismic action, or a beam end supported on a girder at a distance from a joint of the girder with a vertical member, cannot develop large seismic moments; so there is no beam 'critical region' in those cases.
3. A beam section where the hogging moment from the analysis for the seismic design situation attains its maximum value along the span; often that section is at the beam end or nearby and the 'critical region' coincides with one of those described under 2 above.

The length of 'critical regions' prescribed for RC members by Eurocode 8 is given at the top of Tables 5.1, 5.4 and 5.5. These tables give the special detailing rules – mostly prescriptive – that apply in these regions. Sections 5.7.4 to 5.7.5 focus on and elaborate the application of those detailing rules, which have a rational basis.

5.7.2 Material requirements

Ductility depends not only on the detailing of RC members, but also on the ductility and quality of their materials. So the requirements of Eurocode 8 on concrete and steel increase with DC.

Eurocode 8 sets a lower limit of 16 MPa on the nominal cylindrical strength of concrete (concrete class C16/20) in primary elements of DC M buildings, or of 20 MPa (concrete class C20/25) in those of DC H. These limits are at the low end of what is normally used in buildings in Europe. Neither Eurocode 8 nor Eurocode 2 set a lower limit on concrete strength in DC L buildings. All concrete classes foreseen in Eurocode 2 are allowed by Eurocode 8: there is no upper limit for any DC.

The requirements of Eurocode 8 on reinforcing steel are summarised in Table 5.6. The lower limits on the 10%-fractile of the hardening ratio, f_t/f_y, and of the strain at maximum stress (also called tensile strength, f_t), ε_{su}, ensure a minimum extent of the flexural plastic hinge and a minimum curvature ductility, respectively. The aim of the upper limits on f_t/f_y and on the 95%-fractile of the actual yield stress is to avoid flexural overstrength at plastic hinges beyond what is covered by the overstrength factors γ_{Rd} in Equations 5.31, 5.42, 5.44 and 5.54 and avoid jeopardising the capacity design of columns in flexure and of beams, columns or walls in shear, respectively.

The requirements for DC M or L buildings are met by steel bars of Class B or C per Eurocode 2. The requirements for DC H in the last two rows of Table 5.6 are met by steel of Class C of Eurocode 2, but not by class B. That on $f_{yk,0.95}/f_{yk}$ for DC H comes from Eurocode 8 and is not automatically met by steels of class C (let alone B); however, steel types with special ductility produced and used in the most seismic-prone part of Europe do meet this additional requirement. Note that this requirement is sometimes violated, when a quantity of steel, which is originally produced for a certain nominal yield strength but fails the minimum criteria on its f_{yk}-value as 5%-fractile, is then re-classified and marketed as steel of lower nominal yield strength.

Table 5.6 EC8 requirements for reinforcing steel in primary members

Ductility class	DC L or M	DC H
5%-fractile of yield strength (: nominal yield strength), f_{yk}	400 to 600 MPa	
95%-fractile of actual yield strength to nominal, $f_{yk,0.95}/f_{yk}$	–	≤ 1.25
10%-fractile of the ratio of tensile strength (maximum stress) to the yield strength, $(f_t/f_y)_{k,0.10}$	≥ 1.08	≥ 1.15 ≤ 1.35
10%-fractile of strain at maximum stress, $\varepsilon_{su,k,0.10}$	$\geq 5\%$	$\geq 7.5\%$

Example 5.5 illustrates the categorisation of steel according to the Eurocode 8 criteria for DC L, M or H, on the basis of statistics of the steel properties from samples of bars.

The requirements in Table 5.6 for DC L apply throughout the length of any primary element. Strictly speaking, those for DC M or H apply only within 'critical regions'. However, the whole length of a primary member of DC M or H should meet the requirements in the middle column of the table, because its local ductility may not be inferior to that of a DC L member in any respect. The additional requirements in the last column apply only to the 'critical regions' in DC H buildings. However, it is not practical to implement different material specifications in the 'critical region' than over the rest of the element length. So, in practice the requirements on steel in 'critical regions' are applied over the whole of a primary element of DC M or H, including the slab it may be working with (as they apply to the slab bars that are parallel to a primary beam and fall within its effective flange width in tension defined in Section 5.2.2).

5.7.3 Curvature ductility demand in 'critical regions'

Eurocode 8 links the local deformation demand for which a plastic hinge should be detailed to the basic value of the behaviour factor, q_o, applicable to the building's DC and structural system per Table 4.1 in Section 4.6.3. Only in few cases are the values of q_o in Eurocode 8 discrete: for inverted pendulums or torsionally flexible systems and for wall systems of DC M; in all other DC M or H systems, q_o is proportional to α_u/α_1 (see Section 4.6.3), hence takes values in a continuous range. Therefore, it is not feasible to specify discrete values of the curvature ductility factor, μ_φ, for these other structural systems. So, Eurocode 8 gives μ_φ as an algebraic function of q_o (see Equations 5.64). This expression is derived from:

1. The q–μ–T relation between the global displacement ductility factor, μ_δ, the ductility dependent part of the behaviour factor, q_μ, and the period, T, of an SDOF oscillator adopted in Eurocode 8 (Equations 3.119, 3.120 in Section 3.2.3);
2. The approximate equality, $\mu_\theta \approx \mu_\delta$, of μ_δ to the local ductility factor of chord rotation, μ_θ, at those member ends where plastic hinges form in a beam-sway mechanism imposed on the structural system by a stiff/strong spine provided by the walls of wall systems or wall-equivalent dual systems, or by the strong columns of frame systems or frame-equivalent dual systems (see Section 4.5.2 and Figures 2.9b to e).
3. A safe-sided approximation of the curvature ductility factor at the member's end section, μ_φ, in terms of μ_θ, underlain by the Eurocode 2 model for confined concrete and a safe-sided average plastic hinge length, L_{pl}, equal to 18.5% of the shear span (M/V-ratio), L_s, at the end section of a typical RC member in buildings:

$$\mu_\theta = 1 + 3\frac{L_{pl}}{L_s}\left(1 - \frac{L_{pl}}{2L_s}\right)(\mu_\varphi - 1) \approx 1 + 0.5(\mu_\varphi - 1) \rightarrow \mu_\varphi = 2\mu_\theta - 1 \qquad (5.63)$$

4. A safe-sided assumption that the full basic value of the behaviour factor is due to ductility, neglecting overstrength: $q_o = q_\mu$.

By combining 1 to 4 above, Eurocode 8 gives the following relation between q_o and μ_φ:

$$\mu_\varphi = 2q_o - 1 \quad \text{if } T \geq T_C \tag{5.64a}$$

$$\mu_\varphi = 1 + 2(q_o - 1)\frac{T_C}{T} \quad \text{if } T < T_C \tag{5.64b}$$

where T is the period of the first mode in the vertical plane (or close to) where bending of the member being detailed takes place and T_C is the upper corner period of the constant–spectral–acceleration plateau of the elastic spectrum – cf. Equation 5.54b. Equations 5.64 use the basic value, q_o, instead of the final value, q, of the behaviour factor; q may be less than q_o owing to irregularity in elevation, or other features which may reduce the global ductility capacity for given local ductility capacities (e.g. because of non-uniform distribution of ductility in height-wise irregular buildings).

In ductile walls designed to Eurocode 8, the lateral force resistance – which is the quantity directly related to the q-factor – depends only on the moment capacity of the base section. The ratio $M_{Rd,o}/M_{Ed,o}$ captures the wall overstrength (where $M_{Ed,o}$ is the moment at the wall base from the analysis for the design seismic action and $M_{Rd,o}$ the design value of moment resistance under the corresponding axial force from the analysis). So the behaviour factor value utilised by the wall is $q/(M_{Rd,o}/M_{Ed,o})$. As a result, Eurocode 8 allows computing μ_φ at the base of individual ductile walls using in Equations 5.64 the value of q_o divided by the minimum value of the wall $M_{Rd,o}/M_{Ed,o}$-ratio in all combinations of the seismic design situation.

Because a less ductile steel of Class B per Eurocode 2 used as longitudinal reinforcement in the 'critical region' of a primary element (as indeed allowed in DC M, see Table 5.6) may reduce its flexural ductility, Eurocode 8 requires to use for the detailing of members with Class B steel a value of μ_φ increased by 50% over the one resulting from Equation 5.64.

5.7.4 Upper and lower limit on longitudinal reinforcement ratio of primary beams

If the beam cross-section is large, the longitudinal reinforcement may fracture when the concrete cracks, unless it can resist the cracking moment without yielding. In other words, the yield moment should exceed the cracking moment. This condition gives the minimum steel ratio listed at the row 2 of the requirements in Table 5.1 for DC M or H beams (approximately double the minimum ratio for DC L beams per Eurocode 2). Although the minimum steel ratio applies only to the tension side of the beam, it is prudent to implement it at both top and bottom of every section, because the magnitude of seismic moments and their distribution along the beam are very uncertain.

Recalling the lower limit listed in Table 5.6 for the 10%-fractile margin between the tensile strength, f_t, and the yield stress, f_y, of steel and taking into account that the mean yield stress, f_{ym}, normally exceeds the nominal, f_{yk}, by about 15%, the minimum steel ratio for DC M and H beams in Table 5.1 gives a safety margin against potential steel fracture due to overstrength of the concrete in tension (the 95%-fractile of the concrete tensile strength exceeds f_{ctm} by about 30%, but increases with age much less than the compressive strength).

The upper limit to the steel ratio for DC M or H beams listed at the third row of requirements in Table 5.1 aims at ensuring that the value of μ_φ from Equations 5.64 is achieved at the end section. It is derived from:

- The definition of μ_φ as the ratio of: (a) the curvature, φ_u, when the extreme compression fibres reach the ultimate concrete strain per Eurocode 2, $\varepsilon_{cu2} = 0.0035$, to (b) the curvature at yielding, φ_y, taken equal to the semi-empirical value $\varphi_y = 1.54\varepsilon_y/d$ fitted to tests of beams or columns (Fardis 2009)
- The calculation of φ_u as $\varepsilon_{cu2}/(\xi_u d)$, with ξ_u taken from Equation 5.39a in Section 5.4.3 for $\omega_{vd} = 0$, $\nu_d = 0$, $\varepsilon_{cu2} = 0.0035$ and $\varepsilon_{c2} = 0.002$

Example 5.6 at the end of this chapter illustrates the application of Equations 5.64 alongside the calculation of the maximum reinforcement ratio in beams.

The physical meaning behind the maximum top reinforcement ratio is the following: The most likely failure mode of the plastic hinge is crushing of the narrow compression zone at the bottom, in its effort to balance the tension force of the top reinforcement due to hogging bending (see Figure 2.22c). The compression zone is assisted in this task by the bottom reinforcement, with which it shares the force to be balanced. So, the lower the difference between top and bottom reinforcement, the less the burden falling on the concrete (mathematically, the lower the value of ξ_u from Equation 5.39a) and its risk of failure.

The upper limit on the top reinforcement ratio is very restrictive at the supports of DC M and H beams, especially if the value of μ_φ is high (notably for the high q_o-values of DC H). The amount of top steel reinforcement which the beam needs in order to satisfy the ULS in bending at the supports in the seismic design situation and EN1990's 'persistent and transient design situation' (i.e. under factored gravity loads) is fixed. To accommodate it, without excessively increasing the width of the beam to reduce the top steel ratio, the bottom reinforcement ratio, ρ', should preferably be increased beyond the prescriptive minimum values given at the second row of requirements in Table 5.1 and at its last two rows.

5.7.5 Confining reinforcement in 'critical regions' of primary columns

Columns normally have symmetric longitudinal reinforcement: $\omega_{1d} = \omega_{2d}$. Besides, the compression zone also has to resist the compression force, ν_d. So, in a flexural plastic hinge of a column the value of μ_φ from Equations 5.64 cannot be achieved in the same way as in a beam, that is, by reducing ξ_u from Equation 5.39a through a reduction in $(\rho_1 - \rho_2)$. Instead, the extreme concrete fibres are allowed to reach their ultimate strain, $\varepsilon_{cu2} = 0.0035$, and spall; the plastic hinge relies thereafter on the enhanced ultimate strain of the confined concrete core inside the hoops, to provide the required value of μ_φ through confinement.

The effective mechanical volumetric ratio of confining reinforcement, $a\omega_{wd}$, required in plastic hinges of DC M or H columns is given at the last row of Table 5.4 (see also footnotes d through f thereof) and is repeated here for convenience:

$$a\omega_{wd} = 30\mu_\varphi\,\varepsilon_{yd}\nu_d\,\frac{b_c}{b_o} - 0.035 \tag{5.65a}$$

The mechanical volumetric ratio ω_{wd} is defined as $(\rho_h + \rho_b)f_{ywd}/f_{cd}$, with the transverse reinforcement ratios, ρ_h, ρ_b, referring not to the external dimensions of the column section, to the dimensions of the confined core to the centreline of the perimeter hoop:

$$h_o = h_c - 2(c + d_{bw}/2), \qquad b_o = b_c - 2(c + d_{bw}/2) \tag{5.66}$$

where h_c, b_c are the external depth and width of the column section, respectively, c is the concrete cover to the outside of the hoop and d_{bw} is the hoop diameter. The confinement effectiveness factor, a, is a product of:

- One component, a_s, reflecting the variation of confinement along a column with discrete stirrups, and
- a_n, expressing the assumption that there is no confinement over the part of the section outside parabolic arcs emerging from the centres of adjacent vertical bars laterally restrained at tie corners or cross-tie hooks, at an angle of 45° to the chord connecting these two bar centres

For a rectangular section with a perimeter hoop at a centreline spacing of s, the confinement effectiveness factor, a, is

$$a = \left(1 - \frac{s}{2b_o}\right)\left(1 - \frac{s}{2h_o}\right)\left(1 - \frac{\sum b_i^2/6}{b_o h_o}\right) \qquad (5.67)$$

where the product of the first two terms is a_s and the third term is a_n (with b_i denoting the spacing along the perimeter of adjacent laterally restrained vertical bars, the summation extending over all pairs of such bars and the denominator being the area enclosed by the polygonal line connecting the laterally restrained bar centres, see Figure 5.9).

Example 5.7 at the end of this chapter illustrates the application of Equations 5.64, 5.65a, 5.66, 5.67, while Examples 5.8 and 5.9 demonstrate the definition and calculation of a_n in non-rectangular sections. Finally, Example 5.10 shows alternative layouts of confining reinforcement and compares them in terms of cost-effectiveness.

Although they appear so different, the expression for $a\omega_{wd}$ in Equation 5.65a and the maximum steel ratio in beams at the third row of requirements in Table 5.1 are derived similarly; the differences from the two bullet points in Section 5.7.4 are that:

- The full section depth, h, is used, instead of the effective one, d, in non-dimensional values, such as $\xi_u = x_u/h$, $v_d = N_d/(bhf_{cd})$ and so forth, in the semi-empirical expression: $\varphi_y = 1.75\varepsilon_y/h$ and others.

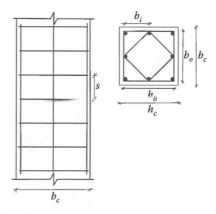

Figure 5.9 Definition of geometric terms for the confinement of a rectangular column.

- φ_u is calculated for the confined core, with the ultimate strain of its extreme compression fibres given as a function of $a\omega_{wd}$ according to the confinement model in Eurocode 2, with $\xi_u = x_u/h$ computed from Equation 5.39a for $v_d = N_d/(bhf_{cd})$, $\omega_{1d} = \omega_{2d}$ and ω_{vd} neglected, compared to v_d.

The confinement reinforcement per Equation 5.65a is required by Eurocode 8 not indiscriminately in every 'critical region' of a column, but only where a plastic hinge may form by design, namely at the base of DC M or H columns – at the connection to the foundation or at the top of a rigid basement. In all other 'critical regions' of DC M columns, only the prescriptive detailing rules in Table 5.4 for the minimum ω_{wd}-value, the maximum spacing, s_w, and the minimum diameter, d_{bw}, of stirrups apply. In DC H buildings, however, Eurocode 8 requires confining reinforcement as per the last row of Table 5.4 in the 'critical regions' at all column ends which are not checked per Equation 5.31 – that is, those falling into the exemptions from Equation 5.31 per Eurocode 8, listed herein in Section 5.4.1 – see footnotes g and i in Table 5.4. Besides, some confining reinforcement is also required even in the 'critical regions' at the ends of DC H columns, which are protected from plastic hinging by meeting Equation 5.31 in both horizontal directions. That confining reinforcement is computed from Equation 5.65a, but for a μ_φ-value (denoted in Table 5.4 by μ_φ^*), which is obtained using in Equations 5.64 two-thirds of the basic q-factor value, q_o, applicable for the design, instead of the full q_o-value (see footnote g in Table 5.4).

Wherever it is required, the confining reinforcement should be computed separately in the two directions of bending, using the values of q_o (and hence of μ_φ) applying to the structural system in these two directions and the most unfavourable (i.e. maximum) value of the axial force from the analysis for the seismic design situation. The largest value from these two separate calculations should be used for ω_{wd}. It should be implemented as the sum of the mechanical reinforcement ratios in both transverse directions, $(\rho_h + \rho_b)f_{ywd}/f_{cd}$, providing, however, approximately equal transverse reinforcement ratios in both: $\rho_h \approx \rho_b$.

If the value of $a\omega_{wd}$ comes out as negative for $b_o = b_c$, then the target value of μ_φ can be achieved by the unspalled section without confinement. In that case, the stirrups in the 'critical region' may just follow the prescriptive detailing rules of the corresponding DC concerning their minimum ω_{wd}-value, maximum spacing s_w, minimum diameter d_{bw}, etc. (see Table 5.4).

For a sample application of this section at the base of the columns of the seven-storey example building, see Section 7.6.2.2.

5.7.6 Confinement of 'boundary elements' at the edges of a wall section

It was pointed out in the second paragraph of Section 5.6.1.2 that the ULS design and the detailing of a wall as RC member differ from those of columns: the moment resistance of a wall is provided by 'tension and compression chords' or 'flanges' at the edges of its section; its shear resistance by the 'web' in-between. The wall's vertical reinforcement is concentrated in 'boundary elements' at the two edges of the section; confinement of concrete is also limited there (see Figure 5.7 and examples in Figures 7.45 to 7.47 of Chapter 7).

Table 5.5 gives in separate sections the detailing rules of Eurocode 8 for the 'boundary elements' and the 'web'. The first part of the 'boundary elements' section refers to the critical region; the second, to the rest of the wall height. The rows before the last one in the first part give prescriptive rules for the geometry, the vertical bars and the confining reinforcement of boundary elements; the last row specifies the effective mechanical volumetric ratio of confining reinforcement, $a\omega_{wd}$, in the boundary elements of DC H or M walls as a function of the value of μ_φ, which corresponds, via Equations 5.64, to the product of q_o times the ratio $M_{Ed,o}/M_{Rd,o}$ at the wall base (see second to last paragraph of Section 5.7.3):

$$a\omega_{wd} = 30\mu_\varphi\, \varepsilon_{yd}\left(\nu_d + \omega_{vd}\right)\frac{b_w}{b_o} - 0.035 \tag{5.65b}$$

b_w appears in Equation 5.65b instead of b_c, but has the same meaning: it is the external width of the compression flange. Equation 5.65b has the same rationale and derivation as Equation 5.65a for columns, but includes also the mechanical ratio of vertical bars in the web, $\omega_{vd} = \rho_v f_{yd}/f_{cd}$, as non-negligible compared to ν_d.

Footnote c in Table 5.5 points out that, under certain conditions often met in practice, Eurocode 8 allows to determine the confining reinforcement of boundary elements in DC M walls according to the rules for walls of DC L. As a matter of fact, if the conditions outlined in that footnote are met, Equation 5.65a most likely gives a negative outcome for $a\omega_{wd}$ when $b_o = b_w$, that is, the target μ_φ-value can be achieved at the unspalled section without confinement (cf. last paragraph in Section 5.7.5).

Above the 'critical region' of DC M or H walls, DC L rules apply for the confining reinforcement of boundary elements, as well as for their geometry and vertical bars. They come from Eurocode 2 and essentially require smaller boundary elements around any edge region of the section in which the vertical bars give a local vertical steel ratio above 2%. Such a region should be enclosed by hoops, following the prescriptive rules in Eurocode 2 concerning hoop diameter and spacing; these rules are very much relaxed compared to the ones of Eurocode 8 for the critical region of DC H or M walls.

The horizontal extent of a confined boundary element in the critical region may be limited to the part of the section where, when the wall reaches its ultimate deformation, the concrete strain exceeds the ultimate strain of unconfined concrete per Eurocode 2, that is, $\varepsilon_{cu2} = 0.0035$. The hoop enclosing a boundary element should have a centreline length of $x_u(1 - \varepsilon_{cu2}/\varepsilon_{cu2,c})$ in the direction of the wall length, $l_w\ (= h_c)$, with the neutral axis depth after concrete spalling, x_u, estimated as

$$x_u = \left(\nu_d + \omega_{vd}\right)\frac{h_c b_c}{b_o} \tag{5.68}$$

and the ultimate strain of unconfined concrete, $\varepsilon_{cu2,c}$, estimated per Eurocode 2 as

$$\varepsilon_{cu2,c} = 0.0035 + 0.1\, a\omega_{wd} \tag{5.69}$$

using the actual value of $a\omega_{wd}$ in the boundary element. In Equation 5.68, b_c, h_c are the same as the wall's b_w, l_w, respectively. The overall length of the confined boundary element includes the concrete cover and the perimeter hoop:

$$l_c \geq x_u\left(1 - \frac{\varepsilon_{cu2}}{\varepsilon_{cu2,c}}\right) + 2\left(c + \frac{d_{bw}}{2}\right) \tag{5.70}$$

The final value of l_c should respect the minimum values at the first row of part (a) 'critical regions' in the 'boundary elements' section of Table 5.5.

For a sample application of this section to the critical height of one wall of the seven-storey example building, see Section 7.6.2.4

5.7.7 Confinement of wall or column sections with more than one rectangular part

Wall or column sections often consist of several rectangular parts: sections with T-, L-, I-, U-shape, walls with 'barbells' at the edges of the section, etc. For such sections ω_{wd} should be determined separately for each rectangular part of the section, which may play the role of a compression flange under any direction of the seismic action. Equations 5.65 should first be applied using the external width of the compression flange at the extreme compression fibres as b_c in Equation 5.65a, or as b_w in Equation 5.65b. This applies also to the normalisation of N_{Ed}, and of the area of vertical reinforcement between the tension and compression flanges, A_{sv}, as $v_d = N_{Ed}/(b_c b_c f_{cd})$, $\omega_{vd} = A_{sv}/(b_c b_c)f_{yd}/f_{cd}$, with h_c denoting the maximum dimension of the unspalled section at right angles to b_c (as if the section were rectangular, with width b_c and depth h_c). For this to apply, the compression zone should be limited within the compression flange, whose width is b_c. To check if this is the case, the neutral axis depth, x_u, at the ultimate curvature after the concrete cover spalls at the compression flange is computed from Equation 5.68. The outcome is then compared to the dimension of the rectangular compression flange at right angles to b_c (i.e. parallel to h_c), after reducing it by $(c + d_{bw}/2)$ for spalling. If this reduced dimension exceeds x_u, the outcome of Equations 5.65 for ω_{wd} is implemented by placing stirrups in the compression flange in question. Approximately equal stirrup ratios should preferably be provided in both directions of this compression flange. However, what mainly counts in this case is the steel ratio of the stirrup legs at right angles to b_c.

If the value of x_u from Equation 5.68 exceeds the dimension of the compression flange at right angles to b_c by much more than $(c + d_{bb}/2)$, there are two practical options:

1. To physically increase the dimension of the rectangular compression flange at right angles to b_c, so that, after its reduction by $(c + d_{bw}/2)$ due to spalling, it exceeds x_u from Equation 5.68.
2. To confine the rectangular part of the section at right angles to the compression flange (the 'web'), instead of the compression flange itself. This is meaningful only if the compression flange for which the neutral axis depth has first been calculated from Equation 5.68 is shallow and not much wider than the 'web'. Equations 5.65 should be then applied with a width b_c or b_w equal to the thickness of the 'web' (also in the normalisation of N_{Ed} and A_{sv} into v_d, ω_{vd}). The outcome of Equations 5.65 for ω_{wd} should be implemented through stirrups in the 'web'. It is consistent with this choice to sacrifice the compression flange by placing in its parts outside the 'web' transverse reinforcement meeting only the prescriptive rules for stirrup spacing and diameter, without confinement requirements. It is more prudent, though, to place in them the same confining reinforcement as in the 'web'.

Example 5.11 at the end of this chapter demonstrates the relationship between confining steel at key points of a wall section and the available ductility factor as per Eurocode 8.

What has been said so far in this section covers both walls and columns with composite section. For walls of this type, Eurocode 8 requires a rigorous approach, namely to go to the fundamentals (equilibrium, σ-ε laws for steel and confined concrete per Eurocode 2, etc.), to provide the required value of $\mu_\varphi = \varphi_u/\varphi_y$. The level of safety provided by Equations 5.65 should be maintained.

5.8 DIMENSIONING FOR VECTORIAL ACTION EFFECTS DUE TO CONCURRENT SEISMIC ACTION COMPONENTS

5.8.1 General approaches

Beams are dimensioned in flexure or shear for scalar internal forces/action effects, that is, bending moment or shear force, respectively. By contrast, columns and walls are dimensioned for uniaxial (or biaxial) bending with axial force and for uniaxial shear with axial force, that is, for two or three concurrent internal forces. Let us consider these internal forces as arranged in a vector: $[M_y, M_z, N]^T$ for biaxial bending with axial force, $[M, N]^T$, or $[V, N]^T$, for uniaxial bending or shear with axial force, respectively. If it was the result of a single seismic action component, that vector would be added to and subtracted from its counterpart due to the quasi-permanent actions; dimensioning or verification would be carried out separately for the vector sum and the vector difference. The question is how do we combine the vectors of peak responses predicted through linear analysis for the individual seismic action components, notably the horizontal ones X and Y, when we know that these peak responses are not simultaneous?

Let us consider biaxial bending, with the vectors of seismic action effects produced by the horizontal components X and Y denoted as

$$E_X = [M_{y,X}, M_{z,X}, N_X]^T, \quad E_Y = [M_{y,Y}, M_{z,Y}, N_Y]^T \tag{5.71}$$

Equation 3.99 gives the expected peak values of individual internal force components due to concurrent horizontal seismic action components, X and Y:

$$\pm M_{y,max} = \pm\sqrt{M_{y,X}^2 + M_{y,Y}^2}, \pm M_{z,max} = \pm\sqrt{M_{z,X}^2 + M_{z,Y}^2}, \pm N_{max} = \pm\sqrt{N_X^2 + N_Y^2} \tag{5.72}$$

Their counterparts from Equation 3.100 are:

$$M_{y,max} = \pm max\Big[\big(|\ M_{y,X}\ | + \lambda\ |\ M_{y,Y}\ |\big); \big(\lambda\ |\ M_{y,X}\ | + |\ M_{y,Y}\ |\big)\Big] \tag{5.73a}$$

$$M_{z,max} = \pm max\Big[\big(|\ M_{z,X}\ | + \lambda\ |\ M_{z,Y}\ |\big); \big(\lambda\ |\ M_{z,X}\ | + |\ M_{z,Y}\ |\big)\Big] \tag{5.73b}$$

$$N_{max} = \pm max\Big[\big(|\ N_X\ | + \lambda\ |\ N_Y\ |\big); \big(\lambda\ |\ N_X\ | + |\ N_Y\ |\big)\Big] \tag{5.73c}$$

It is physically implausible and over-conservative to assume that the maxima of the three internal forces in Equation 5.72 or Equation 5.73 take place concurrently as

$$E = [\pm M_{y,max}, \pm M_{z,max}, \pm N_{max}]^T \tag{5.74}$$

where $M_{y,max}$, $M_{z,max}$, N_{max} are given by Equations 5.72 or 5.73. Nevertheless, Equation 5.74 is commonly used in practice. Alternative, more plausible combinations are described in Fardis (2009); they depend on the type of linear analysis carried out. Another question concerns the permutations of signs among the three internal forces. Modal response spectrum analysis always gives positive results, taken with plus and minus sign. By contrast, the lateral force method gives results with signs; so, when the sign of the seismic action is reversed, all internal forces change sign: internal forces with the same sign keep having the same sign; those with opposite signs stay with opposite signs.

A simple approximation is suggested below as an alternative to the eight combinations of Equation 5.74. Strictly speaking, it does not have a rigorous basis, but is rational and gives reasonable results, close to those from the rigorous approaches highlighted in Fardis (2009). For the general case of biaxial bending, Equations 5.71, this alternative includes 16 combinations:

$$[\pm M_{y,max}, \pm \lambda M_{z,max}, \pm N_{max}]^T \tag{5.75a}$$

$$[\pm \lambda M_{y,max}, \pm M_{z,max}, \pm N_{max}]^T \tag{5.75b}$$

The vector of internal forces due to the translational seismic action components, X and Y, is superimposed to the vector due to the torques produced by the accidental eccentricities of both horizontal components (see Section 3.1.8). As this latter vector is computed via static analysis, its components have signs, which may be reversed all together but not individually. If the combination of components retains the signs of individual action effects, the superposition takes place with signs, such that the internal force which is maximised in the vector due to the translational components is superimposed to its counterpart due to the torques from accidental eccentricities with the same sign; the signs of the other components of the vector due to the accidental eccentricities follow suit, so that they are the same or opposite to each other, in line with how they came out from the static analysis. This is illustrated in Example 5.12.

5.8.2 Implications for the column axial force values in capacity design calculations

The value of the column moment resistance, $M_{Rd,c}$, used in capacity-design calculations should be based on a safe-sided, yet meaningful value of the column axial force, N, within the range of values from the analysis for the combination of the design seismic action with the quasi-permanent loads. More specifically:

1. For the strong column–weak beam capacity design of Equation 5.31, the minimum compressive or maximum tensile axial force in the column should be used.
2. For the capacity design shear of beams, Equations 5.42, we use the maximum compressive axial force in the columns connected to the beam.
3. For the capacity design shear of the column itself, Equation 5.44, we are interested both in the maximum compressive and the maximum tensile (or minimum compressive) axial force in the column.
4. For the capacity design of the foundation system and the bearing capacity verification of the soil, both the maximum compressive and the maximum tensile (or minimum compressive) force in the column are of interest (see Section 6.3.2).

The maximum or minimum compressive axial forces in the seismic design situation come from the maximum compressive and maximum tensile axial force, respectively, from the analysis for the seismic action.

In principle, the value used for N should be consistent with the sense of action (sign) of $M_{Rd,c}$. As an example, for response dominated by the first mode in a given horizontal direction, flexural plastic hinges at the base of columns normally have tension at the 'windward' side of the column and compression at the opposite; the reverse normally holds in plastic hinges at column tops. On the other hand, the first mode dominated response induces tensile axial forces at the top and bottom of exterior columns of the 'windward' side and compressive ones in those of the 'leeward' side.

The controlling moment component is the one for which a plastic hinge forms, let us say M_y. The other moment component is not of interest. If $M_{Rd,c}$ is conventionally taken as positive, M_y is considered positive if it has the same sense of action as $M_{Rd,c}$. N is taken positive if it is compressive. Depending on which one of the two options in Section 3.1.7 is used to combine the effects of the two seismic action components, the extreme values of N may be estimated as follows:

1. E_X, E_Y are combined via Equation 3.99: The maximum compressive or tensile seismic force is

$$\pm \max N_E = \pm \sqrt{\left(N_X^2 + N_Y^2\right)} \tag{5.76}$$

These values are used if the sense (sign) of the bending moments does not make a difference to the value of $\Sigma M_{Rd,b}$. If it does, assuming that the plastic hinge forms in the direction of M_y, the outcome of Equation 5.76 is multiplied by $M_{Rd,c}/M_{y,\max}$. Another physically meaningful option is to take the magnitude of N as $\sqrt{\left(N_X^2 + N_Y^2\right)}$ and use the sign of the axial force in the mode with the largest contribution to the moment in the direction of $M_{Rd,c}$ when that contribution has the same sense (sign) as $M_{Rd,c}$.

2. E_X, E_Y are combined through Equation 3.100. If modal analysis is used, or, if the lateral force method is applied but the sense (sign) of bending moments does not make a difference for the beam moment resistance sums, $\Sigma M_{Rd,b}$, the maximum compressive force in the column is

$$\max N_E = \max[(|N_X| + \lambda |N_Y|);\ (|N_Y| + \lambda |N_X|)] \tag{5.77}$$

The maximum tensile force is given by the same expression but with a minus sign. If E_X, E_Y are computed separately by the lateral force method of analysis, and, in addition, the sense (sign) of bending moments makes a difference to the beam moment resistance sums, $\Sigma M_{Rd,b}$, then the maximum compressive force in the column is taken as

$$\max N_E = \max[(\text{sign}(M_{y,X}N_X)N_X + \text{sign}(M_{y,Y}N_Y)\lambda N_Y);\\ (\text{sign}(M_{y,Y}N_Y)N_Y + \text{sign}(M_{y,X}N_X)\lambda N_X)] \tag{5.78a}$$

while the maximum tensile (minimum compressive) force is

$$\min N_E = \min[(\text{sign}(M_{y,X}N_X)N_X + \text{sign}(M_{y,Y}N_Y)\lambda N_Y);\\ (\text{sign}(M_{y,Y}N_Y)N_Y + \text{sign}(M_{y,X}N_X)\lambda N_X)] \tag{5.78b}$$

The column axial forces due to the accidental eccentricities of both horizontal components (see Section 3.1.8) are added to the extreme seismic axial force determined as highlighted above, with the same sign (i.e. as tensile for minimum N, or compressive for maximum N).

5.9 'SECONDARY SEISMIC ELEMENTS'

5.9.1 Special design requirements for 'secondary' members and implications for the analysis

The contribution of 'secondary' members to lateral stiffness is meant to be neglected in the seismic response analysis from which the seismic action effects for the verification of 'primary' members are computed. On the other hand, Eurocode 8 imposes two special requirements on 'secondary' members, which require special calculations and verifications:

1. The total contribution to lateral stiffness of all 'secondary' members must be less than or equal to 15% of that of all 'primary' ones.
2. 'Secondary' members must remain elastic under the displacements and deformations imposed on them in the seismic design situation.

To check condition no. 1, but also to estimate the deformations imposed on 'secondary' members in the seismic design situation, the designer needs to carry out two linear analyses per horizontal component of the seismic action:

a. One including the contribution of 'secondary' members to lateral stiffness, and
b. Another one neglecting it

For condition no. 1 to be met, the (inter)storey drifts computed from analysis (b) should be less than 1.15 times those from analysis (a). Note that it is on the basis of the results of analysis (b) that 'primary' members are designed and that all the verifications per Eurocode 8, which do not concern 'secondary' members are carried out (including the damage limitation checks on the basis of inter-storey drifts due to the damage limitation seismic action, see Section 1.3.2). On the other hand, a structural model which includes the contribution of 'secondary' members to lateral stiffness is essential for the design of these members against combinations of actions which include other lateral loadings, for example, if the building is also designed for wind. Besides, the same model can be used for the analysis under factored gravity loads ('persistent and transient design situation'). Finally, the results of an analysis of type (a) can be used to estimate the deformations imposed on 'secondary' members in the seismic design situation (see next section). So, for several reasons, it is indeed necessary to perform both types of analysis, (a) and (b).

5.9.2 Verification of 'secondary' members in the seismic design situation

According to Eurocode 8, the design moment and shear resistances of 'secondary' members at the ULS per Eurocode 2, M_{Rd} and V_{Rd}, may not be less than the internal forces (bending moments and shears) derived for these members from the deformations imposed by the rest of the system in the seismic design situation, in a seismic response analysis that neglects the contribution of 'secondary' members to lateral stiffness. These internal forces are to be derived from the imposed seismic deformations using the cracked stiffness of 'secondary' members (i.e. 50% of the gross, uncracked section stiffness). At an extreme limit case, 'secondary' members must be designed for seismic action effects derived with a q-factor of $1/1.15 = 0.87$! To meet this onerous requirement, the lateral stiffness of 'secondary members' should indeed be very low and the global stiffness of the system of 'primary' members and its connectivity to the 'secondary' ones should be such that seismic deformations imposed on the latter are small.

The seismic deformation demands imposed on 'secondary' members in the seismic design situation are determined according to the equal displacement rule through a multi-step procedure:

I. The elastic deformation demands in the 'secondary' members due to the design seismic action are estimated from a linear seismic analysis of type (a) in the previous section, that is, by including the 'secondary' members in the model. The design spectrum is used, that is, the one divided by the behaviour factor, q, but its deformation results are back-multiplied by q, to estimate the displacements as though the structure were elastic.

II. The outcome of Step I for storey i is multiplied by the ratio of inter-storey drifts in that storey from a type (b) linear analysis to those from a type (a) linear analysis. The result is the deformation estimate we seek; it is multiplied by the cracked stiffness of the 'secondary' member in order to estimate its internal forces, to be compared to M_{Rd} and V_{Rd} (see Equation 1.1).

5.9.3 Modelling of 'secondary' members in the analysis

In the structural model for the analysis which neglects the contribution of 'secondary' members to lateral stiffness (type (b) analysis in Section 5.9.1), 'secondary' members should be included only with those of their properties that are essential for their gravity-load-bearing function:

- 'Secondary' vertical elements may be included with their axial stiffness only and with zero flexural rigidity, or with moment releases (i.e. hinges) introduced between their ends and the joint they frame into. Such an approximation is acceptable, so long as the seismic axial forces in these members are small. This precludes vertical elements on the perimeter from such modelling (anyway, it is not sound engineering practice to consider such members as 'secondary').
- 'Secondary' beams directly supported on vertical elements and continuous over two or more spans should be modelled with their flexural stiffness as prescribed by Eurocode 8 for 'primary' members (i.e. 50% of the uncracked, gross section stiffness). Their connectivity with the vertical elements depends on whether the latter are also 'secondary' or not; if they are, zero flexural rigidity of these 'secondary' vertical members, or moment releases (hinges) at their connections with the beam–column joint are satisfactory also for the 'secondary' beams supported on them. If the vertical elements are 'primary', then two separate nodes may be introduced at interior beam–column joints, with pin connection between them: one node on the beam and another on the vertical element; the beam and the vertical element that continue past the joint will resist the gravity loads or the seismic action, respectively, with their flexural stiffness as per Eurocode 8 (50% of the uncracked, gross section stiffness); moment releases (hinges) in the beam may be used at joints where the beam terminates (this includes single-span 'secondary' beams).
- 'Secondary' beams not directly supported on vertical elements (e.g. supported on girders) may be included in the model with their full flexural stiffness and connectivity, because their seismic action effects are negligible anyway.

Note that using different structural models in the analyses of type (a) and (b) is inconvenient, if analysis and design take place in an integrated computational environment; the design modules will have to receive analysis results for the same or different members from the two types of analysis, and combine/modify them appropriately. The alternative, namely to use a single model that neglects the contribution of 'secondary' members to lateral stiffness (for a type (b) analysis), does not allow checking Condition 1 in Section 5.9.1, nor designing the building for other lateral actions, for example, wind. Moreover, as the chord rotations at the ends of 'secondary' members due to the seismic action are not computed from an analysis of type (a), the internal forces in 'secondary members' due to their seismic deformation demands can be estimated only by ad-hoc, approximate and onerous procedures (Fardis 2009), most likely by hand or with spreadsheets.

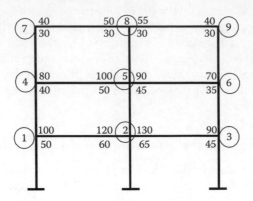

Figure 5.10 Frame of Example 5.1.

EXAMPLE 5.1

The design values of moment resistance of the beams, $M_{Rb,d}$, in a three-storey frame are displayed in Figure 5.10 (in kNm) next to the corresponding tension side of the beam (top or bottom). Calculate the minimum design values of moment resistances of the columns, $M_{Rc,d}$, to meet Equation 5.31, assuming that the columns have symmetric section and reinforcement and that, if the cross-section and the reinforcement above and below a joint are the same, the higher axial load at the column section below the joint increases the moment resistance by 10% compared to the section above.

Answer

Below the joint: $M_{Rc,d1} \geq 1.10 \times 1.3 (\Sigma M_{Rb,d})/2.1$
Above the joint: $M_{Rc,d2} \geq 1.3 (\Sigma M_{Rb,d})/2.1 = 1.3 (\Sigma M_{Rb,d}) - M_{Rc,d1}$
Node 1: Below: $M_{Rc,d1} \geq (1.1/2.1) \times 1.3$ max $(100, 50) = 68$ kNm
   Above: $M_{Rc,d2} \geq (1.0/2.1) \times 1.3$ max $(100, 50) = 62$ kNm
Node 2: Below: $M_{Rc,d1} \geq (1.1/2.1) \times 1.3$ max $(120 + 65, 130 + 60) = 129.4$ kNm
   Above: $M_{Rc,d2} \geq (1.0/2.1) \times 1.3$ max $(120 + 65, 130 + 60) = 117.6$ kNm
Node 3: Below: $M_{Rc,d1} \geq (1.1/2.1) \times 1.3$ max $(90, 45) = 61.3$ kNm
   Above: $M_{Rc,d2} \geq (1.0/2.1) \times 1.3$ max $(90, 45) = 55.7$ kNm
Node 4: Below: $M_{Rc,d1} \geq (1.1/2.1) \times 1.3$ max $(80, 40) = 54.5$ kNm
   Above: $M_{Rc,d2} \geq (1.0/2.1) \times 1.3$ max $(80, 40) = 49.5$ kNm
Node 5: Below: $M_{Rc,d1} \geq (1.1/2.1) \times 1.3$ max $(100 + 45, 90 + 50) = 98.7$ kNm
   Above: $M_{Rc,d2} \geq (1.0/2.1) \times 1.3$ max $(100 + 45, 90 + 50) = 89.8$ kNm
Node 6: Below: $M_{Rc,d1} \geq (1.1/2.1) \times 1.3$ max $(70, 35) = 47.7$ kNm
   Above: $M_{Rc,d2} \geq (1.0/2.1) \times 1.3$ max $(70, 35) = 43.3$ kNm

At the nodes of the roof, capacity design per Equation 5.31 is not required and is indeed meaningless. However, it is extended here to these nodes, to show that it sometimes leads to absurdly large column moment capacities:

Node 7: Below: $M_{Rc,d1} \geq 1.3$ max $(40, 30) = 92$ kNm
Node 8: Below: $M_{Rc,d1} \geq 1.3$ max $(50 + 30, 55 + 30) = 110.5$ kNm
Node 9: Below: $M_{Rc,d1} \geq 1.3$ max $(40, 30) = 52$ kNm

EXAMPLE 5.2

For the same design values of moment resistances of the beams, $M_{Rb,d}$, as in the previous example, the design values of moment resistances of the columns, $M_{Rc,d}$, are depicted in Figure 5.11. Estimate the likely location for plastic hinges to form, if the response to the seismic action is from the left to the right or from the right to the left.

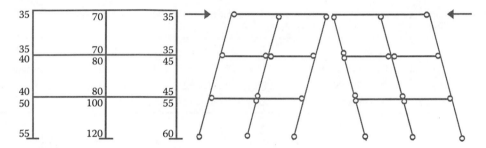

Figure 5.11 Frame of Example 5.2.

Answer

For seismic response from left to right, we have sagging bending at the left end of each beam, and hogging at the right one; the reverse for seismic response from right to left (see Table 5.7).

A storey-sway mechanism ('soft-storey') is not apparent. The closest a storey comes to such a mechanism is when the seismic response is from right to left, with plastic hinges forming at the top and bottom of two columns of the ground storey and the bottom of the third one. Conclusions for the top storey do not change, no matter whether the hinge forms at the top of the column or at the beam end next to it.

When the plastic mechanism is mixed, as in all storeys of both cases herein, a storey-sway mechanism is more likely to happen if at the top nodes of the storey the value of the storey index, $\Sigma(\Sigma M_{Rb})/\Sigma(\Sigma M_{Rc})$, exceeds 1.0 (with the outer sums in numerator and denominator of the index extending over all top nodes of the storey). This is checked in Figure 5.11.

Response from left to right (see Table 5.8).

Table 5.7 Comparison of total beam and column moment resistances around joints in example 5.2

Node	ΣM_{Rc} (kNm)	Response from left to right		Response from right to left	
		ΣM_{Rb} (kNm)	Plastic hinges in:	ΣM_{Rb} (kNm)	Plastic hinges in:
1	90	>50	Beam	<100	Column
2	180	<185	Column	<190	Column
3	100	>90	Beam	>45	Beam
4	75	>40	Beam	<80	Column
5	150	>145	Beam	>140	Beam
6	80	>70	Beam	>35	Beam
7	35	>30	Beam	<40	Column
8	70	<80	Column	<85	Column
9	35	<40	Column	>30	Beam

Table 5.8 Aggregate ratios of total beam and column moment resistances around floor joints in example 5.2 for response from left to right

Floor	$\Sigma(\Sigma M_{Rc})$	$\Sigma(\Sigma M_{Rb})$	$\Sigma(\Sigma M_{Rb})/\Sigma(\Sigma M_{Rc})$
1	90 + 180 + 100 = 370 kNm	50 + 185 + 90 = 325 kNm	0.88 < 1: beam-sway
2	75 + 150 + 80 = 305 kNm	40 + 145 + 70 = 255 kNm	0.84 < 1: beam-sway
3	35 + 70 + 35 = 140 kNm	30 + 80 + 40 = 150 kNm	1.07 > 1: neutral

Table 5.9 Aggregate ratios of total beam and column moment resistances around floor joints in Example 5.2 for response from right to left

Floor	$\Sigma(\Sigma M_{Rc})$	$\Sigma(\Sigma M_{Rb})$	$\Sigma(\Sigma M_{Rb})/\Sigma(\Sigma M_{Rc})$
1	90 + 180 + 100 = 370 kNm	100 + 190 + 45 = 335 kNm	0.91 < 1: beam-sway
2	75 + 150 + 80 = 305 kNm	80 + 140 + 35 = 255 kNm	0.84 < 1: beam-sway
3	35 + 70 + 35 = 140 kNm	40 + 85 + 30 = 155 kNm	1.11 > 1: neutral

Response from right to left (see Table 5.9).

Note the value of the index at the ground storey for response from right to left: despite the hinging at two column tops out of three, the index is still less than 1.0, thanks to the large margin of the total column moment resistance at the top of the third column with respect to the beam connected to it. Values of the index at the top storey larger than 1.0 have no practical impact: it does not matter whether the hinge forms at column tops or in the roof beam.

EXAMPLE 5.3

Calculate the design moment resistance about the strong and the weak axis of a 0.7×0.3 m column section with the reinforcement in Figure 5.12 (corner bars, Ø18; intermediate ones, Ø14), for two extreme values of the axial load: max $N = 1199$ kN, min $N = 852.1$ kN. Concrete strength $f_{ck} = 25$ MPa, reinforcement yield stress $f_{yk} = 500$ MPa. Ties are 8 mm in diameter and have a concrete cover of 35 mm. (Note: this is the base section of column C12 at the top of the basement in the example of Chapter 7).

Answer

$d_1 = 35 + 8 + 14/2 = 50$ mm. For $\gamma_c = 1.5$: $f_{cd} = f_{ck}/\gamma_c = 25/1.5 = 16.67$ MPa.
For $\gamma_s = 1.15$: $f_{yd} = f_{yk}/\gamma_s = 500/1.15 = 434.8$ MPa; $\varepsilon_{yd} = f_{yd}/E_s = 434.8/200{,}000 = 0.00217$.

   a. About the strong axis (y):

$d = 0.7 - 0.05 = 0.65$ m; $b = 0.3$ m. $\delta_1 = d_1/d = 0.05/0.65 = 0.077$.

A_{sv} (including a fictitious half of Ø14 bar at each corner, to give a uniform distribution of A_{sv} between A_{s1} and A_{s2}) = $8 \times 154 = 1232$ mm²; $\omega_{vd} = 1232/(650 \times 300) \times 434.8/16.67 = 0.1648$.

A_{s1} and A_{s2} (subtracting the fictitious half of Ø14 bar at each corner, which counts into A_{sv}): $A_{s1} = A_{s2} = 2 \times 254.5 = 509$ mm²; $\omega_{1d} = \omega_{2d} = 509/(650 \times 300) \times 434.8/16.67 = 0.0681$.

For max $N = 1199$ kN: max $v_d = 1.199/(0.65 \times 0.3 \times 16.67) = 0.369$.

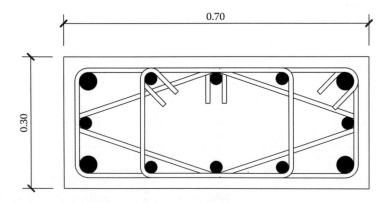

Figure 5.12 Column section for Example 5.3.

For min $N = 852.1$ kN: min $v_d = 0.8521/(0.65 \times 0.3 \times 16.67) = 0.262$.

The left-hand side of Equation 5.37a and right-hand side of Equation 5.37b is
$v_2 = 0.1648/(1 - 0.077) \times (0.077 \times (0.0035 + 0.00217)/(0.0035 - 0.00217) - 1) + 0.077 \times (0.0035 - 0.002/3)/(0.0035 - 0.00217) = 0.044$.

The right-hand side of Equation 5.37a is
$v_1 = 0.1648/(1 - 0.077) \times ((0.0035 - 0.00217)/(0.0035 + 0.00217) - 0.077) + (0.0035 - 0.002/3)/(0.0035 + 0.00217) = 0.528$;

Equation 5.37a is met by both max $v_d = 0.369$ and min $v_d = 0.262$: $v_2 = 0.044 <$ minv_d, max$v_d < v_1 = 0.528$. So, case (i) applies and Equation 5.39a gives:

For max $v_d = 0.369$:
$\xi = [(1 - 0.077) \times 0.369 + (1 + 0.077) \times 0.1648]/[(1 - 0.077) \times (1 - 0.002/(3 \times 0.0035)) + 2 \times 0.1648] = 0.481$, and Equation 5.38a gives:
$M_{Rdy} = 0.3 \times 0.65^2 \times 16667 \times \{0.481 \times [(1 - 0.481)/2 - 0.002/(3 \times 0.0035) \times (0.5 + (0.002/(4 \times 0.0035) - 1) \times 0.481)] + (1 - 0.077) \times 0.0681 + 0.1648/(1 - 0.077) \times [(0.481 - 0.077) \times (1 - 0.481) - (0.481 \times 0.00217/0.0035)^2/3]\} = 447.4$ kNm

For min$v_d = 0.262$:
$\xi = [(1 - 0.077) \times 0.262 + (1 + 0.077) \times 0.1648]/[(1 - 0.077) \times (1 - 0.002/(3 \times 0.0035)) + 2 \times 0.1648] = 0.389$, and Equation 5.38a gives:
$M_{Rdy} = 0.3 \times 0.65^2 \times 16667 \times \{0.389 \times [(1 - 0.389)/2 - 0.002/(3 \times 0.0035) \times (0.5 + (0.002/(4 \times 0.0035) - 1) \times 0.389)] + (1 - 0.077) \times 0.0681 + 0.1648/(1 - 0.077) \times [(0.389 - 0.077) \times (1 - 0.389) - (0.389 \times 0.00217/0.0035)^2/3]\} = 422.4$ kNm

b. About the weak axis (z):
$d = 0.3 - 0.05 = 0.25$ m; $b = 0.7$ m. $\delta_1 = d_1/d = 0.05/0.25 = 0.2$.

A_{sv} (including a fictitious half of Ø14 bar at each corner, to give a uniform distribution of A_{sv} between A_{s1} and A_{s2}) $= 4 \times 154 = 616$ mm²; $\omega_{vd} = 616/(250 \times 700) \times 434.8/16.67 = 0.0918$.

A_{s1} and A_{s2} (subtracting the fictitious half of Ø14 bar at each corner, which counts into A_{sv}): $A_{s1} = A_{s2} = 2 \times 254.5 + 2 \times 154 = 817$ mm²; $\omega_{1d} = \omega_{2d} = 817/(250 \times 700) \times 434.8/16.67 = 0.1218$.

For max$N = 1199$ kN: max $v_d = 1.199/(0.25 \times 0.7 \times 16.67) = 0.411$.

For min$N = 852.1$ kN: min $v_d = 0.8521/(0.25 \times 0.7 \times 16.67) = 0.292$.

The left-hand side of Equation 5.37a and right-hand side one of (5.37b) is:
$v_2 = 0.0918/(1 - 0.2) \times (0.2 \times (0.0035 + 0.00217)/(0.0035 - 0.00217) - 1) + 0.2 \times (0.0035 - 0.002/3)/(0.0035 - 0.00217) = 0.409$.

The right-hand side of Equation 5.37a is:
$v_1 = 0.0918/(1 - 0.2) \times ((0.0035 - 0.00217)/(0.0035 + 0.00217) - 0.2) + (0.0035 - 0.002/3)/(0.0035 + 0.00217) = 0.54$;

Equation 5.37a is met: $v_2 = 0.409 <$ max$v_d = 0.411 < v_1 = 0.54$. So, case (i) applies, and Equation 5.39a gives for $v_d = 0.411$:
$\xi = [(1 - 0.2) \times 0.411 + (1 + 0.2) \times 0.0918]/[(1 - 0.2) \times (1 - 0.002/(3 \times 0.0035)) + 2 \times 0.0918] = 0.528$, and Equation 5.38a gives:
$M_{Rdz} = 0.7 \times 0.25^2 \times 16667 \times \{0.528 \times [(1 - 0.528)/2 - 0.002/(3 \times 0.0035) \times (0.5 + (0.002/(4 \times 0.0035) - 1) \times 0.528)] + (1 - 0.2) \times 0.1218 + 0.0918/(1 - 0.2) \times [(0.528 - 0.2) \times (1 - 0.528) - (0.528 \times 0.00217/0.0035)^2/3]\} = 168.4$ kNm

min$v_d = 0.292$ meets Equation 5.37b: min$v_d < v_2 = 0.409$. So, case (ii) applies and Equation 5.39b becomes:
$\{1 - 0.002/(3 \times 0.0035) + 0.0918 \times (0.0035 + 0.00217)/[2 \times (1 - 0.2) \times (0.0035 \times 0.00217)]\}\xi^2 - [0.292 + 0.1218 \times (1 - 0.0035/0.00217) + 0.0918 \times (1 + 0.2 \times 0.0035/0.00217)/(1 - 0.2)]\xi - [0.1218 - 0.5 \times 0.0918 \times 0.2/(1 - 0.2)] \times 0.2 \times 0.0035/0.00217 = 0$, or $1.0524\xi^2 - 0.3691\xi - 0.03559 = 0 \rightarrow \xi = 0.429$, and Equation 5.38b gives:
$M_{Rdz} = 0.7 \times 0.25^2 \times 16667 \times \{0.429 \times [(1 - 0.429)/2 - 0.002/(3 \times 0.0035) \times (0.5 + (0.002/(4 \times 0.0035) - 1) \times 0.429)] + (1 - 0.2)/2 \times 0.1218 \times [1 + (0.429 - 0.2) \times 0.0035/(0.429 \times 0.00217)] + 0.25 \times 0.0918/(1 - 0.2) \times [0.429 \times (1 + 0.00217/0.0035) - 0.2] \times [1 + 0.0035 \times (0.429 - 0.2)/(0.00217 \times 0.429)] \times [1 - 0.2/3 - 2 \times 0.429 \times (1 + 0.00217/0.0035)/3]\} = 156.6$ kNm

EXAMPLE 5.4

A DC H beam, shown schematically in Figure 5.13, has depth $h = 450$ mm, effective depth $d = 415$ mm and three 14 mm bars as top and bottom reinforcement all along the span; additional top reinforcement consists of one more 14 mm bar at the left support (index: 1) and three more (total six 14 mm bars) at the right one (index: 2), all of 500 MPa steel. The beam spans $L_n = 5.0$ m between the faces of its supporting columns, which are stronger in flexure than the beams around the two beam ends ($\Sigma M_{Rb} < \Sigma M_{Rc}$). The design shear forces at the two ends of the beam are computed by capacity design for two values of the quasi-permanent transverse load: $g + \psi q = 14$ kN/m, and $g + \psi q = 20$ kN/m, considering the possibility that the plastic hinge in positive (sagging) bending may form at some distance from the end section.

Answer

Calculation of the beam moment resistance with the approximation of Equation 5.30c:
$M^+_{Rd} = 0.9 \times 0.415 \times 3 \times 154 \times 500/1.15 = 75{,}000$ Nm = 75 kNm, constant all along the span.
$M^-_{Rd,1} = 4 \times 75/3 = 100$ kNm at the left end (index: 1);
$M^-_{Rd,2} = 6 \times 75/3 = 150$ kNm at the right one (index: 2).

a. For $g + \psi q = 14$ kN/m: $V_{g+\psi q,o,1} = V_{g+\psi q,o,2} = 14 \times 5.0/2 = 35$ kN
   If $\Sigma M_{Rb} < \Sigma M_{Rc}$, Equation 5.42a gives for $\gamma_{Rd} = 1.2$ (DC H):
   $\max V_{d,1} = 1.2 \times (150 + 75)/5 + 35 = 89$ kN, $\max V_{d,2} = 1.2 \times (100 + 75)/5 + 35 = 77$ kN
b. For $g + \psi q = 20$ kN/m: $V_{g+\psi q,o,1} = V_{g+\psi q,o,2} = 20 \times 5.0/2 = 50$ kN
   At first sight, the design shears increase by $50 - 35 = 15$ kN, owing to the increase in $V_{g+\psi q,o}$. This holds for $\max V_{d,1}$, but not for $\max V_{d,2}$, for the following reasons:
   At the instant $\max V_{d,1}$ occurs at end 1, the concurrent shear force at end 2 is:
   $\min V_{d,2} = -1.2 \times (150 + 175)/5 + 50 = 4$ kN > 0; that is, the shear does not change sign between the two ends, implying that there is no local maximum of the sagging bending moment along the beam in this particular case of the seismic design situation. By contrast, when $\max V_{d,2} = 1.2 \times (100 + 75)/5 + 50 = 92$ kN develops at end 2, the value of $V_{d,1}$ at end 1 is $\min V_{d,1} = 1.2 \times (100 + 75)/5 - 50 = -8$ kN; that is, the shear changes sign along the span, going through zero at a point where the bending moment attains its maximum value. As the slope (derivative) of the V-diagram is equal to the transverse load, an estimate of the distance of the maximum moment point to end 1 is $x = |\min V_{d,1}|/(g + \psi q) = 8/20 = 0.4$ m. That maximum moment is equal to the moment at end 1, taken for the present purposes equal to $\gamma_{Rd} M^+_{Rd} = 1.2 \times 75 = 90$ kNm, plus the area under the V-diagram between the maximum moment point and end 1. This area is equal to $|\min V_{d,1}| x/2 = 8 \times 0.4/2 = 1.6$ kNm, giving a maximum moment of $90 + 1.6 = 91.6$ kNm. As expected, this value exceeds the overstrength sagging moment resistance, which is equal to $\gamma_{Rd} M^+_{Rd} = 90$ kNm all along the span. Therefore, the values calculated in the present paragraph, including the capacity design shear of $\max V_{d,2} = 1.2 \times (100 + 75)/5 + 50 = 92$ kN cannot materialise,

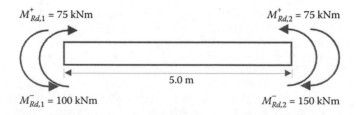

$M^+_{Rd,1} = 75$ kNm $\quad\quad\quad M^+_{Rd,2} = 75$ kNm

5.0 m

$M^-_{Rd,1} = 100$ kNm $\quad\quad\quad M^-_{Rd,2} = 150$ kNm

Figure 5.13 Beam of Example 5.4.

without violating the overstrength sagging moment resistance somewhere along the span; so, they are invalid.

The true value $\max V_{d,2}$ is equal to $20l_x/2 + 1.2 \times (100 + 75)/l_x$, where $l_x = L_n - x$ is the distance from the maximum moment (and zero shear) point to end 2. The value of x estimated in the paragraph above as giving a maximum moment of 91.6 kNm (greater than the overstrength sagging moment resistance of 90 kNm), gives first trial values of $l_x \approx 5.0 - 0.4 = 4.6$ m and $\max V_{d,2} \approx 20 \times 4.6/2 + 1.2 \times (100 + 75)/4.6 = 91.65$ kN. If the maximum moment at a distance $x = 0.4$ m from end 1 is equal to $\gamma_{Rd}M^+_{Rd} = 90$ kNm, the moment at end 1 is equal to that maximum moment minus the area under the V-diagram between the maximum moment point and end 1, that is, to $90 - 0.4 \times 8/2 = 88.4$ kNm. This new moment value at end 1 corresponds to a shear force value at that end equal to $\min V_{d,1} = (1.2 \times 100 + 88.4)/5 - 50 = -8.32$ kN and to a new estimate for the distance of that end to the maximum moment (and zero shear) point of: $x \approx 8.32/20 = 0.416$ m. The new x-value gives a new moment estimate at end 1, equal to: $90 - 0.416 \times 8.32/2 = 88.27$ kNm, which in turn yields an estimate of $\max V_{d,2} = (1.2 \times 100 + 88.27)/5 + 50 = 91.65$ kN, coinciding with the value $20l_x/2 + 1.2 \times (100 + 75)/l_x = 10 \times (5 - 0.416) + 1.2 \times 175/(5 - 0.416) = 91.65$ kN. This is taken as the final value of $\max V_{d,2}$. Note that it could have been computed from the value of $x = 0.4$ m estimated in the paragraph above, without iterations. More importantly, the difference with the outcome of Equation 5.42, namely, $\max V_{d,2} = 1.2 \times (100 + 75)/5 + 50 = 92$ kN, is minor and safe-sided.

A corollary is that Equation 5.42 is safe-sided, even when the plastic hinge in positive (sagging) bending forms at some distance from the end section.

EXAMPLE 5.5

Reinforcing steel is checked against the Eurocode 8 requirements for bars used in primary members of DC L, M or H buildings as steel with nominal yield stress $f_{y,nom} = 500$ MPa. Coupon tests have given the mean values, m, and standard deviation, s, of the yield stress, the strain at maximum stress and the hardening ratio shown in Table 5.10. If the number of samples suffices to consider the underlying probability distribution of material properties to be Normal (Gaussian) and the mean, μ, and standard deviation, σ, of the distribution to be equal to the mean, m, and the standard deviation, s, of the sample, then, the 5%-, 10%- and 95%-fractiles of material property x are equal to: $x_{k,0.05} = m - 1.645s$, $x_{k,0.10} = m - 1.282s$, $x_{k,0.95} = m + 1.645s$. In a second case, the statistics come from only 15 tested samples; then: $x_{k,0.05} = m - 2.33s$, $x_{k,0.10} = m - 1.87s$, $x_{k,0.95} = m + 2.33s$.

Answer

1. If the number of samples is large, then:
   $f_{yk,0.95} = 550 + 1.645 \times 40 = 615.8$ MPa, $f_{yk,0.95}/f_{yk} = 615.8/500 = 1.2316 < 1.25$;
   $(f_t/f_y)_{k,0.10} = 1.26 - 1.282 \times 0.085 = 1.151 > 1.15\%$ and $< 1.35\%$;
   $\varepsilon_{suk,0.10} = 10.6 - 1.282 \times 2.0 = 8.04\% > 7.5\%$.
   The above comparison with the limits in Table 5.6 suggests that the steel may be used for DC H, M or L. However, it cannot be used as S500, because:
   $f_{yk,0.05} = 550 - 1.645 \times 40 = 484.2$ MPa $< f_{y,nom} = 500$ MPa.
   It may be used, though, at least in DC M or L, as of grade S480 or lower. To see if it may be used as S480 in DC H as well, the check $f_{yk,0.95}/f_{yk} < 1.25$ must be carried

Table 5.10 Table of test statistics, Example 5.5

	f_y (MPa)	ε_{su} (%)	f_t/f_y
Sample mean, m	550	10.2	1.26
Sample standard deviation, s	40	2.0	0.085

out; but $f_{yk,0,95}/f_{yk} = 615.8/480 = 1.2829 > 1.25$. So, it may not be used in DC H as of grade S480 or lower.

The conclusion is that the steel may be used in DC M or L, as of grade S480 or lower, but cannot be used in DC H as any grade.

2. For 15 samples only:

$f_{yk,0,95} = 550 + 2.33 \times 40 = 643.2$ MPa, $f_{yk,0,95}/f_{yk} = 643.2/500 = 1.2864 > 1.25$;

$(f_t/f_y)_{k,0,10} = 1.26 - 1.87 \times 0.085 = 1.101 < 1.35\%$ and $> 1.08\%$, but $< 1.15\%$;

$\varepsilon_{suk,0,10} = 10.6 - 1.87 \times 2.0 = 6.86\% > 5\%$, but $< 7.5\%$.

The above comparison with the limits in Table 5.6 shows that the steel may be used in DC M or L, but not in DC H. Again, it cannot be used as S500, because:

$f_{yk,0,05} = 550 - 2.33 \times 40 = 456.8$ MPa $< f_{y,nom} = 500$ MPa.

It may be used, though, as steel of grade S450 or lower, in DC M or L buildings.

EXAMPLE 5.6

Calculate the curvature ductility factor required for the following combinations: $T_1 = 0.5$ s (stiff building) or $T_1 = 0.7$ s (average stiffness building), design with $q_o = 4$ or $q_o = 6$, if $T_C = 0.6$ s. Calculate the corresponding maximum top steel ratio in beams for C25/30 concrete, S500 steel $(f_{cd} = f_{ck}/1.5, f_{yd} = f_{yk}/1.15)$.

Answer

Beam maximum top steel ratio:

$\rho' = \max(0.5\rho_1; \rho_{min})$ with $\rho_{min} = 0.5f_{ctm}/f_{yk} = 0.5 \times 2.6/500 = 0.0026$, and:

$$\rho_{1,max} = \max\left(0.5\rho_{1,max}; \rho_{min}\right) + \frac{0.0018}{\mu_\varphi \dfrac{500}{1.15 \cdot 200,000}} \frac{25 \cdot 1.15}{1.5 \cdot 500}$$

$$= \max\left(0.5\rho_{1,max}; 0.0026\right) + \frac{0.03174}{\mu_\varphi}$$

1. If $T_1 = 0.5$ s $< T_C$:
   - For $q_o = 4$: $\mu_\delta = 1 + (q_o - 1)T_C/T_1 = 4.6$ and $\mu_\varphi = 2\mu_\delta - 1 = 8.2 \rightarrow \rho_1 = 0.0078$
   - For $q_o = 6$: $\mu_\delta = 1 + (q_o - 1)T_C/T_1 = 7$ and $\mu_\varphi = 2\mu_\delta - 1 = 13 \rightarrow \rho_1 = 0.005$
2. If $T_1 = 0.7$ s $> T_C$:
   - For $q_o = 4$: $\mu_\delta = q_o = 4$ and $\mu_\varphi = 2\mu_\delta - 1 = 7 \rightarrow \rho_1 = 0.009$
   - For $q_o = 6$: $\mu_\delta = q_o = 6$ and $\mu_\varphi = 2\mu_\delta - 1 = 11 \rightarrow \rho_1 = 0.0058$.

The resulting values of $\rho_{1,max}$ are low, especially for the stiff building $(T_1 = 0.5$ s) and/or the very ductile one $(q = 6)$. In such cases, $\rho_{1,max}$ may be increased by selecting $\rho' > \max(0.5\rho_1; \rho_{min})$.

EXAMPLE 5.7

Calculate the required confinement reinforcement corresponding to the curvature ductility factors of Example 5.6, at the base of a ground storey column, having 0.4 m square section, 10 mm stirrups, eight 16 mm vertical bars (three per side), cover of stirrup 30 mm and maximum axial force $N_d = 975$ kN (C25/30 concrete, S500 steel).

Answer

Normalised axial force $\nu_d = 975/(0.40^2 \times 25,000/1.5) = 0.365$, $b_o = 0.4 - 2 \times (0.03 + 0.01/2) = 0.33$ m

The 'demand' for confinement is:

$a\omega_{wd} \geq 30\,\mu_\varphi \nu_d \varepsilon_{yd} b_c/b_o - 0.035 = 30\,\mu_\varphi \times 0.365 \times 500/(1.15 \times 200,000) \times 0.4/0.33 - 0.035 = 0.029\,\mu_\varphi - 0.035 = 0.203, 0.342, 0.168$ and 0.284, for $\mu_\varphi = 8.2, 13, 7$ and 11, respectively.

Concerning the 'supply side' of confinement:

For one bar at column mid-side: $b_i = b_o/2$ and $a_n = 1 - 8 \times (1/2)^2/6 = 2/3$.

A diamond-shaped interior tie is placed around the four mid-side bars (as in Figure 5.9), giving steel ratio per transverse direction (with respect to the dimension b_o of the confined core):

$\rho_x = (2 + \sqrt{2})A_{sw}/(b_o s) = (2 + \sqrt{2}) \times 78.5/330s = 0.813/s$, where s is the tie spacing in mm.

So, $\omega_{wd} = 2\rho_x f_{ywd}/f_{cd} = 42.4/s$, $a_s = (1 - s/2b_o)^2 = (1 - s/660)^2$,

Hence the 'supply side' is:

$$a\omega_{wd} = (2/3)(1 - s/660)^2(42.4/s) = (28.26/s)(1 - s/660)^2.$$

Setting this 'supply side' equal to the 'demand' of 0.203, 0.342, 0.168 and 0.284, for $\mu_\varphi = 8.2, 13, 7$ and 11, respectively, we find:

1. If $T_1 = 0.5$ s $< T_C$:
   - For $q_o = 4$: $s \leq 100$ mm
   - For $q_o = 6$: $s \leq 67$ mm, rounded to 65 mm
2. If $T_1 = 0.7$ s $> T_C$:
   - For $q_o = 4$: $s \leq 115$ mm
   - For $q_o = 6$: $s \leq 78$ mm, rounded to 75 mm

EXAMPLE 5.8

Calculate the factor a_n for confinement effectiveness within the cross section of a tie, in an octagonal concrete column with a single octagonal tie engaging the corner bars along the perimeter (Figure 5.14).

Answer

If R is the radius to a corner of the octagonal tie, each tie side has length $b_i = 2R \sin(22.5°)$ and distance from the centre $R \cos(22.5°)$. The area of the confined core inside the tie is $8R\cos(22.5°)b_i/2 = 8R^2\cos(22.5°)\sin(22.5°) = 4R^2\sin(45°) = 2\sqrt{2}R^2$, from which the area outside the 8 parabolic arches based on the tie sides, $8b_i^2/6 = 8 \times (2R\sin(22.5°))^2/6$, is subtracted.

So, $a_n = 1 - (16/3)R^2\sin^2(22.5°)/(2\sqrt{2}R^2) = 1 - 8\sin^2(22.5°)/(3\sqrt{2}) = 0.724$.

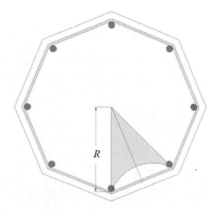

Figure 5.14 Column section for Example 5.8.

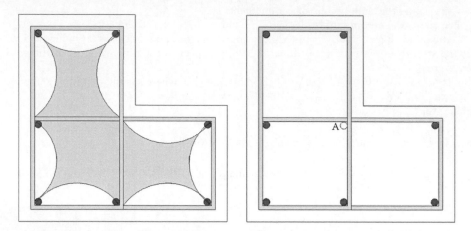

Figure 5.15 Column section for Example 5.9.

EXAMPLE 5.9

An L-shaped column section has one closed rectangular tie extending all along each leg of the L and overlapping over the common part of the two legs next to the corner of the section (Figure 5.15). Calculate: (a) the factor a_n for confinement effectiveness within the section, assuming that the ties are engaged by a vertical bar at each corner of the two stirrups (seven bars in total, considering that the two stirrups share the external corner of the L); and (b) the effect of adding an eighth bar inside the intersection of the ties' interior legs at the re-entrant corner of the section (point A).

Answer

A vertical bar engaging a straight tie leg prevents it from bending outwards under the pressures applied on it by the confined concrete. The two tie legs intersecting each other at the re-entrant corner cannot bend outwards along the part of their length embedded in the concrete: the concrete on one side of that length is 'confined' by the concrete on the other side, with no need of confinement by the tie. So, no matter whether an eighth bar is added at point A, the effectiveness factor within the section, a_n, is equal to $1 - \sum_{i=1-8}\left(b_i^2/6\right)/$(Area inside outline of the two overlapping rectangular ties), with b_i ($i = 1$–8) denoting the distance between successive tie corners on the perimeter, including the intersection of the two straight tie legs at the re-entrant corner.

EXAMPLE 5.10

A square column section has two intermediate vertical bars along each side (Figure 5.16). Compare the cost-effectiveness of a single octagonal tie engaging all eight intermediate bars to that of two rectangular interior ties, each one extending from one side of the section to the opposite and engaging just the two pairs of intermediate bars of these two sides.

Answer

If $3b$ is the length of each side of the perimeter tie, the octagonal one has a total length of $L_8 = 4(1 + \sqrt{2})b$; the two rectangular interior ties have $L_{2x4} = 16b$. If $A_{sw}f_{yw}$ is the yield force of a tie, the octagonal one exerts at right angles to the perimeter a confining force of $A_{sw}f_{yw}/\sqrt{2}$ at each corner, or $F_8 = 8A_{sw}f_{yw}/\sqrt{2}$ in total; the two rectangular interior ties exert confining forces equal to $A_{sw}f_{yw}$ at each corner, or $F_{2x4} = 8A_{sw}f_{yw}$ in total. The ratio of the total confining force to the total volume of each tie is $F_8/A_{sw}L_8 = \sqrt{2}f_{yw}/(1 + \sqrt{2})$ $b = 0.586f_{yw}/b$ for the octagonal, $F_{2x4}/A_{sw}L_{2x4} = 0.5f_{yw}/b$ for the two rectangular interior ties. So, the octagonal tie is about 17% more cost-effective.

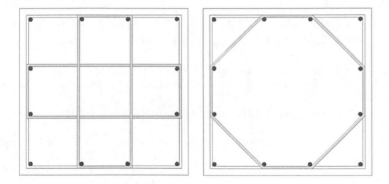

Figure 5.16 Column sections for Example 5.10.

EXAMPLE 5.11

In the U-shaped wall section shown in Figure 5.17, the concrete strength is $f_{ck} = 24$ MPa; the 10 and 12 mm vertical bars ($\Phi10$, $\Phi12$) have $f_{yk} = 520$ MPa and the 8 mm stirrups ($\Phi8$) have $f_{ywk} = 560$ MPa. The outer 0.375 m long parts of each rectangular side of the section are detailed as boundary elements with the minimum length per Eurocode 8 of $1.5b_w$. The concrete cover of the stirrups is ~20 mm, giving a distance of their centreline from the concrete surface equal to 25 mm. A compressive force $N = 2$ MN acts at the centroid of the section. The available curvature ductility factor is computed for bending about centroidal axes normal or parallel to the two sides or 'arms' of the section. Design values of material strengths are used.

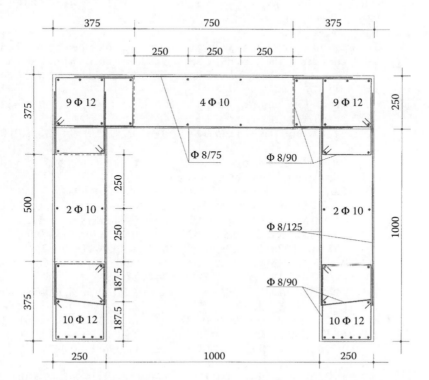

Figure 5.17 Wall section for Example 5.11.

Answer

For $\gamma_c = 1.5$: $f_{cd} = f_{ck}/\gamma_c = 24/1.5 = 16$ MPa. For $\gamma_s = 1.15$: $f_{yd} = f_{yk}/\gamma_s = 520/1.15 = 452$ MPa; $\varepsilon_{yd} = f_{yd}/E_s = 452/200,000 = 0.00226$; $f_{ywd} = f_{ywk}/\gamma_s = 560/1.15 = 487$ MPa.

1. Bending parallel to the web (or 'back') between the two sides or 'arms':
   a. We first neglect any confinement of the boundary elements at the two edges of the side.
      i. Geometry and reinforcement of equivalent rectangular section; neutral axis depth:
         As the central length of the side is unconfined, it is safe-sided to neglect the confinement of its two end regions, that is, to take $a\omega_{wd} = 0$. This implies $\varepsilon_{cu} = 0.0035$ all along the extreme compression fibres of the sides and $b = b_o = 1.25$ m; so $\nu_d = 2/(1.5 \times 1.25 \times 16) = 0.0667$.

         We first consider as tension and compression reinforcement, the reinforcement of each side ('arm') up to slightly over half its thickness: $10\Phi12 + 1\Phi10$ (1210 mm^2) ($\omega_1 = \omega_2$); ω_v comprises the rest: $20\Phi12 + 6\Phi10$ (2733 mm^2); then $\omega_{vd} = 2733/(1250 \times 1500) \times 452/16 = 0.0412$. We estimate the neutral axis depth at ultimate condition from Equation 5.68: $x_u = (0.0667 + 0.0412) \times 1.5 = 0.162$ m, which is less than the thickness of the compression flange, namely 0.25 m. So, the assumption that $b = 1.25$ m is confirmed; but apparently, the compression zone includes $11\Phi12 + 1\Phi10$ (1323 mm^2), at a centroidal distance from the concrete surface $d_1 = 68$ mm.

         Recalculating ω_{vd} for $18\Phi12 + 6\Phi10$ (2507 mm^2) as $\omega_{vd} = 2507/(1250 \times 1500) \times 452/16 = 0.0378$, x_u is revised to $x_u = (0.0667 + 0.0378) \times 1.5 = 0.157$ m, consistent with the new allotment of the section's reinforcement to $\omega_1, \omega_2, \omega_v$.
      ii. Confinement and available curvature ductility factor:
         μ_φ is computed from $a\omega_{wd} = 30 \mu_\varphi(0.0667 + 0.0378) \times 452/200,000 - 0.035 = 0 \rightarrow \mu_\varphi = 4.94$, $\mu_\delta = (\mu_\varphi + 1)/2 = 2.97$.
      iii. Moment resistance:
         The moment resistance at spalling of concrete is computed as per Section 5.4.3, considering the unconfined section as rectangular, with: $h = 1.50$ m, $b = 1.25$ m, $d_1 = 68$ mm, $d = 1500 - 68 = 1432$ mm: $\delta_1 = d_1/d = 68/1432 = 0.0475$:

         $A_{s1} = A_{s2} = 1323$ mm^2: $\omega_{1d} = \omega_{2d} = A_{s1}/(bd) \cdot (f_{yd}/f_{cd}) = 1323/(1250 \times 1432) \times 452/16 = 0.0209$.

         $A_{sv} = 2507$ mm^2: $\omega_{vd} = A_{sv}/(bd) \cdot (f_{yd}/f_{cd}) = 2507/(1250 \times 1432) \times 452/16 = 0.0396$.

         $\nu_d = 2/(1.432 \times 1.25 \times 16) = 0.0698$.

         The right-hand side of Equation 5.37b is:

         $\nu_2 = 0.0396/(1 - 0.0475) \times (0.0475 \times (0.0035 + 0.00226)/(0.0035 - 0.00226) - 1) + 0.0475 \times (0.0035 - 0.002/3)/(0.0035 - 0.00226) = 0.0762 > \nu_d = 0.0667$.
         Case (ii) in Section 5.4.3 applies; Equation 5.39b takes the form:
         $[1 - 0.002/(3 \times 0.0035) + 0.5 \times 0.0396/(1 - 0.0475) \times (0.0035 + 0.00226)^2/(0.0035 \times 0.00226)]\xi^2 - [0.0698 + 0.0209 \times (1 - 0.0035/0.00226) + 0.0396/(1 - 0.0475) \times (1 + 0.0035 \times 0.0475/0.00226)]\xi - [0.0209 - 0.0396 \times 0.0475/(2 \times (1 - 0.0475))] \times 0.0475 \times 0.0035/0.00226 = 0$, that is, $0.8967\xi^2 - 0.103\xi - 0.00146 = 0 \rightarrow \xi = 0.127$. Then Equation 5.38b gives:
         $M_{Rd,c} = 1.25 \times 1.432^2 \times 16,000 \times \{0.127 \times [(1 - 0.127)/2 - 0.002/(3 \times 0.0035) \times (0.5 + (0.002/(4 \times 0.0035)) - 1) \times 0.127)] + 0.5 \times (1 - 0.0475) \times 0.0209 \times [1 + (1 - 0.0475/0.127) \times 0.0035/0.00226] + 0.25 \times 0.0396/(1 - 0.0475) \times [0.127 \times (1 + 0.00226/0.0035) - 0.0475] \times [1 + 0.0035/0.00226 \times (1 - 0.0475/0.127)] \times [1 - 0.0475/3 - 2 \times 0.127/3 \times (1 + 0.00226/0.0035)]\} = 2805$ kNm.

b. We acknowledge that the 0.375 m long boundary elements at the edges of each side are confined:

i. Geometry and reinforcement of equivalent rectangular section; neutral axis depth:

The unconfined, 0.5 m long, central part of each side between the two boundary elements is presumed to fail in compression, once $\varepsilon_c > 0.0035$. Then, what remains is confined and has: $b = 2 \times 0.375$ m $= 0.75$ m, $b_o = 2 \times (0.375 - 0.05) = 0.65$ m; v_d is normalised to $b = 0.75$ m, as $v_d = 2/(0.75 \times 1.5 \times 16) = 0.1111$.

Neglecting the 2Φ10 that fall within the unconfined part presumed to fail, the tension or compression reinforcement determined in (a) above at each side ('arm') is: 11Φ12 (1244 mm²); the rest, that is, 16Φ12 + 4Φ10 (2124 mm²) is then assigned to ω_v: $\omega_{vd} = 2124/(750 \times 1500) \times 452/16 = 0.0533$. This gives $x_u = (0.1111 + 0.0533) \times 1.5 \times 0.75/0.65 = 0.285$ m, and we may consider as tension or compression reinforcement the full reinforcement of each side ('arm') except for the 2Φ10 in the unconfined part presumed to fail, that is, 17Φ12 (1923 mm²). The rest, that is, 4Φ12 + 4Φ10 (767 mm²) is assigned to ω_v: $\omega_{vd} = 767/(750 \times 1500) \times 452/16 = 0.0193$, giving: $x_u = (0.1111 + 0.0193) \times 1.5 \times 0.75/0.65 = 0.225$ m, confirming this last assumption regarding the allotment of reinforcement to ω_1, ω_2, ω_v. However, the neutral axis depth of 0.225 m slightly exceeds the 0.2 m depth of the boundary elements of the side; so, at the confined corner of the side and the 'back', the compression zone extends into the 0.375 m long boundary element of the 'back'.

For the time being, this discrepancy is neglected, owing to the different geometry of the boundary elements at the far edge ('toe') of the side and the corner of the side and the 'back' of the wall; the confined core is taken as rectangular, with width $b_o = 2 \times 0.325 = 0.65$ m, fully lying within one side of the section, that is, not extending into the section's 'back' (or web).

ii. Confinement and available curvature ductility factor:

At both boundary elements of the side:
$a_s = (1 - 90/(2 \times 200))(1 - 90/(2 \times 325)) = 0.668$; $\rho_y = \Sigma A_{swy}/(b_{xo}s) = 2 \times 50/(200 \times 90) = 0.00556$.

For the boundary elements of the far edge ('toe') of the side:
$\rho_x = \Sigma A_{swx}/(b_{yo}s) = 3 \times 50/(325 \times 90) = 0.00513$; 2 min($\rho_x$; ρ_y) = 0.01026; $\omega_{wd} = 0.01026 \times 487/16 = 0.3123$.
$a_n = 1 - (4 \times 162.5^2 + 2 \times 200^2)/(200 \times 325 \times 6) = 0.524$; then $a = 0.668 \times 0.524 = 0.35$;
$a\omega_{wd} = 0.35 \times 0.3123 = 0.1093$.

For the boundary element at the corner of the side and the 'back':
$\rho_x = \Sigma A_{swx}/(b_{yo}s) = 4 \times 50/(325 \times 90) = 0.00684$; 2 min($\rho_x$; ρ_y) = 0.01111; $\omega_{wd} = 0.01111 \times 487/16 = 0.3382$.
$a_n = 1 - (2 \times 125^2 + 4 \times 200^2)/(6 \times 200 \times 325) = 0.510$; then $a = 0.668 \times 0.510 = 0.34$;
$a\omega_{wd} = 0.34 \times 0.3382 = 0.115$.

The most critical of the two edges is the one at the 'toe' of the side, where $a\omega_{wd} = 0.1093$.

Before calculating the available value of μ_φ from $a\omega_{wd} = 0.1093$, the geometric extent of this confinement should be checked: it should extend up to a distance from the extreme compression fibres of at least $(\varepsilon_{cu2,c} - \varepsilon_{cu})/\varepsilon_{cu2,c}$ times the neutral axis depth x_u of 0.225 m, with $\varepsilon_{cu2,c} = 0.0035 + 0.1a\omega_{wd} = 0.01443$ (see Equation 5.69) and $\varepsilon_{cu} = 0.0035$. So, confinement is needed up to a distance of $0.225 \times 0.01093/0.01443 = 0.17$ m to the centreline of the perimeter stirrup; indeed, it is available up to 0.20 m from there. This margin, from 0.20 to 0.17 m, allows disregarding the shortfall of the 0.20 m confined core of the side's toe with respect to the neutral axis depth of 0.225 m, which was estimated as consistent with the present calculation; besides, the other confined core, near the corner of the side and the 'back' of the section, allows the compression zone to partly extend into the 'back'. So, the available value of μ_φ is estimated from: $0.1093 = 30\,\mu_\varphi(0.1111 + 0.0193) \times 0.00226 \times 0.75/0.65 - 0.035 \to \mu_\varphi = 14.15$, $\mu_\delta = (\mu_\varphi + 1)/2 = 7.57$.

iii. Moment resistance:

To check the μ_φ calculation, we estimate the moment resistance, $M_{Rd,o}$, after the presumed loss of the 0.5 m long unconfined part of the side between the two boundary elements, and compare it to the value $M_{Rd,c} = 2805$ kNm computed in I(a) above for the full, unspalled section. Because the ultimate condition is conventionally defined as one where the moment resistance falls below 80% of the peak resistance, a drop from $M_{Rd,c}$ to $M_{Rd,o}$ by less than 20% of $M_{Rd,c}$ means that spalling does not constitute failure and the value $\mu_\varphi = 14.15$ calculated above on the basis of the confined part of the compression zone after complete loss of the unconfined one may be retained as realistic.

The moment resistance of the spalled but confined section is computed per Section 5.4.3, this time considering the confined part of the section only, and indeed as rectangular with: $b_o = 1.45$ m as h, $b_o = 0.65$ m as b, $A_{s1} = A_{s2} = 1923$ mm^2 at $d_1 = 100$ mm from the extreme confined fibres, $d = 1450 - 100 = 1350$ mm ($\delta_1 = d_1/d = 100/1350 = 0.074$), $A_{sv} = 767$ mm^2.

Instead of the unconfined strength, f_{cd}, that of confined concrete, $f_{c,cd}$, from Equation 5.60 is used; it is computed from the average value of $a\omega_{wd}$ in the two boundary elements, $a\omega_{wd} = (0.115 + 0.1093)/2 = 0.1122$: $f_{c,cd} = $ min $[(1 + 2.5 \times 0.1122); (1.125 + 1.25 \times 0.1122)]f_{cd} = 1.265 \times 16 = 20.24$ MPa. This new value is used hereafter as f_{cd}. By the same token, in lieu of $\varepsilon_{cu} = 0.0035$, the confined ultimate strain from Equation 5.69 is used: $\varepsilon_{cu2,c} = 0.0035 + 0.1 \times 0.1122 = 0.0147$, and instead of $\varepsilon_{c2} = 0.002$, the confined strain at ultimate strength is used, from Equation 5.61: $\varepsilon_{c2,c} = 0.002 \times (20.24/16)^2 = 0.0032$.

$\omega_{1d} = \omega_{2d} = 1923/(650 \times 1350) \times 452/20.24 = 0.0489$, $\omega_{vd} = 767/(650 \times 1350) \times 452/20.24 = 0.0195$, $v_d = 2/(0.65 \times 1.35 \times 20.24) = 0.1126$.

The left-hand side of Equation 5.37a and right-hand one of 5.37b is:

$v_2 = 0.0195/(1 - 0.074) \times (0.074 \times (0.0147 + 0.00226)/(0.0147 - 0.00226) - 1) + 0.074 \times (0.0147 - 0.0032/3)/(0.0147 - 0.00226) = 0.0622 < v_d = 0.1126$;

The right-hand side of Equation 5.37a is:

$v_1 = 0.0195/(1 - 0.074) \times ((0.0147 - 0.00226)/(0.0147 + 0.00226) - 0.074) + (0.0147 - 0.0032/3)/(0.0147 + 0.00226) = 0.992$;

Equation 5.37a is met: $v_2 = 0.0622 < v_d = 0.1126 < v_1 = 0.992$; case (i) applies; Equation 5.39a gives:

$\xi = [(1 - 0.074) \times 0.1126 + (1 + 0.074) \times 0.0195]/[(1 - 0.074) \times (1 - 0.0032/(3 \times 0.0147)) + 2 \times 0.0195] = 0.139$, and Equation 5.38a gives: $M_{Rd,o} = 0.65 \times 1.35^2 \times 20240 \times \{0.139 \times [(1 - 0.139)/2 - 0.0032/(3 \times 0.0147) \times (0.5 + (0.0032/(4 \times 0.0147) - 1) \times 0.139)] + (1 - 0.074) \times 0.0489 + 0.0195/(1 - 0.074) \times [(0.139 - 0.074) \times (1 - 0.139) - (0.139 \times 0.00226/0.0147)^2/3]\} = 2460$ kNm $\to M_{Rd,o}/M_{Rd,c} = 0.88 > 0.8$;

So, spalling does not constitute failure and the higher value $\mu_\varphi = 14.2$ ($\mu_\delta = 7.57$) is taken to apply.

iv. Conclusion:

Once the unconfined, 0.5 m long central part of the side fails in compression at $\mu_\varphi = 4.94$ (and $\mu_\delta = 2.97$), confinement is activated in the two boundary elements of the side, allowing the already partially failed section to reach the ultimate condition at $\mu_\varphi = 14.2$ ($\mu_\delta = 7.57$).

II Bending about the centroidal axis which is normal to the two sides (or 'arms') and parallel to the web (or 'back').

IIa Bending induces compression in the web ('back'):

    a. We first neglect any confinement of the boundary elements at the two edges of the 'back':

       i. Geometry and reinforcement of equivalent rectangular section; neutral axis depth:

As the central length of the 'back' is unconfined, it is safe-sided to neglect confinement of its two end regions and take $a\omega_{wd} = 0$, that is, $\varepsilon_{cu} = 0.0035$ all along the extreme compression fibres of the 'back'; then $b = b_o = 1.50$ m, $v_d = 2/(1.25 \times 1.5 \times 16) = 0.0667$ (as in I(a)).

We consider as tension reinforcement $2 \times 6\Phi12$ (1357 mm²) at the very edge of the two 'arms' which are in tension; as compression reinforcement we take $2 \times 4\Phi12 + 2\Phi10$ (1062 mm²). The rest of the section reinforcement counts as A_{sv}: $2 \times 9\Phi12 + 6\Phi10$ (2507 mm²).

As ω_1 and ω_2 are different, $(v_d + \omega_{1d} - \omega_{2d} + \omega_{vd})$ is used in Equations 5.65b – for μ_φ – and Equation 5.68 – for x_u – instead of $(v_d + \omega_{vd})$:

$\omega_{1d} - \omega_{2d} + \omega_{vd} = (1357 - 1062 + 2507) \times 452/(1500 \times 1250 \times 16) = 0.0422$.

The neutral axis depth at ultimate condition is: $x_u \approx (v_d + \omega_{1d} - \omega_{2d} + \omega_{vd})hb/b_o = (0.0667 + 0.0422) \times 1.25 = 0.136$ m, that is, less than the thickness of the compression flange (0.25 m), confirming the validity of b as the full 1.5 m width of the 'back' and the assumed allotment of reinforcement to ω_1, ω_2, ω_v.

      ii. Confinement and available curvature ductility factor:

Equation 5.65b gives for $a\omega_{wd} = 0$: $30\,\mu_\varphi(0.0667 + 0.0422) \times 0.00226 - 0.035 = 0 \rightarrow \mu_\varphi = 4.74$, $\mu_\delta = (\mu_\varphi + 1)/2 = 2.87$.

      iii. Moment resistance:

The moment resistance at spalling of concrete is computed per Section 5.4.3, considering the unconfined section as rectangular, with: $h = 1.25$ m, $b = 1.5$ m, $d_1 = 25$ mm, $d = 1250 - 25 = 1225$ mm: $\delta_1 = d_1/d = 25/1225 = 0.02$; $v_d = 2/(1.5 \times 1.225 \times 16) = 0.068$.

$A_{s1} = 1357$ mm²: $\omega_{1d} = A_{s1}/(bd) \cdot (f_{yd}/f_{cd}) = 1357/(1500 \times 1225) \times 452/16 = 0.0209$;
$A_{s2} = 1062$ mm²: $\omega_{2d} = A_{s2}/(bd) \cdot (f_{yd}/f_{cd}) = 1062/(1500 \times 1225) \times 452/16 = 0.0163$;
$A_{sv} = 2507$ mm²: $\omega_{vd} = A_{sv}/(bd) \cdot (f_{yd}/f_{cd}) = 2507/(1500 \times 1225) \times 452/16 = 0.0385$.

The left-hand side of Equation 5.37a and right-hand side one of 5.37b is:
$v_2 = 0.0163 - 0.0209 + 0.0385/(1 - 0.02) \times (0.02 \times (0.0035 + 0.00226)/(0.0035 - 0.00226) - 1) + 0.02 \times (0.0035 - 0.002/3)/(0.0035 - 0.00226) = 0.0055$.

The right-hand side of Equation 5.37a is:
$v_1 = 0.0163 - 0.0209 + 0.0385/(1 - 0.02) \times ((0.0035 - 0.00226)/(0.0035 + 0.00226) - 0.02) + (0.0035 - 0.002/3)/(0.0035 + 0.00226) = 0.495$;

Equation 5.37a is met: $v_2 = 0.0055 < v_d = 0.068 < v_1 = 0.495$; case (i) applies; Equation 5.39a gives:
$\xi = [(1 - 0.02) \times (0.068 + 0.0209 - 0.0163) + (1 + 0.02) \times 0.0385]/[(1 - 0.02) \times (1 - 0.002/(3 \times 0.0035)) + 2 \times 0.0385] = 0.127$.

Then Equation 5.38a gives:
$M_{Rd,c} = 1.5 \times 1.225^2 \times 16{,}000 \times \{0.127 \times [(1 - 0.127)/2 - 0.002/(3 \times 0.0035) \times (0.5 + (0.002/(4 \times 0.0035) - 1) \times 0.127)] + (1 - 0.02) \times (0.0209 + 0.0163)/2 + 0.0385/(1 - 0.02) \times [(0.127 - 0.02) \times (1 - 0.127) - (0.127 \times 0.00226/0.0035)^2/3]\} = 2440$ kNm.

    b. We acknowledge next the 0.375 m long boundary elements at the two edges of the back as confined:

      i. Geometry and reinforcement of equivalent rectangular section; neutral axis depth:

The unconfined, 0.75 m long, central part of the 'back' between the two boundary elements is presumed to fail in compression, once $\varepsilon_c > 0.0035$. Then, as in I(b) above, we have: $b = 2 \times 0.375 = 0.75$ m, $b_o = 2 \times (0.375 - 0.05) = 0.65$ m.

The tension reinforcement is the same as in IIa (a): $2 \times 6\Phi12$ (1357 mm²) at the tensioned 'toes' of the two 'arms'. We assume the full thickness of the 'back' to be in compression and take as compression reinforcement $2 \times 7\Phi12$ (1583 mm²). The $4\Phi10$ in-between the two boundary elements are neglected, as falling in the part of the section presumed to be lost. The rest of the section reinforcement counts as A_{sv}: $2 \times 6\Phi12 + 2 \times 2\Phi10$ (1671 mm²).

$\omega_{1d} - \omega_{2d} + \omega_{vd}$ and ν_d are normalised to $b = 0.75$ m, as $(\omega_{1d} - \omega_{2d} + \omega_{vd}) = (1357 - 1583 + 1671) \times 452/(1250 \times 750 \times 16) = 0.0435$, $\nu_d = 2/(0.75 \times 1.25 \times 16) = 0.1333$. The neutral axis depth at ultimate is: $x_u \approx (\nu_d + \omega_{1d} - \omega_{2d} + \omega_{vd})hb/b_o = (0.1333 + 0.044) \times 1.25 \times 0.75/0.65 = 0.255$ m, that is, it exceeds the 0.2 m wide confined core within the thickness of the 'back', confirming the assumed allotment of compression reinforcement to ω_1, ω_2, ω_v, but not the validity of b as 2×0.375 m. So, we proceed to an alternative assumption.

c. Only the 0.375 m long boundary elements of the two 'arms' next to the 'back' are taken as confined, neglecting the confined 0.125 m long segments of the back next to them:

  i. Geometry and reinforcement of equivalent rectangular section; neutral axis depth:

  We consider as confined only the 0.375 m long boundary elements of the two sides ('arms') at the corners with the 'back': the acting section is taken as rectangular, with the aggregate width of these two boundary elements: $b = 2 \times 0.25 = 0.5$ m. We neglect, for simplicity, the contribution of the concrete in the confined 0.125 m long segments of the back next to the corners with the sides, but we include that of its reinforcement.

  The tension reinforcement is as in IIa (a): $2 \times 6\Phi12$ (1357 mm²) at the tensioned 'toes' of the 'arms'. Assuming the full thickness within the 'back' to be in compression, but neglecting the $4\Phi10$ bars in-between the two boundary elements, as falling in the part of the section presumed to be lost, the compression reinforcement is $2 \times 7\Phi12$ (1583 mm²). The rest of the section reinforcement counts as A_{sv}: $2 \times 6\Phi12 + 2 \times 2\Phi10$ (1671 mm²). Then $(\omega_{1d} - \omega_{2d} + \omega_{vd}) = (1357 - 1583 + 1671) \times 452/(1250 \times 500 \times 16) = 0.0653$; $\nu_d = 2/(0.5 \times 1.25 \times 16) = 0.2$. The neutral axis depth at ultimate is: $x_u \approx (\nu_d + \omega_{1d} - \omega_{2d} + \omega_{vd})hb/b_o = (0.2 + 0.0653) \times 1.25 \times 0.5/0.4 = 0.415$ m. This value exceeds the 0.375 m length of the confined boundary element. To check if it is consistent with the present assumptions, we calculate the extent of necessary confinement:

  The confining reinforcement of each boundary element encompasses two 0.375 m long stirrup legs along the sides of the section, giving $\rho_y = \Sigma A_{swy}/(b_{xo}s) = 2 \times 50/(200 \times 90) = 0.00556$, and the two cross-legs of the same tie together with the 0.375 m long legs of the rectangular tie along the boundary element of the 'back', which give $\rho_x = \Sigma A_{swx}/(b_{yo}s) = 4 \times 50/(325 \times 90) = 0.00684$. So, its volumetric ratio is $2 \min(\rho_x; \rho_y) = 0.01111$; then $\omega_{wd} = 0.01111 \times 487/16 = 0.3382$ (the same as for the confined corner of the side and the 'back' in I(b) above). We also have: $a_n = 1 - (2 \times 125^2 + 4 \times 200^2)/(200 \times 325 \times 6) = 0.51$, $a_s = (1-90/(2 \times 200))(1-90/(2 \times 325)) = 0.668$, $a = 0.668 \times 0.51 = 0.34$, $a\omega_{wd} = 0.34 \times 0.3382 = 0.115$ (again as at the confined corner of the side and the 'back' in I(b)).

  The depth over which confinement is required is the fraction $(\varepsilon_{cu2,c} - \varepsilon_{cu})/\varepsilon_{cu2,c}$ of the neutral axis depth $x_u = 0.415$ m, with $\varepsilon_{cu2,c} = 0.0035 + 0.1a\omega_{wd}$. So, it is necessary to have confinement over a depth of $0.1a\omega_{wd}x_u/(0.0035 + 0.1 a\omega_{wd}) = 0.0115 \times 0.415/0.015 = 0.318$ m from the centreline of the perimeter stirrup; this depth is indeed inside the 0.325 m length of the confined core.

  ii. Available curvature ductility factor:

  The available value of μ_φ is determined from:
  $0.115 = 30 \mu_\varphi(0.2 + 0.0653) \times 0.00226 \times 0.5/0.4 - 0.035 \rightarrow \mu_\varphi = 6.67$, $\mu_\delta = (\mu_\varphi + 1)/2 = 3.84$.

  iii. Moment resistance:

  Before adopting the higher value $\mu_\varphi = 6.67$ ($\mu_\delta = 3.84$), we check the moment resistance, $M_{Rd,o}$, after the presumed loss of the unconfined part of the 'back', vs. the value $M_{Rd,c} = 2440$ kNm computed in IIa (a) above for the full, unspalled section. A drop from $M_{Rd,c}$ to $M_{Rd,o}$ by less than 20% of $M_{Rd,c}$ means that spalling does not constitute failure; then the value $\mu_\varphi = 6.67$, calculated on the basis of only the confined part of the compression zone after complete loss of the unconfined one, may be retained as realistic.

The moment resistance of the spalled, confined section is computed per Section 5.4.3, considering only the confined part of the section as rectangular, with: $b_o = 1.2$ m as h, $b_o = 0.4$ m as b, $A_{s1} = 1357$ mm^2, $A_{s2} = 1583$ mm^2, at $d_1 = 25$ mm from the extreme confined fibres, $d = 1200 - 25 = 1175$ mm ($\delta_1 = d_1/d = 25/1175 = 0.021$), $A_{sv} = 1671$ mm^2.

Instead of the unconfined strength, f_{cd}, that of confined concrete, $f_{c,cd}$, from Equation 5.60 is used, for $a\omega_{wd} = 0.115$: $f_{c,cd} = \min[(1 + 2.5 \times 0.115); (1.125 + 1.25 \times 0.115)]f_{cd} = 1.26875 \times 16 = 20.3$ MPa. This new value is used hereafter as f_{cd}. Besides, in lieu of $\varepsilon_{cu} = 0.0035$, the confined ultimate strain from Equation 5.69 is used: $\varepsilon_{cu2,c} = 0.0035 + 0.1 \times 0.115 = 0.015$, and instead of $\varepsilon_{c2} = 0.002$, the confined strain at ultimate strength from Equation 5.61: $\varepsilon_{c2,c} = 0.002 \times (20.3/16)^2 = 0.0032$.

$\omega_{1d} = 1357/(400 \times 1200) \times 452/20.3 = 0.0629$, $\omega_{2d} = 1583/(400 \times 1200) \times 452/20.3 = 0.0734$, $\omega_{vd} = 1671/(400 \times 1200) \times 452/20.3 = 0.0775$, $v_d = 2/(0.4 \times 1.2 \times 20.3) = 0.2053$.

The left-hand side of Equation 5.37a and right-hand side of (5.37b) is:

$v_2 = 0.0734 - 0.0629 + 0.0775/(1 - 0.021) \times (0.021 \times (0.015 + 0.00226)/(0.015 - 0.00226) - 1) + 0.021 \times (0.015 - 0.0032/3)/(0.015 - 0.00226) = -0.043$.

The right-hand side of Equation 5.37a is:

$v_1 = 0.0734 - 0.0629 + 0.0775/(1 - 0.021) \times ((0.015 - 0.00226)/(0.015 + 0.00226) - 0.021) + (0.015 - 0.0032/3)/(0.015 + 0.00226) = 0.93$;

Equation 5.37a is met: $v_2 = -0.043 < v_d = 0.2053 < v_1 = 0.93$; case (i) applies; Equation 5.39a gives:

$\xi = [(1 - 0.021) \times (0.2053 + 0.0629 - 0.0734) + (1 + 0.021) \times 0.0775]/[(1 - 0.021) \times (1 - 0.0032/(3 \times 0.015)) + 2 \times 0.0775] = 0.254$.

Then Equation 5.38a gives:

$M_{Rd,c} = 0.4 \times 1.175^2 \times 20,300 \times \{0.254 \times [(1 - 0.254)/2 - 0.0032/(3 \times 0.015) \times (0.5 + (0.0032/(4 \times 0.015) - 1) \times 0.254)] + (1 - 0.021) \times (0.0629 + 0.0734)/2 + 0.0775/(1 - 0.021) \times [(0.254 - 0.021) \times (1 - 0.254) - (0.254 \times 0.00226/0.015)^2/3] = 1960$ kNm; $\rightarrow M_{Rd,o}/M_{Rd,c} = 0.803 > 0.8$; so, be it marginally, spalling does not constitute failure; the higher value $\mu_\varphi = 6.67$ ($\mu_\delta = 3.84$) is taken to apply.

iv. Conclusion:

After the unconfined central part of the 'back' fails in compression at $\mu_\varphi = 4.74$ ($\mu_\delta = 2.87$), confinement is activated in the two boundary elements of the 'sides', allowing the already partially failed section to reach the ultimate condition at $\mu_\varphi = 6.67$ ($\mu_\delta = 3.84$).

IIb Bending induces tension in the web (or 'back') and compression in the 'toe' of each side:

1. Geometry and reinforcement of equivalent rectangular section; neutral axis depth:

The tension and compression reinforcements are reversed: the tension reinforcement is $2 \times 4\Phi12 + 2\Phi10$ (1062 mm^2) at the extreme tension fibres; the compression one is $2 \times 6\Phi12$ (1357 mm^2) at the two 'toes' of the sides ('arms'). The rest of the section reinforcement counts as A_{sv}: $2 \times 4\Phi12 + 2 \times 5\Phi12 + 6\Phi10$ (2507 mm^2).

The total compression flange width is $b = 2 \times 0.25 = 0.5$ m and includes two confined cores, each one of width $b_o = 0.2$ m. So: $v_d = 2/(0.5 \times 1.25 \times 16) = 0.2$ and $\omega_{1d} - \omega_{2d} + \omega_{vd} = (1062 - 1357 + 2507) \times 452/(500 \times 1250 \times 16) = 0.1$. The neutral axis depth is: $x_u \approx (v_d + \omega_{1d} - \omega_{2d} + \omega_{vd})hb/b_o = (0.2 + 0.1) \times 1.25 \times 0.5/0.4 = 0.469$ m.

2. Confinement and available curvature ductility factor:

A value of $a\omega_{wd} = 0.35 \times 0.3122 = 0.1093$ has been computed in I(b) above for the boundary elements of the far edge ('toe') of the side. The depth over which confinement is required is the fraction $(\varepsilon_{cu2,c} - \varepsilon_{cu})/\varepsilon_{cu2,c}$ of the neutral axis depth $x_u = 0.469$ m, with $\varepsilon_{cu2,c} = 0.0035 + 0.1a\omega_{wd} = 0.01443$ and $\varepsilon_{cu} = 0.0035$. This gives a depth of $0.469 \times (0.01443-0.0035)/0.01443 = 0.355$ m to the centreline of the perimeter stirrup, which exceeds the 0.325 m length of the confined core. So, the value of μ_φ is controlled, not by the available value of $a\omega_{wd}$, but by the length over which it is available. We estimate this value of μ_φ by equating the product of $x_u = 0.469$ m and $(\varepsilon_{cu2,c} - \varepsilon_{cu})/\varepsilon_{cu2,c}$ to the length of the confined core, namely 0.325 m. We calculate a new value $\varepsilon_{cu2,c} = 0.0114$, which corresponds, via $\varepsilon_{cu2,c} = 0.0035 + 0.1a\omega_{wd}$, to $a\omega_{wd} = 0.079$. Then, from $0.079 = 30\,\mu_\varphi(0.2 + 0.1) \times 0.00226 \times 0.5/0.4 - 0.035 \rightarrow \mu_\varphi = 4.48$, $\mu_\delta = (\mu_\varphi + 1)/2 = 2.74$.

Summary:

I For bending in a plane parallel to the web (or 'back'): $\mu_\varphi = 14.2$ ($\mu_\delta = 7.57$), after the unconfined, 0.5 m long, central part of the side fails in compression (at $\mu_\varphi = 4.94$ and $\mu_\delta = 2.97$).

II For bending in a plane parallel to the two sides (or 'arms'):
- For compression in the web ('back'): $\mu_\varphi = 6.67$ ($\mu_\delta = 3.84$), after the unconfined central part of the side fails in compression (at $\mu_\varphi = 4.74$ and $\mu_\delta = 2.87$).
- For tension in the web ('back'): $\mu_\varphi = 4.48$ ($\mu_\delta = 2.74$), conditioned by the length of the confined boundary elements at the 'toes' of the two sides.

Whenever the compression flange has a sizeable width, the ductility supply is satisfactory, at least for a DC M wall. However, the boundary elements at the two 'toes' of the section are quite strained, when they alone play the role of the compression zone (for bending at right angles to the web); they make possible a ductility supply barely enough for a DC M wall system, about the same as the one which the wall would provide without further confinement other than that in these boundary elements. The curvature ductility supply is controlled not only by the amount of confinement reinforcement in these boundary elements, but also by their extent (length) along the sides of the section.

EXAMPLE 5.12

Specify the vectors of biaxial moments and axial force due to concurrent seismic action components X and Y to be used in design/verifications from the analysis results of the individual seismic action components given in Table 5.11.

Answer

The numerical results are listed in Table 5.12. Each line of results corresponds to two triplets of internal forces: one with the upper set of signs, another with the lower ones.

Lateral force analysis with the linear combination of E_X, E_Y as per Equation 3.100 gives just 4 combinations in total. All other combinations of analysis methods with either Equation 3.99 or 3.100 give 8 or 16 combinations.

Table 5.11 Analysis results for individual seismic action components for Example 5.12

	$M_{y,X}$ (kNm)	$M_{z,X}$ (kNm)	N_X (kN)	$M_{y,Y}$ (kNm)	$M_{z,Y}$ (kNm)	N_Y (kN)
Lateral force analysis	±100 (main)	∓50	∓10	±20	±80 (main)	±20
Modal response spectrum	100	50	10	20	80	20
Accidental eccentricity	∓10	±2	±1	∓1	∓10	∓2

Table 5.12 Internal force vectors for concurrent seismic action components

	Analysis method	M_y (kNm)	M_z (kNm)	N (kN)
Linear combination of E_x, E_y, Equation 3.100	Lateral force, before accidental eccentricity	$\pm(100 + 0.3 \times 20) = \pm106$ $\mp0.3 \times 100 \pm 20 = \mp10$	$\mp50 \pm 0.3 \times 80 = \mp26$ $\pm(50 \times 0.3 + 80) = \pm95$	$\mp10 \pm 0.3 \times 20 = \mp4$ $\pm0.3 \times 10 \pm 20 = \pm23$
	Lateral force, with accidental eccentricity	$\pm(110 + 0.3 \times 21) = \pm116.3$ $\mp0.3 \times 110 \pm 21 = \mp12$	$\mp52 \pm 0.3 \times 90 = \mp25$ $\pm(52 \times 0.3 + 90) = \pm105.6$	$\mp11 \pm 0.3 \times 22 = \mp4.4$ $\pm0.3 \times 11 \pm 22 = \pm25.3$
	Modal response spectrum analysis, without accidental eccentricity; all components taken with their maximum values, Equations 5.73, 5.74	$\pm(100 + 0.3 \times 20) = \pm106$ ±106 ±106 ±106 $\pm(0.3 \times 100 + 20) = \pm50$ ±50 ±50 ±50	$\mp(50 + 0.3 \times 80) = \mp74$ ∓74 ∓74 ∓74 $\pm(0.3 \times 50 + 80) = \pm95$ ∓95 ∓95 ±95	$\pm(10 + 0.3 \times 20) = \pm16$ ±16 ∓16 ∓16 $\pm(0.3 \times 10 + 20) = \pm23$ ±23 ∓23 ∓23
	Modal response spectrum analysis, with accidental eccentricity; all components taken with their maximum values, Equations 5.73, 5.74	$\pm(110 + 0.3 \times 21) = \pm116.3$ ±116.3 ±116.3 ±116.3 $\pm(0.3 \times 110 + 21) = \pm54$ ±54 ±54 ±54	$\mp(52 + 0.3 \times 90) = \mp79$ ∓79 ∓79 ∓79 $\pm(0.3 \times 52 + 90) = \pm105.6$ ∓105.6 ∓105.6 ±105.6	$\pm(11 + 0.3 \times 22) = \pm17.6$ ±17.6 ±17.6 ±17.6 $\pm(0.3 \times 11 + 22) = \pm25.3$ ±25.3 ∓25.3 ±25.3
SRSS, Equation 3.99; all components with their maximum values, Equations 5.72, 5.74	Lateral force method or modal response spectrum analysis, no accidental eccentricity	$\pm\sqrt{(100^2 + 20^2)} = \pm102$ ±102 ±102 ±102	$\pm\sqrt{(50^2 + 80^2)} = \pm94.3$ ∓94.3 ∓94.3 ±94.3	$\pm\sqrt{(10^2 + 20^2)} = \pm22.4$ ±22.4 ∓22.4 ∓22.4
	Lateral force method or modal response spectrum analysis with accidental eccentricity	$\pm\sqrt{(110^2 + 21^2)} = \pm112$ ±112 ±112 ±112	$\pm\sqrt{(52^2 + 90^2)} = \pm103.9$ ∓103.9 ∓103.9 ±103.9	$\pm\sqrt{(11^2 + 22^2)} = \pm24.6$ ±24.6 ∓24.6 ±24.6

continued

Table 5.12 (continued) Internal force vectors for concurrent seismic action components

	Analysis method	M_y (kNm)	M_z (kNm)	N (kN)
Equation 5.75: SRSS, Equation 3.99, for the maxima of individual components; orthogonal component from Equation 3.100	Lateral force method or modal response spectrum analysis, before accidental eccentricity	$\pm\sqrt{(100^2 + 20^2)} = \pm102$ ±102 ±102 ±102 $\pm0.3\sqrt{(100^2 + 20^2)} = \pm30.6$ ±30.6 ±30.6 ±30.6	$\pm0.3\sqrt{(50^2 + 80^2)} = \pm28.3$ ∓28.3 ∓28.3 ±28.3 $\pm\sqrt{(50^2 + 80^2)} = \pm94.3$ ∓94.3 ∓94.3 ±94.3	$\pm\sqrt{(10^2 + 20^2)} = \pm22.4$ ±22.4 ∓22.4 ∓22.4 ±22.4 ±22.4 ∓22.4 ∓22.4
	Lateral force method or modal response spectrum analysis, with accidental eccentricity	$\pm\sqrt{(110^2 + 21^2)} = \pm112$ ±112 ±112 ±112 $\pm0.3\sqrt{(110^2 + 21^2)} = \pm33.6$ ±33.6 ±33.6 ±33.6	$\pm0.3\sqrt{(52^2 + 90^2)} = \pm31.2$ ∓31.2 ∓31.2 ±31.2 $\pm\sqrt{(52^2 + 90^2)} = \pm103.9$ ∓103.9 ∓103.9 ±103.9	$\pm\sqrt{(11^2 + 22^2)} = \pm24.6$ ±24.6 ∓24.6 ∓24.6 ±24.6 ±24.6 ∓24.6 ∓24.6

The internal forces due to the accidental eccentricity are superimposed at the outset to those of the translational components, with signs such that the magnitude of the component, which is taken equal to its maximum value per Equations 5.72 or 5.73, increases in absolute value due to the contribution of the accidental eccentricity.

The combinations per Equation 5.75, listed near the bottom, give results close to those listed at the top, namely from lateral force analysis with linear combination per Equation 3.100.

QUESTION 5.1

One end of beam B7 in Example 4.7 (Figure 4.18) is indirectly supported on beam B4. How would you take that into account in the calculation of the capacity design shear force at the other end of B7 (the one connected to column C3)?

QUESTION 5.2

An old concrete frame (see Figure 5.18) has all columns the same: 0.25 m wide and 0.4 m deep, with the strong direction in the plane of the frame and only one 18 mm dia. bar at each corner. All the beams have a depth of 0.5 m and a width of 0.25 m; they have two 14 mm dia. bars at top and bottom, continuous across all spans, plus two additional 14 mm top bars over the interior supports on the columns. The quasi-permanent gravity load, $g + \psi_2 q$, is 8 kN/m² (all inclusive) and is applied over the 4.0 m wide tributary floor strip of the frame. Gravity loads go to the column which is nearest in plan.

Concrete is C30/37 and steel S500, with a 25 mm concrete cover.

The frame is evaluated under a seismic action represented by a system of horizontal forces on the floors, with inverted triangular height-wise distribution: $f_i = 0.1V_b i$, where i indexes the storeys (from bottom to top) and V_b is the total seismic base shear. The overturning moment due to these seismic loads induces axial forces only in the two outer columns; seismic axial forces in the interior columns may be neglected.

1. Given that the size and the reinforcement of members is the same in all storeys and that only the column axial force changes from storey to storey, around which beam–column joint is the strong column–weak beam criterion $\sum M_{Rd,c} > \sum M_{Rd,b}$ most likely to be met and around which one is it least likely? Provide separate answers for interior

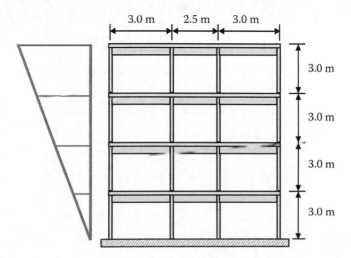

Figure 5.18 Four-storey frame of Question 5.2.

and exterior columns, excluding the top floor and taking into account any effects of the overturning moment. For the two interior and the two exterior joints expected to be most or least likely to fulfil this criterion, identify where the plastic hinges will form around these joints by checking numerically the criterion $\Sigma M_{Rd,c} > \Sigma M_{Rd,b}$. On the basis of the outcome, identify the most likely plastic hinge pattern and plastic mechanism in the frame under lateral seismic loading.

2. On the same basis as in (1), identify the beam span and the interior or exterior column in the frame with the largest capacity design shear force according to Eurocode 8. For the beam span and the interior and exterior columns with the expected highest capacity design shear, calculate its value. You may calculate any effects of the overturning moment using a seismic base shear equal to 20% of the weight.

3. Estimate the maximum horizontal force resistance that the frame can develop at the base, from the shear forces that can develop in its four columns when plastic hinges form at the base of these columns and around their top joint. Express it as a fraction of the weight of the frame.

QUESTION 5.3

A three-storey RC frame with storey height $H = 3$ m has two bays, each one with span length $L = 5$ m (Figure 5.19). The central column is 0.4 m square; the outer ones 0.35 m square. The beams have width $b_w = 0.3$ m and depth $h_b = 0.5$ m and are connected on both sides to a 150 mm thick slab. Design is for a ground motion with design peak ground acceleration (on rock) of 0.30 g and type 1 spectrum per Eurocode 8 on ground type C. Ductility Class (DC) is medium (M).

The moment, M, and the axial force, N, diagrams shown in Figure 5.20 over the clear member length (joints are considered rigid) are obtained from linear analysis for the quasi-permanent gravity loads, $G + \psi_2 Q$, with $\psi_2 = 0.3$, and for the design seismic action. For the latter, the full quasi-permanent gravity loads are taken to produce inertia forces (without reduction for the calculation of masses). The lateral force method is used, but, since the fundamental period, T_1, is not known yet, the M- and N-diagrams have been constructed assuming that T_1 is in the constant-acceleration range, that is, shorter than the corner period $T_C = 0.6$ s on type C ground. The column axial forces at the base give the total weight and

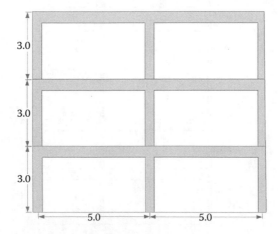

Figure 5.19 Three-storey frame of Question 5.3.

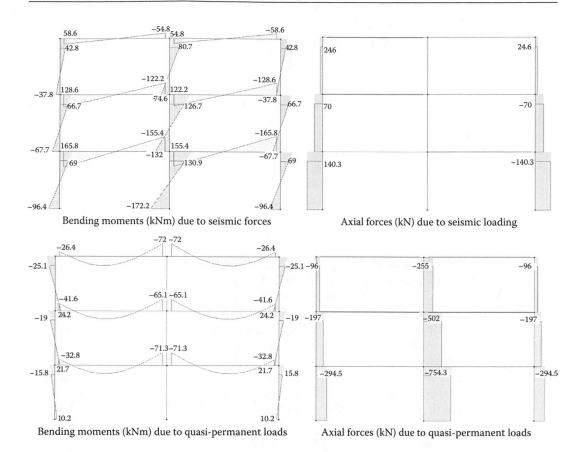

Figure 5.20 Moment and axial force diagrams for Question 5.3.

hence the mass of the frame which corresponds to the quasi-permanent gravity loads; its distribution to the floors is obtained from the storey shears. Columns are taken as fixed at the top of the footing.

Concrete grade is C30/37 and steel is of Class C with 500 MPa nominal yield stress; the concrete cover to reinforcement is $c = 25$ mm. Importance Class is II (ordinary).

1. Calculate from the moment diagram the lateral forces, f_i, and the resulting floor displacements, u_j; use these values to calculate the fundamental period of the frame through the Rayleigh quotient, Equation 3.109; correct the moment and axial force diagrams in Figure 5.20 to be consistent with the computed value of T_1.
2. Calculate the inter-storey drifts under the damage limitation seismic action and the sensitivity coefficient for second-order effects.
3. Dimension the longitudinal bars of the beams in floors 1 and 2, taking into account the 'persistent and transient design situation' for the combinations of Equations 6.10a, 6.10b of EN 1990 (the most unfavourable of $(1.35\xi)G_k$ '+' $1.5Q_k$ or $1.35G_k$ '+'$(1.5\psi_0)$ Q_k, with $\xi = 0.85$ and $\psi_0 = 0.7$); to this end, you may assume that the ratio of permanent-to-imposed nominal loads, G_k-to-Q_k, is 3.
4. Dimension the vertical reinforcement of the central and outer columns in storeys 1 and 2 to meet the strong column–weak beam capacity design rule, Equation 5.31.

5. Calculate the capacity design shears at the end sections of the first and second storey beams and columns from Equations 5.42, 5.44.
6. Dimension the transverse reinforcement of the first storey beams.
7. Dimension and detail the transverse reinforcement of the first storey columns, including confinement at the base.

QUESTION 5.4

For the building shown in Figure 5.21:

- The seismic action in direction X is considered to be resisted by the two exterior 3-bay frames alone. Seismic forces are applied at floor levels and at the lowest level of the roof; they are derived from the masses and a presumed inverted triangular pattern of horizontal displacements. The two interior columns of these frames have twice the moment of inertia of the corner ones and take twice as large seismic shears as the corner columns; hence the seismic moments at the two ends of the beams of that frame are numerically equal across all three spans of a floor. The columns of the two X-direction frames may be considered to develop zero seismic moment (inflection point) at storey mid-height. At the top they are fixed against rotation within the X-direction vertical plane, because the sloping roof works with the type $B1$, $B2$ perimeter beams as a very wide, inclined flange, imparting to these beams very high stiffness and flexural resistance for bending in the plane of the X-direction frames; for that reason, column $C3$ and the like cannot escape from plastic hinging at the top under strong seismic action in direction X.
- The pitched roof is supported by beams only along the perimeter. Its ridge is a non-deflecting support of the two roof slabs on either side. These slabs are one-way and, by in-plane action, transfer to the perimeter beams, which are parallel to the ridge,

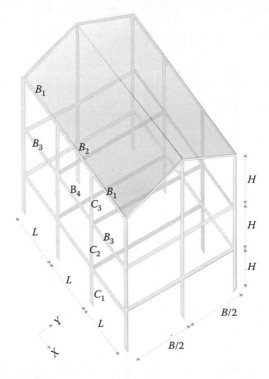

Figure 5.21 Building of Question 5.4.

the vertical reaction to gravity loads that would normally go to the support along the ridge. So, the full gravity load of the roof goes to beams B1 and B2. Under a uniform load p (kN/m), these beams develop bending moments at the interior supports equal to $0.1pL^2$, and zero at the end supports.

- Gravity loads go to the closest column in plan, but may be taken to induce no bending moments to columns. Floor slabs are one-way and can be taken to be supported only on their long sides in plan: floor beams B3, B4 and the like are considered as unloaded by gravity loads.

1. Estimate the seismic moments and axial forces in the members due to the seismic action in direction X.
2. Dimension the longitudinal reinforcement at the end sections of the second floor beams B3 and B4 and of the roof beams B1 and B2.
3. Dimension the vertical reinforcement of the third and second floor columns, C3 and C2, to meet the strong column–weak beam capacity design rule around their joint with beams B3, B4 and to resist at the top the seismic moments from the analysis for earthquake in direction X.
4. Calculate the capacity design shears of the second storey beam B4 and of the third-storey column C3 in the plane of the exterior X-direction frame.

- Ductility Class H (High)
- Base shear in direction X: 25% of the weight of the building
- Bay lengths: $L = 5.0$ m, $B = 11$ m. Storey height $H = 3.6$ m. Roof slope to the horizontal: 12°
- Concrete C25/30, S500 steel. Cover of reinforcement 30 mm
- Permanent loads (all inclusive): for the roof, 8 kN/m² per m² of horizontal projection; for the floors, 9 kN/m²
- Live loads: 2 kN/m² on the floors; zero on the roof
- $\psi_2 = 0.30$
- Beam width 0.3 m and depth 0.5 m. Slab thickness 0.16 m
- Interior columns: 0.6 m square; corner ones: 0.5 m square
- Curvature ductility demand for detailing: $\mu_\varphi = 2q_o - 1$, where q_o is the behaviour factor appropriate for the building

QUESTION 5.5

The building in Figure 5.22 has many similar four-bay, two-storey frames in direction X. Column tops are connected in direction Y, through beams of type B3, into five parallel Y-direction frames, each one with practically infinite, similar bays. There is a diaphragm only at roof level.

Simplifying assumptions:

- The self-weight of beams and columns is neglected for all purposes.
- The roof comprises one-way slabs, supported only on the Y-direction beams B3; beams of type B1 may be taken as not loaded by the roof slabs.
- Under gravity loads, beams of type B3 are considered as fixed at the end section against rotation.
- Bending of columns due to gravity loads is ignored.
- The seismic action is considered to produce horizontal forces only at the roof level.
- The seismic action components in direction X and Y are taken to act separately, not concurrently.

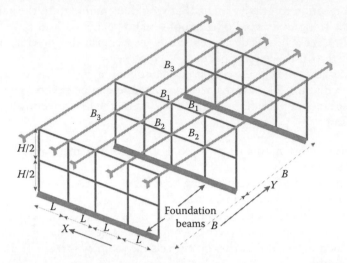

H/2

H/2

L   L   L   L

X

Foundation
beams B

Y

B

B₃ B₁ B₁ B₂ B₂ B₃

Figure 5.22 Three-storey building of Question 5.5.

- Columns take the horizontal seismic forces, as well as the gravity loads acting on the roof, in proportion to their tributary area in plan. Exterior columns have one-half the moment of inertia of interior ones; so, their share of the forces may indeed be assumed to be about half of that of interior columns.
- Under the seismic action, the inflection point (zero moment) of the columns is at the following fraction of the storey height from the base of the column in the storey:
- In the one-storey frames along the Y-direction: $(6k_Y + 1)/(12k_Y + 1)$, where $k_Y = (EI)_{B3}/(EI)_{CY}(H/B)$, with $(EI)_{B3}$ denoting the rigidity of beam B3, $(EI)_{CY}$ denoting that of an interior column for bending within a plane parallel to Y (strong axis) and B, H, denoting the length of these elements.
- In the two-storey frames along the X-direction:
  - At the lower storey: $(3k_{X2} +1)(12k_{X1} +1)/[(6k_{X2} +1)(12k_{X1} +1) - 1/2]$
  - At the upper storey: $[6k_{X2}(6k_{X1} +1) + 1/2]/[(6k_{X2} +1)(12k_{X1} +1) - 1/2]$
  where $k_{X1} = (EI)_{B1}/(EI)_{CX}(H/2L)$, $k_{X2} = (EI)_{B2}/(EI)_{CX}(H/2L)$, with $(EI)_{B1}$, $(EI)_{B2}$ denoting the rigidity of beams B1 and B2, $(EI)_{CX}$ that of an interior column for bending within a plane parallel to X (weak axis) and L, H/2 the length of these elements.
- The inflection points of the beams under seismic loading are always at mid-span.
- The effective flange width of roof beams B1 and B3 may be taken as per Eurocode 2: on each side of the web where there is a slab: 10% of the distance of the beam from the nearest parallel beam (but not greater than 7% of the beam span) plus another 7% of the beam span.

1. What is the value of the behaviour factor, q, of the building in directions X and Y according to Eurocode 8 for Ductility Class High (DC H)?
2. Calculate the fundamental periods of the building in directions X and Y, after establishing the stiffness of the corresponding single-degree-of-freedom (SDOF) system.
3. Using the outcomes of (1) and (2), compute the floor seismic forces for the design of the building in directions X and Y.
4. Calculate the inter-storey drifts under the design seismic action in directions X and Y and use them to estimate the sensitivity coefficients to second-order effects and the inter-storey drifts under a damage limitation earthquake equal to 50% of the design seismic action.

5. What is the use of X-direction beams of type B2 at building mid-height, since there are no slabs or seismic forces at that level?
6. Dimension the longitudinal reinforcement of interior beams B1, B2 and B3 at the supports.
7. Dimension the vertical reinforcement of an interior column, separately in directions X and Y, on the basis of the analysis results for the seismic forces in (3).
8. Calculate the capacity design shears at the ends of interior beams B1, B2, B3 and at both storeys of an interior column, in directions X and Y.
9. Dimension and detail the shear reinforcement of beams B1, B2 and B3.
10. Dimension and detail the transverse reinforcement of an interior column.

- Type 1 spectrum of Eurocode 8 for ground type E and design ground acceleration 0.42 g.
- Ductility Class H (High).
- Bay lengths: $L = 3.0$ m, $B = 10$ m.
- Height to mid-depth of roof slab, where the seismic forces are applied: $H = 7$ m.
- Concrete C35/45, S500 steel. Cover of reinforcement 25 mm.
- The roof slab is 160 mm thick and has only permanent loads: $g = 6.5$ kN/m².
- Beams B1, B2: width 0.3 m; depth 0.40 m; beams B3: width 0.3 m; depth 0.50 m.
- Interior column: 0.35 m in direction X, 0.60 m in Y; Exterior column: 0.30 m in X, 0.50 m in Y.

QUESTION 5.6

An elevated concrete silo, 8 m in diameter, is supported on four concrete columns at a 5 m square arrangement (see Figure 5.23). The columns are 0.6 m square and have a clear height of 4.5 m, with double fixity at top and bottom. The silo may be considered as rigid, with a centre of mass 3 m above the top of the supporting columns. According to Part 4 of Eurocode 8 ('Silos, tanks and pipelines'), the seismic design of the columns and their foundation follows Part 1 of Eurocode 8, except that the q-factor is reduced by 30% owing to the irregularity in elevation. The design peak ground acceleration (on rock) is 0.3 g and the Eurocode 8 spectrum for ground type B applies. Ductility Class medium (DC M) is chosen. The total weight of the silo and its contents is 3000 kN and may be taken as permanent

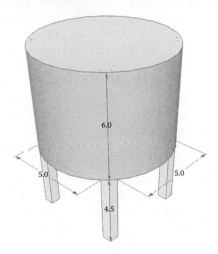

Figure 5.23 Silo of Question 5.6.

load. Concrete grade is C35/45 and steel is of Class C with 500 MPa nominal yield stress; concrete cover to reinforcement is $c = 30$ mm. Importance Class is II (ordinary).

1. Verify the columns against the Eurocode 2 slenderness limits for negligible second-order effects.
2. Considering the structure as an SDOF system, calculate its period and compute its design base shear and the horizontal displacements under the design seismic action. Calculate the sensitivity coefficient to second-order effects. Compute the correlation coefficient of the two natural modes in the horizontal seismic action components X and Y and consider the implications for the complete quadratic combination (CQC) rule. Consider the case of a horizontal seismic action component acting along the diagonal of the column section (including the implications for the column axial forces, as calculated from the overall overturning moment at column mid-height).
3. Calculate the accidental eccentricity per Part 1 of Eurocode 8 and its effects on column internal forces, for concurrent horizontal seismic action components X and Y. Discuss the implications of the correlation of the modes in the context of accidental eccentricity.
4. Dimension the vertical reinforcement of the columns.
5. Calculate the capacity design shears of the columns.

QUESTION 5.7

Design the columns and beams of the perimeter frame of Question 3.3, for the design seismic action specified in Question 3.3 and the Ductility Class and q-factor value chosen in Question 4.7. Consider an accidental eccentricity of 5% of the dimension of the perimeter at right angles to the seismic action component. Apply the linear combination of the effects of the two seismic action components, Equation 3.100. Concrete C20/25, steel S500; cover to reinforcement 30 mm.

- The columns bear the gravity loads acting in their tributary length along the perimeter; their bending moments due to gravity loads may be neglected.
- The horizontal seismic forces are distributed to the columns in proportion to their contribution to lateral stiffness.
- For seismic loading, the inflection point (zero moment) of the columns is at a distance from the column base equal to $(6k + 1)H/(12k + 1)$, where $k = (EI)_b/(EI)_c(L/B)$, with $(EI)_b$ denoting the rigidity of the beam and $(EI)_c$ that of an interior column for bending in the plane of the frame and L, H, the span and height of these elements.
- The corner columns in a frame have (approximately) half the section stiffness of interior columns; hence their seismic shear may be taken as half that of interior columns.
- The inflection point of beams under seismic loading is always at mid-span.
- The long-side beams develop hogging moments at the face of interior columns having depth b, equal to their own distributed quasi-permanent load times $(1 - 3b/L)L^2/12$. Moreover, in order to distribute the quasi-permanent loads of the roof to the long-side intermediate columns, these beams develop moments (hogging over the first and third interior columns, sagging over the second and fourth ones) equal to the roof load within the tributary areas of the column in plan, times one-quarter the clear bay length, $(L - b)$.
- The short-side beams develop hogging moments at the face of their central supporting column (with depth b), equal to their own distributed quasi-permanent load times $(1 - 2.5b/L)L^2/8$.

Chapter 6

Design of foundations and foundation elements

6.1 IMPORTANCE AND INFLUENCE OF SOIL–STRUCTURE INTERACTION

It is common in most seismic design codes to neglect soil–structure interaction (SSI) during seismic design of ordinary buildings. This is equivalent to assuming that the supporting ground is infinitely stiff and that the structural response is not affected by the fact that the ground is not explicitly modelled; in the analysis this is called a 'fixed base' model. The rationale behind this practice is the belief that SSI is always beneficial. Simple arguments to demonstrate this statement can be developed based on the smooth response spectra specified in building codes. Referring to Figure 6.1a and assuming that the natural period of the fixed base building, T_{fb}, corresponds, for instance, to a spectral acceleration associated with the plateau of the response spectrum (point A), the effect of SSI is to lengthen this period by introducing additional flexibility, that is, ground flexibility, at the foundation level. Therefore, the fixed base period will increase from T_{fb} to T_{SSI} and the spectral acceleration will at most remain constant; it is more likely that it will decrease, as the response shifts from the plateau of the spectrum to the constant velocity branch (point B). Furthermore, modelling SSI has an additional effect on the response: the energy imparted to the building by the incoming motion at its base will not be trapped in the building; instead, a good part of it will be diffracted back into the soil medium. The overall effect is an energy dissipation mechanism in the structure, which may be considered as additional damping. Consequently, point B will move, at the same period, from the initial 5% damped response spectrum to a response spectrum with higher damping (point C). The overall effect of SSI on the structural response, represented by a shift from point A to point C, is a decrease in the spectral acceleration; therefore, there is a reduction of the inertia forces for which the structure has to be designed. Of course, although SSI may be beneficial for forces, on the downside there is an increase in displacements. This is highlighted in Figure 6.1b, which presents the same situation as before, but on the displacement response spectrum.

Although the reasoning seems rather straightforward, it must be kept in mind, as pointed out by Mylonakis and Gazetas (2000), that it is entirely based on the smooth spectral shapes defined in codes. Spectra of real earthquakes do not exhibit such smooth shapes; they normally have peaks and troughs. If the shift towards longer periods on the spectrum moves the dominant periods of the response from a trough up a peak, it may even increase spectral accelerations; however, this cannot be foreseen in design, which works with smooth, average spectra.

Another important aspect of SSI, which is sometimes overlooked, is the modification of the base motion to which the structure is subjected. In a fixed base analysis the free-field motion is directly applied to the foundation. When SSI is considered, this motion may be significantly different both in amplitude and in frequency content: for example, large foundation rafts

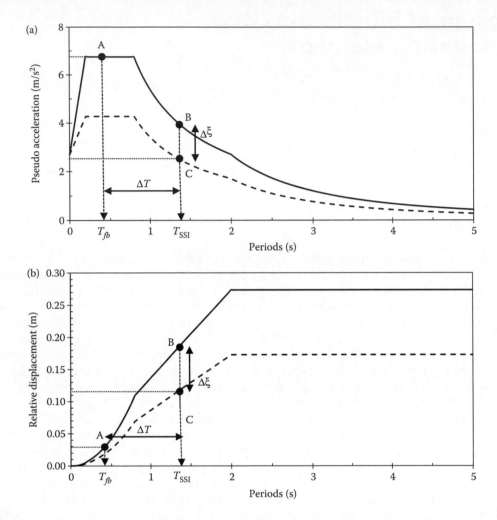

Figure 6.1 Impact of soil–structure interaction on response: (a) accelerations, (b) displacements.

filter the high frequency content of the incoming motion and smooth the effect of incoherent motions. In addition, even for a horizontal free-field motion, rocking motion may be induced at the foundation level. This rocking component may be important for slender structures, like masts and chimneys. Figure 6.2 illustrates an example of modifications of the free-field motion due to SSI. The foundation of a bridge pier is composed of 35 large diameter piles (2.5 m) crossing a 11 m thick very soft mud layer, with shear wave velocity of the order of 100 m/s. The piles penetrate a residual soil layer with V_S of 250–400 m/s and reach the competent rock formation at a depth of 25 m. The free-field ground response spectrum determined from a site-specific response analysis has a smooth shape; the kinematic interaction motion, that is, the motion of the piled foundation without the superstructure, exhibits a marked peak at 0.5 s and a significant de-amplification with respect to the free-field motion between 0.8 and 3.0 s. This phenomenon is due to the inability of the piled foundation to follow the ground motion. This is because of the stiffness of piles. Were the piles more flexible (i.e. of smaller diameter), the kinematic interaction motion would not be very different from the free-field one. An illustrative example of the consequences of kinematically induced rocking motions is depicted in Figure 6.3. This picture was taken in Mexico City after the

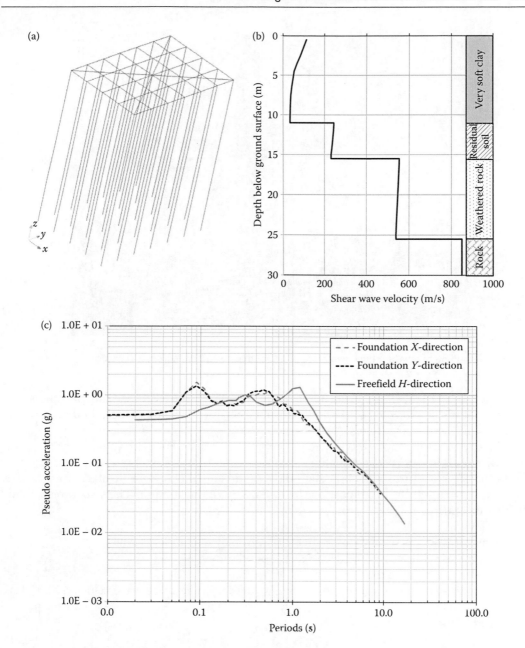

Figure 6.2 Modification of the piled foundation motion due to soil–structure interaction: finite element mesh (a); shear wave velocity profile (b); freefield and foundation response spectra (c).

1985 Michoacán Guerrero earthquake; two adjacent buildings originally of the same height, experienced severe rocking movements because of the very low stiffness of the Mexico lake deposits. The separation joint between the buildings was small and pounding occurred, causing a structural failure with the loss of three storeys of the building on the left. Without SSI, that is, if the buildings had been founded on rock, the rocking movements would have been negligible and the buildings may have survived the earthquake. Rocking motions are particularly significant for embedded caissons due to the interaction between the soil and the caisson

Figure 6.3 Pounding of adjacent buildings in Mexico City due to soil–structure interaction.

along the vertical embedded faces: the caisson is usually much stiffer than the surrounding soil and, when subjected to the soil deformations, it experiences a rigid body rotational motion as depicted in Figure 6.4.

Part 5 of Eurocode 8 follows the general approach presented earlier and does not call for SSI analyses except in special situations, namely those that are likely to cause detrimental effects on the structural response, such as a significant increase in displacement, or are known to have a significant impact on the foundation input motion. The structures explicitly identified are as follows:

- Structures where second order effects $(P - \Delta)$ play a significant role
- Structures with massive or deep-seated foundations
- Slender, tall structures
- Structures supported on very soft soils, with an average shear wave velocity, $V_{S,30}$, less than 100 m/s
- Piled foundations

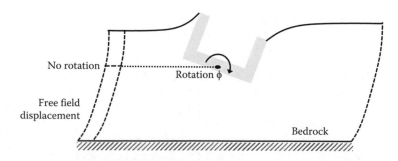

Figure 6.4 Effect of soil–structure interaction: rocking motion of an embedded caisson.

A rigorous treatment of SSI in the linear range is based on elasto-dynamic theories. A direct (or complete) interaction analysis, in which both the soil and the structure are modelled with finite elements, is very time-demanding and not well suited for design, especially in 3D. Substructuring divides the problem into more tractable stages. Moreover, should modifications occur in the superstructure, only certain stages of the analysis will need to be repeated. Substructuring is of great mathematical convenience and rigor, which stems, in linear systems, from the superposition theorem (Kausel and Roesset 1974). This theorem states that the seismic response of the complete system can be computed in two stages (Figure 6.5):

1. Determination of the kinematic interaction motion, involving the response to base acceleration of a system, which differs from the actual one in that the mass of the superstructure is equal to zero
2. Calculation of the inertial interaction effects, referring to the response of the complete soil–structure system to forces associated with base accelerations equal to the accelerations arising from the kinematic interaction

The second step is further divided into two subtasks:

1. Computation of the dynamic impedances at the foundation level; the dynamic impedance of a foundation represents the reaction forces acting under the foundation, when it is directly loaded by harmonic forces
2. Analysis of the dynamic response of the superstructure supported on the dynamic impedances and subjected to the kinematic motion, also called effective foundation input motion

For a more in-depth exposition of the theory, the reader is referred to Pecker (2007).

For rigid foundations, the dynamic impedances can be viewed as sets of frequency-dependent springs and dashpots lumped at the underside of the footing. For rigid shallow foundations with six degrees of freedom, the complex-valued impedance matrix in the most general situation is a full 6×6 matrix. However, for regular geometries, with two axes of symmetry and horizontally layered soil profiles, off-diagonal terms are equal to zero.

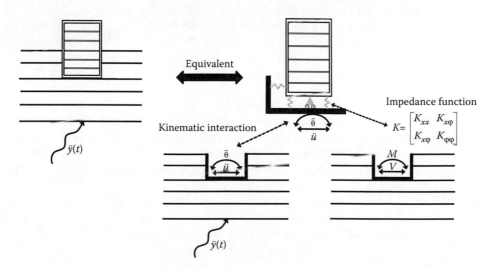

Figure 6.5 Substructuring approach for soil–structure interaction.

Analytical expressions for the different terms of the impedance matrix can be found in the literature for shallow footings of various geometries (Gazetas 1983, 1991). Each term is a complex number, which can be written as

$$k_{ij} = k_{ij}^s \left(k_{ij}^d + ia_0 c_{ij}^d \right) = k_{ij}^s \left(k_{ij}^d + i \frac{\omega B}{V_S} c_{ij}^d \right) \tag{6.1}$$

where
  k_{ij}^s is the static component of the impedance
  the terms in parenthesis correspond to the dynamic contribution to the impedance; k_{ij}^d
    and c_{ij}^d are frequency-dependent parameters
  $a_0 = \omega B / V_S$ is a dimensionless frequency
  ω is the circular frequency of excitation
  B is some characteristic dimension of the foundation (radius, width, etc.)
  V_S is the soil shear wave velocity
  $i^2 = -1$

The terms in Equation 6.1 have the following physical interpretation: $k_{ij}^s k_{ij}^d$ represents a spring and $(B/V_S)k_{ij}^s c_{ij}^d$ represents a dashpot. The equivalent damping ratio of the system is given by

$$\xi_{ij} = \frac{\omega B}{V_S} \frac{c_{ij}^d}{2k_{ij}^d} \tag{6.2}$$

Although the substructure approach described is rigorous for the treatment of linear SSI, its practical implementation is subject to several simplifications:

- Full linear behaviour of the system is assumed; it is well recognised that this assumption is a major one, since non-linearities occur in the soil and at the soil–foundation interface (sliding, uplift, etc.). Soil non-linearities can be partly accounted for, as recommended in Eurocode 8 – Part 5, by choosing for the calculation of the impedance matrix reduced soil properties that reflect the soil non-linear behaviour in the free field. This implicitly assumes that additional non-linearities taking place at the soil–foundation interface do not contribute significantly to the overall seismic response.
- Kinematic interaction is usually not considered; this means that the input motion used for the dynamic response of the structure is simply the free-field motion. Although that assumption is exact for shallow foundations subject to the vertical propagation of body waves, and partly exact for shallow foundations in a more complex seismic environment or for flexible piles, it becomes strongly inaccurate for embedded caissons or very stiff piled foundations, as discussed earlier.
- The frequency-dependent terms of the stiffness matrix are approximated by constant values. Except in a homogeneous soil profile (a rare case and a very restrictive assumption), this condition is far from being met. A fair approximation for the constant value of the stiffness term can be obtained by choosing a value corresponding to the frequency of the SSI mode.

Based on the concept of springs and dashpots to model the effect of the soil on the foundation, a wide class of, so-called, Winkler models has also been used extensively for modelling SSI. They consist of using distributed springs (and dashpots) across the interface of the footing and the soil, to represent their interaction. Although, conceptually, the soil reaction forces are still represented by the action of springs and dashpots, it must be realised that, unlike for the impedance matrix, there is no rational or scientifically sound method to define those springs and dashpots. Their values, but more importantly their distribution across the foundation, vary with frequency. Besides, there is no unique distribution that preserves the global foundation stiffness for all degrees of freedom. For instance, a uniform distribution may be chosen for the vertical stiffness, although this is highly questionable; however, in that case the rocking stiffness will not be correctly matched. Therefore, Winkler-type models, although attractive, should not be preferred, in view of all the uncertainties underlying the choice of their parameters.

6.2 VERIFICATION OF SHALLOW FOUNDATIONS

6.2.1 Three design approaches in EN 1990 and EC7

EN 1990 (CEN 2002) and Eurocode 7 provide three alternatives for the ultimate limit state (ULS) verification of the foundation in the 'persistent and transient design situation' (termed also 'fundamental combination' of actions). They are called design approaches (DAs) and are considered as nationally determined parameters (NDP); the choice is left to the national authorities through the National Annexes. The difference between them lies in the application:

1. Of partial factors on the action effects, γ_F, for the calculation of the design action effect, E_d, and
2. Of partial factors, γ_M, on the characteristic material properties, for the calculation of the design value of the resistance, R_d, or (as an alternative)
3. Of a global factor, γ_R, on the characteristic value of resistance, R_k, which is based on characteristic values of the material properties

Unlike what applies in the ULS verification of structural elements, where γ_F and γ_M (or γ_R) are applied simultaneously to the corresponding side of the verification inequality $E_d \le R_d$, in geotechnical verifications in general γ_F and γ_M take turns, applying only to one side at a time.

The differences between the three DAs reflect, to a certain extent, deep-rooted design traditions in different European countries, as well as a singular feature of geotechnical problems: that the soil often appears on both sides of the verification inequality (as, e.g. in the stability verifications of retaining structures); so, consistent factoring of its properties on both sides by a partial factor, γ_M, normally affects both in the same direction; therefore, it is not clear whether it is safe-sided or not.

The three design approaches for the case of shallow foundations are summarised in Table 6.1. This is done for completeness (as the foundation should be verified at the ULS, not only in the seismic design situation, but for the 'fundamental combination of actions as well) and for the additional reason that Part 5 of Eurocode 8 has implicitly made a choice between them for the seismic design situation (see footnote d in Table 6.1).

Table 6.1 Design approaches for ULS Verification of shallow foundations for the 'fundamental combination' of actions, and values of partial factors recommended in EN 1990 and EC7

| | Action partial factor, γ_F | | | | Resistance partial factors, γ_M | | | Resistance global factor, γ_R | |
| | Unfavourable | | Favourable | | Effective friction, $\tan\varphi'$ | Effective cohesion, c' | Undrained shear strength, c_u | Bearing capacity (Vertical) | Sliding (Horizontal) |
Design approach	Permanent G_k	Variable Q_k	Permanent G_k	Variable Q_k					
DA1[a] DA1-1	1.35[b], or 1.35ξ[c]	1.5[b], or 1.5ψ0[c]	1.0	0	1.0	1.0	1.0	–	–
DA1-2[d]	1.0	1.3	1.0	0	1.25	1.25	1.4	–	–
DA2	1.35[b], or 1.35ξ[c]	1.5[b], or 1.5ψ0[c]	1.0	0	–	–	–	1.4	1.1
DA3 On structural actions only	1.35[b] or 1.35ξ[c]	1.5[b], or 1.5ψ0[c]	1.0	0	1.25	1.25	1.4	–	–
Geotechnical actions[e]	1.0	1.3	1.0						

a   If DA1 is chosen, both DA1-1 and DA1-2 should be checked and verified.
b   These values apply, if Equation 6.10 in EN 1990 is chosen over Equations 6.10a and 6.10b
c   If Equations 6.10a and 6.10b in EN 1990 are chosen over Equation 6.10; the most unfavourable of $((1.35\xi)G_k +1.5Q_k$ or $1.35G_k +(1.5\psi_0)Q_k$ applies; $\xi = 0.85$ and – for buildings – $\psi_0 = 0.7$ (except in storage areas, where $\psi_0 = 1.0$).
d   This design approach is essentially chosen in Part 5 of EC8 for the seismic design situation, which does not apply partial factors, γ_F on actions G_k and $\psi_2 Q_k$.

6.2.2 Verifications in the 'seismic design situation'

The design verifications in general include verification of earthquake-induced settlements, sliding capacity and seismic bearing capacity. The sliding capacity and the bearing capacity of the foundation are verified for the forces acting on the foundation, which are computed as design action effects for the 'seismic design situation'. Theoretically, the foundation settlements are also a function of these design forces but, owing to the complexity of the analysis needed for their calculation, which would require full-fledged non-linear soil–structure modelling and analysis, settlements are usually estimated for free-field conditions.

The design forces acting on the foundation in the seismic design situation comprise the effects of quasi-permanent actions, of other potential actions concurrent with the seismic one and of the seismic action itself. The design action effects on the foundation shall be computed in accordance with the design of the superstructure. For non-dissipative structures, that is, those designed for a q-factor of 1.5, the action effects are the ones obtained from the analysis of the structure. For dissipative structures, capacity-design principles are in general applied instead, accounting for the potential overstrength. However, the so-computed seismic action effects need not exceed the action effects calculated, assuming elastic behaviour of the superstructure (i.e. for $q = 1.5$). Details are given in Section 6.3.2.

According to Eurocode 8 – Part 1, for dissipative structures the design values of the action effect on the foundation are given by

$$E_{Fd} = E_{F,G} + \gamma_{Rd} \, \Omega \, E_{F,E} \tag{6.3}$$

where γ_{Rd} is the overstrength factor, equal to 1.0 for a behaviour factor q less or equal to 3, and equal to 1.2 otherwise;

$\Omega = R_{di}/E_{di} \leq E_{di}$ for the dissipative zone or element i of the structure, which has the highest influence on the effect E_F under consideration; R_{di} is the design resistance of element i and E_{di} is the design value of the action effect on element i in the seismic design situation;

$E_{F,G}$ is the action effect of the quasi-permanent loads and $E_{F,E}$ is the effect of the seismic action.

6.2.3 Estimation and verification of settlements

The magnitude of settlements caused by the earthquake should be addressed when there are extended layers or thick lenses of loose, unsaturated cohesionless materials at shallow depths. Excessive settlements may also occur in very soft clays because of cyclic degradation of their shear strength under ground shaking of long duration, or in saturated sands upon dissipation of earthquake-induced pore water pressures. If the settlements caused by densification or cyclic degradation appear capable of affecting the stability of the foundations, ground improvement should be considered.

Earthquake-induced settlement in unsaturated sands can be estimated using empirical relationships between volumetric strain, SPT N-values (standard penetration test blow count values corrected for overburden) and cyclic shear strain. The peak shear strain computed from the one-dimensional response analysis at a point in the soil and the corresponding SPT corrected N-value are entered into the Tokimatsu and Seed chart (Figure 6.6) to estimate the volumetric strain. The total settlement can then be obtained by integrating these strains over depth.

For saturated layers, the post-liquefaction volumetric strain is a function of the safety factor against liquefaction and of the initial relative density of the layer. The smaller the safety factor, the larger is the volumetric strain. Ishihara (1993) has proposed the chart shown in Figure 6.7 where the initial density is related either to the SPT blow count or to the CPT

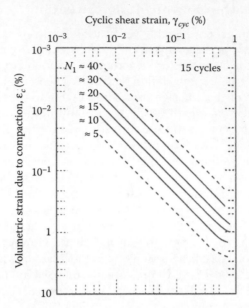

Figure 6.6 Volumetric strains under cyclic loading of unsaturated sands.

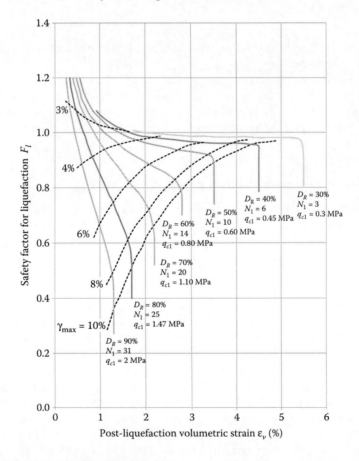

Figure 6.7 Post-liquefaction volumetric strains.

(static cone penetration test) point resistance (both quantities being normalised for overburden). The factor of safety against liquefaction is computed from a separate analysis under free-field conditions, that is, without the presence of the structure. The volumetric strains in liquefied saturated sands are an order of magnitude larger than the volumetric strains in dry sands for the same density and applied stress. Typical post-liquefaction volumetric strains reach values of several percent (typically 4%–6%), as soon as the safety factor drops below 1.1. If the safety factor is larger than 1.25, a typical value specified in seismic design codes, the volumetric strain is less than 1%.

6.2.4 Verification of sliding capacity

For the capacity of the foundation against sliding to be verified, the total horizontal driving force, V_{Ed}, calculated according to Section 6.2.2, should be less than the maximum resisting horizontal force. The horizontal resisting force is provided by friction under the base of the footing and, for embedded foundations, by friction along the lateral sides and soil reaction on the front face. Note that, although full friction on the base and the lateral sides of the foundation can be mobilised, the soil reaction on the front face of the foundation should be considered with caution. Eurocode 8 does not allow relying on more than 30% of the full passive resistance:

$$V_{Ed} \le F_{H1} + F_{H2} + 0.3F_B \tag{6.4}$$

where

F_{H1}: Friction under the base of the footing, equal to $N_{Ed}\tan(\delta)/\gamma_M$
F_{H2}: Friction over the lateral sides (for embedded foundations)
F_B: Ultimate passive resistance
N_{Ed}: Vertical design force acting on the foundation
δ: Friction angle between the foundation and the soil
γ_M: Partial factor for friction

The rationale for this limitation is that mobilisation of the full passive resistance requires a significant amount of displacement to take place, which is inconsistent with the usual performance goal for building foundations. Under certain circumstances, though, sliding may be accepted, because it is an effective means to dissipate energy. Furthermore, numerical simulations generally show that the amount of sliding is limited. For this situation to be acceptable, the ground characteristics should remain unaltered during seismic loading and sliding should not affect the functionality of lifelines. Since soil under the water table may be prone to pore pressure build-up, which will affect its shear strength, sliding should ideally only be tolerated when the foundation is located above the water table. The second condition simply recognises that buildings are not isolated facilities, but are connected to lifelines. So, the designer should make sure that displacements imposed by buildings on lifelines will not damage either the connection, or the lifelines themselves. For instance, during the Loma Prieta earthquake (1989), liquefaction in the Marina district caused severe lateral spreading, which did not really damage the buildings, but caused failure of the gas pipelines. It must be further pointed out that the prediction of the foundation displacements when sliding is allowed, strongly depends on the friction coefficient δ, which in turn depends on the surface material and the construction method. If reliable estimates are necessary, *in situ* tests are generally warranted.

6.2.5 Foundation uplift

Until the 1990s, it has been common practice to impose an upper limit on the eccentricity of the vertical load at the base of the footing with respect to its centre, with a value between one-sixth of the footing's parallel dimension – which is equivalent to saying that uplift is not allowed – and one-third – which means uplift over at most half the foundation plan. Nowadays, uplift seems to be more commonly tolerated. It is recognised that rocking of the foundation reduces the forces entering the structure and therefore protects it. However, rocking may be allowed only if the soil conditions are sufficiently good to avoid yielding of the soil under the loaded edge, which may produce permanent settlement and tilting of the foundation. To evaluate this aspect of the behaviour, it is recommended to carry out non-linear static (pushover) analysis, considering material non-linearity of the soil (soil yielding) and geometric non-linearity (uplift of the footing). The analysis results of prime interest are the moment–rotation characteristics of spread footings reflecting the non-linear effects. These results not only provide rotational stiffness parameters, but also depict the geotechnical mode which gives the ultimate moment capacity. Note that the foundation cannot develop overturning moments that are higher than the ultimate moment capacity. In spread footings, the geometric non-linearity (uplift) is the most important source of non-linearity.

6.2.6 Bearing capacity of the foundation

In addition to the verification of sliding, the seismic bearing capacity of the foundation shall be checked, taking into consideration the inclination and eccentricity of the force acting on the ground, as well as the effects of the inertia forces developed in the soil medium by the passage of the seismic waves. A general expression is provided in Annex F of Part 5 of Eurocode 8, derived from theoretical limit analyses of a strip footing (Paolucci and Pecker 1997; Pecker 1997; Salençon and Pecker 1995a,b).

The verification condition of the foundation against bearing capacity failure simply expresses that the design forces N_{Ed} (design vertical force), V_{Ed} (design horizontal force), M_{Ed} (design overturning moment) and the soil seismic forces should lie within the surface depicted in Figure 6.8. The analytical expression of the surface is

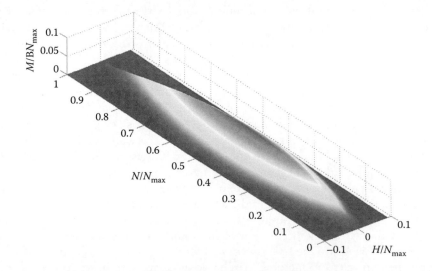

Figure 6.8 Surface of ultimate loads for the foundation bearing capacity.

$$\frac{\left(1 - e\overline{F}\right)^{c_T} \left(\beta\overline{V}\right)^{c_T}}{\left(\overline{N}\right)^{a}\left[\left(1 - m\overline{F}^{k}\right)^{k'} - \overline{N}\right]^{b}} + \frac{\left(1 - f\overline{F}\right)^{c'_M} \left(\gamma\overline{M}\right)^{c_M}}{\left(\overline{N}\right)^{c}\left[\left(1 - m\overline{F}^{k}\right)^{k'} - \overline{N}\right]^{d}} - 1 = 0 \qquad (6.5)$$

where the coefficients in lower case (a, b, etc.) are numerical values that depend on the soil type according to Annex F of Eurocode 8 – Part 5 (see Table 6.2), and

$$\overline{N} = \frac{\gamma_{Rd}N_{Ed}}{N_{max}}, \quad \overline{V} = \frac{\gamma_{Rd}V_{Ed}}{N_{max}}, \quad \overline{M} = \frac{\gamma_{Rd}M_{Ed}}{BN_{max}}, \quad \overline{F} = \left\{ \begin{array}{c} \dfrac{\gamma_{Rd}\rho a B}{C_u} \\ \dfrac{\gamma_{Rd}a}{g\tan\phi} \end{array} \right\} \qquad (6.6)$$

where

N_{max} is the ultimate concentric bearing capacity of the foundation under a vertical load; it may be estimated with any reliable method (strength parameters according to Eurocode 7, empirical correlations from field tests, etc.).

B is the foundation width.

$\overline{F}$ is the dimensionless soil inertia force.

$a = Sa_g$ is the peak acceleration at the top of the ground.

γ_{Rd} is the model factor.

ρ is the soil mass density.

C_u is the soil undrained shear strength, for cohesive soils.

φ is the soil friction angle, for cohesionless soils.

Note that a is meant to be the peak acceleration at the top of the ground accompanying the maximum force acting on the foundation; Sa_g is a safe-sided convenient estimate of it.

Recent studies have shown that the same expression is still valid for a circular footing, provided that the ultimate vertical force under a vertical concentric load, N_{max}, entering Equation 6.5 is computed for a circular footing and the footing width is replaced by the footing diameter in Equation 6.6 when the dimensionless moment $\overline{M}$ is computed and by the footing radius when the dimensionless force $\overline{F}$ is determined (Chatzigogos et al. 2007).

Although Equation 6.5 does not look familiar to geotechnical engineers accustomed to the 'classical' bearing capacity formulae with correction factors for load inclination and eccentricity, it reflects the same aspect of foundation behaviour. It is similar to the interaction diagrams commonly used in structural engineering for the design or the verification of a concrete section under combined axial force and bending moment.

The model factor γ_{Rd} is introduced to reflect the uncertainties in the theoretical model; as such, it should be larger than 1.0. Nevertheless, it also reflects that a certain (limited) magnitude of permanent foundation displacement may be tolerated, in cases when the

Table 6.2 Parameters in the foundation bearing capacity expression Equation 6.5

Soil	a	b	c	d	e	f	m	k	k'	c_T	c_M	c'_M	β	γ
Cohesive	0.70	1.29	2.14	1.81	0.21	0.44	0.21	1.22	1	2	2	1	2.57	1.85
Cohesionless	0.92	1.25	0.92	1.25	0.41	0.32	0.96	1.0	0.39	1.14	1.01	1.01	2.90	2.80

Table 6.3 Model partial factor γ_{rd} for Equation 6.5

Soil	γ_{Rd}
Medium dense to dense sand	1.00
Loose dry sand	1.15
Loose saturated sand	1.50
Non-sensitive clay	1.00
Sensitive clay	1.15

point representing the forces acting on the foundation lies outside the surface described by Equation 6.5. To account for this aspect, γ_{Rd} may be less than 1.0. The values in Annex F of Part 5 of Eurocode 8 intend to combine both effects; as shown in Table 6.3, for the most sensitive soils (loose saturated sands) the model factor is higher than for stable ones (medium dense sand).

6.3 DESIGN OF CONCRETE ELEMENTS IN SHALLOW FOUNDATIONS

6.3.1 Shallow foundation systems in earthquake-resistant buildings

The defining role of the foundation is to transfer the gravity loads from the (vertical members of the) structure to the ground. From that viewpoint, the natural choice for the foundation of a concrete column is to widen its base to adapt the section area, through which the vertical load passes, to the ground bearing capacity – lower than that of the concrete section. Each one of the resulting footings – normally concentric and square – is often connected to neighbouring ones via horizontal tie-beams with rectangular section (see Figure 4.11a). As pointed out in Section 4.4.5, the main role of tie-beams is to reduce the magnitude and impact of differential settlements of adjacent footings, due to large imbalances between their vertical loads and/or variations in the underlying soil. If a more interconnected foundation is essential, instead of placing a number of tie-beams and isolated footings in a row, a foundation beam is used: a deep beam with an inverted-T or L section, which transfers vertical loads through the underside of its bottom flange to the ground all along its length, not just around the column base (see Figures 4.11b and 6.9a). If the overall weight of the building and its contents is so large that the soil bearing capacity needs to be mobilised over most of its footprint area, it is normally more cost-effective to combine all foundation elements into a raft, which acts as a single footing under the entire building, transferring vertical loads to the ground throughout its plan area (Figure 6.9b). We thus have the following three types of shallow foundation systems for buildings, listed in ascending order of cost and effectiveness in transferring the gravity loads to the soil:

1. Isolated footings (or 'pads'), with or without tie-beams
2. Two-way foundation beams
3. Foundation rafts

In modern construction, these foundation systems are always made of concrete, even when the superstructure is built of another material.

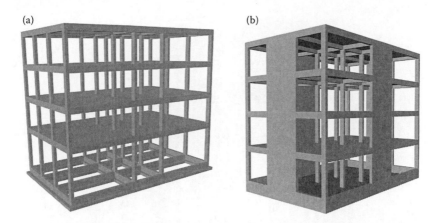

Figure 6.9 Foundation systems: (a) two-way foundation beams in frame building; (b) raft foundation for the columns and perimeter foundation beams.

If the building is subjected to significant lateral loads, notably seismic, the foundation elements should be designed to transfer to the ground the large bending moments that develop at the base of each vertical element, as well as the action effects of the overall overturning moment due to the lateral loads, notably the uplifting of the windward side of the building and the additional vertical compression of the leeward one. In foundation systems of type 1, the vertical force that a footing is called to transfer to the ground acts at a large eccentricity (ratio of moment to vertical force) with respect to the centre of the footing's underside, especially when the vertical force due to the overall overturning moment is tensile. To accommodate the eccentricity, footings may have to be oversized in plan and/or connected via stiff tie-beams, which work in counter-flexure to reduce the moment transferred to the ground by the underside of the footing (Figure 6.10). It goes without saying that, to take into account in the design the role of tie-beams against the eccentricities of footings and the rotations they cause, as well as to dimension the tie-beams for their seismic internal forces, the seismic analysis model should include both the tie-beams and the rotational compliance of the soil under the footing (usually by means of one rotational spring connected to the centre of the footing's underside).

Tie-beams have another important role in design against lateral loads, to prevent differential horizontal slippage of footings, a role, though, which Eurocode 8 allows to delegate to a horizontal slab between them, not integral to them and not included in the analysis model (Figure 6.11). The soffit of tie-beams or a slab connecting different footings should be below the top of the footings, to avoid creating a squat concrete member, which is very vulnerable in shear.

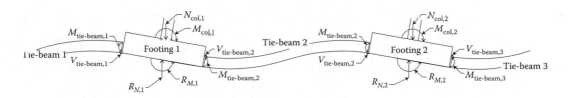

Figure 6.10 Tie-beams reduce eccentricity, R_M/R_N, of the vertical soil reaction on footings, R_N, under seismic actions.

Figure 6.11 Replacement of tie-beams by horizontal slab between footings.

For the reasons mentioned, isolated footings may not be cost-effective for a building in seismic regions, be it on competent ground. If the tie-beams have to be very deep and stiff to alleviate the eccentricity problem of footings, they may well be merged with them into a foundation beam (see Figures 4.11b and 6.9b). That beam works with its overall length as a single integral member, absorbing into its own bending moment diagram the bending moments applied at its top flange by the columns and efficiently transferring to the ground the overall overturning moment applied to the beam by the columns it supports thanks to its long overall length, with very limited uplift (if any) of its windward end. So, for high seismicity, foundations of type 2 are the system of choice for buildings, especially tall ones, almost regardless of the competence of the subsoil. If such a system is used, it is strongly recommended (but not essential) to include in the analysis model the vertical compliance of the soil: either by modelling the part of a foundation beam between adjacent joints with the supported columns as a single 'Beam-on-elastic-foundation' element, having the subgrade reaction modulus, k_s, as one of its properties; or by splitting each stretch of the foundation beam between adjacent columns into a number of prismatic beam sub-elements separated by intermediate nodes, each node supported on the soil through a vertical Winkler spring with a vertical stiffness of $k_s(b\Delta x)$, where b is the width of the bottom flange of the foundation beam and Δx is the node's tributary length along the beam. This is, for instance, how the compliance of the soil is modelled along the basement-deep foundation beams along the perimeter of the example building in Chapter 7 (see Section 7.2.3).

Unlike two-way foundation beams, a foundation raft (Figure 6.9b) does not offer additional advantages for seismic design. Moreover, its analysis and design are demanding even for gravity loads: the raft should be discretised with a fairly fine mesh of plate finite elements, each node being supported on the soil through a vertical Winkler spring of vertical stiffness $k_s\Delta A$, where ΔA is the node's tributary plan area. So, there is no special reason to choose a raft over a two-way system of foundation beams for the purposes of earthquake resistance.

The ideal foundation system for earthquake resistant buildings – especially tall ones – is a box extending throughout the building footprint area and comprising a wall all around the perimeter, working as a deep foundation beam, plus two rigid horizontal diaphragms: one at the top level of the perimeter wall and another at the bottom. Naturally, such a system is more suitable and convenient for buildings with basement(s), where the cover slab of the (upper) basement serves as top diaphragm (see Section 7.1 for the example building in Chapter 7); the bottom diaphragm may consist of a raft, if one has been chosen for the transfer of gravity loads to the ground, or of a two-way system of tie-beams or foundation beams connecting the bases of interior vertical elements among themselves and to the bottom of the wall-cum-foundation beam around the perimeter. The role of a rigid diaphragm at the bottom can be played even by a non-integral horizontal slab among the interior isolated footings and between them and the bottom of the perimeter wall, which may be used in lieu of the two-way tie-beams (cf. Figure 6.11 for the building of Figure 4.11a).

Such a box foundation system transfers the full seismic base shear and overturning moment to the ground through the perimeter wall-cum-foundation beam and its strip footing. Moreover, it ensures that the base sections of all vertical elements, interior ones or on the perimeter, rotate the same in a vertical plane parallel to the horizontal component of the earthquake; therefore, these elements may be taken as fixed at the top level of the box foundation, developing their plastic hinges just above that level. Note that, thanks to the shear rigidity of the perimeter wall-cum-foundation beam, the horizontal drift between the top and bottom of the box is negligible, protecting the stretch of interior vertical members within the depth of the box foundation system from large flexural deformations. Therefore, Eurocode 8 allows taking that stretch as remaining elastic during the earthquake and to design/detail it as such. With the exception of the seismic shear in that stretch of an interior wall, which is exceptionally high due to the large bending moment of the wall at the top of the box, the seismic moments and shears in the stretch of interior elements within the depth of the box and at their connection to their own foundation element are low; if footings are used for the interior elements, they are not penalised by a large eccentricity of the vertical force and do not need stiff tie-beams against it.

Examples 4.8 and 4.9 at the end of Chapter 4 raised and addressed issues of conceptual design of the foundation. Example 6.1, at the end of this chapter, also deals with similar issues.

6.3.2 Capacity design of foundations

The foundation is of prime importance for the stability of the structure as a whole. It is also hard to access for inspection and even harder to repair after an earthquake. Moreover, it is technically impossible to reverse settlements or other soil deformations due to an earthquake. Last but not least, the uncertainty concerning the properties of the soil and its likely response and behaviour in case of an earthquake is greater than that for the superstructure and the materials it is made of, no matter how extensive the soil investigation is. So, we have every reason to be more cautious and conservative in the seismic design of the foundation. As pointed out in Section 6.2, one way of doing this is by capacity-designing the foundation system and the underlying soil to remain elastic, till and after a plastic mechanism develops in the superstructure.

In the simplest case of an isolated footing under a vertical element – column or wall – in a building designed to Eurocode 8 for ductility (i.e. of DC M or H), it is checked that the footing and the soil underneath do not reach their design resistance even when the seismic action exhausts the moment resistance of the vertical element, M_{Rd}, at its connection to the footing. Note that, once the moment acting at that section reaches M_{Rd}, a plastic hinge forms there. Large inelastic rotations may develop in that hinge at very little further increase of the acting moment (and at any rate, below the 20% margin provided by Eurocode 8). The plastic

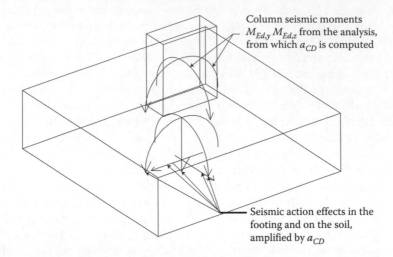

Column seismic moments
$M_{Ed,y}$ $M_{Ed,z}$ from the analysis,
from which a_{CD} is computed

Seismic action effects in the
footing and on the soil,
amplified by a_{CD}

Figure 6.12 Capacity design of isolated footing.

hinge acts as a fuse, putting a ceiling on the magnitude of forces which can be transmitted through the footing from the vertical element to the ground or vice versa. So, the footing and the soil will not fail, if they can safely resist that ceiling, indeed increased in general by the 20% margin imposed by Eurocode 8.

The capacity design of the footing highlighted earlier is implemented as follows (Figure 6.12): A capacity-design magnification factor is computed at the section of the vertical element where it is connected to the footing (cf. Equation 6.3 in Section 6.2, where the more general symbol Ω is used):

$$a_{CD} = \gamma_{Rd} M_{Rd} / M_{Ed} \leq q \qquad (6.7)$$

where

γ_{Rd} is equal to 1.0 if q is less or equal to 3.0 (as in wall systems of DC M), or to 1.2 otherwise

M_{Rd} is the design value of moment resistance of the element's section, computed for the value of the element's axial force in the seismic design situation

M_{Ed} is the bending moment at the element section in the seismic design situation

q is the value of the behaviour factor used in the linear analysis for the design seismic action (its use as an upper limit to a_{CD} corresponds to elastic response to the design seismic action).

The factor a_{CD} from Equation 6.7 multiplies all seismic action effects in the footing and the soil from the linear analysis for the design seismic action (see Equation 6.3 in Section 6.2, where the more general symbol Ω is used in lieu of a_{CD}), namely:

- The force reactions (the vertical and the two horizontal ones) and the moment reactions (the two reacting moments in the vertical planes parallel to the footing's sides) at the centre of the underside of the footing
- The internal forces (vertical shear force and bending moment in a vertical plane) of tie-beams at their end section at the face of the footing

The so amplified action effects in the elastic footing and the surrounding soil cannot be exceeded in the seismic design situation, because they are controlled by the moment resistance of the plastic hinge at the base of the vertical element.

Equation 6.7 is applied separately to each case considered in the linear analysis for the seismic design situation (see Sections 5.8.1 and 7.6.2.5); its outcome multiplies the analysis results of that case.

One further complication comes from the fact that every seismic analysis in 3D generally gives bending moment components at the base section of the vertical element, $M_{Ed,y}$ and $M_{Ed,z}$, about both local axes y and z of the section, respectively. For simplicity, we always work in the direction of these two axes, not in the most adverse oblique direction between y and z. So, it is natural to presume that the plastic hinge will form in the direction where M_{Ed} is numerically closer to M_{Rd}. Therefore, in the general case:

$$a_{CD} = \min[a_{CDy}; a_{CDz}], a_{CDy} = \min[\gamma_{R,dy} M_{Rd,y} / M_{Ed,y}; q_y], a_{CDz} = \min[\gamma_{Rd,z} M_{Rd,z} / M_{Ed,z}; q_z]$$

$$(6.7a)$$

The design values of moment resistance, $M_{Rd,y}$ and $M_{Rd,z}$, are computed for the value of the axial force in the particular case of the seismic design situation considered. In principle, they should be determined not for uniaxial bending of the section, but for biaxial, with a ratio of resisting moment components, $M_{Rd,y}/M_{Rd,z}$, taken equal to that of the acting ones, $M_{Ed,y}/M_{Ed,z}$. However, normally one of the two ratios in Equation 6.7a governs clearly and it is safe-sided to neglect the biaxiality effect due to the other moment component.

Note that Equations 6.7 or 6.7a are applied at one location, namely at the section of the vertical element where it is connected to the footing, but its outcome multiplies the seismic action effects elsewhere, notably at the interfaces of the same footing with the ground and the tie-beams (Figure 6.12). The internal forces in the footing itself are computed from equilibrium with the effects of quasi-permanent loads and the amplified-by-a_{CD} seismic ones acting on the boundary of the footing.

Example 6.2 at the end of this chapter illustrates the calculation of the capacity-design amplification factor of Equation 6.7a in an isolated footing.

This procedure cannot be applied if a single foundation element (namely a foundation beam or a system thereof, a spread footing, a raft, a box-type foundation system) supports more than one ('primary') vertical element. In such cases, in buildings designed for ductility (i.e. of DC M or H), Eurocode 8 permits to use a universal value of

$$a_{CD} = 1.4 \tag{6.8}$$

which multiplies the values of all seismic action effects (internal ones and on the boundary) in the jointly used foundation element obtained from the linear analysis for the seismic action. This amounts to designing the foundation element and the supporting soil for a q-factor reduced by 40% relative to that used for the superstructure. At any rate, it is a major simplification of the seismic design of the foundation elements and the underlying soil.

The overdesign of foundation elements through the given capacity-design procedure protects them from plastic hinging and allows them to stay elastic after a plastic mechanism develops in the superstructure – theoretically forever. Therefore, it is not necessary to detail them for ductility or to protect them from pre-emptive brittle failure through capacity design in shear. Indeed, Eurocode 8 allows to dimension them in shear for the action effects derived from the linear analysis and to apply to them the detailing rules of Eurocode 2, in lieu of those of Eurocode 8. In fact, this is the only practical option for footings or rafts (both being slab-type elements) or deep foundation beams, as Eurocode 8 does not cover detailing of such members for ductility, nor does capacity design in shear work in them.

In buildings designed for DC M or H, Eurocode 8 allows to apply the capacity-design magnification factor of Equations 6.7a and 6.8 only in the verification of the ground and to use the unmagnified seismic action effects from the linear analysis to dimension or verify the foundation elements themselves, provided that they are (capacity-) designed and detailed for ductility. Strictly speaking, this exception can only be applied to tie- or foundation beams, alongside the rules for detailing and capacity design in shear provided in Eurocode 8 for beams of the corresponding ductility class. Practically speaking, it can be applied to advantage mainly in small- to moderately sized tie- or foundation beams. The minimum reinforcement of large ones for crack-control as per Eurocode 2 gives them sufficient resistance to sustain the seismic action effects from the linear analysis magnified by 40% as per Equation 6.8.

Another simplification allowed by Eurocode 8 is to use a q-factor of 1.5 for the design of the foundation and the verification of the soil. This makes sense from the economic point of view only for buildings whose superstructure is designed for ductility with a q-factor less than 1.5 times the outcome of Equations 6.7a or 6.8, which is indeed a very rare case.

6.3.3 Design of concrete foundation elements: Scope

Every foundation element is first dimensioned in plan, by verifying that, with the chosen plan dimensions, the ground has a design value of bearing capacity which is sufficient against the following system of forces, depicted in Figure 6.13, alongside the notation used here for the geometric parameters and the action effects in the column:

- The vertical reaction, R_N, including the weight, W_f, of the foundation element and the overlying soil; R_N may be eccentric with respect to the centre of the horizontal interface of the foundation element and the ground; the eccentricity is reflected in the biaxial reacting moments, R_{Mx}, R_{My}, with respect to two orthogonal horizontal axes of symmetry of the footing's base, y and x, or, often, via the corresponding biaxial eccentricities:

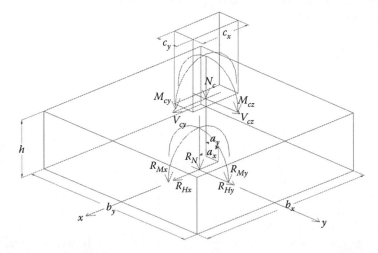

Figure 6.13 Notation for footing geometry, action effects in the column and reactions acting on the soil at the centre of the footing's base.

$$e_x = R_{Mx}/R_N \tag{6.9a}$$

$$e_y = R_{My}/R_N \tag{6.9b}$$

Note that for consistency with the eccentricities e_x (measured along the x-axis) and e_y (measured along the y-axis), the moments R_{Mx}, R_{My} are indexed as per the vertical plane of bending and the rotation they cause, and not according to the direction of their vector.

- Horizontal reactions R_{Hx}, R_{Hy}, along the two axes of symmetry x and y, often expressed in terms of the inclinations to the horizontal of the force resultant in the vertical plane through the x-axis: R_N/R_{Hx}, and in the vertical plane through the y-axis: R_N/R_{Hy}.

The verification of the bearing capacity of the ground may take place according to the more advanced method in Annex F of Part 5 of Eurocode 8, taking into account the magnitude and impact of the stresses in the soil itself due to the seismic waves and the associated shear strains (see Section 6.2, Equations 6.5 and 6.6). If these stresses and strains are low, a simplified static verification may be carried out (i.e. one where the effect of the acceleration on the soil is neglected, as is done in the pseudo-static verifications in Part 5 of Eurocode 8). More specifically, according to Informative Annex D of Part 1 of Eurocode 7:

- A rectangular horizontal interface of the foundation element and the ground, having dimensions b_x and b_y along the x- or y-axis, respectively, and acted upon by a vertical reaction R_N with corresponding eccentricities e_x, e_y, is verified, taking R_N as concentric on an effective loaded area with sides $(b_x - 2e_x)$ and $(b_y - 2e_y)$.
- The uniform pressure $R_N/(b_x - 2e_x)(b_y - 2e_y)$ is compared then to the soil bearing capacity, itself a function of R_{Hx}, R_{Hy}, $(b_x - 2e_x)$ and $(b_y - 2e_y)$.

The depth and any other dimensions of the foundation element in elevation, and its reinforcement, are all dimensioned so that they satisfy the ULS of the foundation element itself, considered as a RC element in bending, shear or punching shear, as relevant (see Sections 6.3.5 to 6.3.8).

6.3.4 Distribution of soil pressures for the ULS design of concrete foundation elements

For the structural verification of the foundation element itself at the ULS, Part 1 of Eurocode 7 allows to take the soil pressure at a point of its interface with the ground proportional to the local settlement of the ground at that point. Presuming that the size of the foundation element in plan is such that the soil bearing capacity is verified, and indeed with the large partial safety factors typically involved, the soil under the footing is reasonably expected to be everywhere on the elastic branch of its σ–ε law. Hence, the elastic soil hypothesis is employed for the structural design of the foundation elements. This hypothesis allows to use in linear analysis the subgrade reaction modulus, k_s, to model soil impedance under a foundation beam or a raft at a point-by-point basis. It further allows to dimension them as a concrete beam or slab using directly the moment and shear diagrams from the linear analysis (with their parts due to the seismic action multiplied by $a_{CD} = 1.4$). There is not much more to say concerning structural design of this type of foundation elements at the ULS in the seismic design situation.

Footings are commonly chosen deep enough to be considered as rigid relative to the soil underneath. Unlike flexible footings, which should be modelled like small rafts, with plate finite elements on a bed of Winkler springs, a rigid footing may have the soil impedance lumped into three springs connected to the centroid of its underside: a rotational spring in each one of the two orthogonal vertical planes parallel to horizontal axes x and y and the sides b_x and b_y of the footing and a vertical spring. Note that shallow foundations are normally modelled without horizontal springs between the soil and foundation elements; instead, the nodes at the underside of these elements are fully restrained (fixed) in both horizontal directions. For the seismic overturning moment at the level of the foundation to translate into a non-uniform distribution of vertical reactions (soil pressures), such restraints should be at the same horizontal level throughout the building plan (see also Example 6.1); if they are not, the analysis produces fictitious horizontal reactions at the restrained nodes to equilibrate the seismic overturning moment through the couples produced by the horizontal reactions.

To reflect the footing's rigidity, the nodes on its boundary where it is connected to the centroidal axes of the vertical element and the tie-beams framing into the footing are connected to the node at the centroid of its underside through rigid links. Moreover, since a rigid footing settles and rotates like a rigid block, the settlement of its underside is linear in x and y; the dependence of soil pressure with x and y follows suit, thanks to the presumed proportionality of the soil pressure to the local settlement. The linear – in x and y – variation of soil pressures under the footing is the basis for the calculation of internal forces in the footing for its own ULS design according to Sections 6.3.5 to 6.3.7.

A rigorous criterion for a footing to be considered as rigid relative to the soil may be formulated by considering it as a beam-on-elastic-foundation (e.g. an 'elastic length' less than 1.0). Instead, a practical rule-of-thumb is in common use, namely, to consider the footing as rigid if it does not protrude from the vertical element in any horizontal direction by more than twice its depth.

In a foundation element taken as rigid relative to the soil (normally a footing, but possibly a foundation beam or raft) and having a rectangular in plan contact area with sides b_x and b_y, the soil pressure $p(x,y)$ (positive for compression) varies linearly with x and y:

1. Suppose first that the biaxial eccentricities of the vertical reaction R_N (including the weight, W_f, of the foundation element and its soil cover), e_x and e_y, along the horizontal axes of symmetry x and y, respectively, fall into the kernel of the $b_x \times b_y$ plan area, by satisfying the condition:

$$\frac{|e_x|}{b_x} + \frac{|e_y|}{b_y} \leq \frac{1}{6} \tag{6.10}$$

Then, the distribution of soil pressure is

$$p(x,y) = \frac{R_N}{b_x b_y}\left(1 + 12\frac{e_x}{b_x}\frac{x}{b_x} + 12\frac{e_y}{b_y}\frac{y}{b_y}\right) \geq 0 \tag{6.11}$$

2. If Equation 6.10 is not met, uplift from the ground takes place over a part of the plan area. The soil pressure is zero over that part, but is still linear in x and y over the contact area; it is given by a set of complicated expressions, which depend on the part of the plan area where the eccentricities e_x and e_y fall. For the sake of simplicity, these

expressions are not presented here. They are behind the expressions in Sections 6.3.5 to 6.3.7 for the action effects in the footing, which apply whenever Equation 6.10 is not met.

Eurocodes 7 or 8 do not set a limit to the magnitude of the eccentricity. However, uplift over a major part of the plan area may lead to large non-linear rotations of the foundation element, even on elastic ground. Such non-linearities are not accounted for in linear analysis and may cast doubt on the accuracy of its results. For this reason, care is exercised in ordinary design to keep the eccentricity low, for example, less than 20% to 30% of the corresponding size of the foundation element.

6.3.5 Verification of footings in shear

Footings are slabs. As such, they may be constructed without shear reinforcement (shear links through the footing depth), unless their acting design shear, V_{Ed}, exceeds the design shear resistance as per Eurocode 2 for members without shear reinforcement, $V_{Rd,c}$. In kN, $V_{Rd,c}$ is

$$V_{Rd,c} = \left\{ \max\left[\frac{180}{\gamma_c}(100\rho_1)^{1/3};\; 35\sqrt{1 + \sqrt{\frac{0.2}{d}}}\; f_{ck}^{1/6} \right]\left(1 + \sqrt{\frac{0.2}{d}} \right)f_{ck}^{1/3} \right\} b_w d = v_{Rd,c} b_w d \quad (6.12)$$

(cf. Equation 5.61) where
   d is the effective depth of the footing (m)
   b_w is the width of the web at the control section (m)
   γ_c is the partial factor for concrete
   f_{ck} is the concrete strength grade (MPa)
   ρ_1 is the ratio of the tension reinforcement that extends past the control section by at least d plus the bar anchorage
   $v_{Rd,c} = V_{Rd,c}/(b_w d)$ is the design value of the vertical shear stress resistance per unit area of the control section (MPa or N/mm²), for no shear reinforcement

A footing works as a two-way, inverted, double cantilever. Therefore, its critical sections in shear are the vertical ones through the depth of the footing at the face(s) of the vertical element. According to Eurocode 2, loads (presently the soil pressure) applied at one side of the member (in this case the underside) up to a distance d from the critical section and supported by a reaction at the opposite side (in the present case, by the base section of the vertical element) do not require shear reinforcement. Such reinforcement may be determined from the loads acting beyond a distance d from the face of the vertical element: the control sections for the footing's shear reinforcement are the vertical ones at distance d from the face of the vertical element. In general, there are two such sections which are normal to the x-axis; their acting and resisting shears are indexed by x; the corresponding shears at the two control sections normal to the y-axis are indexed by y. One of the two sections in each pair is closer to the side of the footing where the peak soil pressure is applied; the other is further away; a prime is used to distinguish this latter section from the former, as both have the same index. So, the two control sections normal to the x-axis are at distances from the centre of the footing in plan (Figure 6.14a):

$$s_{vx} = \min(b_x/2;\; a_x + 0.5c_x + d) \quad (6.13a)$$

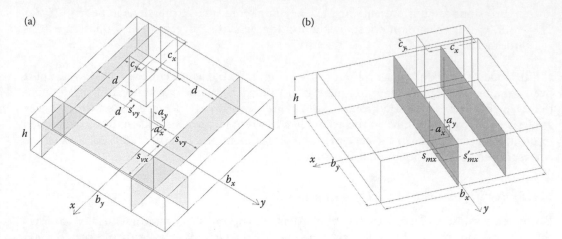

Figure 6.14 Location in plan of control sections: (a) for the dimensioning of the footing in shear, as RC element without shear reinforcement; (b) for the dimensioning of the footing reinforcement parallel to horizontal axis *x*.

$$s'_{vx} = \min(-b_x/2;\ a_x - 0.5c_x - d) \qquad (6.13b)$$

where

a_x is the x-eccentricity of centroidal axis of the vertical element with respect to the centre of the footing in plan, taken positive if it lies on the same side with respect to the centre as the eccentricity e_x and the peak soil pressure ($a_x = 0$ in a concentric footing).

c_x is the cross-sectional size of the vertical element along the x-direction of the footing.

The upper limit of $b_x/2$ in Equation 6.13a and the lower one of $-b_x/2$ in Equation 6.13b are activated if the vertical element is closer than d from the nearest edge of the footing.

The acting shear forces at the two control sections, at distances s_{vx} and s'_{vx}, respectively, from the centre of the footing in plan, are the resultants of all vertical forces or distributed loads acting on the body of the footing between the control section and the footing edge parallel to it:

- If Equation 6.10 and case 1 in Section 6.3.4 apply (i.e. for full contact with the ground), then

$$V_{Ed,x} = \left[\left(1 + \frac{3\,|e_x|}{b_x}\left(1 + \frac{2s_{vx}}{b_x}\right)\right)R_N - W_f\right]\left(\frac{1}{2} - \frac{s_{vx}}{b_x}\right) + \sum_{x=s_{vx}}^{x=b_x/2} V_{\text{tie-beam}} \qquad (6.14a)$$

$$V'_{Ed,x} = \left[\left(1 - \frac{3\,|e_x|}{b_x}\left(1 - \frac{2s'_{vx}}{b_x}\right)\right)R_N - W_f\right]\left(\frac{1}{2} + \frac{s'_{vx}}{b_x}\right) + \sum_{x=s'_{vx}}^{x=-b_x/2} V_{\text{tie-beam}} \qquad (6.14b)$$

- If Equation 6.10 is not met, and case 2 of Section 6.3.4 applies instead (loss of contact over part of the plan area), then

$$V_{Ed,x} = \left[\frac{\left(2.5 - (6\,|e_x|\,/b_x) + (s_{vx}/b_x)\right)}{\left((1/2) - (|e_x|\,/b_x)\right)^2} \frac{R_N}{9} - W_f \right]\left(\frac{1}{2} - \frac{s_{vx}}{b_x} \right) + \sum_{x=s_{vx}}^{x=b_x/2} V_{\text{tie-beam}} \qquad (6.15a)$$

$$V'_{Ed,x} = \frac{R_N}{9}\left(\frac{\max\left(0;\, 1 - (3\,|e_x|\,/b_x) + (s'_{vx}/b_x)\right)}{(1/2) - (|e_x|\,/b_x)} \right)^2$$

$$- W_f\left(\frac{1}{2} + \frac{s'_{vx}}{b_x} \right) + \sum_{x=s'_{vx}}^{x=-b_x/2} V_{\text{tie-beam}} \qquad (6.15b)$$

Figures 6.15a and 6.15b depict the forces and pressures from which the counterparts of Equations 6.14 and 6.15, respectively, are derived, for the verification in shear at the two control sections normal to the y-axis (replacing x in Equations 6.13 to 6.16 by y and vice versa). W_f in Equations 6.14 and 6.15 and Figure 6.15 is the weight of the foundation element and its soil cover (included also in R_N). The sum in each last term comprises the shears of all tie-beams which frame into the footing between the control section and the footing edge parallel to it; tie-beams parallel to either x or y are included; these shears are taken as positive, if they act upwards on the footing (as they are expected to do in the tie-beams connected to the footing on its sides where the peak soil pressure is exerted, i.e. on the same side as the eccentricity e_x); if they act downwards, they should be taken as negative. For a footing rotating in the sense in which the moment $R_{Mx} = e_x R_N$ acts on the ground, these shears are expected to act upwards and be positive in Equations 6.14a and 6.15a and downwards, that is, negative, in Equations 6.14b and 6.15b. Strictly speaking, if i indexes the end of a tie-beam at the footing in question and j indexes the opposite end, its shear cannot exceed the capacity values: $(M^+_{Rb,d,i} + M^-_{Rb,d,j})/l_n$ for an upwards acting shear and $(M^-_{Rb,d,i} + M^+_{Rb,d,j})/l_n$ for one acting downwards on the footing.

The sections normal to the x-axis do not require shear reinforcement, if, in every single case of the seismic design situation (in general, in all the cases of Section 5.8.1), the following condition is met:

$$\max(|V_{Ed,x}|;\, |V'_{Ed,x}|) \le V_{Rd,cx} \qquad (6.16)$$

where $V_{Rd,cx}$ is obtained from Equation 6.12, with b_w taken equal to b_y and using as ρ_1 the ratio of the bottom reinforcement in the x-direction (typically the minimum reinforcement as per Eurocode 2 in the main direction of a slab). Note that $V'_{Ed,x}$ very rarely governs in Equation 6.16: unless the footing is strongly eccentric in the x-direction and unless s_{vx} is equal to $b_x/2$, it is not worth computing $V'_{Ed,x}$ from Equations 6.14b and 6.15b at all.

Normally, a footing which is thick enough to be considered as rigid relative to the soil (cf. third paragraph of Section 6.3.4) does not need shear reinforcement. If it does, d should be increased, but proportionally much less than the shortfall of shear resistance, $V_{Rd,c}$, with respect to the acting shear, V_{Ed}: an increase in d not only increases $V_{Rd,c}$ as per Equation 6.12, but also reduces V_{Ed}, thanks to the larger distance of the control sections from the centre of the footing – cf. Equation 6.13 – and to the reduced eccentricities due to the increase in R_N effected by the heavier footing – cf. Equation 6.9.

Note that the eccentricity e_y does not appear in Equations 6.14 and 6.15. Indeed, it does not affect the value of $V_{Ed,x}$ if Equation 6.10 is met (i.e. for full contact with the ground).

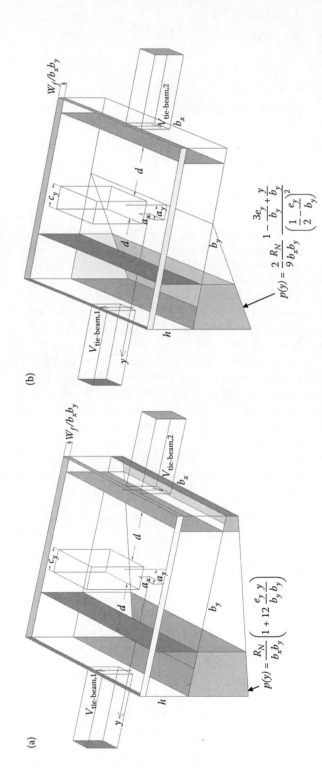

Figure 6.15 Forces and pressures on footing for the calculation of the design shear force on vertical control sections normal to horizontal axis y, for (a) uniaxial eccentricity $e_y \leq b_y/6$ (full contact of the underside with the ground); (b) uniaxial eccentricity $e_y > b_y/6$ (partial uplift from the ground).

Moreover, even in case 2 of Section 6.3.4, the value of e_y does not affect that of $V_{Ed,x}$, provided that it meets the condition:

$$\frac{(1/2) - (|e_x|/b_x)}{(1/2) - (s_{vx}/b_x)} \geq \frac{1}{3} \cdot \frac{1 + \sqrt{1 - 12(e_y/b_y)^2} + 12(e_y/b_y)^2}{1 + \sqrt{1 - 12(e_y/b_y)^2} - 6(|e_y|/b_y)} \tag{6.17}$$

Similarly, the value of e_x does not affect that of $V_{Ed,y}$, provided that Equation 6.10 is met, or case 2 of Section 6.3.4 applies and a condition obtained by substituting x in Equation 6.17 for y and vice versa is satisfied. In all other cases of practical interest, the value of e_x impacts mainly $V_{Ed,x}$ and very little $V_{Ed,y}$, while e_y affects primarily $V_{Ed,y}$ but almost not at all $V_{Ed,x}$.

6.3.6 Design of the footing reinforcement

Working as two-way, inverted, double cantilevers, footings in buildings normally have two-way reinforcement at the bottom face, dimensioned at vertical sections through the depth of the footing at the face(s) of the vertical element. In general, there are two such sections normal to the x-axis and two normal to the y-axis; they are indexed by x and y, respectively; their dimensioning gives reinforcement parallel to the corresponding axis. As in Section 6.3.5, a prime denotes the control section further away from the side of the footing where the peak soil pressure takes place, to distinguish it from the other control section, with which it shares the same index.

The distances of the two control sections which are normal to the x-axis from the centre of the footing in plan are (Figure 6.14b):

$$s_{mx} = a_x + 0.5c_x \tag{6.18a}$$

$$s'_{mx} = a_x - 0.5c_x \tag{6.18b}$$

where the notation is as for Equation 6.13. The acting moment at the section normal to the x-axis is the moment resultant with respect to the section due to all moments, vertical forces, or distributed loads acting on the body of the footing between the control section and the footing edge parallel to it:

- If Equation 6.10 and case 1 of Section 6.3.4 apply (i.e. for full contact with the ground), then

$$M_{Ed,x} = \left[\left(1 + \frac{4 |e_x|}{b_x} \left(1 + \frac{s_{mx}}{b_x} \right) \right) R_N - W_f \right] \frac{b_x}{2} \left(\frac{1}{2} - \frac{s_{mx}}{b_x} \right)^2$$
$$+ \sum_{x=s_{mx}}^{x=b_x/2} \left(M_{\text{tie-beam},x} + V_{\text{tie-beam}} (x_{\text{tie-beam}} - s_{mx}) \right) \tag{6.19a}$$

$$M'_{Ed,x} = \left[\left(1 - \frac{4 |e_x|}{b_x} \left(1 - \frac{s'_{mx}}{b_x} \right) \right) R_N - W_f \right] \frac{b_x}{2} \left(\frac{1}{2} + \frac{s'_{mx}}{b_x} \right)^2$$
$$- \sum_{x=s'_{mx}}^{x=-b_x/2} \left(M_{\text{tie-beam},x} + V_{\text{tie-beam}} (s'_{mx} - x_{\text{tie-beam}}) \right) \tag{6.19b}$$

- If Equation 6.10 is not met, but case 2 of Section 6.3.4 applies instead (loss of contact over part of the plan area), then

$$
M_{Ed,x} = \left[\frac{2}{27} \cdot \frac{4 - \left(9\ |e_x|/b_x \right) + \left(s_{mx}/b_x \right)}{\left((1/2) - \left(|e_x|/b_x \right) \right)^2} R_N - W_f \right] \frac{b_x}{2} \left(\frac{1}{2} - \frac{s_{mx}}{b_x} \right)^2
$$

$$
+ \sum_{x=s_{mx}}^{x=b_x/2} \left(M_x + V_{\text{tie-beam}} \left(x_{\text{tie-beam}} - s_{mx} \right) \right) \tag{6.20a}
$$

$$
M'_{Ed,x} = \frac{R_N b_x}{27} \frac{\max\left(0; \left(1 - \left(3\ |e_x|/b_x \right) + \left(s'_{mx}/b_x \right) \right)^3 \right)}{\left((1/2) - \left(|e_x|/b_x \right) \right)^2} - \frac{W_f b_x}{2} \left(\frac{1}{2} + \frac{s'_{mx}}{b_x} \right)^2
$$

$$
- \sum_{x=s'_{mx}}^{x=-b_x/2} \left(M_{\text{tie-beam},x} + V_{\text{tie-beam}} \left(s'_{mx} - x_{\text{tie-beam}} \right) \right) \tag{6.20b}
$$

Figures 6.16a and 6.16b depict the forces and pressures from which the counterparts of Equations 6.14 and 6.15, respectively, are derived, for dimensioning of the reinforcement which is parallel to y at the two control sections normal to the y-axis (replacing x in Equations 6.18 to 6.20 by y and vice versa).

As in Equations 6.14 and 6.15, the sums in the last term of Equations 6.19 and 6.20 refer to all tie-beams which frame into the footing between the control section and the footing edge close and parallel to it. Only tie-beams in the x-direction apply a moment M_x; the ones in the y-direction contribute to the sums only through their shear force. The x-axis is positive in the direction of the eccentricity e_x. The sign convention for the shear in a tie-beam is the same as in Equations 6.14 and 6.15: positive if it acts upwards on the footing, negative if it acts downwards. Moments M_x are considered in Equations 6.19 and 6.20 as positive, if they act on the footing in the sense in which the reacting moment $R_{Mx} = e_x R_N$ does. For the way the footing is expected to rotate due to a positive eccentricity e_x and moment R_{Mx}, the end of a tie-beam connected to it is expected to be sagging near the edge where the peak soil pressure is exerted (i.e. in Equations 6.19a and 6.20a) and hogging at the opposite edge (in Equations 6.19b and 6.20b); if it is not, its moment should enter these expressions as negative. Moments M_x cannot exceed the corresponding capacities, $M^+_{Rb,d}$ for sagging, $M^-_{Rb,d}$ for hogging. Moreover, as in Equations 6.14 and 6.15, if i indexes the end of a tie-beam at the footing in question and j the opposite end, the absolute value of its shear cannot exceed $(M^+_{Rb,d,i} + M^-_{Rb,d,j})/l_n$ if it acts upwards on the footing, or $(M^-_{Rb,d,i} + M^+_{Rb,d,j})/l_n$ if it acts downwards.

Similar to Equations 6.14 and 6.15, the eccentricity e_y does not appear in Equations 6.19 and 6.20, because it does not affect the value of $M_{Ed,x}$:

- When Equation 6.10 is met (i.e. for full contact with the ground), or
- In case 2 of Section 6.3.4, under the condition that e_y meets an expression similar to Equation 6.17, but with s_{mx} at the denominator of the left-hand side, in lieu of s_{vx}.

Bottom bars parallel to x cross the two vertical control sections through the faces of the vertical element that are normal to the x-axis and have width b_y and effective depth d. They

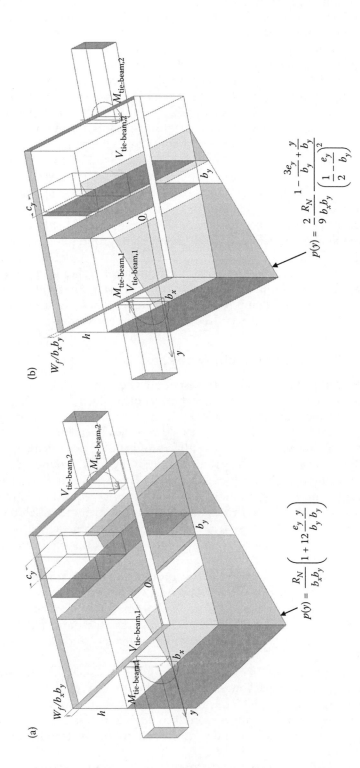

$$p(y) = \frac{R_N}{b_x b_y} \left(1 + 12 \frac{e_y}{b_y} \frac{y}{b_y} \right)$$

$$p(y) = \frac{2}{9} \frac{R_N}{b_x b_y} \frac{1 - \dfrac{3e_y}{b_y} + \dfrac{y}{b_y}}{\left(\dfrac{1}{2} - \dfrac{e_y}{b_y} \right)^2}$$

Figure 6.16 Moments, forces and pressures on footing for the calculation of the design moments on the vertical control sections normal to horizontal axis y, for: (a) uniaxial eccentricity $e_y \leq b_y/6$ (full contact of the underside with the ground); (b) uniaxial eccentricity $e_y > b_y/6$ (partial uplift from the ground).

are controlled by the maximum of the moments $M_{Ed,x}$ and $M'_{Ed,x}$ from Equations 6.19 and 6.20 among all cases of the seismic design situation (defined for multi-component action effects in Section 5.8.1, depending on the way the effects of the two horizontal components are combined). Their dimensioning for the ULS in flexure may take place considering the footing as an inverted corbel, because its effective depth, d, is normally larger than the shear span of the vertical control section considered. That shear span is equal to the controlling value of $M_{Ed,x}$ or $M'_{Ed,x}$ from Equations 6.19 or 6.20, divided by the shear computed from Equations 6.14 or 6.15, but using s_{mx} or s'_{mx} instead of s_{vx} or s'_{vx}; that shear comes from the case of the seismic design situation which gives the controlling – that is, the maximum – value of $M_{Ed,x}$ or $M'_{Ed,x}$. Normally $M'_{Ed,x}$ is much smaller than $M_{Ed,x}$; however, a negative-valued $M'_{Ed,x}$ means tension at the top face of the footing and may require top bars dimensioned for the value of $|M'_{Ed,x}|$ from that case of the seismic design situation which gives the algebraically minimum $M'_{Ed,x}$ value. Note that longitudinal bars of tie-beams crossing these control sections of the footing at the level of its top or bottom reinforcement and fully anchored past them, may count as part of the footing reinforcement. All other bars of the footing should extend to its edge and be anchored there with a 90° bent, as appropriate to a corbel-type element like the footing.

Although a footing does not meet the Eurocode 2 definition of a slab as a 2D element with a depth less than 5-times the minimum horizontal dimension, it is natural to apply the Eurocode 2 rules for slabs for its minimum reinforcement ratio. These are the same as in beams of DC L, listed under DC L in Table 5.1 (second row of requirements).

6.3.7 Verification of footings in punching shear

Being a two-way slab loaded at the bottom with distributed soil pressures and at the top with a concentrated reaction (at the base section of the vertical element), a footing should, in principle, be verified for punching shear. Moreover, it should not require punching shear reinforcement, that is, vertical links around the base of the vertical element.

Punching shear is two-way shear in the footing all around the vertical element, arising from a vertical load, R_N, transferred to the ground fairly uniformly around the base of the element. However, footings designed for earthquake resistance normally transfer R_N to the ground with a large uniaxial or biaxial eccentricity, which equilibrates the large seismic moments. So, for (approximately) the same R_N, such footings are seldom more critical in punching shear all around the vertical element than in eccentric one-way shear at one of its sides. For the same reason, punching shear is rarely an issue in footings of walls.

For completeness, this section highlights the verification of footings without shear reinforcement at the ULS of punching shear as per Eurocode 2. For the reasons explained, it focusses on (practically) concentric footings of columns with small-to-moderate eccentricity of the vertical force resultant, R_N, and activation of a contact area all around the column for the transfer of R_N to the ground. In such cases, Equation 6.10 is met and case 1 in Section 6.3.4 most likely applies.

The design value of the vertical shear stress resistance, $v_{Rd,c}$, in a slab without punching shear reinforcement is still given by Equation 6.12, but applies on a vertical cylindrical surface ('control section') through the slab, which is nowhere closer than a to the perimeter of the column. For footings, the distance a ranges from 0 to $2d$, where d is the mean effective depth of the footing in the two directions, x and y, at the column perimeter; the same value of d is used in Equation 6.12.

The perimeter of the control section on a horizontal plane ('control perimeter') should be convex and as short as possible. Around a polygonal column (including the ones with

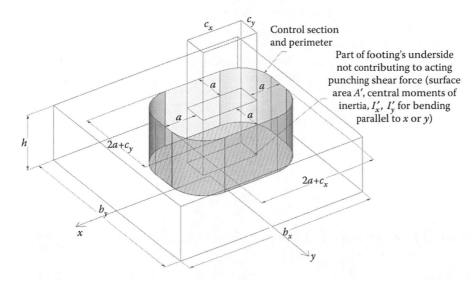

Figure 6.17 'Control surface' in punching shear of footing around a rectangular column and area of the footing's underside which does not contribute to the acting punching shear force.

re-entrant corners, e.g. L-, T-, C-sections) it consists of a sequence of circular arcs, having radius $a \leq 2d$ and its centre at one of the (non-reentrant) corners, and of straight segments tangential to two adjacent arcs (Figure 6.17). The design shear resistance, $v_{Rd,c}$, decreases with increasing closest distance of the 'control perimeter' from the column, $a < 2d$, (units MN, m) as follows:

$$v_{Rd,c}(a) = \frac{2d}{a}\left\{\max\left[\frac{0.18}{\gamma_c}(100\rho_1)^{1/3}; 0.035\sqrt{1 + \sqrt{\frac{0.2}{d}}}\ f_{ck}^{1/6}\right]\left(1 + \sqrt{\frac{0.2}{d}}\right)f_{ck}^{1/3}\right\}$$

$$\leq 0.3\left(1 - \frac{f_{ck}}{250}\right)f_{cd} \tag{6.12a}$$

where the mean effective depth of the footing in x and y is used as d and the geometric mean, $\sqrt{(\rho_x\rho_y)}$, of the tension reinforcement ratios in these directions is used as ρ_1.

For a footing to be without punching shear reinforcement, the value of $v_{Rd,c}$ from Equation 6.12a may not be less than the maximum acting shear stress, v_{Ed}, along any control perimeter at a closest distance from the column equal to $a < 2d$. The value of v_{Ed} is estimated as

$$v_{Ed}(a) = \beta(a)\frac{V_{Ed,red}(a)}{u(a)d} \tag{6.21}$$

where

$V_{Ed,red}(a)$ is that part of R_N, which passes through the cylindrical 'control section' on its way from the ground to the column (loads applied on the footing inside the 'control section' are transferred to the column directly, by inclined compression).

$u(a)$ is the length of the control perimeter around the column, equal to $2(\pi a + c_x + c_y)$ if it surrounds a rectangular column with cross-sectional sides c_x and c_y in the x- and y-directions of the footing.

$\beta(a)$ is the 'eccentricity factor', which converts the mean of the acting shear stress, represented by the fraction in Equation 6.21, to its peak value around the control perimeter due to that part of the biaxial reacting moments, R_{Mx}, R_{My} (with respect to axes of symmetry y and x), which passes through the 'control section' on its way from the ground to the column base.

Denoting by $M_{Ed,x,red}(a)$, $M_{Ed,y,red}(a)$ the parts of R_{Mx}, R_{My}, respectively, going to the column base via the 'control section', their effect is expressed through their eccentricities:

$$e_{x,red}(a) = M_{Ed,x,red}(a)/V_{Ed,red}(a) \qquad (6.22a)$$

$$e_{y,red}(a) = M_{Ed,y,red}(a)/V_{Ed,red}(a) \qquad (6.22b)$$

For uniaxial eccentricity along direction x (i.e. for $e_{y,red}(a) = 0$), the 'eccentricity factor' is

$$\beta(a) = 1 + ke_{x,red}(a)\frac{u(a)}{W_x(a)} \qquad (6.23)$$

where
k is the fraction of $M_{Ed,x,red}(a)$ transferred through vertical shear stresses on the cylindrical 'control section'; for a rectangular column with cross-sectional sides c_x and c_y in the x- and y-directions of the footing, respectively, k is equal to:

- For $c_x/c_y \leq 0.5$: $k = 0.45$ (6.24a)
- For $c_x/c_y = 1.0$: $k = 0.60$ (6.24b)
- For $c_x/c_y = 2.0$: $k = 0.70$ (6.24c)
- For $c_x/c_y \geq 3.0$: $k = 0.80$ (6.24d)

with interpolation in-between;
$W_x = \oint_{u(a)} |x|\,ds$ is the 'plastic modulus' of the control perimeter, corresponding to vertical shear stresses equilibrating $kM_{Ed,x,red}(a)$, which are uniformly distributed over each half of the cylindrical 'control section' on each side of its centroidal axis parallel to y; for a rectangular column with cross-sectional sides c_x and c_y along x and y:

$$W_x(a) = c_x^2/2 + c_x c_y + 2ac_y + \pi ac_x + 4a^2 \qquad (6.25)$$

For uniaxial eccentricity in y (i.e. for $e_{x,red}(a) = 0$), x is replaced by y in Equations 6.23 to 6.25 and vice versa. For full-fledged biaxial eccentricity (i.e. when neither one of $e_{x,red}(a)$, or $e_{y,red}(a)$ is negligible compared to the other), there is no simple expression for $\beta(a)$; for that case, Eurocode 2 gives the following approximation for rectangular columns:

$$\beta(a) \approx 1 + 1.8\sqrt{\left(\frac{e_{x,red}(a)}{c_y + 2a}\right)^2 + \left(\frac{e_{y,red}(a)}{c_x + 2a}\right)^2} \qquad (6.26)$$

Expressions are given below for $V_{Ed,red}(a)$ and $M_{Ed,x,red}(a)$ – that is, the parts of R_N and $R_{Mx} = e_x R_N$, respectively, which are transferred to the column base via the 'control section'.

Strictly speaking, these expressions apply under conditions of small-to-moderate eccentricity, notably when Equation 6.10 is met and case 1 in Section 6.3.4 applies. As explained in the first two paragraphs of this section, these conditions are normally met in footings that are more critical in punching shear all around the column than in one-way shear induced at one of its four sides by a large eccentricity. In such footings, even when Equation 6.10 is not met but case 2 in Section 6.3.4 applies instead, the expressions given represent a satisfactory approximation of the fairly complex exact alternatives holding then.

$$V_{Ed,red}(a) = \left(1 - \frac{A'}{b_x b_y}\right)(R_N - W_f) - \frac{12A'}{b_x b_y}\left(\frac{a_x e_x}{b_x^2} + \frac{a_y e_y}{b_y^2}\right)R_N + \sum_{\text{tie-beams}} V_{\text{tie-beam}} \quad (6.27a)$$

$$M_{Ed,x,red}(a) = e_x R_N - \frac{12 R_N}{b_x b_y}\left(\frac{I_x' e_x}{b_x^2} + \frac{I_{xy}' e_y}{b_y^2}\right) + \sum_{\text{tie-beams}}\left(M_{\text{tie-beam},x} + V_{\text{tie-beam}}(x_{\text{tie-beam}} - a_x)\right)$$

$$(6.28a)$$

where (see Figure 6.17):

b_x, b_y are the dimensions of the footing in plan along the x- or y-axis, respectively

e_x, e_y are the eccentricities of the vertical reaction R_N along the x- or y-axis, as per Equation 6.9

a_x, a_y are the eccentricities of the column's centroidal axis with respect to the centre of the footing in plan in the x- or y-direction, respectively, taken positive if on the same side with respect to the centre as the eccentricities e_x, e_y and the peak soil pressure ($a_x = 0$, $a_y = 0$ in concentric footings).

W_f is the weight of the footing and of its soil cover (included in R_N).

A', I_x', I_{xy}' are the surface area, central moment of inertia for bending parallel to the x-axis (moment of inertia about a centroidal axis parallel to y) and cross-moment of inertia about centroidal axes parallel to x and y, respectively, of the part of the footing underside enclosed by the 'control section'. If the column is rectangular, with cross-sectional sides c_x and c_y along the x- or y-axis, respectively, then

$$A' = \pi a^2 + c_x c_y + 2a(c_x + c_y) \quad (6.29a)$$

$$I_x' = \frac{\pi a^2(c_x^2 + a^2)}{4} + \frac{a c_x^3}{6} + \frac{c_y(2a + c_x)^3}{12} \quad (6.29b)$$

$$I_{xy}' = 0 \quad (6.29c)$$

Strictly speaking, Equations 6.27a and 6.28a apply only if the footing does not uplift, that is, in case 1 of Section 6.3.4 (i.e. if Equation 6.10 is met). If there is uplift, the counterpart of these expressions is relatively simple only for essentially uniaxial eccentricity, e_x ($e_y = 0$):

$$V_{Ed,red}(a) = \left(1 - \frac{2}{9}\frac{A'}{b_x b_y}\frac{\left(1 - (3e_x/b_x)\right)\left(1 + (a_x/b_x)\right)}{\left((1/2) - (e_x/b_x)\right)^2}\right)R_N - \left(1 - \frac{A'}{b_x b_y}\right)W_f$$

$$+ \sum_{\text{tie-beams}} V_{\text{tie-beam}} \quad (e_x > b_x/6, e_y = 0) \quad (6.27b)$$

$$M_{Ed,x,red}(a) = \left(e_x - \frac{2}{9} \frac{I'_x}{\left((1/2) - (e_x/b_x) \right)^2 b_x^2 b_y} \right) R_N$$

$$+ \sum_{\text{tie-beams}} \left(M_{\text{tie-beam},x} + V_{\text{tie-beam}}(x_{\text{tie-beam}} - a_x) \right) \quad (e_x > b_x/6, e_y = 0) \quad (6.28\text{b})$$

Moreover, Equations 6.29 apply, so long as the uplift area does not overlap with the part of the footing underside enclosed by the 'control section', that is, as long as the distance of the control section from the column perimeter satisfies the condition:

$$b_x - 3e_x - c_x/2 + a_x \geq a \quad (6.30)$$

with the eccentricity of the column axis with respect to the centre of the footing, a_x, being positive if it is in the same direction as the eccentricity of the loading, e_x.

The sums in the last term of Equations 6.27 and 6.28 encompass all tie-beams which frame into the footing in x or y from any side. Only tie-beams parallel to the x-axis apply a moment M_x; those in the y-direction contribute through their shear force alone. The sign convention is the same as in Equations 6.14, 6.15, 6.19, 6.20: shears are positive if they act upwards on the footing; moments M_x are positive if they act on the footing in the sense in which the reacting moment $R_{Mx} = e_x R_N$ does (i.e. if they induce tension in the bottom face of the footing on the side where bearing pressures are larger due to the eccentricity, or compression on the opposite side). For the way the footing is expected to rotate under a positive eccentricity e_x and moment R_{Mx}, M_x is expected to be positive both at the connecting ends of tie-beams near the edge where the peak soil pressure is exerted (sagging end), as well as at the opposite edge (hogging); by contrast, tie-beam shears are expected to be positive in the former case, negative in the latter. As in Equations 6.14, 6.15, 6.19, 6.20, if i indexes the end of a tie-beam at the footing in question and j indexes the opposite end, the magnitude of its shear force cannot exceed the capacity values: $(M^+_{Rb,d,i} + M^-_{Rb,d,j})/l_n$ at a sagging end (upwards acting shear), or $(M^-_{Rb,d,i} + M^+_{Rb,d,j})/l_n$ at a hogging one.

An expression for $M_{Ed,y,red}(a)$ is obtained from Equations 6.28 and 6.29 by replacing x with y and vice versa.

For a footing to be allowed without punching shear reinforcement, its design shear resistance, $v_{Rd,c}$, from Equation 6.12a should be at least equal to the maximum acting shear stress, v_{Ed}, from Equations 6.21 to 6.29, for all values of a from 0 to $2d$. Note that v_{Ed} increases with decreasing a, and so does $v_{Rd,c}$, but only up to an upper bound value given by the last part of Equation 6.12a, which corresponds to failure of the concrete in diagonal compression. The verification that v_{Ed} does not exceed that upper bound to $v_{Rd,c}$ should always be carried out at the lateral perimeter of the column, that is, for $a = 0$. Other than that, the most critical value of a is when the ratio $v_{Ed}/v_{Rd,c}$ attains its maximum value.

Note that, while a sweeps through the range from 0 to $2d$, the control perimeter may intersect the edge of the footing. However, a truncated control perimeter is not physically plausible as critical in punching shear: apart from the fact that some of the expressions in the calculation of v_{Ed} do not apply then, one-way shear in the x- or the y-direction at one side of the column will most likely be more critical than punching shear around two or three sides of the column.

Example 6.3 at the end of this chapter illustrates a complete design of an isolated footing.

6.3.8 Design and detailing of tie-beams and foundation beams

Eurocode 8 considers the minimum cross-sectional dimensions of tie-beams, foundation beams or foundation slabs used instead of tie-beams, as NDPs; the recommended values are

   i. For tie-beams and foundation beams: width, $b_w \geq 0.25$ m; depth, $h_b \geq 0.4$ m in build-ings with up to three storeys above the basement; $h_b \geq 0.5$ m in taller ones.

   ii. For foundation slabs connecting individual footings: thickness, $t \geq 0.2$ m.

The minimum longitudinal steel ratio all along these elements is also an NDP; the values recommended for it are

- For tie-beams and foundation beams, 0.4%, separately at top and bottom
- For foundation slabs connecting individual footings, 0.2%, separately at top and bottom

Eurocode 2 specifies a minimum downward load on tie-beams, to represent the effects of compaction machinery. Its value is an NDP, with a recommended value of 10 kN/m. By its nature, it may be classified as imposed (live) load with a quasi-permanent value (ψ_2 factor) of 0.

According to Eurocode 8, a postulated axial force should be considered at the ULS in flexure of a tie-beam (or a tie-zone in a foundation slab replacing the tie-beams), alongside the moments from the analysis for the seismic design situation. It is more unfavourable to take that force as tensile. Its magnitude, N_{tb}, is specified in Eurocode 8 as a fraction of the mean value, $N_{m,c}$, of the design axial forces of the connected vertical elements in the seismic design situation. This fraction is equal to the design ground acceleration in g's, αS, times $\lambda = 0.3$, 0.4 or 0.6 for ground type B, C or D, respectively.

$$N_{tb} = \lambda(\alpha S)N_{m,c} \tag{6.31}$$

This axial force is meant to account for the effects of horizontal relative displacements between foundation elements not accounted for explicitly in the analysis for the seismic design situation. It may be neglected for ground type A, as well as over ground type B but only in low seismicity cases (recommended as those cases where $\alpha S \leq 0.1$). To take into account $N_{Ed} = N_{tb}$ (positive for tension) in dimensioning the longitudinal reinforcement of the tie-beam, instead of M_{Ed} in Equation 5.13 of Section 5.3.1, the net internal moment with respect to the level of the tension reinforcement is used, $M_{sd} = M_{Ed} - y_{s1}N_{Ed}$, with y_{s1} denoting the distance of the tension reinforcement from the centroidal axis of the tie-beam. The dimensionless axial force $v_d = N_{Ed}/(b_{eff}df_{cd})$ is also added to the right-hand side of Equations 5.15 for the mechanical ratio of tension reinforcement.

According to the second to last paragraph of Section 6.3.2, the designer has two options:

1. To multiply the seismic action effects for the tie-beam and foundation beam from the linear analysis by the magnification factor of Equations 6.7a and 6.8. Then
   a. The capacity design of these beams is not carried out; the so-magnified seismic shear forces are used instead.
   b. Section 5.5.3 does not apply to the dimensioning of the beam in shear, even though DC H may have been chosen for the superstructure; Section 5.5.2 applies instead.
   c. The minimum requirements at the beginning of the present Section are supplemented with the non-conflicting rules for DC L in Tables 5.1, 5.3 and 5.6.
2. To dimension the tie- or foundation beams for the ULS in flexure, using the unmagnified seismic action effects from the linear analysis. From that point on, these beams are designed and detailed as the primary ones in the superstructure:
   a. The capacity-design shears are computed from Equation 5.42a in Section 5.5.1. In foundation beams, the possibility of plastic hinging is considered, either at the

base of the vertical element or in the foundation beam itself. In contrast, tie-beams may always be taken to form the plastic hinges themselves (i.e. 1.0 is taken as the minimum in the term multiplying $M_{Rd,b}$ at each tie-beam end).

b. If DC H has been chosen for the superstructure, Section 5.5.3 applies for the dimensioning of the beam in shear.

c. The minimum requirements at the beginning of this section are supplemented by the non-conflicting rules in Tables 5.1, 5.3 and 5.6 for the ductility class of the building.

Example 6.4 at the end of this chapter illustrates the effect of tie-beams on the design of an isolated footing, as well as the design of tie-beams.

EXAMPLE 6.1

Pros and cons of the alternative foundation schemes (a) to (d) (Figure 6.18) for earthquake resistance in a building on a steep slope; alternative scheme (far right), with justification.

Answer

In options (a) to (c), there is no assurance that all footings will be subjected to the same horizontal seismic displacement time-history: it may differ among footings, owing to the incident seismic waves or the seismic response of the superstructure. In that respect, the uncertainty concerning the seismic response of the building or its parts is higher than in option (d). If all footings have the same horizontal seismic displacements, then, owing to the rigid diaphragm above them, each footing, and the length of the vertical element immediately above it, develops elastic shears (about) proportional to the stiffness of that element. In cases (a) and (d), that stiffness is inversely proportional to the cube of the clear length of that element; in (b), almost the entire shear goes to the footing on the left. In case (c), the total shear is more uniformly shared by the four columns, because the connection to a transverse beam at about mid-height increases their lateral stiffness. The leftmost column above the footing in cases (a), (c), (d), the second one from the left in (b) and the lower part of all columns in (c) are squat, hence vulnerable to shear.

None of these options is appropriate for earthquake resistance. A suitable one is depicted on the right; it ensures common displacement of all footings, avoids squat columns above the footings or excessive excavation to bring all footings to the same horizontal level. Yet, it considerably increases the volume of excavation compared to (a)–(d); it suits a building with a basement under the part of the building on the right.

EXAMPLE 6.2

Calculation of a_{CD} at the base of the column of Example 5.3, for the following behaviour factor and γ_{Rd} values in the strong and the weak direction of bending, respectively: $q_y = 3.6$, $q_z = 3.0$, $\gamma_{Rd,y} = 1.2$, $\gamma_{Rd,z} = 1.0$, and for the two extreme values of the column

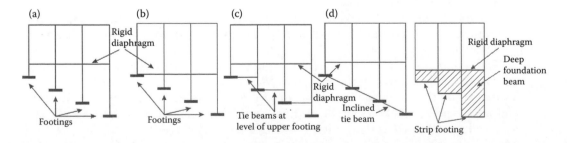

Figure 6.18 (a–d) Alternative foundation schemes for Example 6.1.

Table 6.4 Example 6.2: Action effects from analysis at the column base

Actions and combinations	N (kN)	Strong direction (y)		Weak direction (z)	
		M_{cy} (kNm)	V_{cy} (kN)	M_{cz} (kNm)	V_{cz} (kN)
$G + \psi_2 Q$	1025.5	17.75	7.85	21	12.75
$\pm E$	± 173.5	± 135.15	± 56.15	± 26.1	± 11.55
$G + \psi_2 Q \pm E$	1199/852	152.9/ −117.4	64.0/ −48.3	47.1/ −5.1	24.3/1.2

axial load in the seismic design situation. Calculation of the design action effects at the centre of the base of the 0.7 m deep concentric footing of the column, which is horizontally connected to other foundation elements through a concrete slab, not through tie-beams. The moments, shears and axial force at the base section of the column in the seismic design situation, as obtained from modal response spectrum analysis, are listed in Table 6.4.

Answer

The design value of the moment capacities at the column's base section, $M_{Rd,y}$, $M_{Rd,z}$, have been found in Example 5.3 for the two extreme values of the column's axial load in the seismic design situation, $\max N = 1199$ kN, $\min N = 852.1$ kN. The capacity-design factor for the design of the column's footing and the verification of the foundation soil is computed in Table 6.5. The values shown for each individual direction are those of $\gamma_{Rd} M_{Rd}/M_{Ed}$, without the corresponding upper limits of q. Those limits are applied in the end, when Equation 6.7a is used.

As the footing is concentric and not connected to tie-beams, the design action effects at the centre of its base can be found by equilibrium from those at the top, as follows:

- Each horizontal reaction is equal to the corresponding shear at the base of the column, V.
- The reacting moment is equal to the bending moment at the column's base, M, plus the corresponding shear, V, times the footing's depth, $h = 0.7$ m.
- The vertical reaction is equal to the column's axial force, N, plus the weight of the footing and of the overlying soil.

The part of M, V, N due to the seismic action is multiplied by $a_{CD} = 3.0$. Table 6.6 summarises the calculations and the outcomes, without including yet in N the weight of the footing and that of the overlying soil. The potentially critical M, V, N values are underlined. Both values of N should be considered with each M and V combination.

Table 6.5 Example 6.2: Capacity-design multiplier for footing

Combination of actions	$M_{Rd,y}$ (kNm)	$\|M_{Ed,y}\|$ (kNm)	a_{CDy}	$M_{Rd,z}$ (kNm)	$\|M_{Ed,z}\|$ (kNm)	a_{CDz}	a_{CD}
$M_{Ed,y} > 0, M_{Ed,z} > 0, \max N$	447.4	152.9	3.51	168.4	47.1	3.58	3.0
$M_{Ed,y} > 0, M_{Ed,z} < 0, \max N$	447.4	152.9	3.51	168.4	5.1	33	3.0
$M_{Ed,y} < 0, M_{Ed,z} > 0, \max N$	447.4	117.4	4.57	168.4	47.1	3.58	3.0
$M_{Ed,y} < 0, M_{Ed,z} < 0, \max N$	447.4	117.4	4.57	168.4	5.1	33	3.0
$M_{Ed,y} > 0, M_{Ed,z} > 0, \min N$	422.4	152.9	3.32	156.6	47.1	3.32	3.0
$M_{Ed,y} > 0, M_{Ed,z} < 0, \min N$	422.4	152.9	3.32	156.6	5.1	30.7	3.0
$M_{Ed,y} < 0, M_{Ed,z} > 0, \min N$	422.4	117.4	4.32	156.6	47.1	3.32	3.0
$M_{Ed,y} < 0, M_{Ed,z} < 0, \min N$	422.4	117.4	4.32	156.6	5.1	30.7	3.0

Table 6.6 Example 6.2: Design action effects at the base of the footing

Actions and combinations	N (kN)	Column Strong Direction (y)				Column Weak Direction (z)		
		M_{cy} (kNm)	V_{cy} (kN)	$M_{cy}+V_{cy}h$ (kNm)	M_{cz} (kNm)	V_{cz} (kN)	$M_{cz}+V_{cz}h$ (kNm)
$G+\psi_2 Q$	1025.5	17.75	7.85	23.25	21	12.75	29.9
$\pm E$	±173.5	±135.15	±56.15	±174.45	±26.1	±11.55	±34.2
$G+\psi_2 Q$ $\pm a_{CD}E$	1546/505	423.2/−387.7	176.3/−160.6	546.6/−500.1	99.3/−57.3	47.4/−21.9	132.5/−72.7

EXAMPLE 6.3

Design of a 2.4 m × 2.4 m, 0.7 m deep concentric footing for the column of Examples 5.3 and 6.2. The column cross-sectional dimensions are: $c_x = 0.7$ m, $c_y = 0.3$ m, with the local axes x and y of the footing corresponding to bending in the strong (y) and weak (z) direction of the column. Concrete is C25/30, the steel grade is S500. The soil is saturated clay, with characteristic value of undrained shear strength $c_u = 300$ kPa in the 'persistent and transient design situation', reduced to 270 kPa in the 'seismic design situation'; the design values are 215 and 195 kPa, respectively. The footing is horizontally connected to the other foundation elements through a foundation slab without tie-beams; it is not covered by soil, but the surcharge pressure at the level of its base amounts to $q_{sur} = 14$ kPa (due to a soil with a unit weight of 20 kN/m³). The design values of

- The horizontal reactions R_{Hx} (:V_x,), R_{Hy} (:V_y)
- The reacting moments, R_{Mx} (:M_x,), R_{My} (:M_y)
- The vertical reaction, R_N

at the centre of the footing's base are given in Table 6.7. R_N includes the effect of the footing's weight, $W_f = 25 \times 0.7 \times 2.4 \times 2.4 = 101$ kN, which gives a contribution of $1.35 \times 0.85 \times 101 = 116$ kN to the fundamental combination as per Equation 6.10b of EN1990, or of $1.35 \times 101 = 136$ kN to that as per Equation 6.10a. In the seismic design situation, both values of N are considered for each M and V. The design peak ground acceleration Sa_g is 0.25 g.

Answer

The verification of the footing is, in general, carried out not only for the seismic design situation but also for the combination of persistent and transient actions ('fundamental combination'), and indeed both as per Equation 6.10a in EN1990:2002 and Equation 6.10b. In these cases, that 'fundamental combination' which gives the most unfavourable outcome out of the two applies. If it is not clear *a priori* which one it is, both have to be checked.

Table 6.7 Example 6.3: Design action effects at the base of the footing

Combination of actions	R_N (kN)	Footing direction x		Footing direction y	
		R_{Hx} (kN)	R_{Mx} (kNm)	R_{Hy} (kN)	R_{My} (kNm)
Fundamental: Equation 6.10a in EN1990	1676	11.8	35	19.15	44.9
Fundamental: Equation 6.10b in EN1990	1600	11.1	33	18.1	42.4
Seismic design situation: $G+\psi_2 Q \pm a_{CD}E$	1647/606	176.3	546.6	47.4	132.5

1. Verification of the footing in plan on the basis of the bearing capacity of the soil.
   The bearing capacity of the footing is estimated first as per Informative Annex D in Part 1 of Eurocode 7. According to it, the bearing capacity over clay in undrained conditions is

$$q_{ud} = q_{sur} + (\pi + 2)c_{ud}\{1 + 0.2\min[b_x - 2e_x; b_y - 2e_y]/\max[b_x - 2e_x; b_y - 2e_y]\}i_c$$

with $i_c = \{1 + (2\theta/\pi)\sqrt{(1 - R_{Hx}/V_{cu,d})} + (1 - 2\theta/\pi)\sqrt{(1 - R_{Hy}/V_{cu,d})}\}/2$
where $\tan\theta = R_{Hx}/R_{Hy}$; $V_{cud} = c_{ud}(b_x - 2e_x)(b_y - 2e_y)$ is the undrained shear resistance of the interface of the base of the footing with the soil, whose exceedance triggers sliding failure of the footing.

Design Approach DA3 is adopted in the 'persistent and transient design situation' ('fundamental combination'). The design forces at the centre of the footing's base and the associated eccentricities from Equation 6.9 are listed in Table 6.8, alongside the resulting uniform soil bearing pressures $R_N/(b_x - 2e_x)(b_y - 2e_y)$. The combination in the last row, where the seismic vertical force is tensile, induces a very large eccentricity along the strong direction of the column, which well exceeds one-third of the footing length. However, thanks to the reduction of the net vertical reaction by the uplifting seismic vertical force, the soil bearing capacity is not compromised. Note also that, although the large margin between the pressure and the soil bearing capacity gives the impression that a reduction in the size of the footing, for example, to 2.3 m-square, is feasible, this is not the case: in such a footing R_{Hx} exceeds the value of $V_{cud} = c_{ud}(b_x - 2e_x)(b_y - 2e_y)$ at the last row and the first square-root term in the i_c term does not have a real value.

The bearing capacity of the footing in the seismic design situation is also verified as per Annex F in Part 5 of Eurocode 8 (see Section 6.2). Note that Annex F is for strip footings, but it can also be used for square or circular footings with the appropriate choice of N_{max}, replacing the width of the strip footing by the diameter of a footing with the same area in plan as the one considered (in the dimensionless soil force F, by its radius). That diameter in the present case is $2.4\sqrt{(4/\pi)} = 2.7$ m. The strength of a concentric, vertically loaded footing is: $q_{ud} = q_{sur} + 1.2c_{ud}(\pi + 2) = 14 + 1.2 \times 193 \times (\pi + 2) = 1200$ kPa, hence $N_{max} = 2.4^2 \times 1200 = 6900$ kN.

The seismic horizontal force and moment are taken as the resultant of the two components: $V_{Ed} = 183$ kN, $M_{Ed} = 563$ kNm. Then, Equation 6.6 gives, for $\gamma_{Rd} = 1.0$:

$$\bar{V} = \frac{183}{6900} = 0.0265, \quad \bar{M} = \frac{563}{2.7 \times 6900} = 0.0302, \quad \bar{F} = \frac{20 \times 0.25 \times 1.35}{193} = 0.035$$

$$\bar{N} = \frac{1647}{6900} = 0.239 \text{ for max } N, \quad \bar{N} = \frac{606}{6900} = 0.088 \text{ for min } N$$

The left-hand side of Equation 6.5 is equal to −0.87 or −0.31, for maxN or minN, respectively, confirming stability and the smaller margin for maxN than for

Table 6.8 Example 6.3: Verification of soil bearing capacity as per EC7

Combination of actions	R_N (kN)	R_{Mx} (kNm)	e_x/b_x	R_{My} (kNm)	e_y/b_y	R_{Hx} (kN)	R_{Hy} (kN)	$\theta = \tan^{-1}$ (R_{Hx}/R_{Hy}) (rad)	Soil bearing Pressure (kPa)	Soil bearing Capacity (kPa)
DA3 & Equation 6.10a	1676	35	0.009	44.9	0.011	11.8	19.15	0.552	301	1330
DA3 & Equation 6.10b	1600	33	0.009	42.4	0.011	11.1	18.1	0.55	289	1330
$G + \psi_2 Q \pm a_{CD}E$, maxN	1647	546.6	0.138	132.5	0.034	176.3	47.4	1.308	424	1095
$G + \psi_2 Q \pm a_{CD}E$, minN	606	546.6	0.376	132.5	0.091	176.3	47.4	1.308	519	825

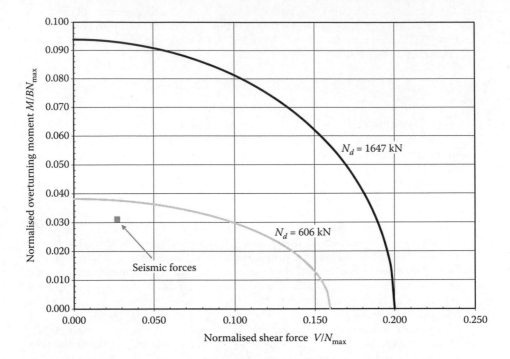

Figure 6.19 Example 6.3: Point of forces acting on the foundation in the 3D force space vis-a-vis the cross sections of the ultimate load surface for maxN and minN.

minN. The location of the point corresponding to the forces acting on the foundation is compared in Figure 6.19 to the cross sections of the ultimate load surface for maxN and minN; it is indeed well inside them.

2. Verification of the depth of the footing on the basis of the ULS in shear.

After the bearing capacity of the soil is verified as mentioned (and indeed with a large margin), it is checked if the effective depth of the footing, $d = h - 0.05 = 0.65$ m, meets the ULS of the footing in shear and punching shear without shear reinforcement.

Equation 6.13 – and their counterparts in the y-direction – give the distance of the control sections in shear from the centre of the footing:

$$s_{vx} = 0.7/2 + 0.65 = 1.0 \text{ m}; \quad s'_{vx} = -1.0 \text{ m};$$
$$s_{vy} = 0.3/2 + 0.65 = 0.8 \text{ m}, \quad s'_{vy} = -0.8 \text{ m}$$

As the footing is concentric, only the sections at s_{vx} and s_{vy} are of interest. Equation 6.14a or 6.15a – and their counterparts in the y-direction – apply, if the eccentricity is less than or exceeds, respectively, one-sixth of the footing length.

Table 6.9 Example 6.3: Design shear forces for ULS verification of footing in shear as per EC2

Combination of actions	R_N (kN)	e_x/b_x	$V_{Ed,x}$ (kN)	e_y/b_y	$V_{Ed,y}$ (kN)
EN1990 Equation 6.10a	1676	0.009	137	0.011	275.4
EN1990 Equation 6.10b	1600	0.009	128.6	0.011	258.7
$G + \psi_2 Q \pm a_{CD}E$, maxN	1647	0.138	233	0.034	304
$G + \psi_2 Q \pm a_{CD}E$, minN	606	0.376	233	0.091	130

Table 6.10 Example 6.3: geometric and loading variables for ULS verification of the footing in punching shear as per EC2, for different distances of the control perimeter from the column base

a (m)	u(a) (m)	A'(a) (m²)	$V_{Ed,red}/(R_N-W_f)$	$W_x(a)$ (m²)	$W_y(a)$ (m²)	$I'_x(a)$ (m⁴)	$M_{Edx,red}/R_{Mx}$	$I'_y(a)$ (m⁴)	$M_{Edy,red}/R_{My}$
0.8	7.03	3.82	0.337	5.254	4.689	0.918	0.668	0.771	0.721
0.6	5.77	2.54	0.559	3.574	3.10	0.446	0.839	0.327	0.882
0.4	4.51	1.51	0.738	2.215	1.832	0.189	0.932	0.111	0.960
0.3	3.88	1.093	0.810	1.655	1.318	0.113	0.959	0.0566	0.980
0.2	3.257	0.7356	0.8723	1.176	0.8835	0.0613	0.978	0.0249	0.991

The outcomes of their application, listed in Table 6.9 as $V_{Ed,x}$, $V_{Ed,y}$, are well below the value of shear resistance, $V_{Rd,cx} = V_{Rd,cy} = 529.2$ kN, obtained from Equation 6.12 by using the values of b_y, b_x, respectively, as b_w and computing ρ_1 from a presumed bottom reinforcement of Ø12/125 mm, which gives the minimum slab reinforcement ratio as per Eurocode 2.

3. Verification of the depth of the footing on the basis of the ULS in punching shear.

The control perimeter for punching shear sweeps the range from the furthest-most complete perimeter surrounding the column from all sides – at a distance to it equal to $a = 0.8$ m – to one as close to the column as $a = 0.2$ m. Table 6.10 lists the values of geometric properties and loading variables of the control perimeter at selected values of a.

Punching shear is more critical when the vertical load is large and its eccentricity low, than the other way round. So, only the combination of actions as per Equation 6.10a of EN1990 and the one giving maxN in the seismic design situation are examined in detail here. In fact, for the combination of actions in the last row of Table 6.11 uplift occurs and, more importantly, Equation 6.30 is violated for any value of a; so, the verification of punching shear for that combination is meaningless.

At a distance $a = 2d$ the shear stress resistance is the same as for the ULS in shear: $v_{Rd,c} = 339.2$ kN; it then increases as inversely proportional to a, to a maximum of $0.3(1 - f_{ck}/250)f_{cd} = 4500$ kPa, attained at $a = 339.2 \times 2 \times 0.65/4500 = 0.098$ m. The verifications in Table 6.11 show that, even when one of the eccentricities is dominant (as the one along x is in the seismic design situation), the use in Equation 6.21 of the approximation as per Equation 6.26 gives higher

Table 6.11 Example 6.3: ULS verification of footing in punching shear as per EC2

Combination of actions	R_N (kN)	R_{Mx} (kNm)	R_{My} (kNm)	a (m)	$V_{Ed,red}$ (kN)	$e_{x,red}$ (m)	$e_{y,red}$ (m)	Equations 6.21 and 6.25 $\beta_x(a)$	$maxv_{Edx}$ (kPa)	$\beta_y(a)$	$maxv_{Edy}$ (kPa)	Equation 6.26 $\beta(a)$	$maxv_{Ed}$ (kPa)	$2v_{Rd}d/a$ (kPa)
EN1990 equation 6.10a	1676	35	44.9	0.8	525.7	0.044	0.062	1.043	120	1.041	120	1.063	122	551
				0.6	072	0.033	0.045	1.039	242	1.038	241	1.058	246	735
				0.4	1151	0.028	0.037	1.042	409	1.041	409	1.064	418	1102
				0.3	1264	0.026	0.034	1.045	523.5	1.045	524	1.07	536	1470
				0.2	1361	0.025	0.033	1.051	675.5	1.054	677.5	1.084	697	2205
$G + \psi_2 Q + a_{CD}E$ maxN	1647	546.6	132.5	0.8	521	0.701	0.183	1.688	192.5	1.123	128	1.679	191.5	551
				0.6	864	0.531	0.135	1.629	375	1.113	256.5	1.65	380	735
				0.4	1141	0.446	0.111	1.666	648.5	1.123	437	1.742	678	1102
				0.3	1252	0.419	0.104	1.720	854	1.138	565	1.85	918.5	1470
				0.2	1349	0.396	0.097	1.804	1150	1.161	740	2.031	1294	2205

Table 6.12 Example 6.3: Action effects for the dimensioning of the footing in flexure as per EC2

Combination of actions	R_N (kN)	e_x/b_x	$M_{Ed,x}$ (kNm)	$M'_{Ed,x}$ (kNm)	e_y/b_y	$M_{Ed,y}$ (kNm)	$M'_{Ed,y}$ (kNm)
$G + \psi_2 Q \pm a_{CD}E$, maxN	1647	0.138	390	76	0.034	410	300
$G + \psi_2 Q \pm a_{CD}E$, minN	606	0.376	320	−15	0.091	170	62

maximum acting shear stress, $\max v_{Ed}$, on the control perimeter than the more rigorous uniaxial expressions, Equation 6.25. The margin between the shear stress resistance and the maximum acting shear stress is the lowest, for a distance a of the control perimeter from the column around 0.3 m. The seismic design situation giving maxN is more critical than the fundamental combination of actions, Equation 6.10a in EN1990.

4. Dimensioning of two-way reinforcement at the bottom face on the basis of the ULS in flexure.

The critical vertical sections of the footing in bending pass through the lateral faces of the column; according to Equation 6.18 (and their counterparts in the y-direction), they are at a distance from the centre of the footing: $s_{mx} = 0.7/2 = 0.35$ m; $s'_{mx} = -0.35$ m; $s_{my} = 0.3/2 = 0.15$ m, $s'_{my} = -0.15$ m. For a concentric footing, the sections at s_{mx} and s_{my} and the values of $M_{Ed,x}$, $M_{Ed,y}$ are of interest for the bottom reinforcement. If the values of $M'_{Ed,x}$, $M'_{Ed,y}$ at the sections with $s'_{mx} = -0.35$ m and $s'_{my} = -0.15$ m are negative, the footing may need top reinforcement as well.

Equation 6.19 or 6.20 – and their counterparts in the y-direction – apply, if the eccentricity is less than or exceeds, respectively, one-sixth of the footing length. The maximum of the computed values of $M_{Ed,x}$, $M_{Ed,y}$ for all combinations of actions, alongside the minimum steel of slabs, determine the two-way reinforcement at the bottom surface of the footing (see Table 6.12). As $M_{Ed,x}$, $M_{Ed,y}$ and even $M'_{Ed,x}$, $M'_{Ed,y}$ attain their highest values when the eccentricity is large, only the combinations of actions in the seismic design situation are examined in detail here: for maxN, that combination gives approximately the same vertical force as the 'fundamental combination' as per Equations 6.10a and 6.10b of EN1990, but at larger eccentricities.

Under minN, the bending moment at the backside with respect to the eccentricity is negative, but small compared to the cracking moment, which is equal to 502 kNm for the mean tensile strength of concrete, 351 kNm for the characteristic. Note that an isolated footing is not subjected to any significant imposed deformations which may cause vertical cracking, while the quasi-permanent gravity loads induce compressive stresses to its top side. So, the top surface of the footing will face the design seismic action uncracked and does not need top reinforcement against the small negative moments that may develop next to the column.

The maximum bending moment of 409.8/2.4 = 170.75 kNm/m is resisted by the minimum slab reinforcement ratio as per Eurocode 2, translated to a two-way bottom reinforcement of Ø12/125 mm.

EXAMPLE 6.4

The footing of Example 6.3 is connected to the adjacent foundation elements, in both horizontal directions, through tie-beams:

i. On one side, the footing is connected to a similar footing of a similar column, subjected to similar seismic action effects.

ii. On the other side and in the same horizontal direction, it is connected to a rigid foundation element, which fixes the other end of the tie-beam against rotation

The tie-beams have the minimum dimensions recommended in Eurocode 8: width of 0.25 m, depth of 0.5 m. The action effects at the top of the footing are the same as in Example 6.3. Positive are those moments and shears which increase the bearing pressures under the footing edge connected to the tie-beams indexed by 2 (see following text). Owing to the restraint of the rotation of the footing about the horizontal axes by the tie-beams, the net moment transferred to the ground at the base of the footing is reduced, at the expense of elastic moments in the tie-beams. Specifically, if the moment, M_c, and the shear, V_c, at the base section of the column are transferred to the footing's base over its depth h as a moment $M_c + hV_c$ and a shear V_c, the moment transferred to the ground is $k_\varphi(M_c + hV_c)/[k_\varphi + \Sigma_i(k_{tb,i})]$ ($i = 1, 2$ indexes the two tie-beams in the plane of action of M_c and V_c), where

- $k_\varphi = 625$ MNm/rad is the rotational spring stiffness at the centre of the base of the footing, modelling the rotational impedance of the ground.
- $k_{tb,i}$ is the rotational stiffness of tie-beam i, referring to the centre of the footing's base. If EI_{tb} is the stiffness of the tie-beam (taken as 50% of the uncracked gross-section stiffness), b_f is the size of the footing in the direction of the tie-beam and L_{cl} is the tie-beam's clear span, then
  - For a tie-beam connected to a similar footing of a similar column (case i), $k_{tb,i} = 6EI_{tb}(1 + b_f/L_{cl,i})^2/L_{cl,i}$, but only for seismic action effects.
  - $k_{tb,i} = 6EI_{tb}[2/3 + b_f/L_{cl,i} + (b_f/L_{cl,i})^2/2]/L_{cl,i}$ in all other cases (including a tie-beam connecting two similar footings of similar columns, but for gravity actions).

The moment at the end section of tie-beam i is $k_{tb,i}(M_c + hV_c)/\{[k_\varphi + \Sigma_i(k_{tb,i})]\cdot[1 + 0.5b_f/L_{s,i}]\}$, where $L_{s,i}$ is the shear span (M/V ratio) of that section for the particular case of loading. In case (i) $L_{s,i} = 0.5L_{cl,i}$; in case (ii), $L_{s,i} = 0.5L_{cl,i}(4/3 + b_f/L_{cl,i})/(1 + b_f/L_{cl,i})$.

The axial distance of the column to the adjacent vertical element is 6 m in the x-direction (strong direction of the column), or 7 m in the y-direction; for the clear span L_{cl}, that axial distance is reduced by $0.5b_f$ at an end connected to a footing, or by 0.5 m at an end connected to a rigid foundation element (case ii).

The tie-beams are designed below and their impact on the design of the footing is evaluated.

Answer

In view of the beneficial effect of the tie-beams, the size of the footing is reduced to 2 m × 2 m. Its weight is $W_f = 25 \times 0.7 \times 2.0 \times 2.0 = 70$ kN, giving a contribution of $1.35 \times 0.85 \times 70 = 80$ kN to the vertical reaction R_N in the fundamental combination as per Equation 6.10a of EN1990.

1. Stiffness of tie-beams and impact on the distribution of action effects among the components.

    The tie-beam connecting the footing to a rigid foundation element (case ii) is indexed with $i = 1$; the opposite one connecting to a similar footing (case i) is indexed with $i = 2$. Then
    a. In the x-direction (strong direction of the column), $L_{cl,1} = 6 - 0.5 - 1 = 4.5$ m, $L_{cl,2} = 6 - 1 - 1 = 4.0$ m
    b. In the y-direction; $L_{cl,1} = 7 - 0.5 - 1 = 5.5$ m, $L_{cl,2} = 7 - 1 - 1 = 5.0$ m
    Tie-beam properties: $EI_{tb} = 0.5 \times 31,500 \times 0.25 \times 0.5^3/12 = 41$ MNm².
    In the x-direction:
    a. $k_{tb,1} = 6 \times 41 \times [2/3 + 2.0/4.5 + (2.0/4.5)^2/2]/4.5 = 66$ MNm/rad
    b. $k_{tb,2} = 6 \times 41 \times [2/3 + 2.0/4.0 + (2.0/4.0)^2/2]/4.0 = 79.5$ MNm/rad   for gravity actions
    c. $k_{tb,2} = 6 \times 41 \times (1 + 2.0/4.0)^2/4.0 = 138.4$ MNm/rad for seismic actions

d. Sharing of $(M_c + hV_c)$ by the soil and the tie-beams:
- For gravity actions:
  - Soil: $k_\varphi/(k_\varphi + \Sigma_i(k_{tb,i})) = 625/(625 + 66 + 79.5) = 81.1\%$
  - Tie-beam 1: $k_{tb,1}/(k_\varphi + \Sigma_i(k_{tb,i})) = 66/(625 + 66 + 79.5) = 8.6\%$
  - Tie-beam 2: $k_{tb,2}/(k_\varphi + \Sigma_i(k_{tb,i})) = 79.5/(625 + 66 + 79.5) = 10.3\%$
- For seismic actions:
  - Soil: $k_\varphi/(k_\varphi + \Sigma_i(k_{tb,i})) = 625/(625 + 66 + 138.4) = 75.3\%$
  - Tie-beam 1: $k_{tb,1}/(k_\varphi + \Sigma_i(k_{tb,i})) = 66/(625 + 66 + 138.4) = 8\%$
  - Tie-beam 2: $k_{tb,2}/(k_\varphi + \Sigma_i(k_{tb,i})) = 138.4/(625 + 66 + 138.4) = 16.7\%$

In the y-direction:
a. $k_{tb,1} = 6 \times 41 \times [2/3 + 2.0/5.5 + (2.0/5.5)^2/2]/5.5 = 49$ MNm/rad
b. $k_{tb,2} = 6 \times 41 \times [2/3 + 2.0/5.0 + (2.0/5.0)^2/2]/5.0 = 56.4$ MNm/rad for gravity actions
c. $k_{tb,2} = 6 \times 41 \times (1 + 2.0/5.0)^2/5.0 = 96.4$ MNm/rad for seismic actions
- For gravity actions:
  - Soil: $k_\varphi/(k_\varphi + \Sigma_i(k_{tb,i})) = 625/(625 + 49 + 56.4) = 85.6\%$
  - Tie-beam 1: $k_{tb,1}/(k_\varphi + \Sigma_i(k_{tb,i})) = 49/(625 + 49 + 56.4) = 6.7\%$
  - Tie-beam 2: $k_{tb,2}/(k_\varphi + \Sigma_i(k_{tb,i})) = 56.4/(625 + 49 + 56.4) = 7.7\%$
- For seismic actions:
  - Soil: $k_\varphi/(k_\varphi + \Sigma_i(k_{tb,i})) = 625/(625 + 49 + 96.4) = 81.1\%$
  - Tie-beam 1: $k_{tb,1}/(k_\varphi + \Sigma_i(k_{tb,i})) = 49/(625 + 49 + 96.4) = 6.4\%$
  - Tie-beam 2: $k_{tb,2}/(k_\varphi + \Sigma_i(k_{tb,i})) = 96.4/(625 + 49 + 96.4) = 12.5\%$

Table 6.13 lists the action effects at the level of the footing's base from Example 6.2 for the seismic action, or 6.3 for the 'fundamental combination' as per Equation 6.10a of EN1990 (revised for the different weight of the footing and modified as explained later due to the imposed load on tie-beams for the effects of compaction machinery). It also distributes the resultant moments at the central node of the base to the soil and the tie-beams according to the fractions given. The vertical and horizontal forces are not distributed: they go directly to the ground under the base. For tie-beams, the moments are given at the node at the centre of the footing's base, as well as at the beam's end section, at the face of the footing. The value listed at each end section of a tie-beam for the 'fundamental combination' as per Equation 6.10a is the so-obtained moment plus the fixed-end moment due to the 10 kN/m imposed load recommended in Eurocode 2 for the effects of compaction machinery

Table 6.13 Example 6.4: Design action effects at footing's base and at end sections of tie-beams

					Footing direction x/column strong direction y							Footing direction x/column weak direction z				
					Moments[a] (kNm)							Moments[a] (kNm)				
						Tie-beam 1		Tie-beam 2					Tie-beam 1		Tie-beam 2	
Actions	R_N (kN)	$V_y = R_{Hx}$ (kN)	$M_y + V_y h$	Soil R_{Mx}	Node	End[a]	Node	End[a]	$V_z = R_{Hy}$ (kN)	$M_z + V_z h$	Soil R_{My}	Node	End[a]	Node	End[a]
Equation 6.10a	1740	11.8	35	25.4	2.7	−19.7[b]	3.2	−11.7[b]	19.2	44.9	34.5	2.1	−28.6[b]	3.1	−19.5[b]
$G+\psi_2 Q$	1095.5	7.85	23.2	18.8	2.0	−1.5	2.4	1.7	12.8	29.9	25.6	2.0	−1.5	2.3	1.7
$\pm E$	±173.5	56.2	±174.5	±131.4	14	∓10.3	29.1	±19.4	11.5	±34.2	±27.7	2.2	∓1.7	4.3	±3.1
$G+\psi_2 Q \pm a_{CD}E$	1546/505	176.3	546.6	±413	44	−32.4/29.4	89.7	59.9/−56.5	47.4	132.5	±108.7	8.6	−6.3/3.6	15.2	11/−7.6

[a] The moment at the end section of a tie-beam is negative for tension at the top side, positive for compression.
[b] The end moments include the tie-beam fixed-end moment due to the imposed load of 10 kN/m recommended in EC2 for the effect of compaction machinery; the ones at the central node do not.

on tie-beams (see Section 6.3.8). That load is not reflected in the action effects due to gravity loads taken from Examples 6.2 and 6.3; so, the fixed-end shears it produces in the tie-beams ($1.35 \times 0.85 \times 10 \times (4.0 + 4.5 + 5.0 + 5.5)/2 = 100$ kN in total) are added to the vertical reaction R_N as per Equation 6.10a. Moreover, the fixed end moments this load produces in the tie-beams are transferred to the node at the centre of the footing's base; the unbalanced moment (i.e. their difference: $1.35 \times 0.85 \times 10 \times (4.0^2 - 4.5^2)/12 = -3.7$ kNm in x, $1.35 \times 0.85 \times 10 \times (5.0^2 - 5.5^2)/12 = -4.6$ kNm in y) is superimposed to the one transferred from the base of the column. It is the net moment from the 'fundamental combination' which is distributed to the soil and the tie-beams.

Option 1 in Section 6.3.8 is chosen for the seismic design of the tie-beams, namely to multiply their seismic action effects by the same capacity-design factor as the footings, $a_{CD} = 3.0$.

2. Design of tie-beams for the ULS in flexure.

For the axial force in tie-beams as per Equation 6.31, the seismic design parameters of the example in Chapter 7 are used: $a_g = 0.25$ g, Ground type B, $S = 1.2$. The most critical tie-beam is no. 2 in direction x, which connects the footing to one of a similar column with similar seismic action effects. So, the maximum axial force of the column in the seismic design situation, namely 1546 kN including the factor $a_{CD} = 3.0$, is taken as mean value for both columns. Then, $N_{tb} = 0.3 \times 0.25 \times 1.2 \times 1546 = 139$ kN. For $d = 0.45$ m, $y_{s1} = 0.2$ m, we have $M_{sd} = 59.9 - 0.2 \times 139 = 32.1$ kNm and $v_d = 139/(0.25 \times 0.45 \times 16,667) = 0.074$. Equations 5.13 to 5.15, modified for the axial load as highlighted in Section 6.3.8, give then $A_{s1} = 486$ mm². The minimum recommended steel ratio of 0.4% gives 500 mm² at top and bottom; 2Ø18 (509 mm²) is chosen.

3. Verification of the footing in plan on the basis of the bearing capacity of the soil.

The moments and shears due to gravity actions are positive (see Table 6.4); this has been defined in this example to mean that they increase the bearing pressures under the footing edges connected to tie-beams indexed by 2. So, it is those sides of the footing that are more critical in terms of the bearing capacity of the soil and shear or flexure in the footing. Therefore, in part 3 herein and 4 to 6 we will focus on that side.

Table 6.14 summarises the eccentricities of the total axial force and the verification of the bearing capacity of the soil, for the same strength parameters as in Example 6.3. In this case there is practically no margin in soil capacity.

The soil bearing capacity in the seismic design situation is also verified according to Annex F in Part 5 of Eurocode 8 (see Section 6.2). The diameter of the footing with the same area in plan as the present footing is $2.0\sqrt{(4/\pi)} = 2.25$ m. The strength of a concentric, vertically loaded footing is: $q_{ud} = q_{sur} + 1.2c_{ud}(\pi + 2) = 14 + 1.2 \times 193 \times (\pi + 2) = 1200$ kPa, hence $N_{max} = 2.0^2 \times 1200 = 4800$ kN. The seismic horizontal force and moment are the resultants of the two components: $V_{Ed} = 183$ kN, $M_{Ed} = 427$ kNm. Then, Equation 6.6 gives, for $\gamma_{Rd} = 1.0$:

Table 6.14 Example 6.4: Verification of soil bearing capacity as per EC7

Combination of actions	R_N (kN)	R_{Mx} (kNm)	e_x/b_x	R_{My} (kNm)	e_y/b_y	R_{Hx} (kN)	R_{Hy} (kN)	$\theta = tan^{-1}$ (R_{Hx}/R_{Hy}) (rad)	Soil bearing Pressure (kPa)	Soil bearing Capacity (kPa)
EN1990 Equation 6.10a	1740	25.4	0.007	34.5	0.010	11.8	19.15	0.552	450	1330
$G + \psi_2 Q \pm a_{CD}E$, maxN	1616	413	0.128	108.7	0.034	176.3	47.4	1.308	583	1075
$G + \psi_2 Q \pm a_{CD}E$, minN	575	413	0.359	108.7	0.095	176.3	47.4	1.308	629	650

Table 6.15 Example 6.4: Design shears for footing and tie-beams

Combination	R_N (kN)	e_x/b_x	$V^a_{tb1,x}$ (kN)	$V^a_{tb2,x}$ (kN)	$V_{Ed,x}$ (kN)	e_y/b_y	$V^a_{tb1,y}$ (kN)	$V^a_{tb2,y}$ (kN)	$V_{Ed,y}$ (kN)
EN1990 Equation 6.10a	1740	0.007	−24.6[b]	−19.7[b]	24.6	0.010	−29.5[b]	−25.2[b]	150.2
$G + \psi_2 Q \pm a_{CD}E$ maxN	1616	±0.128	−11.7/10.6	30.1/−28.1	30.1	±0.034	−1.8/1.1	4.5/−3.0	188.8
$G + \psi_2 Q \pm a_{CD}E$ minN	575	±0.359	−11.7/10.6	30.1/−28.1	30.1	±0.095	−1.8/1.1	4.5/−3.0	84.5

[a] The tie-beam shear is positive if it acts upwards on the footing, negative if it is downwards.
[b] Value includes the end shear due to the imposed load of 10 kN/m recommended in EC2 for the effects of compaction machinery.

$$\overline{V} = \frac{183}{4800} = 0.039, \ \overline{M} = \frac{427}{2.25 \times 4800} = 0.0395, \ \overline{F} = \frac{20 \times 0.25 \times 1.125}{193} = 0.029$$

$$\overline{N} = \frac{1616}{4800} = 0.336 \text{ for maxN}, \ \overline{N} = \frac{575}{4800} = 0.120 \text{ for minN}$$

The left-hand side of Equation 6.5 is equal to −0.85 or −0.32 for maxN or minN, respectively, confirming stability, but with a smaller margin for minN.

4. Verification of the depth of the footing in shear.

The verification of the depth of the footing in shear and the calculation of its reinforcement (see part 6) take into account further effects of the tie-beams, reflected in the last term of Equations 6.14, 6.15, 6.19, 6.20, 6.27 and 6.28. The elastic values of the tie-beam action effects are used there, because tie-beams are designed for seismic action effects multiplied with the capacity-design factor a_{CD} applying to the footings (option 1 in Section 6.3.8). The elastic tie-beam moments are those given in Table 6.13 at the end section of the tie-beam. Apart from the shears as per footnote b of Table 6.15 (those in the 'fundamental combination' as per Equation 6.10a due to Eurocode 2's postulated load of 10 kN/m), the end shears, V_{tbi}, are found by dividing the corresponding end moment by the pertinent shear span, $L_{s,i}$; they are given in Table 6.15.

The shear forces $V_{Ed,x}$, $V_{Ed,y}$ in Table 6.15 are calculated from Equations 6.14 and 6.15. The critical sections are at the same distance from the column as in part 2 of Example 6.3, but in the x-direction they coincide with the lateral face of the footing; so, only the tie-beam shears contribute to $V_{Ed,x}$, $V'_{Ed,x}$. The value listed as $V_{Ed,x}$ is the absolutely maximum of $V_{Ed,x}$, $V'_{Ed,x}$ from Equations 6.14 and 6.15. All acting shears come out much less than the shear resistance, $V_{Rd,cx} = V_{Rd,cy} = 529.2$ kN. The shear verification is much more favourable than in Example 6.3, thanks to the reduced size of the footing made possible by the tie-beams.

Table 6.16 Example 6.4: Geometric variables for ULS verification of footing in punching shear as per EC2, for different distances of the control perimeter from the column's base

a (m)	$1-A'/b_x b_y$	$1-12I'_x / b_y b_x^3$	$1-12I'_y / b_x b_y^3$	$u(a)$ (m)	$W_x(a)$ (m²)	$W_y(a)$ (m²)
0.6	0.365	0.6655	0.7548	5.77	3.574	3.10
0.4	0.6225	0.8582	0.9168	4.51	2.215	1.832
0.3	0.7268	0.9152	0.9576	3.88	1.655	1.318
0.2	0.8161	0.954	0.9813	3.257	1.176	0.8835

Table 6.17 Example 6.4: Action effects on footing for ULS verification in punching shear as per EC2

Combination	R_N (kN)	R_{Mx} (kNm)	$V^a_{tb1,x}$ (kN)	$V^a_{tb2,x}$ (kN)	$M^b_{tb1,x}$ (kNm)	$M^b_{tb2,x}$ (kNm)	R_{My} (kNm)	$V^a_{tb1,y}$ (kN)	$V^a_{tb2,y}$ (kN)	$M^b_{tb1,y}$ (kNm)	$M^b_{tb2,y}$ (kNm)
Equation 6.10a	1740	25.4	−24.6	−19.7	19.7	−11.7	34.5	−29.5	−25.2	28.6	−19.5
$G + \psi_2 Q \pm a_{cD}E$ maxN	1616	±413	−11.7/10.6	30.1/ −28.1	32.4/ −29.4	59.9/ −56.5	±108.7	−1.8/ 1.1	4.5/ −3.0	6.3/ −3.6	11/ −7.6

[a] Tie-beam shears are positive if they act upwards on the footing, negative if downwards.
[b] Tie-beam end moments are positive if they induce tension to the bottom face of the footing on the side where bearing pressures are larger due to the eccentricities, or compression on the opposite side; otherwise they are negative.

5. Verification of the depth of the footing in punching shear.

   Complete control perimeters surrounding the column from all sides start from a distance to it of $a = 0.6$ m and are considered to decrease to $a = 0.2$ m. The values of geometric properties of the control perimeter are listed in Table 6.16, followed in Table 6.17 by the action effects that are needed for the calculation of the reduced shears and moments from Equations 6.27a and 6.28a.

   The verification in Table 6.18 leads to similar conclusions as in Example 6.3; the margin from the shear stress resistance and to maximum acting shear stress is again the lowest when the distance a of the control perimeter from the column is around 0.3 m; it is about the same as in Example 6.3.

6. Dimensioning of two-way reinforcement at the bottom face for the ULS in flexure.

   The moments $M_{Ed,x}$, $M_{Ed,y}$, $M'_{Ed,x}$, $M'_{Ed,y}$ in Table 6.19 are calculated from Equations 6.20 and 6.21 at critical sections coinciding with the column faces.

   The maximum bending moment of $345.4/2.0 = 172.7$ kNm/m is about the same as in Example 6.3; it is again covered by the minimum slab reinforcement as per Eurocode 2 of Ø12/125 mm. The negative bending moment at the backside under minN is larger than in Example 6.3, but still small, compared to the cracking moment of 418 kNm for the mean tensile strength of concrete, 292.5 kNm for the characteristic.

Table 6.18 Example 6.4: ULS verification of footing in punching shear as per EC2 for different distances of the control perimeter from the column's base

Actions	a (m)	$V_{Ed,red}$ (kN)	$M_{Ed,x,red}$ (kNm)	$e_{x,red}$ (m)	$M_{Ed,y,red}$ (kNm)	$e_{y,red}$ (m)	Equations 6.21 and 6.25 $\beta_x(a)$	$maxv_{Edx}$ (kPa)	$\beta_y(a)$	$maxv_{Edy}$ (kPa)	Equation 6.26 $\beta(a)$	$maxv_{Ed}$ (kPa)	$2v_{Rd}d/a$ (kPa)
Equation 6.10a	0.6	507	29.8	0.059	39.4	0.078	1.07	144.5	1.065	144	1.102	149	735
	0.4	934.4	34.7	0.037	45.0	0.048	1.055	337	1.053	335.5	1.084	345.5	1102
	0.3	1107.5	36.1	0.033	46.4	0.042	1.057	464	1.056	463.5	1.088	478	1470
	0.2	1256	37.1	0.030	47.3	0.038	1.06	629	1.062	630.5	1.10	652	2205
$G + \psi_2 Q$ $\perp a_{cD}E/maxN$	0.6	585.1/ −583.7	409/ −399.5	0.699/ −0.684	105.6/ −97.3	0.180/ −0.167	1.828/ 1.81	285.3/ 281.7	1.151/ 1.14	179.5/ 177.5	1.856/ 1.836	289.7/ 285.7	735
	0.4	983.5/ −981.8	488.5/ −479	0.497/ −0.488	123.3/ −115	0.125/ −0.117	1.742/ 1.729	584.5/ 579	1.138/ 1.13	382/ 378.5	1.827/ 1.811	613/ 606.5	1102
	0.3	1145/ −1143	512/ −502.5	0.447/ −0.440	127.7/ −119.4	0.112/ −0.104	1.768/ 1.756	803/ 796	1.148/ 1.138	521.5/ 515.5	1.907/ 1.906	866/ 863.5	1470
	0.2	1283/ −1281	528.1/ −518.6	0.412/ −0.405	130.3/ −122	0.102/ −0.095	1.837/ 1.823	1113/ 1103	1.169/ 1.158	708.5/ 700.5	2.072/ 2.053	1256/ 1244	2205

Table 6.19 Example 6.4: Action effects for the dimensioning of the footing in flexure as per EC2

Combination	R_N (kN)	e_x/b_x	$V^a_{tb1,x}$ (kN)	$V^a_{tb2,x}$ (kN)	$M^b_{tb1,x}$ (kNm)	$M^b_{tb2,x}$ (kNm)	$M_{Ed,x}$ (kNm)	$M'_{Ed,x}$ (kNm)	e_y/b_y	$V^a_{tb1,y}$ (kN)	$V^a_{tb2,y}$ (kN)	$M^b_{tb1,y}$ (kNm)	$M^b_{tb2,y}$ (kNm)	$M_{Ed,y}$ (kNm)
$G + \psi_2 Q \pm a_{CD}E$ maxN	1616	0.128	−11.7/	30.1/	32.4/	59.9/	345.4	20.6	0.034	−1.8/	4.5/	6.3/	11/	336.7
$G + \psi_2 Q \pm a_{CD}E$ minN	575	0.359	10.6	−28.1	−29.4	−56.5	285.7	−43.7	0.095	1.1	−3	−3.6	−7.6	148.5

a The tie-beam shear is positive if it acts upwards on the footing, negative if it acts downwards.

b Tie-beam end moments are positive if they induce tension to the bottom face of the footing on the side where bearing pressures are larger due to the eccentricities, or compression on the opposite side; otherwise they are negative.

QUESTION 6.1

A 20 m thick soil profile is composed of clay with two embedded sand layers. The first one is 1 m thick between 3 and 4 m depth and the second one 2 m thick between 7 and 9 m. The normalised SPT N blow count numbers are, respectively, $N_1 = 10$ and 20. The response of the soil profile is approximated by its fundamental mode of vibration with a displacement pattern given by: $u = (4/\pi)d_0 \sin(\pi z/2H)$ where $4/\pi$ is the mode participation factor, d_0 the surface displacement, equal to 30 mm, and the profile thickness $H = 20$ m. Calculate the surface settlement assuming that the sand layers are unsaturated and that the seismic motion is represented by 15 cycles of constant amplitude shear strain.

QUESTION 6.2

Repeat Question 6.1 for saturated sand strata for which the safety factors against liquefaction are respectively 0.95 and 1.05.

QUESTION 6.3

Circular concrete columns with reinforcement uniformly distributed around the perimeter at a distance to it not more than 10% of the column diameter, D, and with a mechanical reinforcement ratio, ω_{tot}, have dimensionless moment resistance, $\mu_d = M_{Rd}/(A_c D f_{cd})$, given by the following empirical expression, as a function of the mechanical steel ratio, ω_{tot}, and the dimensionless axial load ratio, $\nu_d = N_d/(A_c f_{cd})$:

$$\mu_d = M_{Rd}/(A_c D f_{cd})$$
$$= \left[\omega_{tot} - \left(1.37\nu_d^2 - 1.228\nu_d - 0.083\right)\right]\Big/\left[3.758 + 1.138\nu_d + 3.468\nu_d^2 - 5.781\nu_d^3\right]$$

The column in question has $D = 1.2$ m, is constructed of concrete C25/30, has axial load $N_d = 3265$ kN and 25,000 mm² of 500 MPa steel reinforcement uniformly distributed around its perimeter. The seismic moment and shear at its base from the analysis are: $M_E = 4155$ kNm, $V_E = 1385$ kN. The behaviour factor is equal to 3.6 and the design peak ground acceleration at the top of the soil 0.30 g. The soil is clay with design value of undrained shear strength $c_{ud} = 300$ kPa and unit weight $\gamma_{soil} = 20$ kN/m³. The top of the footing is at grade level.

1. Calculate the design value of the moment resistance and the capacity-design magnification factor of the footing.
2. Calculate the design action effects at the centre of the base of a concentric isolated footing of the column, for a footing depth of 1.5 m.
3. Choose and verify the dimensions of the footing in plan.
4. Choose and verify the footing depth for the ULS in shear.
5. Verify the footing depth for the ULS in punching shear.

QUESTION 6.4

Design the footings of the central and the outer columns of the 3-bay frame of Question 5.3. The soil is clay with design value of undrained shear strength $c_{ud} = 95$ kPa and unit weight $\gamma_{soil} = 21$ kN/m³. The top of the footing is at ground surface (no overburden of the footing by soil).

QUESTION 6.5

Design the foundation of the building in Question 5.5. Each X-direction frame has its own foundation beam, whose length is 4L plus one column width. There are no tie-beams between column bases in direction Y: a concrete slab between the foundation beams provides non-monolithic connection. The foundation beam is designed as such in direction X; in direction Y, it works as an isolated footing, with plan dimensions equal to the length of the foundation beam and the width of its strip footing. The foundation beam should be chosen with cross-sectional moment of inertia, I, large enough to consider the beam as almost rigid relative to the ground. To this end, its elastic length (defined as its real length times $[k_s b/(4EI)]^{1/4}$, where k_s is the subgrade reaction modulus and b the width of the strip footing) should be less than 2.0. The soil is clay with design value of undrained shear strength $c_{ud} = 100$ kPa, unit weight $\gamma_{soil} = 21$ kN/m³ and subgrade reaction modulus, k_s, equal to $100/b$ (MPa/m), where b(m) is the width of the strip footing. The top of the foundation beam is at grade level.

QUESTION 6.6

Consider alternative foundation systems for the four columns of the elevated silo in Question 5.6 and design them. The soil is clay, with design value of undrained shear strength $c_{ud} = 260$ kPa, unit weight $\gamma_{soil} = 20$ kN/m³ and subgrade reaction modulus, k_s, equal to $200/b$ (MPa/m), where b(m) is the minimum dimension in plan of the foundation element. The top of the foundation elements is 0.5 m below grade.

QUESTION 6.7

Assume that you have chosen a square 6.6×6.6 m raft with thickness of 1.65 m for the foundation of the four columns of the elevated silo in Questions 5.6 and 6.6. The soil is clay, with design value of undrained shear strength $c_{ud} = 260$ kPa, unit weight $\gamma_{soil} = 20$ kN/m³, shear wave velocity $V_S = 400$ m/s and Poisson's ratio $\nu = 0.45$. The design peak ground acceleration is 0.30 g and the ground type is B. The stiffness of the foundation for horizontal translation and rocking are

$$k_h = \frac{8Gr}{2 - \nu}, \quad k_\theta = \frac{8Gr^3}{3(1 - \nu)}$$

where r is the equivalent foundation radius $r = \sqrt{(A_f/\pi)}$ (radius of a circular footing with the same area) and G is the soil shear modulus. Calculate the periods of vibration and compute the design base shear and bending moment under the design seismic action taking into account soil–structure interaction. Neglect the vertical component of the seismic action and assume a uniform damping ratio of 5% for the whole system. Furthermore, as a first approximation, you may neglect the rotation of the foundation. Compare the foundation forces with the results without SSI.

Chapter 7

Design example

Multistorey building

7.1 GEOMETRY AND DESIGN PARAMETERS

As a capstone of the book, a multistorey concrete building is fully designed in this chapter, as per Eurocodes 2 and 8. It is a real-life, regular building of Importance Class II (ordinary). The building has six storeys above ground and two basements, extending in one direction beyond the plan of the superstructure (Figures 7.1 to 7.3). The ground storey has a height of 4.0 m; all others, including the two basements, are 3.0 m tall.

The slabs are 0.18 m thick. Perimeter columns have a 0.30×0.70 m section, with the exception of the corner ones, which are 0.30×0.60 m. All interior columns are 0.50 m square. All beams have width $b_w = 0.30$ m and depth $h_b = 0.50$ m. Two rectangular walls, W1 and W2, with dimensions 0.3×4.0 m, are placed at the middle of the exterior frames in direction Y. Two more rectangular walls, W3 and W4, near the centre in plan, with dimensions 0.25×4.0 m, flank the staircase; a U-shaped wall (W5), with outside dimensions $1.8 \text{ m} \times 3.6$ m and thickness of 0.25 m, houses the elevator shaft. The cross-section of columns and beams and the slab thickness are the same as in the upper storeys. A 0.30 m thick retaining wall runs all along the basement's perimeter, serving as a deep foundation beam for the vertical elements of the perimeter.

The plan of the foundation is shown in Figure 7.4. All interior columns have isolated footings, $2.0 \times 2.0 \times 0.7$ m (width × length × depth). Columns C8, C9 and walls W3, W4, W5 share a spread footing of $7.0 \times 9.0 \times 0.7$ m. The perimeter walls have a strip footing, 1.0 m wide and 0.30 m deep. Instead of a system of two-way tie-beams, horizontal connection of the footings and the foundation strip of the perimeter walls is effected by a foundation slab, right below the top of the footings and the perimeter foundation strip; it serves as a floor to the lower basement and completes the box foundation system, together with the perimeter walls and the top slab of the upper-level basement. The top of each footing (including the strip footing of the perimeter wall) is flush with the floor of the basement at Level -2; so, each footing is embedded only over its own depth.

The soil corresponds to Ground type B, for the definition of the seismic action at the top of the ground. It is a saturated clay, with the:

- Design value of undrained shear strength $c_{ud} = 300$ kPa in the 'persistent and transient design situation', reduced by 10% to 270 kPa in the 'seismic design situation'
- Design values of friction angle $\varphi_d = \delta_d = 20°$ ($\tan\delta_d = 0.364$) and of drained cohesion $c_d = 50$ kPa, applying under slow loading in the 'persistent and transient design situation'

These values, for use in Design Approaches DA1-2 or DA3 (see Section 6.2.1), include the partial factors, γ_M.

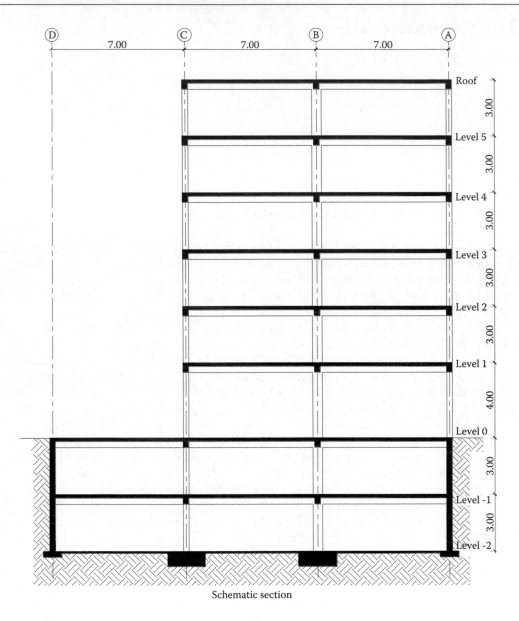

Schematic section

Figure 7.1 Section in Y direction.

C25/30 concrete and S500 steel of Class C as per Eurocode 2 are used. Environmental exposure class is XC3, for which the nominal concrete cover is taken as 35 mm.

The design actions are:

- Self-weight, calculated from a concrete density of 25 kN/m³; dead loads for partitions and finishings amount to an additional 2 kN/m².
- A live load of 2 kN/m² for Categories A (domestic/residential use), B (office) and F (parking and light traffic) of EN 1991-1. The quasi-permanent value of live loads is computed with the value $\psi_2 = 0.3$ recommended for Categories A and B in Annex A of EN 1990. The storeys are considered as independently occupied.

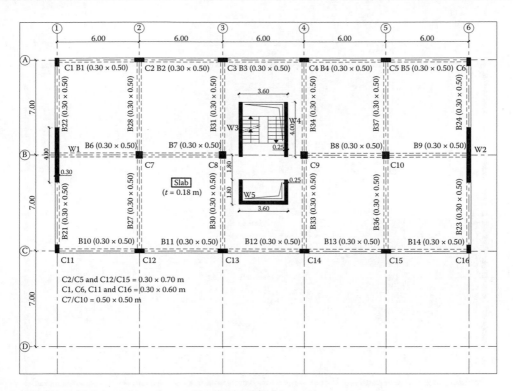

Figure 7.2 Typical framing plan – floors above ground level.

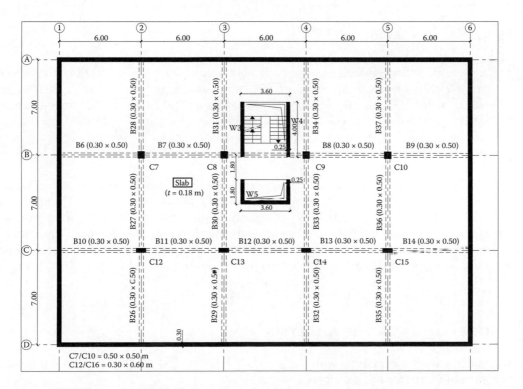

Figure 7.3 Framing plan – basement storeys.

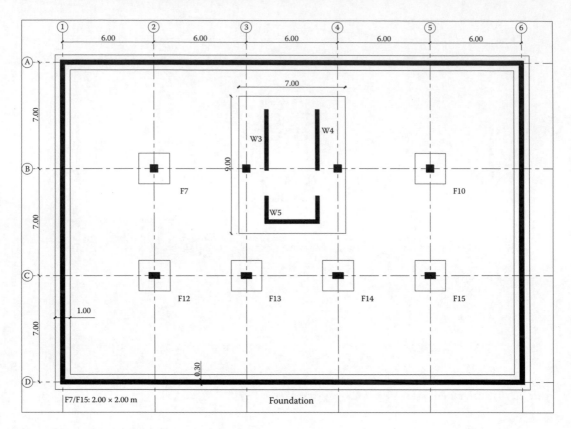

Figure 7.4 Foundation plan.

- The seismic action has reference peak ground acceleration (on rock) $a_{gR} = 0.25$ g; the Type 1 elastic spectrum of Eurocode 8 applies for Ground type B, with the recommended values ($S = 1.2$, $T_B = 0.15$ s, $T_C = 0.5$ s, $T_D = 2.0$ s). For Importance Class II (ordinary), the design peak ground acceleration is $a_g = \gamma_I \, a_{gR} = 1.0 \times 0.25$ g $= 0.25$ g. The conditions for ignoring the vertical component are met.
- Seismic design as per Eurocode 8 is for DC M (Medium).

The building has developed from the example prepared by the authors for the Workshop 'Eurocode 8: Seismic design of buildings', organised in Lisbon by the European Commission in February 2011 (Bisch et al. 2012). A wide range of modifications have been introduced in the geometry and the modelling, while the scope of analysis now includes the lateral-force method. Another version of the Lisbon example was the basis of the Workshop on Eurocode 2, organised in Brussels by the European Commission in February 2013 (Biasoli et al. 2014).

7.2 MODELLING FOR THE ANALYSIS

A 3D model is built for analysis with the software ETABS (CSI 2002). Modelling followed the guidance in Eurocode 8, as relevant.

7.2.1 General modelling

Every member of the superstructure, or interior ones of the basement, is modelled as a linear prismatic element, in a one-to-one correspondence: its two nodes coincide with its connections to other members.

- The elastic flexural and shear stiffness of an element are taken to be equal to 50% of those of the uncracked concrete element (see Section 3.1.11). Its elastic torsional stiffness is taken as 10% of that of the uncracked concrete element.
- As vertical elements have large cross-sectional dimensions (normally larger than the beams), the length of beams that falls within a physical joint with a column or a wall is taken as rigid; that of vertical elements within a physical joint with a beam is taken as elastic.
- The effective flange width on each side of the web of a T- or L-beam is taken in the model according to Eurocode 2; it comes out as 20%–70% of the beam span, that is, 0.85 or 1.0 m for the beams in the X- or Y-direction, respectively.
- Concrete floors are considered as rigid diaphragms; floor masses are lumped at a master node at the centroid of the floor in plan (at the intersection of the floor diagonals), alongside the associated rotary moment of inertia.
- Masses are computed for gravity loads, $G + \psi_E Q$, where $\psi_E = \varphi \psi_2$, with $\psi_2 = 0.3$ and $\varphi = 1$ at the roof, $\varphi = 0.5$ at all other levels (see Section 1.3.1); floor masses are listed in Table 7.1.
- Uniform surface loads on slabs due to gravity are assigned to the nearest beam or wall among those surrounding the slab; the resulting trapezoidal line load on beams is then uniformised along the beam; loads assigned to walls W3, W4, W5, which are not connected to beams at floor levels, are applied to their end node at the floor as a concentrated load.

7.2.2 Modelling of the foundation and the soil

The foundation elements are not considered as fixed in the vertical direction; the compliance of the foundation soil is explicitly included in the model. This allows computing the internal forces (moments and shears) in the deep foundation beams around the basement. Although they also play the role of perimeter walls, the mechanical behaviour of these perimeter elements is controlled by their longer dimension, that is, the horizontal. So, they are modelled here as deep foundation beams on Winkler springs.

Table 7.1 Storey masses, forces for the lateral-force method and torsional moments due to accidental torsion

Storey	z_i (m)	m_i (ton)	$z_i m_i$ (tonm)	f_i/V_b	f_{Xi} (kN)	f_{Yi} (kN)	$e_{aXi} f_{Xi}$ (kNm)	$e_{aYi} f_{Yi}$ (kNm)	M_{SRSS} (kNm)
6	19	373.2	7090	0.27	622	933	444	1413	1482
5	16	390.6	6249.8	0.23	548	822	392	1246	1306
4	13	390.6	5078.1	0.19	446	668	318	1012	1061
3	10	390.6	3906.2	0.15	343	514	245	778	816
2	7	390.6	2734.4	0.10	240	360	171	545	571
1	4	402.7	1610.8	0.06	141	212	101	321	337
Sum:		2338.3	26,669.2	1	2340	3510			

Note: The mass of Levels 0 and –1, amounting to 1442 tons, is not considered.

It is important to stress from the outset that modelling of the vertical soil compliance is not essential for a building with such a deep box-type basement as the present one; it may even be uncommon in practice. Indeed, to reflect this reality, the model of the original version of the building in Bisch et al. (2012) did not consider the vertical compliance of the ground. However, in buildings on normal-depth foundation beams (even, in some cases, with storey-deep perimeter basement walls) or footings connected with tie-beams, or combinations thereof, it is essential to model vertical soil compliance, in order to:

- take into account the impact of differential settlements or rotations at different points of the foundation on the superstructure
- calculate the internal forces in the foundation beams and the tie-beams, which control the amount of their reinforcement or the verification of their dimensions

under all types of design actions (gravity, seismic, etc.). To illustrate one way of doing this in more general cases and to show how the analysis results are used in the design of these foundation elements as per Eurocode 2, vertical soil compliance is included, although in this case it could very well be neglected.

As already indicated and stressed further in Section 7.2.3, the main point of including soil compliance is to capture the vertical deflection of the perimeter walls working as deep foundation beams on a continuous elastic support. The only practical way of doing this is by considering them on a bed of Winkler springs. The subgrade reaction modulus of these springs is taken here to be equal to $k_s = 250/b$ (MPa/m), where b(m) is the width (smaller dimension in plan) of the foundation element. This value is consistent with the soil parameters and the categorisation of the soil as type B per Eurocode 8; for simplicity, the same value is taken to apply both for static and for seismic loading. Vertical springs are introduced at the support nodes at mid-width of the underside of the strip footing under the perimeter wall-cum-foundation beam; their vertical stiffness per linear meter of the footing length is $k_v = k_s b = 250$ MN/m/linear meter.

Soil compliance is also taken into account under the footings, including the spread footing shared by columns C8, C9 and walls W3, W4, W5; in this way, the vertical soil reactions under the footings can be realistically estimated, taking into account their vertical displacement and rotations with respect to the ground. As the footings are thick enough to be considered as rigid, the bed of distributed vertical Winkler springs under each footing may be replaced by a single vertical spring and two rotational springs about the two horizontal directions, all connected to a node at the centre of the underside of each individual footing (cf. Sections 6.1 and 6.3.1). The stiffness of the vertical spring and those for rotation about the x or y axis are taken as (ASCE 2007):

$$K_v = \frac{G_s b_x}{1 - v}\left[1.55\left(\frac{b_y}{b_x}\right)^{3/4} + 0.8\right]\left[1 + \frac{2}{21}\frac{t}{b_x}\left(1 + 1.3\frac{b_x}{b_y}\right)\right]\left[1 + 0.32\left(\frac{d(b_y + b_y)}{b_x b_y}\right)^{2/3}\right] \quad (7.1)$$

$$K_{\varphi x} = \frac{G_s b_x^3}{1 - v}\left[0.47\left(\frac{b_y}{b_x}\right)^{2.4} + 0.034\right]\left[1 + 1.4\left(\frac{d}{b_y}\right)^{0.6}\left(1.5 + 3.7\frac{(d/b_y)^{1.9}}{(d/t)^{0.6}}\right)\right] \quad (7.2)$$

$$K_{\varphi y} = \frac{G_s b_x^3}{1 - v}\left[0.4\frac{b_y}{b_x} + 0.1\right]\left[1 + 2.5\frac{d}{b_x}\left(1 + \left(2d/\sqrt{b_x b_y}\right)/(d/t)^{0.2}\right)\right] \quad (7.3)$$

In Equations 7.1 to 7.3, G_s is the secant shear modulus of the soil at the expected shear strain level and ν is its Poisson ratio; b_x is the smaller of the two plan dimensions of the footing and b_y is the larger one ($b_x < b_y$); the x-axis is parallel to b_x and the y-axis is parallel to b_y; d is the effective embedment depth and t is the soil depth from the surface to the underside of the footing.

For consistency with the modelling of the soil impedance along the foundation-beam-cum-perimeter-wall via the subgrade reaction modulus, k_s, the vertical spring stiffness, expressed as:

$$K_v = k_s b_x b_y \tag{7.4}$$

is set equal to the value from Equation 7.1. In this way, a footing-specific relationship between k_s and $G_s b_x/(1-\nu)$ is established; to derive this relationship, we set the effective embedment depth, d, equal to half the depth of the footing. (Because full lateral contact is presumed to develop over d, capable of both sidewall friction and passive earth pressure, d is normally less than the depth of the footing). We use in Equations 7.1 to 7.3 the value of $G_s b_x/(1-\nu)$ determined in this way, namely as:

$$k_s \approx \frac{3.17G}{b_x(1-\nu)} \text{ for the isolated footings under single columns} \tag{7.5a}$$

$$k_s \approx \frac{2.25G}{b_x(1-\nu)} \text{ for the spread footing under walls W3, W4, W5}$$
$$\text{and columns C8, C9} \tag{7.5b}$$

Note that Equations (7.5), which are based on recent analytical developments, give significantly higher values than the old expression often used for k_s (Horvath 1983):

$$k_s \approx \frac{1.3G}{b_x(1-\nu)} \tag{7.6}$$

By back-calculating the value of $G_s b_x/(1-\nu)$ from the value $k_s = 250/b$ (MPa/m) and replacing in Equations 7.2 and 7.3, the stiffness of the springs connected to the node at the underside of a standard isolated footing of a single column are obtained:

$K_v = 500$ MN/m,
$K_{\varphi x} = K_{\varphi y} = 630$ MNm/rad;

Similarly, the vertical and rotational spring stiffness at the node at the centre of the base of the spread footing shared by walls W3, W4, W5 and columns C8, C9, are:

$K_v = 2250$ MN/m,
$K_{\varphi x} = 44{,}000$ MNm/rad, $K_{\varphi y} = 26{,}000$ MNm/rad,

for rotation about an axis parallel to X or Y, respectively.

The base node of every interior vertical element is placed at the top of its footing and connected to a node at the centre of the underside of the footing through fairly rigid linear elements. The soil springs are connected to that latter node at the underside of the footing. For the spread footing shared by walls W3, W4, W5 and columns C8, C9, the base node of each one of these elements is connected to the node at the centre of the underside of the footing through fairly rigid linear elements.

All nodes of the foundation are taken at the horizontal level of the underside of the strip footing of the perimeter walls; the small difference with the elevation of the underside of the interior footings is ignored. All these foundation nodes are restrained against translation in both horizontal directions and for rotation about the vertical axis.

7.2.3 Modelling of perimeter foundation walls

As the dimension governing the behaviour of the perimeter basement-wall-cum-foundation-beam is the long one, that is, the horizontal, this wall is modelled as a horizontal prismatic beam resting on a bed of vertical springs according to Section 7.2.2. It has an asymmetric I section, with the same depth and thickness as those of the basement wall, and top and bottom flanges roughly consistent with the effective width of the top slab of the basement and the strip footing of the perimeter wall. The centroidal axis of the horizontal elements modelling the perimeter basement wall as a deep foundation beam is placed at Level 0 (top of the basement). Nodes on that axis are connected to the nodes of the foundation at the underside of the wall's strip footing by fictitious vertical elements running through the depth of the basement wall. These elements are rigid in the axial direction, but not in the lateral one, within the plane of the perimeter wall; their aggregate stiffness in that plane reproduces the overall horizontal stiffness of the perimeter wall in shear. The main problem arising from concentrating that stiffness to few fictitious vertical elements, each one right underneath a vertical member of the superstructure, is that the moment diagram of the horizontal beam modelling the perimeter wall as a deep foundation beam shows a major discontinuity at the connecting node. So, a curtain of fictitious vertical elements running through the depth, $H = 6.3$ m, of the basement wall is introduced, by placing nodes every 1.0 m along its axis and connecting each one to a soil node underneath (at Level -2) via a fictitious vertical element running through the depth of the basement. The cross-section of each element is such that its lateral stiffness as a vertical cantilever fixed at Level 0 (i.e. at the centroidal axis of the horizontal elements modelling the perimeter wall), smeared per unit length of the wall, is equal to the shear stiffness, Gt/H, of the perimeter wall per unit length (t being the wall thickness). The vertical soil spring at the tip of a fictitious vertical element has a vertical stiffness representing 1 m of the length of the strip footing, that is, $250 \times 1 = 250$ MN/m. Were a special 'Beam-on-Elastic-Foundation' prismatic element included in the library of the analysis software used, a single horizontal element of that type would have been used to connect the base nodes of adjacent vertical elements of the perimeter directly at level 0, avoiding the curtain of fictitious vertical elements through the depth H of the basement wall; in that case, the shear flexibility of the 'Beam-on-Elastic-Foundation' element should reflect the in-plane lateral stiffness of the basement walls. Note that these particular walls could have been modelled, instead, with shell-type finite elements. However, those elements are not always available in commonly used analysis software. Moreover, their dimensioning at the Ultimate and Serviceability Limit States is not fully covered by Eurocode 2 and their connectivity with the linear elements of the superstructure is delicate. Last, but not least, the modelling adopted here is the only option for the less deep foundation beams normally used in earthquake resistant buildings.

As noted in Section 7.2.2, the model of the original version of the building in Bisch et al. (2012) neglected the vertical compliance of the ground, as inconsequential. Then, modelling the vertical flexibility of the perimeter wall is irrelevant. So, in that model the perimeter wall was modelled as the combination of:

a. A set of practically rigid horizontal prismatic elements connecting the bases of adjacent vertical members at Level 0; this was the counterpart of the horizontal prismatic elements used here at the same level, but using here the moment of inertia of the vertical section of the wall, plus its effective top and bottom flanges.

b. A vertical continuation of each perimeter vertical element from Level 0 down to -2; but with a strong-axis rigidity, EI, for bending parallel to the plane of the perimeter wall equal to the total rigidity of that side of the perimeter wall (from-corner-to-corner) divided by the number of vertical elements replacing it in the model (five on the X direction sides, three on the Y-direction ones, with the corner elements counting as half); this was the counterpart of the curtain of vertical elements used here at a 1 m spacing to reproduce the overall in-plane stiffness of the perimeter wall in shear.

That model was considered here as a benchmark, and sensitivity studies were carried out for the original building geometry in Bisch et al. (2012) and the present model. Different locations of the axis were tried (at mid-depth, or at the bottom of the basement, Levels -1 and -2, respectively) and different stiffness values of the horizontal prismatic elements modelling the foundation beam-cum-basement wall were used (including practically infinite stiffness). Their conclusion was that the chosen location gives seismic moments at the base of the perimeter vertical elements (including the two exterior walls W1, W2) in good agreement with the benchmark model in Bisch et al. (2012). Another key criteria for the model used here are the modal periods and participating masses it produces for the original building vs those of the benchmark in Bisch et al. (2012). As shown in more detail in Section 7.3.5, the two lower modal periods per horizontal direction, which capture nearly the full mass of the superstructure, differ little between the two models: in direction X, by 1% in the first mode and by 3% in the second; along Y, by less than 5% in the first mode and by less than 10% in the second. Sensitivity studies concerning the stiffness of the soil or the basement itself show also that the differences in modal periods are exclusively due to the compliance of the soil, not to the different ways of modelling the basement. Overall, the model used for the soil and the basement impacts much less the participating masses and the forces on the superstructure in the two lower modes per direction than the periods of these modes. Therefore, the conclusion of the sensitivity analyses is that the model used here does not materially alter the attributes of the response which are of prime importance for the design of the superstructure, while it allows to estimate the internal forces in the foundation elements themselves.

At any rate, if the flexibility of the soil under a box-type foundation is included in a global soil-foundation-superstructure model, it is strongly recommended to carry out sensitivity studies of the type highlighted here and in Section 7.3.5 concerning the results in the superstructure.

7.3 ANALYSIS

7.3.1 Fraction of base shear taken by the walls: Basic value of behaviour factor

A preliminary static analysis with the Eurocode 8 lateral-force method is carried out per Section 3.1.6, separately in each one of the two horizontal directions, X, Y, in order to estimate the fraction of the elastic seismic base shear, V_b, taken by the walls and to classify the building accordingly as a wall system, a wall-equivalent dual or a frame-equivalent one. Only the relative magnitude of the storey lateral loads used in this analysis per Equation

3.98 matters. So, they are taken equal to $z_i m_i$, that is, as if the design spectral acceleration, S_d, were equal to 1 g (see Table 7.1). The outcomes of these analyses are:

- $V_{wall,X}/V_{bX} = 63.7\%$ in direction X
- $V_{wall,Y}/V_{bY} = 91.4\%$ in direction Y

Therefore, the system is a wall-equivalent dual in X and a wall system in Y (cf. Section 4.4.4.1).

A side benefit of resisting at least 50% of the base shear in both directions with walls is that, according to Eurocode 8, masonry infills may be ignored in the seismic analysis and design.

As we will see in Section 7.3.3, the building is not torsionally flexible. Then, its basic q-factor value is (see Table 4.1 and Section 4.6.3):

- $q_{oX} = 3\alpha_u/\alpha_1$ in direction X
- $q_{oY} = 3.0$ in direction Y

The regularity of the building in plan determines the default value of α_u/α_1 in X (Section 4.6.3); regularity in elevation or lack thereof determines if there is a 20% reduction of the q-factor in X or Y (see Sections 4.3.5 and 4.6.3).

7.3.2 Possible reduction of behaviour factor due to irregularity in elevation or squat walls

Mere inspection of the drawings shows that the Eurocode 8 criteria for regularity in elevation regarding the variation of mass, stiffness and plan dimensions from storey to storey are met (cf. Section 4.3.4). Note that the large difference in stiffness and plan dimensions between the rigid basements and the superstructure is not relevant for the characterisation of the building as regular in elevation. This is clear from the fact that regularity in elevation is considered by Eurocode 8 as a prerequisite for the applicability of the lateral-force analysis method, while the clauses in Eurocode 8 concerning how the method is applied (e.g. the reference level from which z_i is measured) often refer to buildings with a rigid basement (i.e. they are considered within the scope of this analysis method).

The prevailing aspect ratio of the walls, for the determination of the k_w-factor which reflects the prevailing failure mode of walls (see Section 4.6.3), is $\alpha_o = \Sigma H_{wi}/\Sigma l_{wi}$, where H_{wi}, l_{wi} are the height and length of wall i, respectively (cf. Section 4.6.3):

- In direction X, for wall W5: $\alpha_{oX} = 25/3.6 = 6.9 > 2$, so $k_{wX} = 1$
- In direction Y and for all the walls: $\alpha_{oY} = 6 \times 25/[2 \times (4.0 + 4.0 + 1.8)] = 7.7 > 2$, so $k_{wY} = 1$

There is no reduction of the q-factor due to irregularity in elevation or due to the failure mode of the walls.

7.3.3 Torsional flexibility and regularity in plan: Final value of the behaviour factor

At a glance, the compact rectangular shape in plan (aspect ratio of 30.3/14.3 = 2.12), the perfect symmetry with respect to the Y direction, the near symmetry with respect to the X direction and the absence of large cut-outs or re-entrant corners suggest a building that is regular in plan. This visual impression should be supplemented with the check of the magnitude of the torsional radii, r_X, r_Y, with respect to the radius of gyration of the mass, l_s, and the eccentricity between the centres of stiffness and mass along direction Y (the one in X is zero).

Because in a frame–wall system the frames and the walls inherently have very different vertical patterns of drifts under lateral loading, the location of the centre of stiffness, x_{ck}, y_{ck}, and the torsional radii of a storey cannot be readily determined on the basis of the moments of inertia, I_{Xi}, I_{Yi}, and the co-ordinates, x_i, y_i, of the vertical elements alone. However, as the walls dominate the response, mainly in Y but by-and-large in X, the location of the stiffness centre and of the torsional radii is estimated here neglecting the frames, as well as the stiffness of the rectangular walls in the weak direction. This is safe-sided because, compared to the walls, the (ignored) frames are overall more symmetric and closer to the perimeter.

The origin of the X, Y axes is taken at the centre in plan:

The cross-section of W5 has:

Area: $A = 0.25 \times (2 \times 1.8 + 3.1) = 1.675$ m²; centroid $x_{cg} = 0$, $y_{cg} = -2.7 - 0.775 \times 0.25 \times 3.1/1.675 = -3.06$ m

- For response parallel to X:

$$\Sigma I_X = I_{W5,X} = 2 \times 1.8 \times 0.25^3/12 + 0.25 \times 3.1^3/12 + 2 \times 1.8 \times 0.25 \times 1.675^3 = 3.15 \text{ m}^4$$

- For response parallel to Y:

$$I_{W1,Y} = I_{W2,Y} = 0.3 \times 4^3/12 = 1.6 \text{ m}^4, \quad I_{W3,Y} = I_{W4,Y} = 0.25 \times 4^3/12 = 1.333 \text{ m}^4$$

$$I_{W5,Y} = 2 \times 0.25 \times 1.8^3/12 + 3.1 \times 0.25^3/12 + 3.1 \times 0.25 \times 0.775^2 - 1.675 \times (3.06 - 2.7)^2$$
$$= 0.4972 \text{ m}^4$$

$$\Sigma I_Y = 2 \times 1.6 + 2 \times 1.333 + 0.4972 = 6.363 \text{ m}^4$$

Equation 4.3: $x_{ck} = 0$, $y_{ck} = -3.06$ m

Torsional stiffness: $2 \times 1.6 \times 15^2 + 2 \times 1.333 \times 1.675^2 = 722.7$ m⁶

Torsional radii with respect to the stiffness centre, compared to the radius of gyration of the floor plan (cf. Equations 4.4, 4.7 and 4.8):

- $r_X = \sqrt{(722.7/6.363)} = 10.65 \text{ m} > l_s = \sqrt{[(14.3^2 + 30.3^2)/12]} = 9.67 \text{ m}$
- $r_Y = \sqrt{(722.7/3.15)} = 15.15 \text{ m} > l_s = 9.67 \text{ m}$

Eccentricity between mass and stiffness centres, compared to 30% of torsional radii with respect to the stiffness centre (Equation 4.1): $e_{oY} = -y_{ck} = 3.06 \text{ m} < 0.3 \times 15.13 = 4.54$ m.

The building is, therefore, torsionally stiff, even when the contribution of frames (mainly on the perimeter) is neglected. In fact, Bisch et al. (2012) used a rigorous but rather cumbersome approach, which takes accurately into account both the walls and the frames and their interaction, to compute the torsional radii of the original building of the Lisbon example, which was very similar to the present one but had slightly more flexible frames. The r_X and r_Y values computed there were about 25% or 40% larger, respectively, than the present approximate ones. The eccentricity in Y was also 30% smaller than estimated herein. So, the present approximation is safe-sided.

All the conditions for regularity in plan are met. Therefore, according to Section 4.6.3:

- In direction X: $\alpha_u/\alpha_1 = 1.2$ and $q_{oX} = 3\alpha_u/\alpha_1 = 3.6$.
- In direction Y: $q_{oY} = 3.0$.

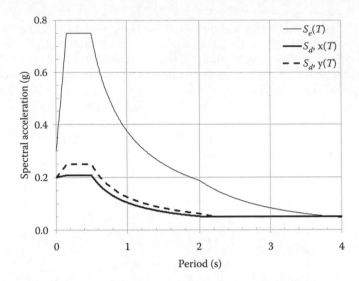

Figure 7.5 5% Damped elastic spectrum of design seismic action; design spectra in X and Y.

The elastic response spectra and the design spectra in the two directions are shown in Figure 7.5.

7.3.4 Lateral-force analysis procedure

As the building is regular in elevation, the analysis for the seismic action may be carried out with the lateral-force method (cf. Section 4.3.5).

The storey lateral forces for use in this procedure were indeed established from the floor masses listed in Table 7.1, assuming a design spectral acceleration, S_d, of 1 g, in order to carry out a lateral-force analysis and estimate from it the fraction of the seismic base shear taken by the walls (see Section 7.3.1 and Table 7.1). The storey drifts from the same analysis can be used to estimate the fundamental period of vibration, T_1, in directions X and Y through the Rayleigh quotient, Equation 3.109:

- $T_{1X} = 0.85$ s
- $T_{1Y} = 0.68$ s

If $T_C = 0.5$ s $\leq T_1 \leq T_D = 2$ s, the design spectral acceleration is $S_d(T_1) = 2.5(Sa_g)(T_C/T_1)/q$. So:

- In X: $S_d(T_{1X}) = 0.25$ g $\times 1.2 \times 2.5 \times (0.50/0.85)/3.6 = 0.12$ g
- In Y: $S_d(T_{1Y}) = 0.25$ g $\times 1.2 \times 2.5 \times (0.50/0.68)/3.0 = 0.18$ g

The fundamental periods in X and Y are less than $2T_C = 1$ s and the building has more than two storeys above the top of a rigid basement. So, according to Section 3.1.6, Eurocode 8 allows to reduce by 15% the base shear of the SDOF system corresponding to T_1, $mS_d(T_1)$, to account for a participating mass less than the total mass, m. So, the base shear at the top of the basement is:

- $V_{bX} = 0.85 \times 2338.3 \times 0.12 \times 9.81 = 2340$ kN
- $V_{bY} = 0.85 \times 2338.3 \times 0.18 \times 9.81 = 3510$ kN

These base shears are distributed to superstructure storey forces as per Equation 3.98 according to the relevant columns of Table 7.1.

A separate linear lateral-force analysis is carried out for the lateral forces in the two horizontal directions, X, Y. The outcomes of these analyses are deemed to represent the peak effects of the two horizontal seismic action components, in X and Y. For simultaneous action of these components, the results of these two separate analyses are combined via the 100%–30%, 30%–100% approximation as per Equations 3.100.

Detailed internal force results from the lateral-force method are given in Section 7.4.

7.3.5 Multi-modal response spectrum analysis: Periods, mode shapes, participating masses

The modal periods and corresponding participating masses for the (ten) vibration modes needed to capture at least 90% of the full mass of the superstructure and the basement are listed in Table 7.2 (cf. Section 3.1.5.3). As depicted in Figure 7.6, mode 1 is primarily translational in X, with a certain twist, due to the sizeable eccentricity in Y between the mass and stiffness centres, e_{oY}; mode 2 is purely translational along Y; mode 3 (like mode 6) is purely torsional. This sequence is consistent with the torsional radii estimated in Section 7.3.3, which were found larger than the radius of gyration of the floor mass in plan. Modes 4 and 5 are the second translational modes in X and Y, respectively. Note that the participating mass ratios refer to the total mass of the superstructure and the basement. As the basement mass amounts to almost 40% of that total mass (see footnote of Table 7.1), these first two modes per direction account for almost the full mass of the superstructure. This is in full agreement with the modal analysis results of the original structure in Bisch et al. (2012) with the benchmark model, which has the foundation nodes vertically restrained at Level -2 and apportions differently the lateral stiffness of the basement wall between vertical elements and a horizontal one at Level 0. Modes 7 in direction X and 8 along Y are third translational modes for the superstructure, but involve also significant lateral deformation of the basement. Two more translational modes in X are necessary in order to achieve more than 90% of the total mass in that direction. However, as pointed out, if only the dynamic response of the superstructure were of interest, that goal would have been achieved with just modes 1, 2, 4 and 5.

The 'Complete Quadratic Combination' of modal responses, Equations 3.88 to 3.90 in Section 3.1.5.3, is adopted.

Table 7.2 Modal periods and participating mass ratios with respect to the sum of superstructure and basement mass

Mode	Period (s)	m_x (%)	m_y (%)
1	0.86	53.3	0.0
2	0.69	0.0	53.5
3	0.49	0.1	0.0
4	0.22	11.4	0.0
5	0.16	0.0	21.1
6	0.12	0.3	0.0
7	0.10	6.2	0.0
8	0.08	0.0	17.8
9	0.07	15.9	0.0
10	0.06	3.8	0.0
	Sum:	91.1	92.3

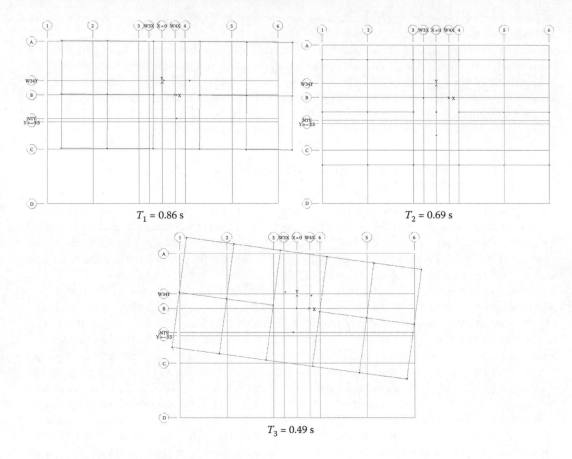

$T_1 = 0.86$ s $T_2 = 0.69$ s

$T_3 = 0.49$ s

Figure 7.6 First three natural modes of the building.

A single modal response spectrum analysis was carried out for the two horizontal components of the seismic action, using the different design response spectra shown in Figure 7.5. The maximum seismic action effect for simultaneously acting horizontal components is obtained by applying the SRSS combination, Equation 3.99, during that single modal response spectrum analysis.

7.3.6 Accidental eccentricity and its effects

The accidental eccentricities of masses at storey i, postulated by Eurocode 8 to be equal to $e_{ai} = 0.05L_i$, with L_i the floor dimension at right angles to the horizontal seismic action component, are in all storeys equal to $e_{aX} = 0.715$ m for the horizontal component along X and $e_{aY} = 1.515$ m for the one along Y (see Section 3.1.8). Their effects are determined by applying static torques $M_{ai} = e_{ai}f_i$ about a vertical axis passing through the centre of mass of each storey i in the superstructure, where f_i is the horizontal force on storey i used in the analysis with the lateral-force method of Section 7.3.4.

Since the fundamental method to combine the peak seismic action effects of the two simultaneous horizontal components is the SRSS rule of Equations 3.99, and, as such, has been adopted in Section 7.3.5 in the context of modal response spectrum analysis, it is also adopted here to combine the effects of the accidental eccentricities accompanying the two simultaneous horizontal components. In fact, the SRSS rule may be conveniently used from the outset, notably to combine the torque loadings in the two directions as $M_{SRSS,i} = \sqrt{[(e_{aXi}f_{Xi})^2 + (e_{aYi}f_{Yi})^2]}$. These SRSS torques, listed at the last column of Table 7.1, are used as the loading for a single static analysis. The action effects from that analysis are superimposed to those obtained from the lateral-force method as per Section 7.3.4, or the modal response spectrum analysis as per Section 7.3.5 for the two simultaneously acting translational components in X and Y.

7.4 SEISMIC DISPLACEMENTS FROM THE ANALYSIS AND THEIR UTILISATION

7.4.1 Inter-storey drifts under the damage limitation seismic action

Columns 2 and 3 in Table 7.3 list the elastic displacements in X and Y, $u_{e,i}$, obtained at the storey centre of mass via modal response spectrum analysis using the design spectra in Figure 7.5. Columns 4 and 5 give the estimate of the inelastic storey displacements under the design seismic action, obtained as per Eurocode 8 as $u_i = qu_{e,i}$, using the q-factor value pertinent to the corresponding horizontal direction ('equal displacement rule', Equation 3.116 in Section 3.2.2). The inter-storey drift, Δu_i, is obtained as $(u_i - u_{i-1})$, and normalised by the corresponding storey height into an inter-storey drift ratio, $\Delta u_i/h_i$. The inter-storey drift ratios listed in the last two columns of Table 7.3 and depicted in Figure 7.7 include the factor ν for the conversion of displacement demands due to the design seismic action to those expected under the 'damage limitation' earthquake; the value $\nu = 0.5$ recommended in Eurocode 8 for buildings of Importance Class II is adopted (see Section 1.3.2). These inter-storey drift ratios meet, at every storey, the limit of 0.5% specified in Eurocode 8 for buildings with brittle non-structural elements attached to the structure.

7.4.2 Second-order effects

According to Section 3.1.12, the inter-storey drift sensitivity index of storey i is defined as $\theta_i = N_{tot,i}\Delta u_i/(V_{tot,i}h_i)$, where $N_{tot,i}$ is the total gravity load at and above storey i in the seismic design situation (computed from the storey masses, m_i, in Table 7.1), Δu_i the inter-storey drift at the storey centre of mass under the design seismic action (computed per Section 7.4.1, but not listed in Table 7.3), $V_{tot,i}$ the total seismic storey shear from modal response spectrum analysis, and h_i the height of storey i (see Equation 3.110).

Table 7.3 Storey displacements and inter-storey drift ratios

Storey, i	$u_{eX,i}$ (m)	$u_{eY,i}$ (m)	$u_{X,i}$ (m)	$u_{Y,i}$ (m)	$\nu\Delta u_{X,i}/h_i$ (%)	$\nu\Delta u_{Y,i}/h_i$ (%)
6	0.031	0.031	0.110	0.098	0.258	0.277
5	0.026	0.026	0.095	0.082	0.294	0.288
4	0.021	0.020	0.077	0.064	0.306	0.293
3	0.016	0.015	0.059	0.047	0.324	0.272
2	0.011	0.010	0.039	0.030	0.300	0.235
1	0.006	0.005	0.021	0.016	0.243	0.152
0	0.001	0.001	0.002	0.004	0.012	0.032
−1	0.000	0.001	0.001	0.002	0.018	0.037

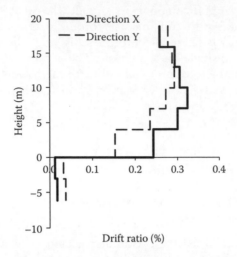

Figure 7.7 Inter-storey drift ratio under the damage limitation seismic action.

The values of θ_i listed in the last two columns of Table 7.4 are all well below the limit of 0.10, beyond which first-order effects should be divided by $(1 - \theta_i)$ to account for second-order ones.

7.5 MEMBER INTERNAL FORCES FROM THE ANALYSES

7.5.1 Seismic action effects

Figures 7.8 to 7.25 depict the effects of the two translational components of the seismic action without the torsional effects of accidental eccentricity per Section 7.3.6, as obtained from modal response spectrum analysis (with all values displayed as positive), or the lateral-force method (retaining the signs of action effects obtained for the particular sense of application of the lateral forces). The presented results are supposed to be due to both concurrent horizontal seismic action components. Results of modal response spectrum analysis are computed via the SRSS rule of Equation 3.99 in Section 3.1.7. By contrast, the approximate combination: $E_X + 0.3E_Y$, $E_Y + 0.3E_X$ as per Equation 3.100 is adopted for the lateral-force method. In that case, the action effects due to E_X and E_Y are 'added' with the same sign. The axial forces from the lateral force method are depicted for seismic action both in the

Table 7.4 Inter-storey drift sensitivity coefficient for $P - \Delta$ effects

Storey, i	$N_{tot,i}$ (kN)	$V_{tot,X,i}$ (kN)	$V_{tot,Y,i}$ (kN)	$\Delta u_{X,i}$ (m)	$\Delta u_{Y,i}$ (m)	h_i (m)	$\theta_{X,i}$	$\theta_{Y,i}$
6	3661	822	1238	0.0077	0.0083	3.0	0.011	0.008
5	7493	1309	1999	0.0088	0.0086	3.0	0.017	0.011
4	11,325	1701	2532	0.0092	0.0088	3.0	0.020	0.013
3	15,157	1995	2988	0.0097	0.0082	3.0	0.025	0.014
2	18,989	2258	3367	0.0090	0.0070	3.0	0.025	0.013
1	22,939	2455	3668	0.0097	0.0061	4.0	0.023	0.010
0	30,172	2722	4042	0.0004	0.0010	3.0	0.001	0.002
−1	36,225	3002	4311	0.0005	0.0011	3.0	0.002	0.003

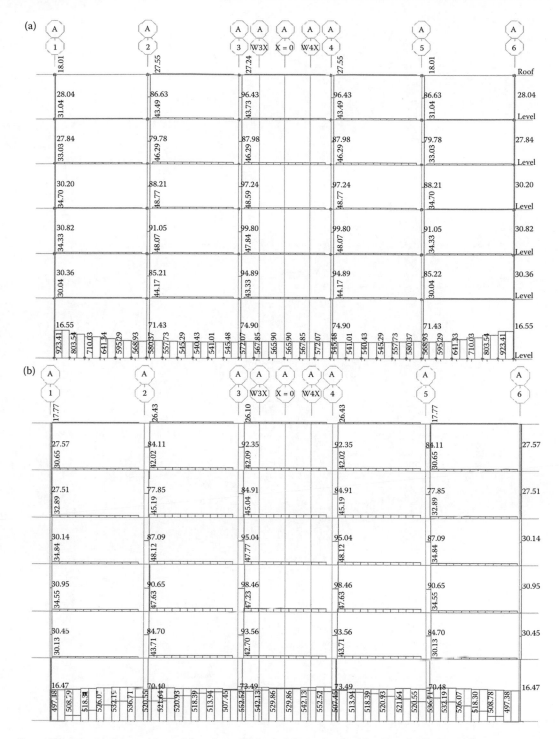

Figure 7.8 Seismic shears: (a) modal analysis; (b) lateral-force method for $E_X + 0.3E_Y$. Frame A.

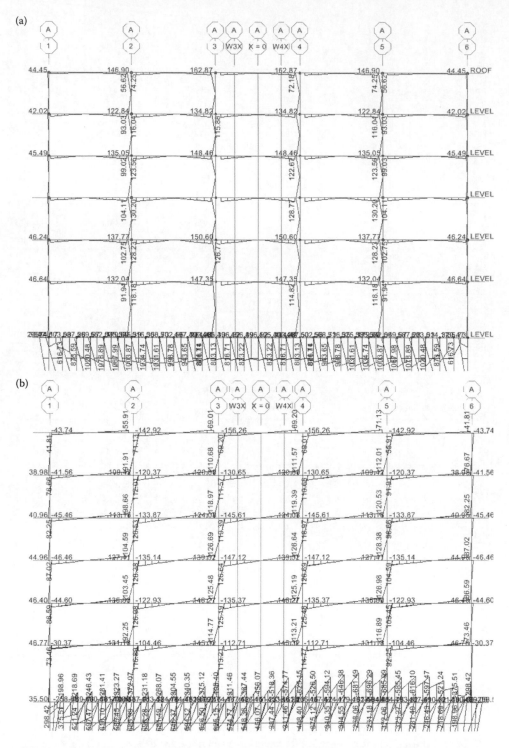

Figure 7.9 Seismic moments: (a) modal analysis; (b) lateral-force method for $E_X + 0.3E_Y$. Frame A.

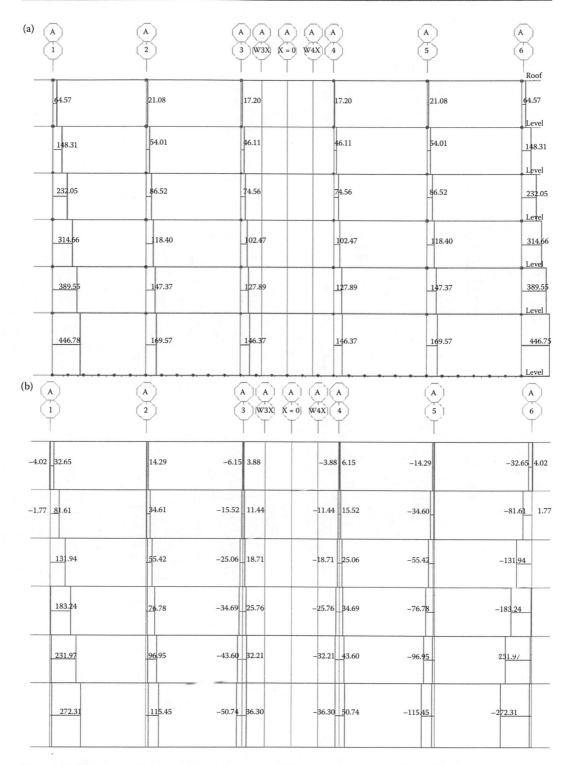

Figure 7.10 Seismic axial forces: (a) modal analysis; (b) lateral-force method, $E_X + 0.3E_Y$. Frame A.

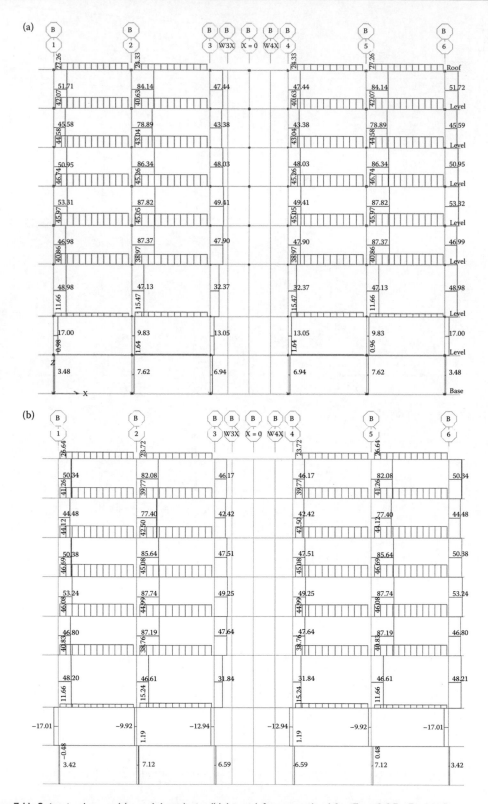

Figure 7.11 Seismic shears: (a) modal analysis; (b) lateral-force method for $E_X + 0.3E_Y$. Frame B.

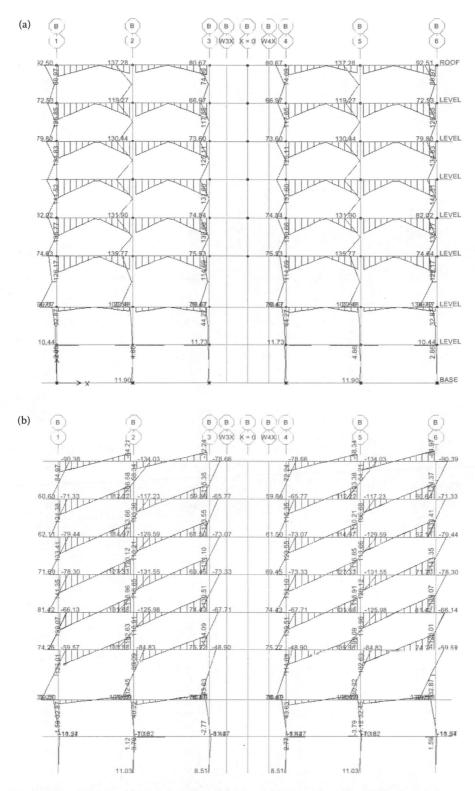

Figure 7.12 Seismic moments: (a) modal analysis; (b) lateral-force method, $E_X + 0.3E_Y$. Frame B.

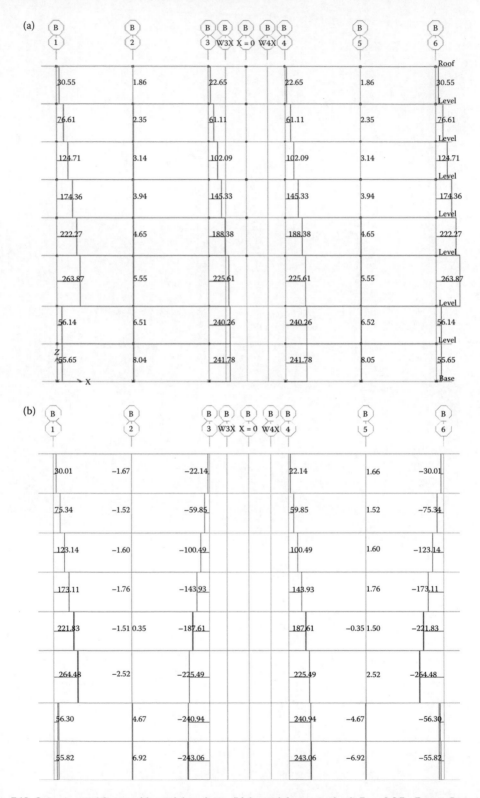

Figure 7.13 Seismic axial forces: (a) modal analysis; (b) lateral-force method, $E_X + 0.3E_Y$. Frame B.

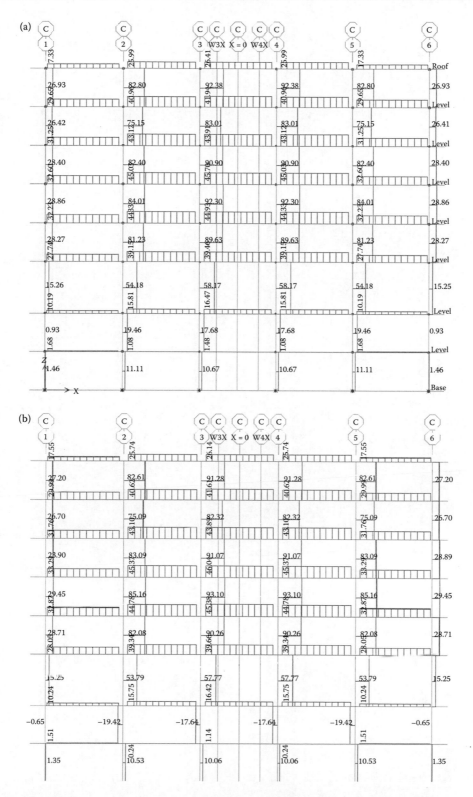

Figure 7.14 Seismic shears : (a) modal analysis; (b) lateral-force method for $E_X + 0.3E_Y$. Frame C.

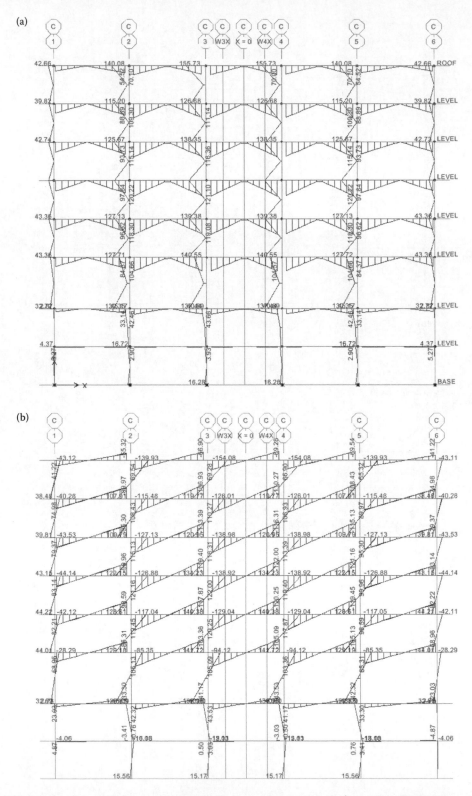

Figure 7.15 Seismic moments: (a) modal analysis; (b) lateral-force method, $E_X + 0.3E_Y$. Frame C.

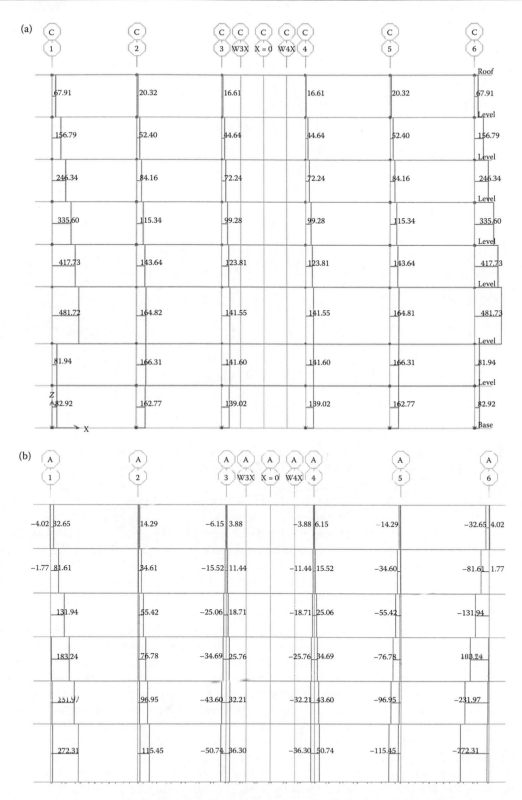

Figure 7.16 Seismic axial forces: (a) modal analysis; (b) lateral-force method, $E_X + 0.3E_Y$. Frame C.

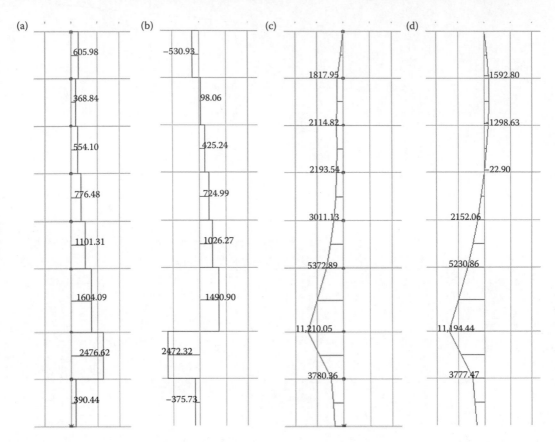

Figure 7.17 (a), (b): Seismic shears; (c), (d): seismic moments; (a) and (c): from modal analysis; (b) and (d): from lateral-force method for $E_X + 0.3E_Y$. W5 – X-plane.

plus and in the minus directions of the axes; for moments and shears, only one direction is shown. The scales used for the two types of analysis are not exactly the same.

Results are presented for frames A, B, C in the X-direction (Figures 7.8 to 7.16), frames 1 and 2 in the Y-direction (Figures 7.18 to 7.23) and walls W3 (Figure 7.24) and W5 (Figures 7.17 and 7.25 for bending in a plane parallel to X or Y, respectively). Frame 1 includes wall W1 (Figures 7.18 to 7.20). Results for frame 3 (not shown) do not differ much from those of frame 2 (Figures 7.20 to 7.23).

At the bottom of frames A (Figures 7.8 and 7.9) and 1 (Figures 7.18, 7.19) is the 6.3 m deep foundation beam which models the perimeter walls of the basement. Since the centroidal axis of that beam has been chosen to be at Level 0 (of the ground), its action effects are depicted at that level. Below Level 0 are depicted the action effects in the 6.3 m tall vertical elements, at a spacing of 1 m, which connect the centroidal axis of the foundation beam to the ground nodes at Level -2. Only the axial forces of these elements are meaningful and of interest. As a matter of fact, they are equal to the vertical reactions from the soil, which, if divided by the tributary contact area with the ground (i.e. the width of the strip footing times the spacing of these elements), give the average bearing pressure at the underside of the strip footing. As the strip footing is 1 m wide and the spacing of these elements is 1 m, the vertical reaction force of each element is numerically equal to the bearing pressure locally exerted on the soil (without the effect of the weight of the overlying soil, directly borne by the strip footing).

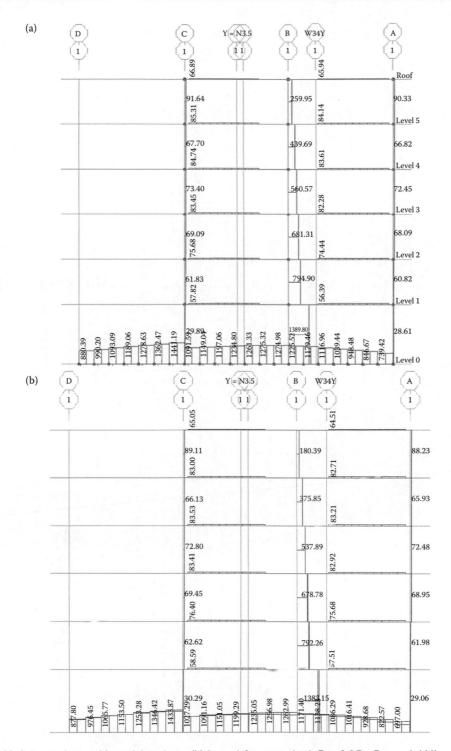

Figure 7.18 Seismic shears: (a) modal analysis; (b) lateral-force method, $E_Y + 0.3E_X$. Frame I, WI.

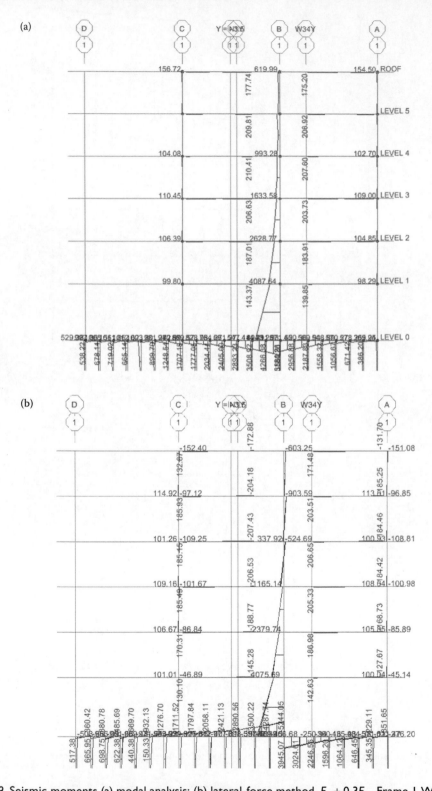

Figure 7.19 Seismic moments (a) modal analysis; (b) lateral-force method, $E_Y + 0.3E_X$. Frame I WI.

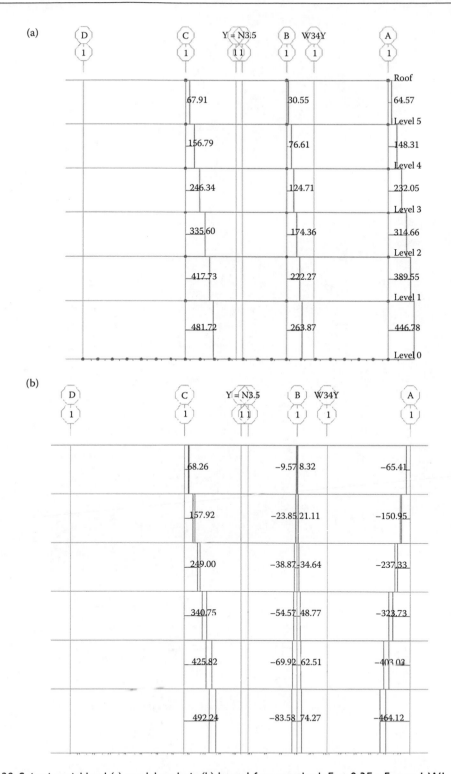

Figure 7.20 Seismic axial load (a) modal analysis (b) lateral-force method, $E_Y + 0.3E_X$. Frame I WI.

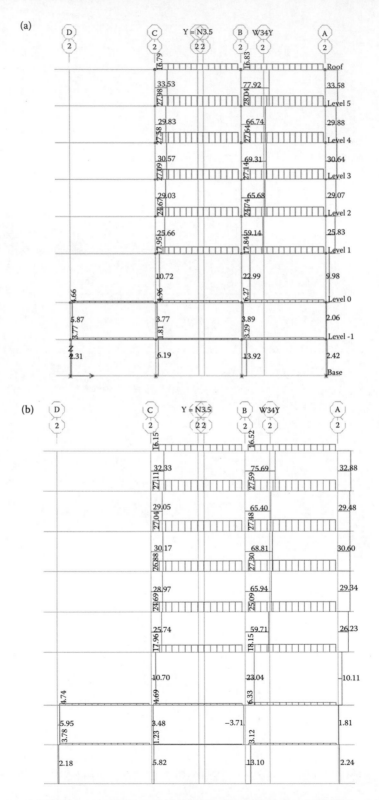

Figure 7.21 Seismic shears: (a) modal analysis; (b) lateral-force method, $E_Y + 0.3E_X$. Frame 2.

(a)

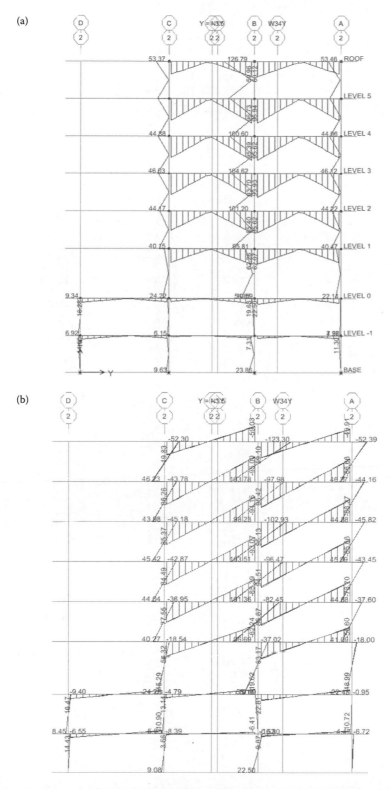

(b)

Figure 7.22 Seismic moments: (a) modal analysis; (b) lateral-force method for $E_Y + 0.3E_X$. Frame 2.

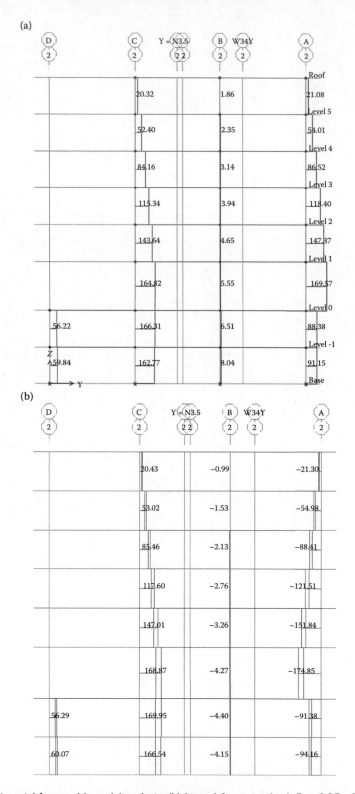

(a)

(b)

Figure 7.23 Seismic axial forces: (a) modal analysis; (b) lateral-force method, $E_Y + 0.3E_X$. Frame 2.

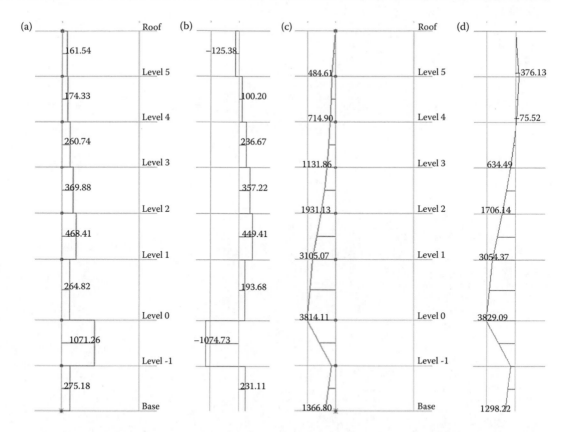

(a)   Roof   (b)   (c)   Roof   (d)

161.54   −125.38   484.61

174.33   Level 5   100.20   714.90   Level 5   −376.13

260.74   Level 4   236.67   1131.86   Level 4   −75.52

369.88   Level 3   357.22   1931.13   Level 3   634.49

468.41   Level 2   449.41   3105.07   Level 2   1706.14

264.82   Level 1   193.68   3814.11   Level 1   3054.37

1071.26   Level 0   −1074.73   3814.11   Level 0   3829.09

275.18   Level -1   231.11   1366.80   Level -1   1298.22

                                          Base   Base

Figure 7.24 (a), (b): Seismic shears; (c), (d): seismic moments; (a) and (c): from modal analysis; (b) and (d): from lateral-force method for $E_Y + 0.3E_X$. Wall W3.

The seismic action effects in the deep foundation beam of Frame A (Figures 7.8 and 7.9) are partly due to the seismic action component orthogonal to Frame A (i.e. in the Y-direction). Most of the overturning moment due to that component is transferred to the ground through bearing pressures distributed fairly uniformly along that beam (as well as counterpart pressures along the foundation beam of Frame D). By contrast, the seismic action effects along the deep foundation beam of Frame 1 (Figures 7.18 and 7.19) are almost fully due to the in-plane seismic action component (in the Y-direction) and are controlled by the transfer of the large moment of wall W1 to the ground via that beam.

Witness in Figures 7.11 to 7.15 and 7.21, 7.22 the very low magnitude of seismic moments and shears in the beams and columns of the two basement floors: these floors transfer the seismic base shear and overturning moment from the superstructure to the ground through the perimeter walls, which are almost rigid. Note also, in Figures 7.8, 7.9, 7.11, 7.12, 7.14 and 7.15, that the seismic moments and shears in the different storeys of a column are approximately the same, which is typical of dual systems; beams occupying the same bay at different floors below the roof exhibit the same pattern. Witness also in Figures 7.18, 7.19, 7.21 and 7.22 the notable increase of column and beam seismic moments and shears from the ground to the roof, as if in that direction (Y) the building were a wall system. The maximum seismic moments and shears in the beams or columns of a frame occur in general at roof level; the smallest ones at the ground floor. The beam flexural reinforcement, depicted later in Figures 7.34 through 7.39, follows that trend.

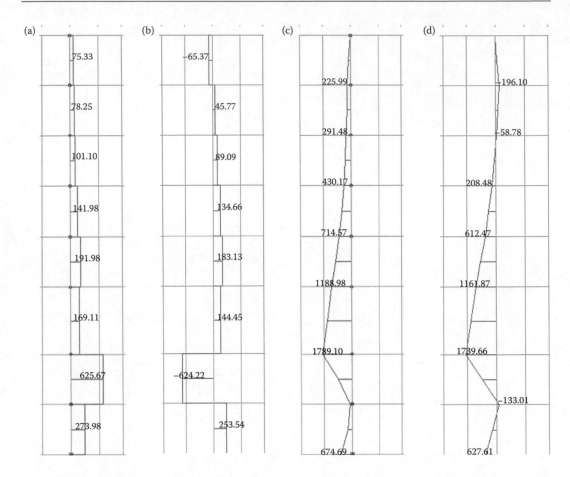

Figure 7.25 (a), (b): Seismic shears; (c), (d): seismic moments; (a) and (c): from modal analysis; (b) and (d): from lateral-force method for $E_Y + 0.3E_X$. Wall W5, Y-plane.

Witness also in Figures 7.17, 7.24 and 7.25 the large seismic shears in the two basement storeys of walls W3 and W5, especially in the upper floor. Consistent with the reversal of the trend in the moment diagrams of these walls at Level 0 (i.e. at the top of the basement), these shears have the opposite sense and sign with respect to the superstructure – a pattern observed also in interior columns of the upper basement storey in Figures 7.11 to 7.15 and 7.21, 7.22. They reflect the horizontal forces exerted on each wall by the basement's diaphragms at Level 0 and at the two levels below; the horizontal forces at these levels are opposite and produce the couple which fixes each interior wall to the box-type foundation.

Seismic action effects from modal response spectrum analysis or from the lateral-force method are very consistent, except possibly at locations where both are relatively low and far from critical. A notable exception is that the lateral-force method predicts much lower seismic axial forces in the corner columns C1, C8, C11, C16 than modal analysis, when these columns are considered as parts of the X-direction frames A and C (Figures 7.10 and 7.16). This discrepancy is fictitious: Figures 7.10b and 7.16b depict the smallest of these columns' axial forces for $E_X + 0.3E_Y$, $E_Y + 0.3E_X$, whereas it is the largest value that should be compared to the outcome of the SRSS rule as per Equation 3.99 in Figures 7.10a and 7.16a. That largest value is indeed depicted in Figure 7.20b for both types of corner columns and is consistent with the outcomes of modal analysis in Figure

7.20a. Note, however, that in order to dimension the corner columns in bending about their weak axis for the peak uniaxial moments in Figures 7.9 and 7.15, it is physically more meaningful and consistent to use the smaller seismic axial forces in part (b) of these figures than the more rigorous peak values in part (a). As a matter of fact, columns are not meant to be dimensioned for the peak seismic biaxial moments and axial forces from the SRSS rule per Equation 3.99, as if they were simultaneous; unless the peak value of each individual component is combined with the likely concurrent values of the other two according to the demanding approach highlighted in Fardis (2009), they may be dimensioned for the simple approximation of Equation 5.75 in Section 5.8.1 (see also right-hand side of Figure 3.10).

The seismic action effects shown in Figures 7.8 to 7.25 are superimposed on their counterparts from the static analyses for the torsional effect of the accidental eccentricities, considered with the same sign. The outcome is superimposed then, with plus and minus sign, to the effects of the quasi-permanent gravity actions, $G + \psi_2 Q$, present in the seismic design situation. These latter gravity action effects are illustrated next.

7.5.2 Action effects of gravity loads

The action effects due to the permanent, G, and the variable, Q, gravity loads are computed from separate static analyses, to be combined:

a. As per Equations 6.10a and 6.10b of EN1990 for the 'persistent and transient design situation'.
b. In the 'quasi-permanent' combination of loads, $G + \psi_2 Q$, considered as concurrent with the design seismic action in the 'seismic design situation'.

Although the uncracked stiffness should be used in the analyses for gravity actions, cracked stiffness values are used here instead, for convenience in combining their results with those of the analyses for the seismic action. If it were not for the modelling of the soil compliance, the use of cracked stiffness would not change the action effects of gravity loads, because the reduction factor on the uncracked stiffness is constant all over the structure (50%).

Analysis results for the 'quasi-permanent' combination are depicted in Figures 7.26 to 7.33. They do have signs and should be superimposed with these signs to the seismic action effects (including the results of the static analyses for torques due to accidental eccentricities per Section 7.3.6); the latter are taken with plus or minus sign.

Witness in Figure 7.26 that the deep foundation beam under frame A works under gravity loading as an inverted continuous beam. Witness also in Figure 7.29 that the deep foundation beam under frame 1 works as if it cantilevers from wall W1 to the left and right. Note that, owing to the model used for the foundation beams, the peaks of moment and shear along them are fictitious. At any rate, their magnitude dwarfs the moments and shears in the superstructure, which, therefore, are not discerned easily in these two figures. The pattern of these moments and shears is very similar to those in Figures 7.28 and 7.30, respectively, but the values differ.

7.6 DETAILED DESIGN OF MEMBERS

7.6.1 Introduction

Design does not have a unique 'solution'. It leaves room to judgement, choice, or even interpretation of code rules. To facilitate construction and supervision, it is common

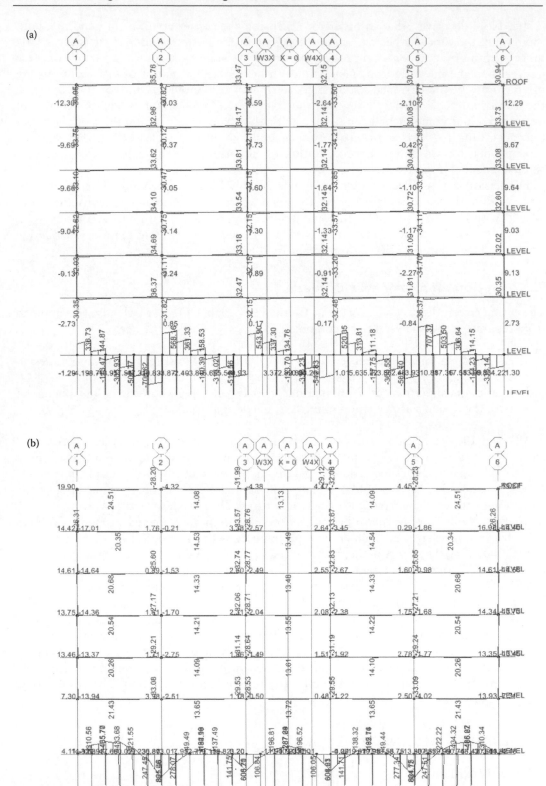

Figure 7.26 Shears (a); and moments (b), for quasi-permanent loads $G + \psi_2 Q$. Frame A.

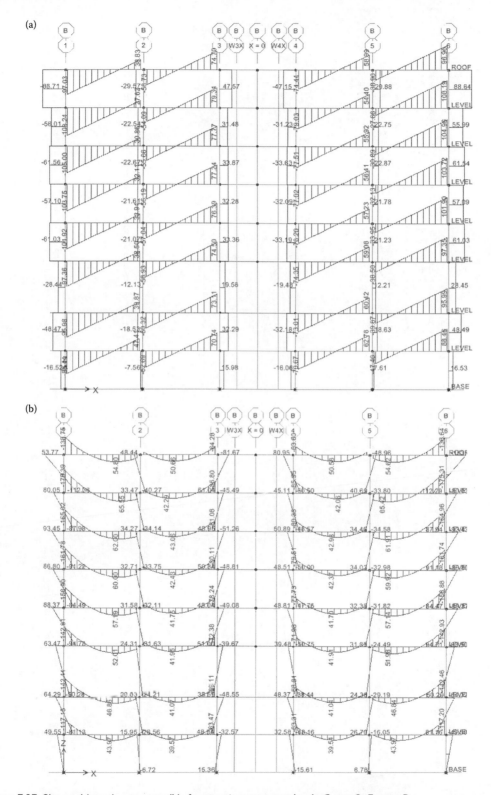

Figure 7.27 Shears (a); and moments (b), for quasi-permanent loads $G + \psi_2 Q$. Frame B.

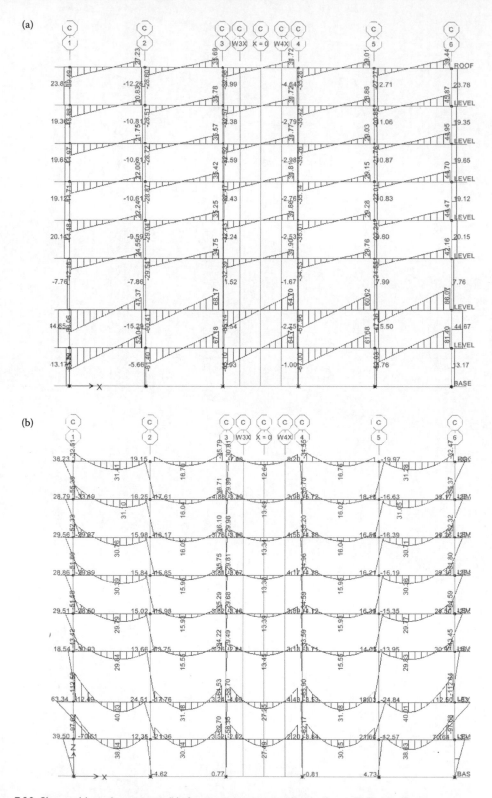

Figure 7.28 Shears (a); and moments (b), for quasi-permanent loads $G + \psi_2 Q$. Frame C.

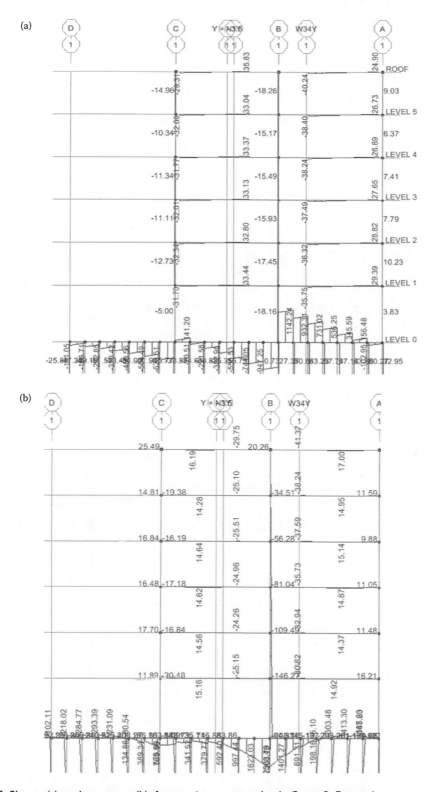

Figure 7.29 Shears (a); and moments (b), for quasi-permanent loads $G + \psi_2 Q$. Frame 1.

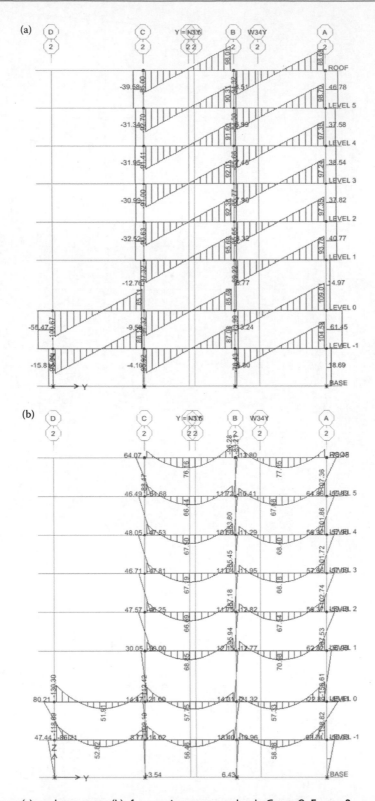

Figure 7.30 Shears (a); and moments (b), for quasi-permanent loads $G + \psi_2 Q$. Frame 2.

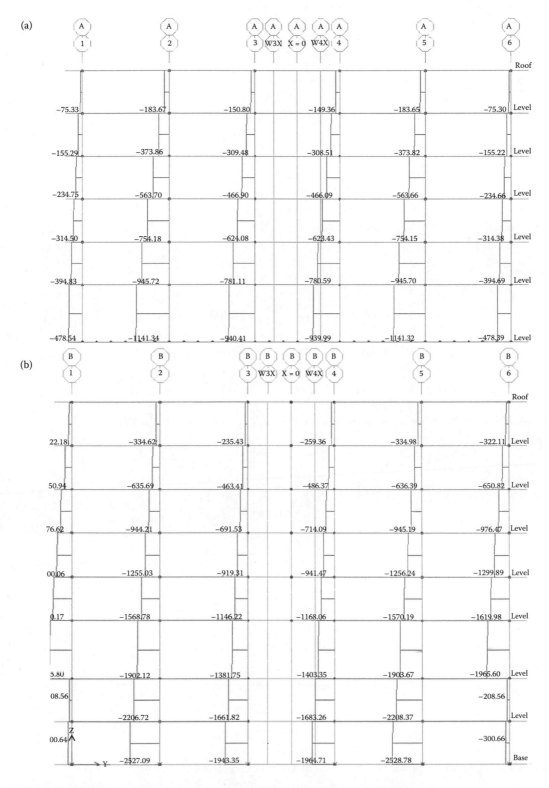

Figure 7.31 Axial forces for quasi-permanent loads $G + \psi_2 Q$. (a) Frame A; (b) Frame B.

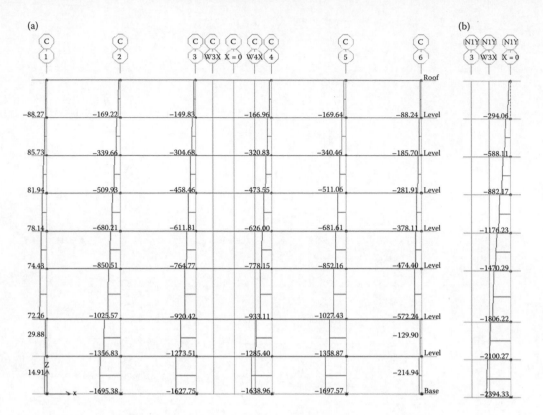

Figure 7.32 Axial forces for quasi-permanent loads $G + \psi_2 Q$. (a) Frame C; (b) Wall W5.

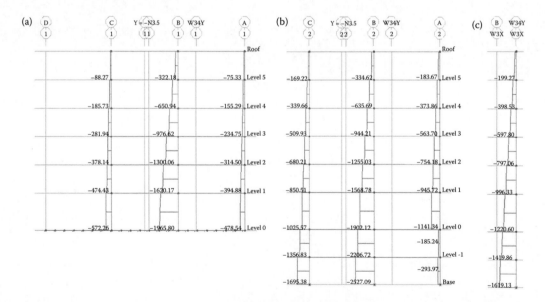

Figure 7.33 Axial forces for quasi-permanent load $G + \psi_2 Q$. (a) Frame 1; (b) Frame 2; (c) Wall W3.

practice to specify in the drawings the same reinforcement within families of geometrically similar beams or columns of the same floor, or across several storeys, or even throughout the building. To be on the safe side, the reinforcement placed is the one of the most adverse cases in the family, producing certain overstrength in all other cases. Of course, the design should fully account for these overstrengths during all capacity design calculations: the strong column–weak beam rule of Equation 5.31 of Section 5.4.1; the capacity design shears of beams or columns from Equations 5.42 and 5.44 of Section 5.5.1, or the magnification factor for footings from Equation 6.7a in Section 6.3.2 and so on. Therefore, these decisions cannot be made and implemented when the drawings for construction are prepared, that is, after dimensioning: they should be made before any subsequent stage of dimensioning that depends on the amount of longitudinal reinforcement of other members. Apart from the increased amount of steel reinforcement and its impact on cost and sustainability, the marked overstrengths that this common practice produces in several members over the demands of the 'seismic design situation' may lead to concentration of inelastic deformations in few locations, instead of spreading them uniformly throughout the building. For this reason, but also to avoid interpretation of subjective choices as the true intent of Eurocode 8, a 'minimalistic' approach is followed in the detailed design of this building: oversized members have been avoided, the reinforcement is tailored to the demands of the analysis for the 'seismic design situation' and the design/detailing rules of Eurocodes 2 or 8 are applied to the limit, without additional safety margins. Overstrengths and optional margins reflecting the choice of the designer, rather than the intent of Eurocodes 2 and 8, are avoided. The resulting large variation of reinforcement at different locations of the building may seem odd to an experienced designer.

After this introductory discourse, it is to be discussed how the results of the analyses are used for the purposes of dimensioning the members. The seismic action effects from modal response spectrum analysis are used. The peak effects of the two horizontal seismic action components are combined via the SRSS rule, Equation 3.99, at the same time as the modal combination is carried out through the CQC rule. We then add to the single seismic action effect for the two horizontal components of the seismic action, the absolute value of its counterpart from the static analysis under storey torques representing the torsional effect of the accidental eccentricities of both horizontal seismic action components. The outcome is superimposed, with a plus or minus sign, to the corresponding action effect of the quasi-permanent loads, $G + \psi_2 Q$, present in the 'seismic design situation'. Concerning the number and composition of cases for multi-component (vectorial) seismic action effects, Section 5.8.1 applies to columns and walls in bending and to footings; however, this case is not fully covered in Section 5.8.1, as its rigorous treatment requires complex expressions, not presented in Section 5.8.1, and calculations not supported by the commonly used analysis software; moreover, these calculations cannot take place in a post-processing module, outside the engine of modal combinations. So, the option chosen here is to list in tabular form the action effects that involve the peak values of all seismic action effects assumed to take place concurrently and taken with all combinations of signs, according to Equations 5.72 and 5.74. This gives eight cases at each section. However, although the 16 combinations per Equation 5.75 are not explicitly listed, they are the ones used, be it implicitly: section dimensioning in bending about the two local axes y and z takes place separately for the two uniaxial combinations $[\pm M_{y,max}, \pm N_{max}]$, $[\pm M_{z,max}, \pm N_{max}]$, tacitly assuming that the orthogonal components omitted, $\lambda M_{z,max}$, $\pm \lambda M_{y,max}$, respectively, are small.

7.6.2 Detailed design sequence

The design process of the full building takes place in a sequence of stages, as follows:

7.6.2.1 Stage 1: Beam longitudinal reinforcement (dimensioning for the ULS in flexure and the SLSs of stress limitation and crack control; detailing per EC2 and EC8)

Design of the beam longitudinal reinforcement is carried out for one multistorey plane frame at a time, storey by storey. The beams of each storey are dimensioned/detailed span-by-span, with continuity of the top and bottom bars from one span to another across a joint, as necessary.

The reinforcement requirements are determined and met at three sections:

- The two end sections, at the face of the supports (top and bottom reinforcement)
- A section at, or near, mid-span (bottom reinforcement only)

The first step is to determine the maximum beam bar diameter, $\max d_{bL}$, that can pass through or terminate at the beam–column joint at each end of the beam, on the basis of Equation 5.3 in Section 5.2.3.3. Input data for Equation 5.3 include the minimum value of the axial force in the column, N_{Ed}, among all combinations of the design seismic action with the quasi-permanent gravity loads ('seismic design situation').

The beams are fully designed:

- For the ULS in flexure, according to Section 5.3.1
- For the SLS of stress limitation in concrete and steel and of crack width (with $w_{max} = 0.3$ mm for environmental exposure XC3) and for the minimum steel area for crack control, according to Section 5.3.3

The minimum reinforcement, $A_{s,min}$, required all along the top and bottom flanges is determined per rows 2 and 4 of Table 5.1; it is implemented in the form of two bars per flange, normally with a diameter equal to the minimum of the two $\max d_{bL}$-values at the beam–column joints at the beam ends (or more than two bars, if this is needed in order to respect the $\max d_{bL}$ values at the beam–column joints of both ends). The other minimum reinforcement conditions in Table 5.1 are respected as well. To meet the maximum steel ratio limit in row 3 of Table 5.1 at the top flange of an end section for given steel area at the top, it is often necessary to add steel area to the bottom flange, in order to increase the compression steel ratio, ρ'.

The top and bottom bars which are continuous all along the span are supplemented with additional bottom bars at mid-length, and top and bottom bars at the support sections, to provide the required steel areas. Additional steel requirements on opposite sides of the same joint are covered, to the extent feasible, by common bars. Additional bottom bars at mid-length are combined with same-diameter additional bottom bars at one or both beam ends, if their ends overlap and so on.

A steel area of 250 mm^2 per metre of effective tension flange width according to Eurocode 8 is included in the top reinforcement area provided at the supports for the ULS in flexure and counts towards the hogging moment resistance there. The contributing slab width extends beyond each side of a supporting column by $4h_f$ at interior joints, or $2h_f$ at exterior ones, where h_f denotes the slab thickness. As this is according to a Eurocode 8 rule mentioned in Section 5.2.2, but there is no similar provision in Eurocode 2, this additional amount of top reinforcement does not count towards meeting the SLS stress or crack width limits, nor the minimum steel area for crack control. For the same reason it is not included in the top reinforcement area at the supports of the beams of the two basement floors, which are designed for the ULS in flexure to Eurocode 2 alone. These latter beams, from Level 0 down, are detailed according to the rules listed in Tables 5.1 and 5.3 under DC L.

Sometimes, the amount of top reinforcement at the support sections is governed by the SLS crack width limits, or the minimum steel area for crack control.

Archived for use in later stages are:

- The design values of beam moment resistances, $M_{Rd,b}$, around joints, to be used in Stage 2 below for the capacity design of columns as per Equation 5.31, and in Stage 3 for the capacity design of beams in shear per Equation 5.42
- The beam longitudinal bar diameters, for use in Stage 3 to determine the maximum stirrup spacing against buckling of these bars – see last row of Table 5.3
- The cracked stiffness, $EI_{b,eff}$, of beams around joints, taking into account their reinforcement and concrete creep per Equation 5.6, for use in Stage 2 for the calculation of the effective buckling length of the columns connected to these beams – see Equations 5.4 and 5.5 in Section 5.2.3.4

A condensed example of Stage 1 is given in Tables 7.5 and 7.6 for the fifth-storey beams of Frame 2. The reinforcement of all storeys is shown at the beam framing plans of Figures 7.34 to 7.41. In these figures, the top and bottom longitudinal reinforcements at the ends of beams and near mid-span are given in the first and second row, respectively. Because the right and left halves of the plan are symmetric, the longitudinal reinforcement is displayed

Table 7.5 Frame 2, storey 5, beams B27, B28: ULS design of longitudinal reinforcement

Location	Max bar Ø (mm)	Compression flange width (m)	MaxM_{Ed} (kNm)	Req. steel area (mm²)	Beam bars Continuous	Added	Prov. steel area (mm²)	Design moment resistance (kNm)
Beam B28, clear length: 6.6 m, section: T, depth h: 0.5 m, width b_w: 0.3 m, flange thickness h_f: 0.18 m								
Left end (C2), top	14	0.30	202.5	1182	2Ø14	4Ø14	1204[a]	213.0
Left end, bottom	14	0.72	−12.2	591	2Ø14	–	616[b]	116.6
Mid-span, bottom	–	2.68	101.7	526	2Ø14	2Ø14	616[b,c]	118.8
Right end (C7), top	24	0.30	164.5	934	2Ø14	2Ø16	1120[a]	193.3
Right end, bottom	24	1.14	44.8	467	2Ø14	–	462[b]	88.7
Beam B27, clear length: 6.6 m, section: T, depth h: 0.5 m, width b_w: 0.3 m, flange thickness h_f: 0.18 m								
Left end (C7), top	24	0.30	185.0	1065	2Ø14	2Ø16	1120[a]	193.3
Left end, bottom	24	1.14	23.8	533	2Ø14	–	616[d,e]	117.7
Mid-span, bottom	–	2.68	100.3	519	2Ø14	2Ø14	616[d]	118.8
Right end (C12) top	14	0.30	183.4	1055	2Ø14	3Ø14	1050[a]	182.6
Right end, bottom	14	0.72	6.5	528	2Ø14	–	616[d]	116.6

[a] Provided top reinforcement includes 250 mm² from the slab per m of an effective tension flange width, which extends beyond each side of a supporting column by $4h_f$ at interior joints or $2h_f$ at exterior ones.
[b] Additional bottom mid-span bars extended: 2Ø14 to the left end; 1Ø14 to right end.
[c] Additional bottom mid-span bars extended across C7 to the left end of beam B27: 1Ø14.
[d] Additional bottom mid-span bars extended: 1Ø14 to the left end; 2Ø14 to right end.
[e] Additional bottom mid-span bars of beam B28 extended across C7 to the left end of beam B27: 1Ø14.

Table 7.6 Frame 2, storey 5, beams B27, B28: SLS checks per EC2: Stress limits; crack width $<w_{max} = 0.3$ mm; Steel area for crack control

Location	For characteristic loads G + Q			For quasi-permanent loads $G + \psi_2 Q$				Steel area/crack control	
	Moment (kNm)	Steel stress/f_{yk}	Concrete stress/f_{ck}	Moment (kNm)	Concrete stress/f_{ck}	Crack spacing (mm)	Crack width (mm)	Minimum (mm²)	Provided (mm²)
Beam B28									
Left end top	125.8	0.628	0.348	107.4	0.297	324.8	0.26	214	923
Mid-span bottom	79.0	0.605	0.084	67.4	0.071	283.2	0.22	75	615
Beam B27									
Left end, top	94.8	0.527	0.271	80.6	0.231	423	0.24	386	709
Mid-span bottom	77.9	0.596	0.083	66.4	0.070	283	0.22	75	615
Right end, top	103.6	0.542	0.291	88.5	0.248	358	0.25	239	769

over the left half; the right half is reserved for the transverse reinforcement of the end and central regions of beams (see Stage 3).

Foundation beams are protected from plastic hinging by being designed with the design seismic action multiplied by a universal a_{CD} factor of 1.4, in order to remain elastic in the seismic design situation (see Equation 6.8 in Section 6.3.2). So, like all other beams in the basements, they follow the Eurocode 2 rules alone and may be dimensioned in shear as early as Stage 1, without awaiting Stage 3. Unlike beams of the superstructure, they are specified as single-storey elements, not as beams at the lowest level of a multistorey plane frame.

The moment (M) and shear (V) diagrams obtained from the analysis for the 1 m long fictitious sub-elements into which the foundation beams have been discretised in order to place the discrete Winkler springs along their underside are consolidated into single M or V diagrams for each 'span' of a foundation beam between vertical elements. The M and V

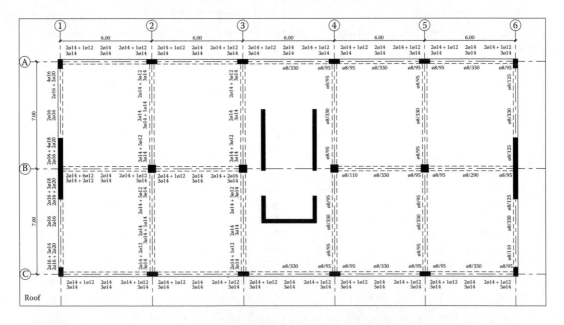

Figure 7.34 Beam framing plan – roof level (longitudinal bars: left-hand side, ties: right-hand side).

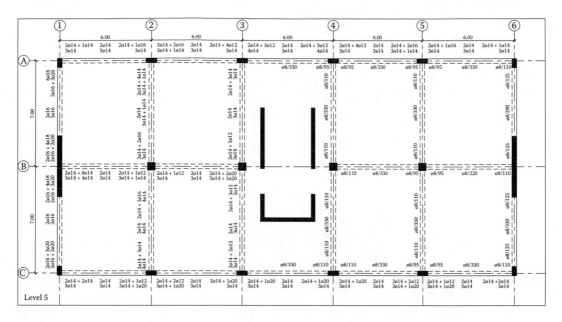

Figure 7.35 Beam framing plan – floor 5 (longitudinal bars: left-hand side, ties: right-hand side).

values needed for the ULS and SLS design of the foundation beams are retrieved from these consolidated diagrams.

The perimeter walls of the basement are also subjected to the static earth pressures at rest (even when the basement moves during the earthquake against the earthfill). They should resist these pressures in out-of-plane bending as one-way slabs spanning vertically between the floor diaphragms (the vertical elements on the perimeter do not restrain laterally the perimeter walls). So, the vertical reinforcement of these walls should meet the minimum

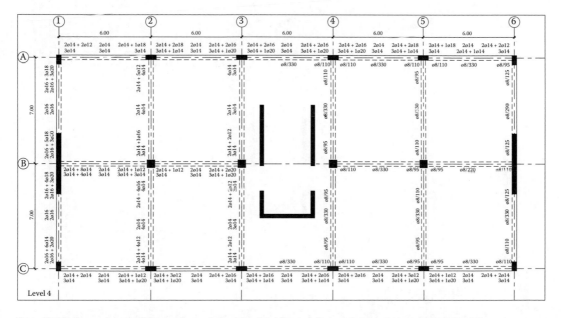

Figure 7.36 Beam framing plan – floor 4 (longitudinal bars: left-hand side, ties: right-hand side).

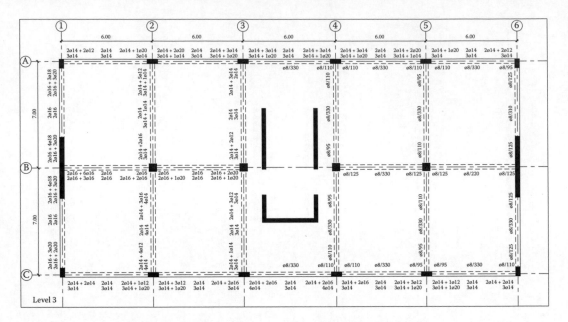

Figure 7.37 Beam framing plan – floor 3 (longitudinal bars: left-hand side, ties: right-hand side).

steel ratio specified in Eurocode 2 for the main direction of slabs; this ratio is the same as the one given at the second row of Table 5.1 for DC L beams, and, as a matter of fact, controls the vertical reinforcement of the foundation beam-cum-perimeter walls. The horizontal reinforcement, placed between the curtains of the vertical bars, is controlled by the goals of limiting the crack width in the web to $w_{max} = 0.3$ mm and of having the skin reinforcement work as minimum steel for crack control, with a value of σ_s which corresponds to the horizontal bars' diameter according to Equation 5.25 (cf. very last part of Section 5.3.3).

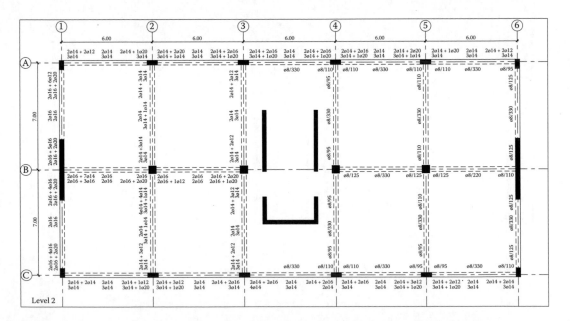

Figure 7.38 Beam framing plan – floor 2 (longitudinal bars: left-hand side, ties: right-hand side).

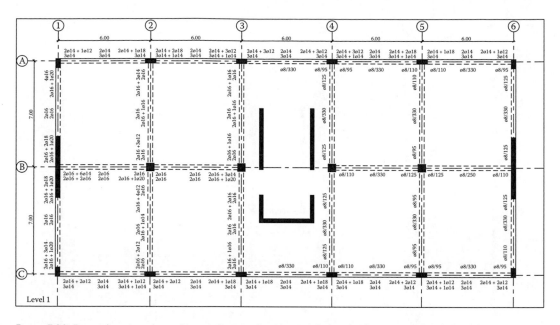

Figure 7.39 Beam framing plan – floor 1 (longitudinal bars: left-hand side, ties: right-hand side).

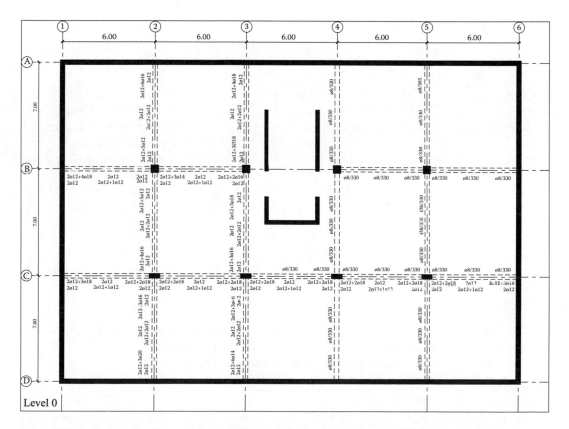

Figure 7.40 Beam framing plan – grade level (longitudinal bars: left-hand side, ties: right-hand side).

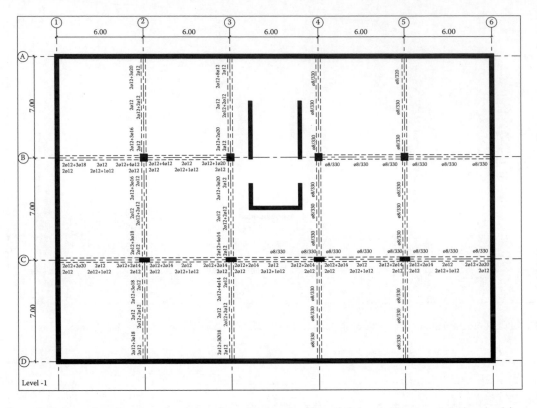

Figure 7.41 Beam framing plan – basement floor (longitudinal bars: left-hand side, ties: right-hand side).

A condensed design example is given in Tables 7.7 to 7.9 for the foundation beam-cum-perimeter wall under frame 1. As shown in the typical vertical section of Figure 7.42, a major difference with the beams of the superstructure is the large amount of horizontal reinforcement at both lateral sides, resulting from the Eurocode 2 requirements for skin reinforcement in deep beams and from its role as minimum steel for crack control. Thanks to this skin reinforcement, the moment resistance of the foundation beam for bending in a vertical plane is several times higher than required for the ULS in flexure under the combined actions from the superstructure and the ground. If the thickness of the perimeter wall were reduced from 300 mm (chosen for easier compaction of concrete and to be flush with the perimeter columns and walls W1 and W2) to 250 mm, the spacing of horizontal bars could increase to 150 mm. Witness also the very low stresses and small crack widths predicted in SLS, owing to the very small service moments, compared to the ULS moments in the 'seismic design situation'.

7.6.2.2 Stage 2: Columns (slenderness check; dimensioning of vertical and transverse reinforcement from the ULSs in flexure and shear with capacity design; detailing per EC8)

This stage is carried out for one multistorey column at a time (from the roof to the foundation).

It is checked and ensured from the outset that the column meets the slenderness limits in Eurocode 2, which allow to neglect second-order effects in the analysis for the 'persistent

Table 7.7 Perimeter frame 1, basement storeys 0 and −1; foundation beams B20, B21, B22; ULS design of longitudinal reinforcement (with multiplier of seismic internal forces: 1.4)

Location	Compression flange width (m)	$MaxM_{Ed}$ (kNm)	Req. steel area (mm²)	Flange bars No. Ø	Flange bars Area (mm²)	Lateral side bars Ø/s mm	Lateral side bars Area (mm²/m)	Design moment resistance (kNm)
Beam B20, clear length: 7.0 m; section: asymmetric *I*; depth *h*: 6.3 m; width b_w: 0.3 m; flange thickness (top) h_f: 0.18 m; (bottom) h_f: 0.30 m								
Left end (D), top	1.00	670	379	4Ø18	1018	Ø10/130	2 × 604	11,420
Left end, bottom	0.72	466	379	8Ø18	2036	Ø10/130	2 × 604	13,792
Mid-length, top	1.49	200	379	2Ø18	509	Ø10/130	2 × 604	10,246
Right end (C11) top	1.00	1157	379	2Ø18	509	Ø10/130	2 × 604	10,245
Right end, bottom	0.51	2697	379	8Ø18	2036	Ø10/130	2 × 604	13,792
Beam B21, clear length: 5.0 m; section: asymmetric *I*; depth *h*: 6.3 m; width b_w: 0.3 m; flange thickness (top) h_f: 0.18 m; (bottom) h_f: 0.30 m								
Left end (C11), top	1.00	400	379	2Ø18	509	Ø10/130	2 × 604	10,245
Left end, bottom	0.72	1551	379	8Ø18	2036	Ø10/130	2 × 604	13,792
Mid-length, bottom	1.28	432	379	8Ø18	2036	Ø10/130	2 × 604	14,038
Right end (W1), top	1.00	2941	379	2Ø18	509	Ø10/130	2 × 604	10,245
Right end, bottom	0.72	4936	379	8Ø18	2036	Ø10/130	2 × 604	13,792
Beam B22, clear length: 5.0 m; section: asymmetric *I*; depth *h*: 6.3 m; width b_w: 0.3 m; flange thickness (top) h_f: 0.18 m; (bottom) h_f: 0.30 m								
Left end (W1), top	1.00	1807	379	2Ø18	509	Ø10/130	2 × 604	10,245
Left end, bottom	0.51	3190	379	8Ø18	2036	Ø10/130	2 × 604	13,573
Mid-length, top	1.49	366.8	379	2Ø18	509	Ø10/130	2 × 604	10,246
Right end (C1), top	1.00	477.5	379	4Ø18	1018	Ø10/130	2 × 604	11,420
Right end, bottom	0.72	278.9	379	8Ø18	2036	Ø10/130	2 × 604	13,792

and transient design situation', that is, for the factored gravity loads per Equations 6.10a and 6.10b of EN1990 (Table 7.10). It is checked first whether the building can be considered as laterally braced or unbraced. To this end, it is checked whether the building's walls satisfy Equation 5.11 in Section 5.2.3.4 at the top of the foundation or of a rigid basement.

Only wall W5, having $I_w = 3.1504$ m⁴, provides bracing to the building in direction X. Then, for $n_{st} = 6$, $H_{tot} = 19$ m, $E_{cd} = E_{cm}/1.2 = 31,000,000/1.2$ (kPa), the right-hand side of Equation 5.11 is equal to 56,000 kN, and exceeds the value of $F_{V,Ed} = 36,925$ kN from Equations 6.10a and 6.10b of EN1990. So, the building can be considered as braced along direction X; the same along direction Y. Had the walls failed to meet Equation 5.11, but by less than a factor of 2.0, we would next check the cracking moment at the base of the walls under the factored permanent and imposed loads in the 'persistent and transient' design situation, which act at the inclination of the building from the vertical postulated in Eurocode 2. If that cracking moment was not exceeded, the right-hand side of Equation 5.11 would then be multiplied by 2 and the bracing condition would be satisfied.

The characterisation of the building as braced or unbraced in a given horizontal direction affects the way the slenderness limit of the column is computed from Equation 5.3 and its value; it affects even more the column's effective buckling length, l_0, from Equations 5.4. The values of restraining stiffness entering Equations 5.4 are calculated from Equation 5.5, using the cracked stiffness values, $EI_{b,eff}$, of the beams framing into the column's top and

Table 7.8 Perimeter frame 1 in basement storeys 0 and −1; foundation beams B20, B21, B22; SLS checks per EC2; stress limits; crack width $<w_{max} = 0.3$ mm; steel area for crack control

Location	For characteristic loads G + Q			For quasi-permanent loads G + ψ₂Q				Steel for crack control	
	Moment (kNm)	Steel stress/f_yk	Concrete stress/f_ck	Moment (kNm)	Concrete stress/f_ck	Crack spacing (mm)	Crack width (mm)	Minimum (mm²)	Provided (mm²)
Beam B20									
Left end, top	114.7	0.006	0.001	102.1	0.001	252.5	0.00	893	1018
Mid-span, top	165.9	0.001	0.001	146.8	0.001	0.0	0.00	174	509
Right end bottom	889.4	0.034	0.010	770.0	0.009	200.2	0.01	1863	2036
Whole span, web						877.8	0.02	1863	2036
Beam B21									
Left end, bottom	671.8	0.026	0.008	575.6	0.006	412.1	0.01	1863	2036
Mid-span bottom	332.0	0.007	0.002	294.9	0.001	200.2	0.01	1867	2036
Right end bottom	1115.3	0.043	0.013	997.4	0.011	200.2	0.01	1863	2036
Whole span, web						877.8	0.03		
Beam B22									
Left end bottom	759.1	0.043	0.013	691.3	0.012	412.1	0.02	1861	2036
Mid-span top	298.1	0.001	0.001	249.2	0.001	0.0	0.00	174	509
Right end top	110.0	0.005	0.001	99.3	0.001	252.5	0.00	893	1018
Whole span, web						877.8	0.03		

Table 7.9 Perimeter frame 1, basement storeys 0 and −1; foundation beams B20, B21, B22; ULS design in shear; dimensioning of transverse reinforcement

| Region | Length (m) | Design shear (kN1) | | Max tie spacing (mm) | Provided ties | | | | Shear resistance (kN) | |
		Seismic	Non-seismic		No.	Ø (mm)	s (mm)	Strut angle	$V_{Rd,s}$	$V_{Rd,max}$
Beam 20										
Left half (D)	3.50	562.5	1897	250	15	10	250	22°	3654	5535
Right half (C11)	3.50	1995	1864	250	15	10	250	22°	3654	5535
Beam 21										
Left half (C11)	2.50	134.7	134.1	250	11	10	250	22°	3654	5535
Right half (W1)	2.50	1960	1567	250	11	10	250	22°	3654	5535
Beam 22										
Left half (W1)	2.50	1705.4	85.0	250	11	10	250	22°	3654	5535
Right half (C1)	2.50	1139.7	−529.7	250	11	10	250	22°	3654	5535

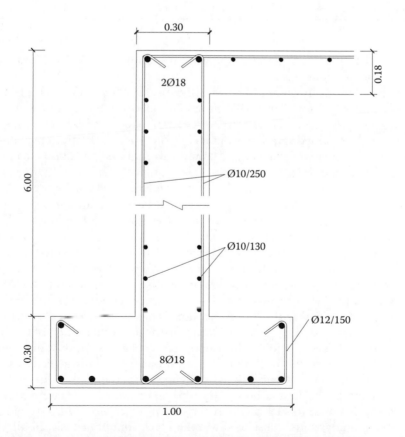

Figure 7.42 Reinforcement and strip footing of basement wall-cum-foundation beam.

Table 7.10 Column C12: Check of column slenderness for negligible second-order effects as per EC2

Storey	Combination of actions per EN1990	Column direction z				Column direction y			
		Slenderness		Column		Slenderness		Column	
		Limit	Actual	Eff. l_0 (m)	Sufficient size (m)	Limit	Actual	Eff. l_0 (m)	Sufficient size (m)
6	Equation 6.10a	184.4	8.0	1.61	0.70	179.8	17.1	1.48	0.30
	Equation 6.10b	186.5	8.0	1.61	0.70	181.9	17.1	1.48	0.30
5	Equation 6.10a	133.2	7.9	1.60	0.70	124.8	17.1	1.48	0.30
	Equation 6.10b	135.2	7.9	1.60	0.70	126.8	17.0	1.48	0.30
4	Equation 6.10a	109.1	7.9	1.60	0.70	109.2	17.2	1.49	0.30
	Equation 6.10b	110.8	7.9	1.60	0.70	111.	17.1	1.48	0.30
3	Equation 6.10a	94.1	7.9	1.61	0.70	94.9	17.3	1.50	0.30
	Equation 6.10b	95.8	7.9	1.61	0.70	96.5	17.3	1.50	0.30
2	Equation 6.10a	83.0	8.0	1.61	0.70	84.9	17.5	1.52	0.30
	Equation 6.10b	84.4	8.0	1.61	0.70	86.5	17.5	1.52	0.30
1	Equation 6.10a	75.0	11.1	2.24	0.70	69.8	23.6	2.04	0.30
	Equation 6.10b	76.6	11.1	2.24	0.70	71.0	23.6	2.04	0.30
0	Equation 6.10a	65.1	8.0	1.62	0.70	67.5	16.6	1.44	0.30
	Equation 6.10b	66.3	8.0	1.62	0.70	68.7	16.6	1.44	0.30
−1	Equation 6.10a	47.1	7.1	1.43	0.70	47.8	15.4	1.33	0.30
	Equation 6.10b	48.0	7.1	1.43	0.70	48.6	15.4	1.33	0.30

bottom joints in the plane of bending considered, as computed in Stage 1 and archived for the purposes of the present check.

The column slenderness limit should be, in principle, checked at every storey and in both lateral directions. Most critical is normally either: (a) the lowest storey above the foundation or the top of a rigid basement, owing to the larger axial load of the column, or (b) the storey above, which is less restrained rotationally at its bottom node. The same calculation may be extended to give an estimate of the required column size, in case the one available turns out in the end not to meet the slenderness limit of Equation 5.3.

Once sufficiency of the column's size is established, the vertical bars are dimensioned for the ULS in flexure with axial force. As the vertical bars of a column serve the column's top section in the storey and the bottom section of the storey above, they are dimensioned for the most adverse design triplet of biaxial moments and axial force, $M_y - M_z - N$, at these two sections. Each of these two sections should be dimensioned for the two triplets from Equations 6.10a and 6.10b of EN1990 ('persistent and transient design situation'), and for the triplet due to the quasi-permanent gravity loads, superimposed to the $M_y - M_z - N$ triplets arising from the two horizontal seismic action components (including the effect of their accidental eccentricities) with all possible signs. The modal response spectrum analysis employed in the present case, alongside the SRSS rule for the combination of the peak effects of the two horizontal seismic action components, gives a signless peak value of each one of M_y, M_z, N. As noted in Sections 5.8.1 and 7.6.1, rigorous, probabilistic methods to estimate the most likely values of M_z, N (with signs) accompanying the peak value of M_y, those of M_y, N accompanying peak M_z, and of M_y, M_z accompanying peak N, are beyond the present scope. Without recourse to such an approach, the peak values of M_y, M_z, N from modal response spectrum analysis are considered to take place simultaneously, in all $2^3 = 8$ combinations of signs. So, the vertical bars of each storey are dimensioned for the most adverse among the 2 + 8 = 10 possible triplets $M_y - M_z - N$, for each one of the two column sections above and below the top joint of the storey. The bars of the top storey of the column are dimensioned for just the top section of that storey; starter bars at the connection with the foundation are dimensioned only for that section. As suggested in Section 5.4.2 with reference to Figure 5.3, this dimensioning is carried out for uniaxial bending, first for the pair $M_y - N$ in each triplet and then for $M_y - N$, superimposing the reinforcement requirements, resulting each time for the corresponding pairs of opposite sides of the section. Table 7.11 follows this procedure, taking as an example column C12.

All columns of the present building are exempted from the strong column–weak beam capacity design rule, Equation 5.31, because the building has a wall system in one direction and a wall-equivalent dual in the other. In a column not exempted from Equation 5.31 within a given vertical plane of bending, the area of vertical bars arranged along the column sides, which are at right angles to that plane, should also be dimensioned at the storey's top joint for the combination of:

1. The average of the minimum N-values in the 'seismic design situation' (axial force due to quasi-permanent actions, minus the peak value of N from modal analysis with SRSS combination of the N-values due to the two horizontal components) above and below the joint.
2. A uniaxial moment of one-half the right-hand side of Equation 5.31, using the design moment resistance values of the beams framing into the column's top joints, as calculated and archived in Stage 1. Table 7.12 shows the values of $\Sigma M_{Rd,b}$ for use in this step in the example case of column C12. If the column section is symmetric, the maximum value of $\Sigma M_{Rd,b}$ for the two possible senses of beam bending around the joint is used; asymmetric columns should be dimensioned for two combinations of this type, using the corresponding value of $\Sigma M_{Rd,b}$ for each sense of beam (and column) bending in the plane considered.

If Equation 5.31 should also be met in the orthogonal plane of bending, the area of vertical bars arranged along the column sides at right angles to that plane is also uniaxially dimensioned in the same way and superimposed to the steel requirements on the two other sides.

With the large column sections which are necessary to provide sufficient depth for bond of beam bars at joints and/or to meet Eurocode 2's slenderness limits for negligible second-order effects, the number and size of vertical bars is normally controlled by minimum requirements. This happens in all columns of the present building, as shown in Figure 7.43 and Table 7.13 and, by way of an example, in Table 7.14 for the case of column C12. Two bar diameters are combined in each column section, in order to meet without margins the detailing rules for: (a) a minimum steel ratio of 1%, (b) the 200 mm maximum spacing of laterally restrained bars along the perimeter applying to DC M, and (c) the minimum of three bars per side. As demonstrated in Table 7.15 for the example for C12, even with this optimisation, the resulting moment resistance values have a large margin over the moment demands from the analysis listed in Table 7.12 and the beam moment resistances around the joints, $\Sigma M_{Rd,b}$, in Table 7.13. Owing to this latter margin, the column capacity design shear, V_{CD}, from Equation 5.44 is controlled by plastic hinging in the beams and the beam moment resistances, $M_{Rd,b}$. The design shears, V_{Ed}, of the columns of the superstructure are equal to the capacity design ones, V_{CD}; by contrast, in the box-type basement, where only Eurocode 2 rules apply, V_{Ed} is the value from the analysis – see Table 7.16 in the example for C12.

The column design shear is constant within a storey, but the detailing rules concerning the diameter, spacing and ratio of transverse reinforcement are much tighter in the 'critical'

Table 7.11 Column C12: Normal stress resultants in the 'seismic design situation' from the analysis, for ULS dimensioning of the vertical reinforcement

	Column base			Column top		
Combination of actions	M_y (kNm)	M_z (kNm)	N (kN)	M_y (kNm)	M_z (kNm)	N (kN)
Storey 6						
EN1990 Equation 6.10a	−26.5	−82.5	252.9	28.9	96.6	231.6
EN1990 Equation 6.10b	−25.2	−78.2	237.9	27.4	91.7	219.8
$G+\psi_2Q+E:+X,+Y$/maxN	99.5	−3.8	190.4	170.2	121.6	174.7
$G+\psi_2Q+E:-X,+Y$/maxN	−134.7	−3.8	190.4	−131.9	121.6	174.7
$G+\psi_2Q+E:+X,-Y$/maxN	99.5	−105.5	190.4	170.2	6.6	174.7
$G+\psi_2Q+E:-X,-Y$/maxN	−134.7	−105.5	190.4	−131.9	6.6	174.7
$G+\psi_2Q+E:+X,+Y$/minN	99.5	−3.8	148.0	170.2	121.6	132.3
$G+\psi_2Q+E:-X,+Y$/minN	−134.7	−3.8	148.0	−131.9	121.6	132.3
$G+\psi_2Q+E:+X,-Y$/minN	99.5	−105.5	148.0	170.2	6.6	132.3
$G+\psi_2Q+E:-X,-Y$/minN	−134.7	−105.5	148.0	−131.9	6.6	132.3
Storey 5						
EN1990 Equation 6.10a	−24.4	−71.7	507.5	24.5	70.1	486.2
EN1990 Equation 6.10b	−23.1	−68.0	477.5	23.2	66.5	459.4
$G+\psi_2Q+E:+X,+Y$/maxN	102.3	1.0	394.6	139.5	94.7	378.9
$G+\psi_2Q+E:-X,+Y$/maxN	−134.6	1.0	394.6	−107.0	94.7	378.9
$G+\psi_2Q+E:+X,-Y$/maxN	102.3	−96.0	394.6	139.5	−1.8	378.9
$G+\psi_2Q+E:-X,-Y$/maxN	−134.6	−96.0	394.6	−107.0	−1.8	378.9
$G+\psi_2Q+E:+X,+Y$/minN	102.3	1.0	284.7	139.5	94.7	269.0
$G+\psi_2Q+E:-X,+Y$/minN	−134.6	1.0	284.7	−107.0	94.7	269.0
$G+\psi_2Q+E:+X,-Y$/minN	102.3	−96.0	284.7	139.5	−1.8	269.0
$G+\psi_2Q+E:-X,-Y$/minN	−134.6	−96.0	284.7	−107.0	−1.8	269.0
Storey 4						
EN1990 Equation 6.10a	−23.9	−72.1	761.9	24.1	72.5	740.7
EN1990 Equation 6.10b	−22.7	−68.4	716.9	22.8	68.7	698.8
$G+\psi_2Q+E:+X,+Y$/maxN	114.3	2.1	598.4	150.1	97.6	582.6
$G+\psi_2Q+E:-X,+Y$/maxN	−146.0	2.1	598.4	−118.1	97.6	582.6
$G+\psi_2Q+E:+X,-Y$/maxN	114.3	−97.7	598.4	150.1	−1.5	582.6
$G+\psi_2Q+E:-X,-Y$/maxN	−146.0	−97.7	598.4	−118.1	−1.5	582.6
$G+\psi_2Q+E:+X,+Y$/minN	114.3	2.1	421.5	150.1	97.6	405.7
$G+\psi_2Q+E:-X,+Y$/minN	−146.0	2.1	421.5	−118.1	97.6	405.7
$G+\psi_2Q+E:+X,-Y$/minN	114.3	−97.7	421.5	150.1	−1.5	405.7
$G+\psi_2Q+E:-X,-Y$/minN	−146.0	−97.7	421.5	−118.1	−1.5	405.7
Storey 3						
EN1990 Equation 6.10a	−24.1	−69.7	1016.3	23.9	70.4	995.1
EN1990 Equation 6.10b	−22.8	−66.2	956.2	22.6	66.8	938.2
$G+\psi_2Q+E:+X,+Y$/maxN	119.4	1.8	801.6	148.6	93.4	785.8
$G+\psi_2Q+E:-X,+Y$/maxN	−151.3	1.8	801.6	−117.0	93.4	785.8
$G+\psi_2Q+E:+X,-Y$/maxN	119.4	−94.3	801.6	148.6	0.0	785.8
$G+\psi_2Q+E:-X,-Y$/maxN	−151.3	−94.3	801.6	−117.0	0.0	785.8
$G+\psi_2Q+E:+X,+Y$/minN	119.4	1.8	558.8	148.6	93.4	543.1
$G+\psi_2Q+E:-X,+Y$/minN	−151.3	1.8	558.8	−117.0	93.4	543.1
$G+\psi_2Q+E:+X,-Y$/minN	119.4	−94.3	558.8	148.6	0.0	543.1
$G+\psi_2Q+E:-X,-Y$/minN	−151.3	−94.3	558.8	−117.0	0.0	543.1

Table 7.11 (Continued) Column C12: normal stress resultants in the 'seismic design situation' from the analysis, for ULS dimensioning of the vertical reinforcement

	Column base			Column top		
Combination of actions	M_y (kNm)	M_z (kNm)	N (kN)	M_y (kNm)	M_z (kNm)	N (kN)
Storey 2						
EN1990 Equation 6.10a	−20.7	−75.4	1270	22.6	71.7	1249
EN1990 Equation 6.10b	−19.6	−71.5	1195	21.5	68.1	1177
$G+\psi_2Q+E:+X,+Y/maxN$	121.8	−6.3	1001	137.5	87.6	986
$G+\psi_2Q+E:-X,+Y/maxN$	−149.3	−6.3	1001	−107.5	87.6	986
$G+\psi_2Q+E:+X,-Y/maxN$	121.8	−93.7	1001	137.5	7.5	986
$G+\psi_2Q+E:-X,-Y/maxN$	−149.3	−93.7	1001	−107.5	7.5	986
$G+\psi_2Q+E:+X,+Y/minN$	121.8	−6.3	699	137.5	87.6	683
$G+\psi_2Q+E:-X,+Y/minN$	−149.3	−6.3	699	−107.5	87.6	683
$G+\psi_2Q+E:+X,-Y/minN$	121.8	−93.7	699	137.5	7.5	683
$G+\psi_2Q+E:-X,-Y/minN$	−149.3	−93.7	699	−107.5	7.5	683
Storey 1						
EN1990 Equation 6.10a	−26.8	−31.6	1531	20.6	45.3	1503
EN1990 Equation 6.10b	−25.4	−30.0	1440	19.6	43.0	1416
$G+\psi_2Q+E:+X,+Y/maxN$	117.4	5.1	1199	102.7	50.1	1178
$G+\psi_2Q+E:-X,+Y/maxN$	−152.9	5.1	1199	−75.4	50.1	1178
$G+\psi_2Q+E:+X,-Y/maxN$	117.4	−47.1	1199	102.7	10.0	1178
$G+\psi_2Q+E:-X,-Y/maxN$	−152.9	−47.1	1199	−75.4	10.0	1178
$G+\psi_2Q+E:+X,+Y/minN$	117.4	5.1	852	102.7	50.1	831
$G+\psi_2Q+E:-X,+Y/minN$	−152.9	5.1	852	−75.4	50.1	831
$G+\psi_2Q+E:+X,-Y/minN$	117.4	−47.1	852	102.7	10.0	831
$G+\psi_2Q+E:-X,-Y/minN$	−152.9	−47.1	852	−75.4	10.0	831
Storey 0						
EN1990 Equation 6.10a	−32.2	−21.1	2029	37.0	21.8	2007
EN1990 Equation 6.10b	−30.6	−20.0	1910	35.1	20.6	1892
$G+\psi_2Q+E:+X,+Y/maxN$	−7.9	−7.8	1531	71.0	19.7	1515
$G+\psi_2Q+E:-X,+Y/maxN$	−34.8	−7.8	1531	−22.0	19.7	1515
$G+\psi_2Q+E:+X,-Y/maxN$	−7.9	−20.2	1531	71.0	9.2	1515
$G+\psi_2Q+E:-X,-Y/maxN$	−34.8	−20.2	1531	−22.0	9.2	1515
$G+\psi_2Q+E:+X,+Y/minN$	−7.9	−7.8	1182	71.0	19.7	1166
$G+\psi_2Q+E:-X,+Y/minN$	−34.8	−7.8	1182	−22.0	19.7	1166
$G+\psi_2Q+E:+X,-Y/minN$	−7.9	−20.2	1182	71.0	9.2	1166
$G+\psi_2Q+E:-X,-Y/minN$	−34.8	−20.2	1182	−22.0	9.2	1166
Storey −1						
EN1990 Equation 6.10a	−7.0	−5.3	2537	18.6	13.2	2516
EN1990 Equation 6.10b	−6.6	−5.1	2391	17.7	12.5	2373
$G+\psi_2Q+E:+X,+Y/maxN$	12.7	6.3	1866	29.8	17.9	1850
$G+\psi_2Q+E:-X,+Y/maxN$	−21.9	6.3	1866	−5.1	17.9	1850
$G+\psi_2Q+E:+X,-Y/maxN$	12.7	−13.4	1866	29.8	−0.4	1850
$G+\psi_2Q+E:-X,-Y/maxN$	−21.9	−13.4	1866	−5.1	−0.4	1850
$G+\psi_2Q+E:+X,+Y/minN$	12.7	6.3	1524	29.8	17.9	1508
$G+\psi_2Q+E:-X,+Y/minN$	−21.9	6.3	1524	−5.1	17.9	1508
$G+\psi_2Q+E:+X,-Y/minN$	12.7	−13.4	1524	29.8	−0.4	1508
$G+\psi_2Q+E:-X,-Y/minN$	−21.9	−13.4	1524	−5.1	−0.4	1508

Table 7.12 Column C12: Sum of design moment resistances of beams around joints with column C12, $\sum M_{Rd,b}$ (kNm), for the capacity design of the column in shear (and for strong column–weak beam design, not required in this building, thanks to its wall or wall-equivalent dual system)

Storey	Location	Direction of M_{Rd} vector			
		+y	−y	+z	−z
6	Top	197.6	197.6	88.0	145.3
	Base	274.4	255.3	116.6	182.6
5	Top	274.4	255.3	116.6	182.6
	Base	293.0	255.3	116.6	181.1
4	Top	293.0	255.3	116.6	181.1
	Base	293.0	255.3	116.6	181.1
3	Top	293.0	255.3	116.6	181.1
	Base	293.0	255.3	88.0	163.5
2	Top	293.0	255.3	88.0	163.5
	Base	245.2	197.6	76.8	160.5
1	Top	245.2	197.6	76.8	160.5
	Base	175.9	175.9	191.4	223.3
0	Top	175.9	175.9	191.4	223.3
	Base	141.8	141.8	216.8	217.0
−1	Top	141.8	141.8	216.8	217.0
	Base	0.0	0.0	0.0	0.0

end regions of the columns than in-between, where Eurocode 2 alone applies (see Table 5.4). As the value of the column's design shear is the same in all these regions, it is used to dimension the transverse reinforcement between the 'critical' regions of the two ends. Table 7.16 in the example of C12 is typical in that the minimum diameter and maximum spacing of ties as per Eurocode 2, used together with a tie layout that meets the Eurocode 2 rules on lateral restraint of a minimum number of bars along each side of the section, give a design shear resistance with a sizeable margin over the design shear force. Normally, that resistance is attained at the minimum value of the strut inclination angle per Eurocode 2 – 22° (see Equation 5.45) – and is controlled by the column resistance in diagonal tension, $V_{Rd,s}$, from Equation 5.46a. In Table 7.16 for the example of column C12, the basement storeys of C12 attain their shear resistance along the strong axis at values of the strut angle above the minimum; in that case the column resistances in diagonal tension and compression, $V_{Rd,s}$ from Equation 5.46a and $V_{Rd,max}$ from Equation 5.48a, are equal.

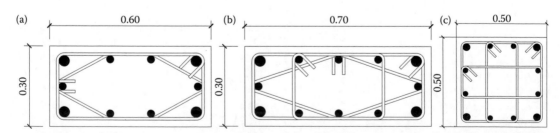

Figure 7.43 Column vertical reinforcement and tie layout: (a) corner columns C1, C6, C11, C16 (vertical bars 4Ø18 + 6Ø14); (b) intermediate columns in frames A and C: C2 to C5, C12 to C15 (vertical bars 4Ø18 + 8Ø14); (c) interior columns C7 to C10 (vertical bars 4Ø18 + 8Ø14).

Table 7.13 Column reinforcement

Column	Vertical bars	Steel ratio (%)	Critical height at column ends: superstructure (m)	In critical height at the column base: level I (mm)	In critical height of all columns except at the column base: level I (mm)	In basement or outside the critical height (mm)	Lapping at floor level (m)
C1, C6	4Ø18 + 6Ø14	1.08	0.6	Ø8/90	Ø6/110	Ø6/170	0.75
C2, C5	4Ø18 + 8Ø14	1.07	0.7	Ø8/90	Ø6/110	Ø6/170	0.75
C3, C4	4Ø18 + 8Ø14	1.07	0.7	Ø8/95	Ø6/110	Ø6/170	0.75
C7, C10	4Ø18 + 8Ø14	1.0	0.7	Ø8/90	Ø6/110	Ø6/170	0.75
C8, C9	4Ø18 + 8Ø14	1.0	0.7	Ø8/110	Ø6/110	Ø6/170	0.75
C11, C16	4Ø18 + 6Ø14	1.08	0.6	Ø8/90	Ø6/110	Ø6/170	0.75
C12, C14	4Ø18 + 8Ø14	1.07	0.7	Ø8/90	Ø6/110	Ø6/170	0.75
C13, C15	4Ø18 + 8Ø14	1.07	0.7	Ø8/95	Ø6/110	Ø6/170	0.75

The prescriptive detailing of the transverse reinforcement is tighter in the two 'critical' end regions than in-between (see Table 5.4). For DC M, there are rules for the volumetric mechanical ratio of confining reinforcement, ω_{wd} and $a\omega_{wd}$, only at the base of the column, at the connection to the foundation (be it of box type). As demonstrated in Table 7.17 for the example of C12, these rules give larger diameter and denser stirrups at that location.

The DC L detailing rules for transverse reinforcement apply in the basement storeys, thanks to the box-type foundation system; ties there have the same spacing, diameter and so on, throughout the storey height.

Table 7.13 lists the details of the transverse reinforcement placed in columns.

All column vertical bars are lap-spliced at floor level. The design value of their basic anchorage length is $l_{bd} = 0.25d_{bL}f_{yd}/(2.25f_{ctd}) = 40d_{bL}$; the concrete cover to the bars is $35 + 6 = 41$ mm. The required lapping comes out equal to 755 mm for the 18 mm dia. bars and to 590 mm for the 14 mm dia. ones (see last row of Table 5.2), but is taken for simplicity as 750 mm for both (see last column of Table 7.13). The bars continue into the foundation

Table 7.14 Column C12: Column geometry and vertical reinforcement

Storey 6: Clear height 2.50 m, h_{cr} 0.70 m, Cross-section rectangular: b_y 0.30 m, b_z 0.70 m
  Vertical steel ratio: 0.0107, Bars: 4Ø18 + 8Ø14 (2Ø18 & 1Ø14 along b_y, 2Ø18 & 3Ø14 along b_z)
Storey 5: Clear height 2.50 m, h_{cr} 0.70 m, Cross-section rectangular: b_y 0.30 m, b_z 0.70 m
  Vertical steel ratio: 0.0107, Bars: 4Ø18 + 8Ø14 (2Ø18 & 1Ø14 along b_y, 2Ø18 & 3Ø14 along b_z)
Storey 4: Clear height 2.50 m, h_{cr} 0.70 m, Cross-section rectangular: b_y 0.30 m, b_z 0.70 m
  Vertical steel ratio: 0.0107, Bars: 4Ø18 + 8Ø14 (2Ø18 & 1Ø14 along b_y, 2Ø18 & 3Ø14 along b_z)
Storey 3: Clear height 2.50 m, h_{cr} 0.70 m, Cross-section rectangular: h_y 0.30 m, b_z 0.70 m
  Vertical steel ratio: 0.0107, Bars: 4Ø18 + 8Ø14 (2Ø18 & 1Ø14 along b_y, 2Ø18 & 3Ø14 along b_z)
Storey 2: Clear height 2.50 m, h_{cr} 0.70 m, Cross-section rectangular: b_y 0.30 m, b_z 0.70 m
  Vertical steel ratio: 0.0107, Bars: 4Ø18 + 8Ø14 (2Ø18 & 1Ø14 along b_y, 2Ø18 & 3Ø14 along b_z)
Storey 1: Clear height 3.50 m, h_{cr} 0.70 m, Cross-section rectangular: b_y 0.30 m, b_z 0.70 m
  Vertical steel ratio: 0.0107, Bars: 4Ø18 + 8Ø14 (2Ø18 & 1Ø14 along b_y, 2Ø18 & 3Ø14 along b_z)
Storey 0: Clear height 2.50 m, h_{cr} 0.70 m, Cross-section rectangular: b_y 0.30 m, b_z 0.70 m
  Vertical steel ratio: 0.0107, Bars: 4Ø18 + 8Ø14 (2Ø18 & 1Ø14 along b_y, 2Ø18 & 3Ø14 along b_z)
Storey -1: Clear height 2.50 m, h_{cr} 0.70 m, Cross-section rectangular: b_y 0.30 m, b_z 0.70 m
  Vertical steel ratio: 0.0107, Bars: 4Ø18 + 8Ø14 (2Ø18 & 1Ø14 along b_y, 2Ø18 & 3Ø14 along b_z)
  Starter bars at the base: 4Ø18 + 8Ø14 (2Ø18 & 1Ø14 along b_y, 2Ø18 & 3Ø14 along b_z)

Table 7.15 Column C12: Design moment resistance of the column, $M_{Rd,c}$ (kNm), values for minN/maxN

Storey	Location	Direction of M_{Rd} vector			
		+y	−y	+z	−z
6	Top	259.3/319.6	−259.3/−319.6	92.3/96.9	−92.3/−96.9
	Base	314.0/322.8	−314.0/−322.8	94.0/98.6	−94.0/−98.6
5	Top	338.3/358.4	−338.3/−358.4	106.7/117.2	−106.7/−117.2
	Base	341.3/361.1	−341.3/−361.1	108.3/118.7	−108.3/−118.7
4	Top	363.0/390.6	−363.0/−390.6	119.7/134.4	−119.7/−134.4
	Base	365.7/392.8	−365.7/−392.8	121.1/135.6	−121.1/−135.6
3	Top	384.9/416.2	−384.9/−416.2	131.3/148.3	−131.3/−148.3
	Base	387.2/417.9	−387.2/−417.9	132.6/149.3	−132.6/−149.3
2	Top	404.1/435.0	−404.1/−435.0	141.7/159.0	−141.7/−159.0
	Base	406.1/436.2	−406.1/−436.2	142.8/159.8	−142.8/−159.8
1	Top	421.0/447.0	−421.0/−447.0	151.0/166.5	−151.0/−166.5
	Base	423.1/448.0	−423.1/−448.0	152.2/169.1	−152.2/−169.1
0	Top	446.5/454.1	−446.5/−454.1	166.1/169.8	−166.1/−169.8
	Base	447.2/454.0	−447.2/−454.0	166.6/169.2	−166.6/−169.2
−1	Top	454.1/445.2	−454.1/−445.2	170.1/156.3	−170.1/−156.3
	Base	454.1/443.8	−454.1/−443.8	169.5/155.7	−169.5/−155.7

Table 7.16 Column C12: Dimensioning of transverse reinforcement between the column end regions, for the ULS in shear (for maxN or minN)

Storey		Design shear, V_{Ed} (kN)		Provided ties				Strut angle		Shear resistance (kN)			
				Ø (mm)	No. legs		Spacing (mm)			$V_{Rd,s}$		$V_{Rd,max}$	
		y	z		y	z		y	z	y	z	y	z
6	For maxN	87	43	6	3[a]	5[b]	170	22°	22°	363	219	545	490
	For minN	87	41					22°	22°	349	214	545	490
5	For maxN	66	24	6	3[a]	5[b]	170	22°	22°	412	236	545	490
	For minN	66	23					22°	22°	382	225	545	490
4	For maxN	66	23	6	3[a]	5[b]	170	22°	22°	461	252	545	490
	For minN	67	22					22°	22°	415	236	545	490
3	For maxN	66	22	6	3[a]	5[b]	170	22°	22°	510	268	545	490
	For minN	66	22					22°	22°	448	247	545	490
2	For maxN	66	20	6	3[a]	5[b]	170	22°	22°	549	284	556	490
	For minN	66	20					22°	22°	482	259	545	490
1	For maxN	39	14	6	3[a]	5[b]	170	22°	22°	523	273	545	490
	For minN	39	14					22°	22°	460	252	545	490
0	For maxN	35	14	6	3[a]	5[b]	170	26°	22°	626	327	626	490
	For minN	35	13					23°	22°	575	298	575	490
−1	For maxN	17	10	6	3[a]	5[b]	170	29°	22°	676	354	676	490
	For minN	17	10					26°	22°	622	325	624	490

[a] The value 3 applies for the number of legs, if a single cross-tie connects the two central bars of the short sides; if a diamond tie is used around all four central bars of the four sides as in Figure 7.43b, instead of orthogonal straight cross-ties, then the number is 3.9.

[b] The value 5 applies for the number of legs, if a single cross-tie connects the two central bars of the long sides; if a diamond tie is used around all four central bars per Figure 7.43b, instead of orthogonal straight cross-ties, then the number is 4.65.

Table 7.17 Column C12: Confinement reinforcement at column ends (for maxN)

					Stirrups									
	Required ω_{wd} (for DC M)		Required $a\omega_{wd}$ (for DC M)		Legs		$\emptyset$ (mm)		Spacing (mm)		Provided ω_{wd}		Provided $a\omega_{wd}$	
Storey	Base	Top	Base	Top	y	z	Base	Top	Base	Top	Base	Top	Base	Top
6	0.00	0.00	0.000	0.000	3ª	5ᵇ	6	6	110	110	0.174	0.174	0.055	0.055
5	0.00	0.00	0.000	0.000	3ª	5ᵇ	6	6	110	110	0.174	0.174	0.055	0.055
4	0.00	0.00	0.000	0.000	3ª	5ᵇ	6	6	110	110	0.174	0.174	0.055	0.055
3	0.00	0.00	0.000	0.000	3ª	5ᵇ	6	6	110	110	0.174	0.174	0.055	0.055
2	0.00	0.00	0.000	0.000	3ª	5ᵇ	6	6	110	110	0.174	0.174	0.055	0.055
1	0.08	0.00	0.136	0.000	3ª	5ᵇ	8	6	90	110	0.379	0.174	0.133	0.055
0	0.00	0.00	0.000	0.000	3ª	5ᵇ	6	6	170	170	0.113	0.113	0.025	0.025
−1	0.00	0.00	0.000	0.000	3ª	5ᵇ	6	6	170	170	0.113	0.113	0.025	0.025

ª See first footnote in Table 7.16.
ᵇ See second footnote in Table 7.16.

down to its lowest level. Even the shallowest footings (with a depth of 0.7 m) more than accommodate the straight anchorage length of $28d_{bL}$ computed for these bars from the last row of Table 5.1, with the product of the first two terms equal to 0.7.

The capacity design factor, a_{CD}, for the verification of the foundation ground and the design of the footing is not computed from the column section at the top of the footing, because the perimeter walls and the box-type action of the basement protect the interior elements of the basement from plastic hinging. Instead, a_{CD} is computed from the moment resistance and the moment from the analysis at the point of C12 nearest to the footing where a plastic hinge may form. This is at Level 0 (the base section of the 1st storey column). The so-computed values of a_{CD} are listed in Table 7.18 for the example of C12.

Archived for use in later stages are:

- The design values of the moment resistances at column end sections, $M_{Rd,c}$, for the maximum and the minimum column axial loads from the analysis for the seismic design situation, to be used next in Stage 3 for the capacity design shears of beams per Equation 5.42.
- The capacity design magnification factors per Equation 6.7a at the connection of the column to the box foundation, for use in Stage 5 for the design of the ground and the foundation elements per Sections 6.3.4 to 6.3.7; they are calculated and archived

Table 7.18 Column C12: Capacity design factor for the design of the column's footing

Combination of actions	M_{Rdy} (kNm)	M_{Edy} (kNm)	a_{CDy}	M_{Rdz} (kNm)	M_{Edz} (kNm)	a_{CDz}	a_{CD}
$G + \psi_2 Q + E: + X, + Y$/maxN	448.0	117.4	4.58	169.1	2.1	33.35	3.0
$G + \psi_2 Q + E: - X, + Y$/maxN	448.0	152.9	3.52	169.1	2.1	33.35	3.0
$G + \psi_2 Q + E: + X, - Y$/maxN	448.0	117.4	4.58	169.1	47.1	3.59	3.0
$G + \psi_2 Q + E: - Y$/maxN	448.0	152.9	3.52	169.1	47.1	3.59	3.0
$G + \psi_2 Q + E: + X, + Y$/minN	423.1	117.4	4.33	152.2	5.1	30.0	3.0
$G + \psi_2 Q + E: - X, + Y$/minN	423.1	152.9	3.32	152.2	5.1	30.0	3.0
$G + \psi_2 Q + E: + X, - Y$/minN	423.1	117.4	4.33	152.2	47.1	3.23	3.0
$G + \psi_2 Q + E: - X, - Y$/minN	423.1	152.9	3.32	152.2	47.1	3.23	3.0

separately for the different directions and sense of action of the design earthquake, producing eight combinations of signs of the column's seismic biaxial moments and axial force.

7.6.2.3 Stage 3: Beams in shear (Capacity design shears; dimensioning of transverse reinforcement for the ULS in shear; detailing per EC8)

The beams and their transverse reinforcement are dimensioned for the ULS in shear one multistorey frame at a time, according to Section 5.5.2. For DC H beams, the special provisions highlighted in Section 5.5.3 apply in addition; they may result in diagonal reinforcement or shear links at ±45° to the beam axis, if the value of $\zeta = \min V_{Ed}/\max V_{Ed}$ at the face of the support satisfies the conditions of Equations 5.49 and 5.50. For DC M beams, the value of the ζ-ratio is immaterial and the full shear reinforcement may always be placed at right angles to the beam axis.

The capacity design shears are computed from Equation 5.42, using the design moment resistances of the beams themselves, $M_{Rd,b}$, and of the columns they are connected to, $M_{Rd,c}$, which have been calculated and archived in Stages 1 and 2, respectively. As different detailing rules apply in the two 'critical' regions at the ends of the beam and in the central region in-between, stirrup spacing is normally constant in each one of these regions. In the 'critical' ones, it is determined from the design shear force at a distance d from the face of the supporting column; in the central part of the beam, the design shears at a distance $z\cot\theta$ from the ends of the two 'critical' regions may be used; in this case, for simplicity, a safe-sided value of d is used for $z\cot\theta$. The design shears due to the 'persistent and transient design situation' (called here 'non-seismic' design shears) should also be considered, because they may be the critical ones; especially in DC M beams, which, unlike DC H ones, do not have more adverse dimensioning rules against 'seismic' design shears than for 'non-seismic' ones. The maximum stirrup spacing in the 'critical' regions depends on the beam longitudinal bar diameters, determined and archived in Stage 1 for use in Stage 3.

The beams of the basement storeys follow the shear design and detailing rules of Eurocode 2 alone; their 'seismic' design shears are obtained from the analysis, not from capacity design, and are very low. So, their shear design does not have to await completion of Stage 2 and takes place in Stage 1, not in a separate Stage 3. For convenience, their end regions of length h (the same as in DC M beams of the superstructure) may be dimensioned in shear separately from the central one, as the 'non-seismic' design shears at distance d from the end section of these two types of region are very different and may produce very different stirrup spacing. Foundation beams are protected from plastic hinging by being overdesigned against the seismic action through multiplication of their seismic action effects by 1.4, per Equation 6.8. So, like all other beams of basements, they follow the Eurocode 2 rules alone and may be dimensioned in shear in Stage 1. The only difference with ordinary beams is that they are specified as one-storey elements, not as the beams at the lowest level of a multistorey plane frame.

An example output of Stage 3 is given in Table 7.19 for the two 5th-storey beams of frame 2, whose design in flexure was presented in Section 7.6.2.1. The transverse reinforcement at the ends and the central part of the beams in all storeys is shown at the right half of Figures 7.34 to 7.41.

7.6.2.4 Stage 4: Walls (Dimensioning of vertical and transverse reinforcement for the ULSs in flexure and in shear; detailing per EC8)

Each wall is fully dimensioned for the ULS in bending and in shear, and detailed for ductility, as a multistorey unit from the roof to the foundation. The linear envelope of the

Table 7.19 Frame 2, storey 5: beams B27, B28; capacity design of beams in shear; ULS dimensioning of transverse reinforcement

Beam	Sums of beam/column design moment resistances around joint, $\sum M_{Rd,b}/\sum M_{Rd,c}$ for maxN (kNm) Beam end and direction of $M_{Rd,b}$ vector:			
	Left end, +y	Left end, −y	Right end, +y	Right end, −y
28	213/220.7 (C2)	116.6/220.7 (C2)	321.6/527.5 (C7)	311/527.5 (C7)
27	321.6/527.5 (C7)	311/527.5 (C7)	116.6/215.8 (C12)	182.6/215.8 (C12)

Beam 28

Left end section: $maxV_{Ed} = 148.9$ kN, $minV_{Ed} = 54.9$ kN, $\zeta = minV_{Ed}/maxV_{Ed} = 0.37$
Right end section: $maxV_{Ed} = 128.1$ kN, $minV_{Ed} = 34.1$ kN, $\zeta = minV_{Ed}/maxV_{Ed} = 0.27$

Region	Length (m)	Design shear (kN)		Max tie spacing, mm	Provided ties			Strut angle	Shear resistance (kN)	
		Seismic	Non-seismic		No.	Ø, mm	s, mm		$V_{Rd,s}$	$V_{Rd,max}$
Left end	0.50	135.8	127.2	112	6	8	110	22°	399.6	416.2
Central	5.60	122.8	107.4	330	18	8	330	22°	133.2	416.2
Right end	0.50	115.0	109.6	112	6	8	110	22°	399.6	416.2

Beam 27

Left end section: $maxV_{Ed} = 142.2$ kN, $minV_{Ed} = 50.6$ kN, $\zeta = minV_{Ed}/maxV_{Ed} = 0.36$
Right end section: $maxV_{Ed} = 132.4$ kN, $minV_{Ed} = 40.8$ kN, $\zeta = minV_{Ed}/maxV_{Ed} = 0.31$

Region	Length (m)	Design shear (kN)		Max tie spacing, mm	Provided ties			Strut angle	Shear resistance (kN)	
		Seismic	Non-seismic		No.	Ø, mm	s, mm		$V_{Rd,s}$	$V_{Rd,max}$
Left end	0.50	138.5	129.2	112	6	8	110	22°	399.6	416.2
Central	5.60	125.5	109.5	330	18	8	330	22°	133.2	416.2
Right end	0.50	111.0	107.6	112	6	8	110	22°	399.6	416.2

moments from the analysis (see Table 7.20) is constructed first, according to Section 5.6.1.1 and Figure 5.6, providing the design moments at the various levels (see Table 7.21). Storey seismic shears from the analysis are multiplied with the amplification factor of Equation 5.56 in DC M walls, or of Equation 5.55 in DC H ones (see Table 7.22). If the wall belongs to a dual system (as, in this case, wall W5 in direction X), the so-amplified shears are replaced in the upper two-thirds of the wall height by a linear diagram as per Figure 5.8. According to Eurocode 8, the design shears in the basement storeys of interior walls are taken equal to the design moment resistance of the wall at the top of the basement times an overstrength factor (equal to 1.1 in DC M or 1.2 in DC H), divided by the height of the basement from the top of the wall's footing to the top of the basement. The diagrams of design moments and shears for walls W1, W3 and W5 (in directions X and Y) are depicted in Figure 7.44. Note that the Eurocode 8 rule for the design shear of interior walls in the basement, devised for single level basements, may fall short of the shear from the analysis in the upper storey of the basement. This happens indeed in the X direction of wall W5 (Figure 7.44d bottom). Although this case is not foreseen in Eurocode 8, the design shear is taken here at least equal to the value from the analysis, be it unmagnified.

The diameter and spacing of the vertical reinforcement in the web is determined from the minimum measures prescribed in Table 5.5 (see Table 7.23 for the example of wall W5). Vertical bars concentrated near the edges of the section supplement the minimum web reinforcement, to meet the moment resistance requirements resulting from the linear diagram of design moments up the wall (see Figure 7.44(top)). Presuming that these bars start at floor levels, their cross-sectional area is computed according to Section 5.6.1.2 from the design moments at these levels and translated into a specific layout of bars at each edge of the wall section. This layout should meet the rules for detailing wall boundary elements in Table 5.5, elaborated in Sections 5.7.6 and 5.7.7 for the particular case of the critical height of the wall. Outside that height, Eurocode 8 does not explicitly require boundary elements, nor prescribes a minimum size for them. However, the vertical steel area necessary to provide the wall moment resistance at a floor level should be spread over a concrete area large enough to limit the steel ratio near the edge to less than the maximum allowed, that is, 4%. That area is de facto configured as a boundary element, by applying the Eurocode 2 and 8 rules for lateral restraint of vertical bars and confinement of concrete wherever the steel ratio exceeds 2% (see the last part of the section of Table 5.5 devoted to boundary elements outside the critical region). Iterations may be needed, for consistency of the bar layout near

Table 7.20 Wall W5: Geometry; M, V at floor level from analysis for the design seismic action; N due to $G + \psi_2 Q$

Cross section: U. flanges: 1.80 m, web: 3.60 m, end stubs: none				
Total/critical height: 25.0 m/3.60 m, flange/web thickness: 0.25 m/0.25 m				
Storey	M_y (kNm)	V_y (kN)	M_z (kNm)	V_z (kN)	N (kN)
6-Top	0	±610	0.0	±75	294
6-Base	±1830	±610	±226	±75	294
5-Base	±2148	±376	±291	±78	588
4-Base	±2257	±564	±430	±101	882
3-Base	±3111	±788	±714	±142	1176
2-Base	±5509	±1113	±1189	±192	1470
1-Base	±11369	±1609	±1739	±169	1806
0-Base	±3832	±2512	±147	±625	2100
−1-Base	±2697	±399	±674	±274	2394

Table 7.21 Wall W5: Design envelopes of moments at floor levels and of magnified shears

Storey		M_y (kNm)	V_y (kN)	M_z (kNm)	V_z (kN)	N (kN)
6-top	$+M_y, +M_z$	−2187	1207	−164	115	0
	$−M_y, +M_z$	2187	−1207	−164	115	0
	$+M_y, −M_z$	−2187	1207	167	−110	0
	$−M_y, −M_z$	2187	−1207	167	−110	0
6-base	$+M_y, +M_z$	−3975	1207	−447	115	294
	$−M_y, +M_z$	3977	−1207	−447	115	294
	$+M_y, −M_z$	−3975	1207	431	−110	294
	$−M_y, −M_z$	3977	−1207	431	−110	294
5-base	$+M_y, +M_z$	−5763	1463	−728	119	588
	$−M_y, +M_z$	5766	−1463	−728	119	588
	$+M_y, −M_z$	−5763	1463	699	−115	588
	$−M_y, −M_z$	5766	−1463	699	−115	588
4-base	$+M_y, +M_z$	−7551	1573	−1010	154	882
	$−M_y, +M_z$	7555	−1572	−1010	154	882
	$+M_y, -M_z$	−7551	1573	966	−149	882
	$−M_y, −M_z$	7555	−1572	966	−149	882
3-base	$+M_y, +M_z$	−9339	1682	−1294	216	1176
	$−M_y, +M_z$	9344	−1682	−1294	216	1176
	$+M_y, −M_z$	−9339	1682	1231	−209	1176
	$−M_y, −M_z$	9344	−1682	1231	−209	1176
2-base	$+M_y, +M_z$	−11127	1670	−1582	292	1470
	$−M_y, +M_z$	11133	−1670	−1582	292	1470
	$+M_y, −M_z$	−11127	1670	1492	−283	1470
	$−M_y, −M_z$	11133	−1670	1492	−283	1470
1-base	$+M_y, +M_z$	−11365	2414	−1792	255	1806
	Corresponding M_{Rd} at base:	11590			6194	at v_d=0.065
	$−M_y, +M_z$	11373	−2414	−1792	255	1806
	Corresponding M_{Rd} at base:	11590			6194	at v_d=0.065
	$+M_y, −M_z$	−11365	2414	1685	−251	1806
	Corresponding M_{Rd} at base:	11590			4279	at v_d=0.065
	$−M_y, −M_z$	11373	−2414	1685	−251	1806
	Corresponding M_{Rd} at base:	11590			4279	at v_d=0.065
0-base	$+M_y, +M_z$	11360	2511	−1387	1135	2100
	$−M_y, +M_z$	11378	−2510	−1387	1135	2100
	$+M_y, −M_z$	−11360	2514	1395	−784	2100
	$−M_y, −M_z$	11378	−2510	1395	−784	2100
-1-base	$+M_y, +M_z$	−6794	2125	−217	1135	2394
	$−M_y, +M_z$	6848	−2125	−217	1135	2394
	$+M_y, −M_z$	−6794	2125	826	−784	2394
	$−M_y, −M_z$	6848	−2125	826	−784	2394

Table 7.22 Wall W5: Design in shear, for diagonal tension and compression (neglecting short-shear-span effects)

Floor	Location	Amplification factor for shear	Design shear $maxV_{Ed}$ (kN)	Horizontal bars					Shear resistance (kN)	
				Ø (mm)	Legs	Spacing s_h (mm)		Strut angle	$V_{Rd,s}$	$V_{Rd,max}$
						Max	Provided			
6	Web	1.5	1208	8	2	200	200	22°	1573	2234
	Flanges	1.5	116	8	2 × 2	200	200	22°	1573	2234
5	Web	1.5	1463	8	2	200	200	22°	1573	2234
	Flanges	1.5	120	8	2 × 2	200	200	22°	1573	2234
4	Web	1.5	1573	8	2	200	200	22°	1573	2234
	Flanges	1.5	154	8	2 × 2	200	200	22°	1573	2234
3	Web	1.5	1683	8	2	200	185	22°	1701	2234
	Flanges	1.5	216	8	2 × 2	200	185	22°	1573	2234
2	Web	1.5	1670	8	2	200	185	22°	1701	2234
	Flanges	1.5	292	8	2 × 2	200	185	22°	1573	2234
1	Web	1.5	2415	8	2	200	115	24°	2428	2428
	Flanges	1.5	256	8	2 × 2	200	115	22°	1573	2234
0	Web	—	2514	8	2	200	105	25°	2516	2516
	Flanges	—	1136	8	2 × 2	200	105	22°	1573	2234
−1	Web	—	2125	8	2	200	145	22°	2170	2234
	Flanges	—	1136	8	2 × 2	200	145	22°	1573	2234

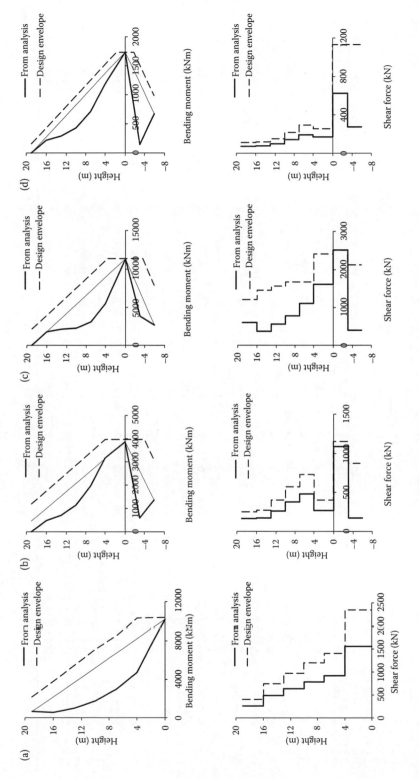

Figure 7.44 Seismic action effects from modal response spectrum analysis versus design action effects: moments (top), shears (bottom) of wall: (a) W1; (b) W3; (c) W5, X-plane; (d) W5, Y-plane.

Table 7.23 Wall W5: vertical, horizontal, hoop reinforcement (whole storey and base of storey above); steel added at construction joints

		Boundary elements: dimensions and reinforcement							Web reinforcement					
			Vertical bars		Hoops		ω_{wd}			Vertical		Horizontal		Added steel @ joint (mm^2)
Floor	Location	Size (m)	Ø (mm)	no	Ø (mm)	s (mm)	Req.	Prov.	Location	Ø (mm)	s_v (mm)	Ø (mm)	s_h (mm)	
6	Corners	0.25 × 0.25	18	4	8	105	0.0	0.32	Web	8	200	8	200	–
	Edges	0.15 × 0.25	18	4	8	105	0.0	0.46	Flanges	8	200	8	200	–
5	Corners	0.25 × 0.25	18	4	8	110	0.0	0.30	Web	8	200	8	200	–
	Edges	0.15 × 0.25	18	4	8	110	0.0	0.43	Flanges	8	200	8	200	–
4	Corners	0.25 × 0.25	20	4	8	110	0.0	0.30	Web	8	200	8	200	–
	Edges	0.15 × 0.25	20	4	8	110	0.0	0.43	Flanges	8	200	8	200	–
3	Corners	0.25 × 0.25	20	5	8	110	0.0	0.30	Web	8	200	8	185	–
	Edges	0.15 × 0.25	20	5	8	110	0.0	0.26	Flanges	8	200	8	185	–
2	Corners	0.25 × 0.25	20	6	8	105	0.0	0.24	Web	8	200	8	185	–
	Edges	0.25 × 0.25	20	6	8	110	0.0	0.24	Flanges	8	200	8	185	–
1	Corners	0.4 × 0.25, 0.55 × 0.25	18	10	8	105	0.0	0.24	Web	8	200	8	115	–
	Edges	0.4 × 0.25	20	6	8	110	0.0	0.24	Flanges	8	200	8	115	–
0	Corners	0.4 × 0.25, 0.55 × 0.25	18	10	8	105	0.0	0.24	Web	8	200	8	105	–
	Edges	0.4 × 0.25	20	6	8	110	0.0	0.24	Flanges	8	200	8	105	–
−1	Corners	0.4 × 0.25, 0.55 × 0.25	18	10	8	105	0.0	0.24	Web	8	200	8	150	–
	Edges	0.4 × 0.25	20	6	8	110	0.0	0.24	Flanges	8	200	8	150	–
−2	Corners	0.4 × 0.25, 0.55 × 0.25	18	10	8	105	0.0	0.24	Web	8	200	8	150	–
	Edges	0.4 × 0.25	20	6	8	110	0.0	0.24	Flanges	8	200	8	150	–

each edge with the distance of the centroid of these bars from the nearest extreme fibres, d_1, and the effective depth, d, used in the algorithm of Section 5.6.1.2.

The approach highlighted in the preceding text gives the layout of vertical reinforcement over the sections of W1, W3 and W5 and up the wall depicted in Figures 7.45 to 7.47. Wall W1 in Figure 7.45 is typical of a gradual reduction of the size of boundary elements up the wall; W3 in Figure 7.46, by contrast, is an example of boundary elements kept constant till the top storey – a common practice, but, strictly speaking, not required by Eurocode 8. In

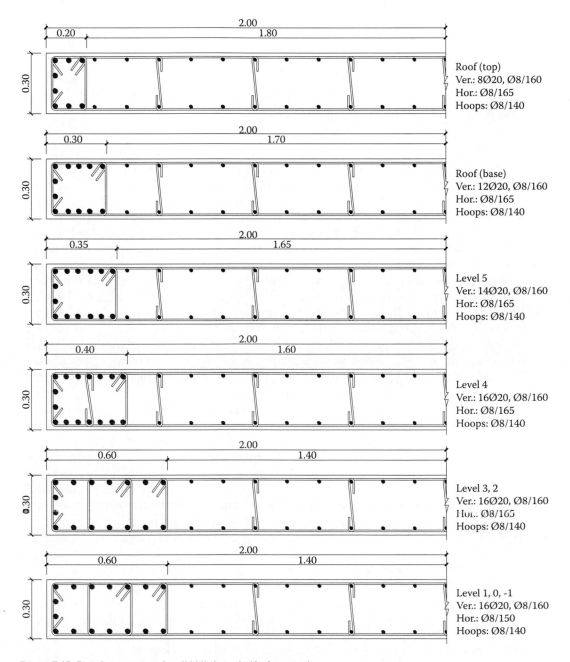

Figure 7.45 Reinforcement of wall WI (one-half of section).

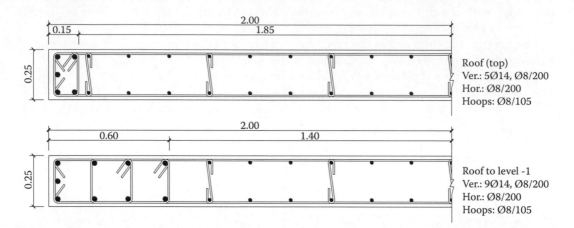

Figure 7.46 Reinforcement of wall W3 (one-half of section).

the channel-shaped section of W5, whose design is presented below as an example, boundary elements are arranged at both ends of each rectangular part of the section (cf. Example 5.11). Above the critical height of the wall, they are limited to the intersection ('corner') of the 'web' with each 'flange' of the section and to the free 'edge' of each flange.

All three types of walls have low axial load ratio at Level 0 (e.g. in the example of W5 presented here, $v_d = 0.065$ at that level), which meets the condition of Note no. (3) in Table 5.5 that allows applying the DC L rules to the confining reinforcement of boundary elements in the critical height. So, the specific values of the absolute or the effective mechanical confining reinforcement ratio, ω_{wd}, $a\omega_{wd}$, listed in the last two rows of the part of Table 5.5 on boundary elements in the critical height region, are not compulsory for these elements. Another point worthy of mention for the channel-shaped W5 in the example here is that, for easier fixing of the bars, the spacing of the horizontal bars in the 'flanges' from Level 3 down is kept the same as in the 'web', according to the verification of the 'web' in shear in the X-direction, although the 'flanges' can resist the Y-direction shear even when their horizontal bars have the maximum allowed spacing of 200 mm.

For the axial force $N = 1806$ kN produced at the base of wall W5 in the 'seismic design situation', the reinforcement layout in Figure 7.47 gives the biaxial moment resistance diagram in Figure 7.48; the biaxial design moments (seismic action effects) at the base of W5 from the analysis in the 'seismic design situation', displayed as asterisks, lie inside or marginally outside this diagram. The small discrepancy is acceptable, because the two components of the seismic moment demand are assumed in Figure 7.48 to take their peak values simultaneously per Equations 5.72 and 5.74, which is a very conservative assumption. The more realistic case of Equation 5.75 corresponds to the full value of one of the two components and 30% of the other, giving biaxial design moments well inside the biaxial resistance diagram.

Archived from this stage should be:

- The capacity design magnification factors at the connection of the wall to the foundation system (in this case at the base of the wall at Level 0), separately for the 8 combinations of signs of the wall's seismic biaxial moments and axial force, for use in Stage 5 for the capacity design of the ground and the foundation elements.

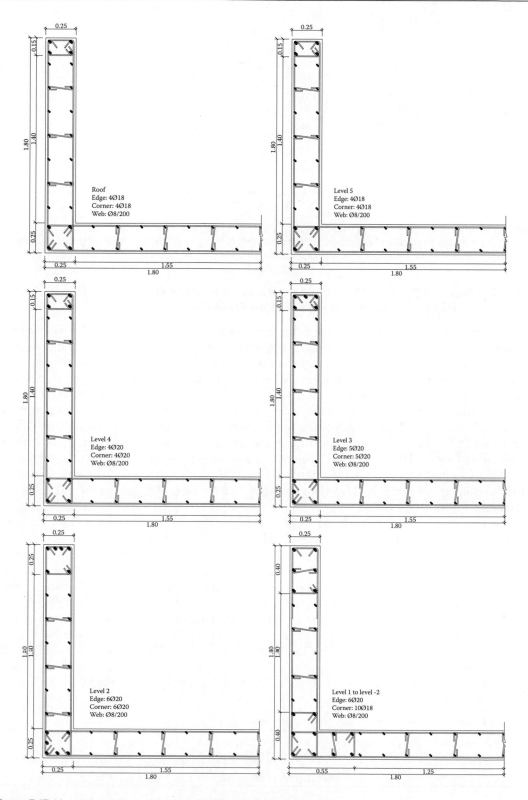

Figure 7.47 Vertical reinforcement of wall W5 (left-half of section).

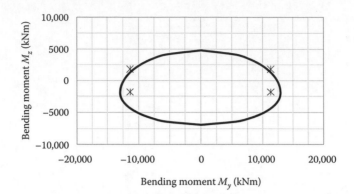

Figure 7.48 Biaxial design moments (seismic action effects) at the base of wall W5 (displayed as asterisks) compared to the biaxial moment resistance diagram (local y-axis is parallel to global Y and corresponds to bending in the X-plane; local z-axis: parallel to global X, bending in the Y-plane).

7.6.2.5 Stage 5: Footings (Bearing capacity; dimensioning for the ULSs in shear; punching shear and flexure)

The bearing capacity of the ground is calculated per Eurocode 7 and checked at the underside of each individual footing, for biaxial eccentricity of the vertical load and bidirectional horizontal forces (bidirectional inclination of the vertical load). Seismic reaction forces and moments at the node connecting the footing to the ground are amplified by the corresponding capacity design magnification factor, a_{CD}, at the base section of the column right above the top of the basement (a different value for the different directions and sense of action of the design earthquake). The depth of the footing is then dimensioned/verified for the ULS in shear and in two-way eccentric punching shear. Its two-way bottom reinforcement is then dimensioned for the ULS in flexure. The verifications of the ground and the ULS dimensioning of the footing are carried out separately for each horizontal direction and sense of action of the design earthquake, and for the persistent and transient design situation (Equations 6.10a and 6.10b in EN 1990).

A design example is given in Tables 7.24 to 7.27 for the footing of column C12. Witness in Tables 7.25 and 7.26 the very low demands posed on it by the 'seismic design situation', thanks to the compelling role of the perimeter wall for the transfer of the (full) seismic action to the ground. Design Approach DA3 is adopted in Table 7.25 for the persistent and transient design situation, because it is more compatible with the verification of the seismic design situation according to Part 5 of Eurocode 8 (see footnote d in Table 6.1).

The strip footings of the foundation beams are designed with a one-way version of the design of individual footings above. This is carried out for the full length of the strip footings of each foundation beam, which may encompass quite a few intermediate nodes and vertical soil springs.

The plan dimensions and reinforcement of the footings are shown in Figure 7.49.

Table 7.24 Footing F12 for column C12: Footing geometry

Footing depth h: 0.70 m; footing plan dimensions: //y b_y = 2.00 m, //z b_z = 2.00 m	
Overburden depth: 0.0 m	Column cross-sectional dimensions: //y c_y = 0.70 m, //z c_z = 0.30 m
	Column axis eccentricity: //y a_y = 0 m, //z a_z = 0 m

Table 7.25 Footing F12 for column C12: Design forces at the centre of the footing's base; soil bearing pressure and capacity per EC7

Combination of actions	Capacity design magnification factor	N_{tot} (kN)	M_y (kNm)	e_y/b_y	V_y (kN)	M_z (kNm)	e_z/b_z	V_z (kN)	Soil bearing Pressure (kPa)	Soil bearing Capacity (kPa)
DA3 EN1990 Equation 6.10a[a]	–	2632	–5	0.001	8	6	0.001	6	661	1276
DA3 EN1990 Equation 6.10b[a]	–	2471	–5	0.001	8	6	0.001	5	620	1276
$G+\psi_2Q+E$:+X,+Y/maxN	3.0	2278	26	0.012	40	56	0.006	23	590	1673[b]
$G+\psi_2Q+E$:–X,+Y/maxN	3.0	2278	33	0.012	29	56	0.007	23	592	1677[b]
$G+\psi_2Q+E$:+X,–Y/maxN	3.0	2278	26	0.010	28	47	0.006	14	588	1674[b]
$G+\psi_2Q+E$:–X,–Y/maxN	3.0	2278	33	0.010	17	47	0.007	14	590	1679[b]
$G+\psi_2Q+E$:+X,+Y/minN	3.0	1252	26	0.023	40	56	0.010	23	334	1669[b]
$G+\psi_2Q+E$:–X+Y/minN	3.0	1252	33	0.023	29	56	0.013	23	336	1674[b]
$G+\psi_2Q+E$:+X,–Y/minN	3.0	1252	26	0.019	26	47	0.010	14	332	1671[b]
$G+\psi_2Q+E$:–X,–Y/minN	3.0	1252	33	0.019	15	47	0.013	14	334	1677[b]

[a] The most unfavourable outcome of the application of Equation 6.10a or 6.10b applies.
[b] The verification per Annex F in Part 5 of EC8 (see Section 6.2.5), with an equivalent footing diameter of $2.0\sqrt{(4/\pi)}=2.25$ m and using as V the resultant of V_y, V_z, and as M that of M_y, M_z, gives a value of the left-hand side of Equation 6.5 between –0.994 and –0.998.

Table 7.26 Footing F12 for column C12: ULS design of footing in shear and punching shear

Combination of actions	Shear stress v_{Ed} and resistance (kPa) Section//y V_{Edy}/b_zd	Section//z V_{Edz}/b_yd	Resistance $v_{Rd.c}$	Punching shear at distance a_v Max stress maxv_{Ed}	Critical distance a_v (m)	Resistance $(2d/a_v)v_{Rd}$ (kPa)
EN1990 Equation 6.10a[a]	12.1	214.4	340.9	679	0.3	1299
EN1990 Equation 6.10b[a]	11.4	201.0	340.9	636	0.3	1299
$G+\psi_2Q+E$:+X,+Y/maxN	10.8	196.0	340.9	598	0.3	1299
$G+\psi_2Q+E$:–X,+Y/maxN	10.9	196.0	340.9	599	0.3	1299
$G+\psi_2Q+E$:+X,–Y/maxN	10.8	194.0	340.9	596	0.3	1299
$G+\psi_2Q+E$:–X,–Y/maxN	10.9	194.0	340.9	597	0.3	1299
$G+\psi_2Q+E$:+X,+Y/minN	5.9	110.8	340.9	327	0.3	1299
$G+\psi_2Q+E$:–X,+Y/minN	6.0	110.8	340.9	328	0.3	1299
$G+\psi_2Q+E$:+X,–Y/minN	5.9	108.7	340.9	325	0.3	1299
$G+\psi_2Q+E$:–X,–Y/minN	6.0	108.7	340.9	326	0.3	1299

[a] The most unfavourable outcome of the application of Equation 6.10a or 6.10b applies.

Table 7.27 Footing F12 for column C12: ULS design of two-way reinforcement at the bottom of the footing

| Maximum bending moments | | | | Reinforcement | | | | |
| Vertical section//b_z | | Vertical section//b_y | | | //b_y | | //b_z | |
M_{Edy}/b_z (kNm/m)	Combination	M_{Edz}/b_y (kNm/m)	Combination	Bar Dia. (mm)	Spacing (mm)	No.	Spacing (mm)	No.
132.0	EN1990 Equation 6.10a	230.5	EN1990 Equation 6.10a	12	130	15	120	16

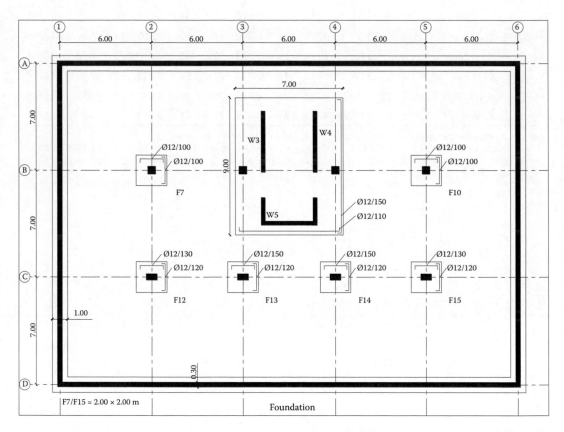

Figure 7.49 Plan and reinforcement of foundation elements; depth of footings: 0.7 m; top face of all foundation elements flush with top surface of RC slab playing the role of tie-beams and bottom diaphragm of the box foundation.

References

Abrahamson, N.N. 2000. State of the practice of seismic hazard evaluation. *GeoEng2000*. Melbourne, Australia.

Akkar, S. and J.J. Bommer. 2010. Empirical equations for the prediction of PGA, PGV, and spectral accelerations in Europe, the Mediterranean region, and the Middle East. *Seismological Research Letters* 81:195–206.

Antoniou, K., G. Tsionis, and M.N. Fardis. 2014. Inelastic shears in ductile RC walls of mid-rise wall-frame buildings and comparison to Eurocode 8. *Bulletin of Earthquake Engineering*. DOI 10.1007/s10518-014-9641-x (published online 11 June 2014).

ASCE. 2007. *ASCE/SEI Standard 41-06, Seismic Rehabilitation of Existing Buildings (including Supplement 1)*. American Society of Civil Engineers, Reston, VA.

Bard, P.Y. and J. Riepl-Thomas. 1999. Wave propagation in complex geological structures and local effects on strong ground motion. In *Wave Motion in Earthquake Engineering*, eds. E. Kausel and G.D. Manolis, 38–95. WIT Press, Southampton, UK. Series Advances in Earthquake Engineering, ISBN 1-85312-744-2, pp. 38–95.

Biasoli, F., G. Manchini, M. Just, M. Curbach, J. Walraven, S. Gmainer, J. Arrieta, R. Frank, C. Morin, and F. Robert. 2014. *Eurocode 2: Background & Applications, Design of Concrete Buildings*. JRC SCIENTIFIC AND POLICY REPORTS, European Commission, Joint Research Centre, Institute for the Protection and Security of the Citizen.

Bisch, P., E. Carvalho, H. Degee, P. Fajfar, M. Fardis, P. Franchin, M. Kreslin et al. 2012. *Eurocode 8: Seismic Design of Buildings—Worked Examples*. JRC SCIENTIFIC AND TECHNICAL REPORTS, European Commission, Joint Research Centre, Institute for the Protection and Security of the Citizen.

Biskinis, D.E. and M.N. Fardis. 2010a. Deformations at flexural yielding of members with continuous or lap-spliced bars. *Structural Concrete* 11(3):127–138.

Biskinis, D.E. and M.N. Fardis. 2010b. Flexure-controlled ultimate deformations of members with continuous or lap-spliced bars. *Structural Concrete* 11(2):93–108.

Biskinis, D.E., G. Roupakias, and M.N. Fardis. 2004. Degradation of shear strength of RC members with inelastic cyclic displacements. *ACI Structural Journal* 101(6):773–783.

Bommer, J.J. and F. Scherbaum. 2008. The use and misuse of logic trees in probabilistic seismic hazard analysis. *Earthquake Spectra* 24:997–1009.

CEN. 2002. *European Standard EN 1990:2002 Eurocode: Basis of Structural Design*. Comite Europeen de Normalisation, Brussels.

CEN. 2003. *European Standard EN 1997-1:2003 Eurocode 7: Geotechnical Design—Part 1: General Rules*. Comite Europeen de Normalisation, Brussels.

CEN. 2004a. *European Standard EN 1998-1:2004 Eurocode 8: Design of Structures for Earthquake Resistance, Part 1: General Rules, Seismic Actions and Rules for Buildings*. Comite Europeen de Normalisation, Brussels.

CEN. 2004b. *European Standard EN 1992-1-1:2004 Eurocode 2: Design of Concrete Structures, Part 1-1: General Rules and Rules for Buildings*. Comite Europeen de Normalisation, Brussels.

CEN. 2004c. *European Standard EN 1998-5:2004 Eurocode 8: Design of Structures for Earthquake Resistance, Part 5: Foundations, Retaining Structures, Geotechnical Aspects*. Comite Europeen de Normalisation, Brussels.

CEN. 2005. *European Standard EN 1998-3:2005 Eurocode 8: Design of Structures for Earthquake Resistance, Part 3: Assessment and Retrofitting of Buildings*. Comite Europeen de Normalisation, Brussels.

Chatzigogos, C.T, A. Pecker, and J. Salençon. 2007. Seismic bearing capacity of circular footing on an heterogeneous cohesive soil. *Soils and Foundations* 47(4):783–797.

Chopra, A. 2007. *Dynamics of Structures* (3rd edition). Pearson Prentice Hall, Upper Saddle River, New Jersey.

Clough, R.W. and J. Penzien. 1993. *Dynamics of Structures* (2nd edition). McGraw-Hill, New York.

Cornell, C.A. 1968. Engineering seismic risk analysis. *Bulletin of the Seismological Society of America* 58(5):1583–1606.

CSI. 2002. *ETABS Manual*. Computers & Structures Inc., Berkeley, California.

Deierlein, G.G., A.M. Reinhorn, and M.R. Wilford. 2010. *Nonlinear Structural Analysis for Seismic Design. A Guide for Practicing Engineers*. NEHRP seismic design technical brief No. 4, NIST GCR 10-917-5, National Institute of Standards and Technology, Gaithersburg, MD, USA.

Der Kiureghian, A. 1981. A response spectrum method for random vibration analysis of MDF systems. *Earthquake Engineering and Structural Dynamics* 9:419–435.

Dolšek, M. 2010. Development of computing environment for the seismic performance assessment of reinforced concrete frames by using simplified nonlinear models. *Bulletin of Earthquake Engineering* 8:1309–1329.

Douglas, J. 2010. Consistency of ground-motion predictions from the past four decades. *Bulletin of Earthquake Engineering* 8:1515–1526.

Eibl, J. and E. Keintzel. 1988. Seismic shear forces in cantilever shear walls. *9th World Conference in Earthquake Engineering*. Tokyo, Kyoto, Japan.

EU. 2004. Directive 2004/18/EC of the European Parliament and of the Council of 31 March 2004, on the coordination of procedures for the award of public works contracts, public supply contracts and public service contracts. Official Journal of the European Union, L134/114 30-4-2004.

EU. 2006. Directive 2006/123/EC of the European Parliament and of the Council of 12 December 2006, on services in the internal market. Official Journal of the European Union, L376/36 27-12-2006.

EU Regulation no 305/2011 of the European Parliament and of the Council of 9 March 2011 laying down harmonised conditions for the marketing of construction products and repealing Council Directive 89/106/EEC. *Official Journal of the European Union* L 88/5, 4.4.2011.

European Commission. 2001. *Guidance Paper L Application and Use of Eurocodes* CONSTRUCT 01/483 Rev.1. Brussels.

European Commission. 2003. *Commission Recommendation on the Implementation and Use of Eurocodes for Construction Works and Structural Construction Products*. Document No. C(2003)4639. Brussels.

Fajfar, P. 1999. Capacity spectrum method based on inelastic demand spectra. *Earthquake Engineering and Structural Dynamics* 28(9):979–993.

Fajfar, P. 2000. A nonlinear analysis method for performance-based seismic design. *Earthquake Spectra* 16(3):573–592.

Fajfar, P. and M. Fischinger. 1987. Non-linear seismic analysis of RC buildings: Implications of a case study. *European Earthquake Engineering* 1:31–43.

Fajfar, P. and M. Fischinger. 1989. N2—A method for non-linear seismic analysis of regular buildings. *9th World Conference on Earthquake Engineering, August 2–9, 1988*. Tokyo, Kyoto, Japan. Proceedings, Vol. 5:111–116. Maruzen, Tokyo.

Fajfar, P., M. Dolšek, D. Marušić, and A. Stratan. 2006. Pre- and post-test mathematical modelling of a plan-asymmetric reinforced concrete frame building. *Earthquake Engineering and Structural Dynamics* 35:1359–1379.

Fajfar, P., D. Marušić,, and I. Peruš. 2005. Torsional effects in the pushover-based seismic analysis of buildings. *Journal of Earthquake Engineering* 9(6):831–854.

Fardis, M.N. 2009. *Seismic Design, Assessment and Retrofitting of Concrete Buildings Based on EN-Eurocode 8*. Springer Science + Business Media, Dordrecht, the Netherlands.

Fardis, M.N., A. Papailia, and G. Tsionis. 2012. Seismic fragility of RC framed and wall-frame buildings designed to the EN-Eurocodes. *Bulletin of Earthquake Engineering* 10(6):1767–1793.

FEMA. 2009. *Quantification of Building Seismic Performance Factors, FEMA P695.* Federal Emergency Management Agency, Washington, DC.

Gazetas, G. 1983. Analysis of machine foundation vibrations: State-of-the-art. *International Journal of Soil Dynamics and Earthquake Engineering* 2(1):2–43.

Gazetas, G. 1991. Foundation vibration. In *Foundation Engineering Handbook*, ed. H.-Y. Fang, Chap. 15 (2nd edition). Kluwer Academic Publishers, Dordrecht, the Netherlands.

Giberson, M.F. 1967. The response of nonlinear multi-story structures subjected to earthquake excitation. PhD thesis, California Institute of Technology, Pasadena, CA.

Goel, R.K. and A.K. Chopra. 2005. Extension of modal pushover analysis to compute member forces. *Earthquake Spectra* 21:125–140.

Gupta, A.K. and M.P. Singh. 1977. Design of column sections subjected to three components of earthquake. *Nuclear Engineering and Design* 41:129–133.

Horvath, J.S. 1983. Modulus of subgrade reaction: New perspective. *ASCE Journal of Geotechnical Engineering Division* 109(GT12):1567–1587.

Idriss, I.M. 1990. Response of soft soil sites during earthquakes. *Proceedings H.B. Seed Memorial Symposium*, 273–290. BiTech Publishers Ltd, Vancouver, BC, Canada.

Idriss, I.M. and R.W. Boulanger. 2008. Soil liquefaction during earthquakes. *Earthquake Engineering Research Institute*, MNO-12, Oakland, CA, 261 pp.

Ishihara, K. 1993. Liquefaction and flow failures during earthquakes. Rankine Lecture. *Géotechnique* 43(3):351–415.

Kausel, E. and J.M. Roesset. 1974. Soil structure interaction for nuclear containment structures. *Proc. ASCE, Power Division Specialty Conference*. Boulder, Colorado.

Keintzel, E. 1990. Seismic design shear forces in reinforced concrete cantilever shear wall structures. *European Journal of Earthquake Engineering* 3(1):7–16.

Kosmopoulos, A. and M.N. Fardis. 2008. Simple models for inelastic seismic analysis of asymmetric multistory buildings. *Journal of Earthquake Engineering* 12(5):704–727.

Krawinkler, H. 2006. Importance of good nonlinear analysis. *The Structural Design of Tall and Special Buildings* 15(5):515–531.

Krawinkler, H. and G.D.P.K. Seneviratna. 1998. Pros and cons of a pushover analysis for seismic performance evaluation. *Engineering Structures* 20(4–6):452–464.

Kreslin, M. and P. Fajfar. 2011. The extended N2 method taking into account higher mode effects in elevation. *Earthquake Engineering and Structural Dynamics* 40(14):1571–1589.

Kreslin, M. and P. Fajfar. 2012. The extended N2 method considering higher mode effects both in plan and elevation. *Bulletin of Earthquake Engineering* 10(2):695–715.

Miranda, E. and V.V. Bertero. 1994. Evaluation of strength reduction factors for earthquake-resistant design. *Earthquake Spectra* 10(2):357–379.

Mitchell, J.K. and F.J. Wentz. 1991. Performance of improved ground during the Loma Prieta earthquake. Earthquake Engineering Research Center UCB/EERC 91–12, Berkeley, CA.

Mylonakis, G. and G. Gazetas. 2000. Seismic soil–structure interaction: Beneficial or detrimental. *Journal of Earthquake Engineering* 4:277–301.

O'Rourke, T.D. 1996. Lessons learned for lifeline engineering from major urban earthquakes. Paper no. 2172. *Eleventh World Conference on Earthquake Engineering*. Acapulco, Mexico.

Paolucci, R. and A. Pecker. 1997. Seismic bearing capacity of shallow strip foundations on dry soils. *Soils and Foundations* 37(3):95–105.

Pecker, A. 1997. Analytical formulae for the seismic bearing capacity of shallow foundations. *International Society of Soil Mechanics and Foundation Engineering. Special volume TC4*, London.

Pecker, A. 2007. Soil structure interaction. In *Advanced Earthquake Engineering Analysis*, ed. A. Pecker, Chapter 3: 33–42, CISM 494. Udine, Italy.

Power, M., B. Chiou, N. Abrahamson, Y. Bozorgnia, T. Shantz and C. Roblee. 2008. An overview of the NGA project. *Earthquake Spectra* 24(1):3–21.

Pyke, R., H.B. Seed, and C.K. Chan. 1975. Settlement of sands under multi-directional shaking. *ASCE Journal of the Geotechnical Engineering Division* 101(4):379–398.

Reid, H.F. 1910. *The Mechanics of the Earthquake. The California Earthquake of April 18. 1906, Report of the State Investigation Commission*, Vol. 2, Carnegie Institution of Washington, Washington, DC.

Salençon, J. and A. Pecker. 1995a. Ultimate bearing capacity of shallow foundations under inclined and eccentric loads—Part I: Purely cohesive soil. *European Journal of Mechanics A/Solids* 14(3):349–375.

Salençon, J. and A. Pecker. 1995b. Ultimate bearing capacity of shallow foundations under inclined and eccentric loads—Part II: Purely cohesive soil without tensile strength. *European Journal of Mechanics A/Solids* 14(3):377–396.

Scordilis, E.M. 2006. Empirical global relations converting MS and mb to moment magnitudes. *Journal of Seismology* 10:225–236.

Smebby, W. and A. Der Kiureghian. 1985. Modal combination rules for multicomponent earthquake excitation. *Earthquake Engineering and Structural Dynamics* 13(1):1–12.

Tokimatsu, K. and H.B. Seed. 1987. Evaluation of settlements in sands due to earthquake shaking. *Journal of Geotechnical Engineering* 113(8):861–878.

Veletsos, A.S. and N.M. Newmark. 1960. Effect of inelastic behavior on the response of simple systems to earthquake motions. *2nd World Conference on Earthquake Engineering*. Tokyo, Japan. Proceedings, Vol. 2: 895–912.

Vidic, T., P. Fajfar, and M. Fischinger. 1994. Consistent inelastic design spectra: Strength and displacement. *Earthquake Engineering and Structural Dynamics* 23:502–521.

Wegener, A. 1915. *Die Entstehung der Kontinente und Ozeane. (On the Origin of Continents and Oceans)*. Friedrich Vieweg & Son, Braunshweig.

Wilson, E.L., A. Der Kiureghian, and E.P. Bayo. 1981. A replacement for the SRSS method in seismic analysis. *Earthquake Engineering and Structural Dynamics* 9:187–194.

Youd, T.L. and I.M. Idriss. 2001. Liquefaction resistance of soils: Summary Report from the 1996 NCEER Workshops on Evaluation of Liquefaction Resistance of Soils. *ASCE Journal of Geotechnical and Geoenvironmental Engineering* 127(10):817–833.

Index